Chemistry: Theory and Practice

Strength, Theory and Practice

Chemistry: Theory and Practice

Editor: Bruce Horak

NY RESEARCH
P R E S S

New York

Published by NY Research Press
118-35 Queens Blvd., Suite 400,
Forest Hills, NY 11375, USA
www.nyresearchpress.com

Chemistry: Theory and Practice
Edited by Bruce Horak

International Standard Book Number: 978-1-63238-583-3 (Hardback)

Cataloging-in-Publication Data

Chemistry : theory and practice / edited by Bruce Horak.
    p. cm.
Includes bibliographical references and index.
ISBN 978-1-63238-583-3
1. Chemistry. I. Horak, Bruce.
QD31.3 .C44 2018
572--dc23

Contents

Permissions

List of Contributors

Index

Preface

The branch of science which studies the structure and composition of matter is known as chemistry. Some of the principles of modern chemistry are bonding, energy, ions and salts, equilibrium, chemical laws, redox, etc. Chemistry can be categorized into a number of branches such as inorganic chemistry, materials chemistry, analytical chemistry, biochemistry, etc. This book will provide interesting topics for research which readers can take up. The aim of this book is to present researches that have transformed this discipline and aided its advancement. The readers would gain knowledge that would broaden their perspective about the subject.

The researches compiled throughout the book are authentic and of high quality, combining several disciplines and from very diverse regions from around the world. Drawing on the contributions of many researchers from diverse countries, the book's objective is to provide the readers with the latest achievements in the area of research. This book will surely be a source of knowledge to all interested and researching the field.

In the end, I would like to express my deep sense of gratitude to all the authors for meeting the set deadlines in completing and submitting their research chapters. I would also like to thank the publisher for the support offered to us throughout the course of the book. Finally, I extend my sincere thanks to my family for being a constant source of inspiration and encouragement.

Editor

FeCl₃-catalyzed Synthesis of Dehydrodiisoeugenol

Athina Mardhatillah[1,2], Mutakin Mutakin[1], Jutti Levita[1]

[1]Faculty of Pharmacy, Universitas Padjadjaran, Jl. Raya Bandung-Sumedang Km.21 Jatinangor, West Java 45363, Indonesia

[2]Faculty of Pharmacy, University of Jenderal Achmad Yani, Jl. Terusan Jenderal Sudirman Cimahi, West Java, Indonesia

Correspondence: Jutti Levita, Faculty of Pharmacy, Universitas Padjadjaran, Jl. Raya Bandung-Sumedang Km.21 Jatinangor Sumedang, West Java 45363, Indonesia. E-mail: jutti.levita@unpad.ac.id

Abstract

Dehydrodiisoeugenol (DDIE) synthesis has been performed by modifying a method recommended by Leopold with a different ratio of isoeugenol and FeCl₃ (1.9:1). FeCl₃ was chosen as catalyst due to its efficiency and environment-friendly property. This modification yielded 22.93 % of product. The product, a white crystalline form, was characterized using thin layer chromatography, melting point, UV, IR, HRMS, and NMR spectroscopy, as well as HPLC, employing pure DDIE as the standard. TLC chromatogram showed Rf 0.32 using n-hexane/ethyl acetate (8:2). The crystals melted at 138-139 °C, while its UV maximum was detected at λ 273 nm. IR spectrum showed a specific broad O-H stretch at 3437.15 cm⁻¹, C-H aromatic and C-H alkene at 3163.26 and 3024.38 cm⁻¹, C-H alkyl stretch at 2951.09 and 2927.94 cm⁻¹. An overtone peak of aromatic was detected at 2100 to 1700 cm⁻¹. C-O peak was detected at 1126.43 cm⁻¹. HPLC showed that this compound was eluted at 11.886 minutes after it was injected to a C18 column 250 x 4 mm using a mixture of methanol and double distilled water (73:27) for mobile phase. HRMS spectra predicted that the molecular structure is C₂₀H₂₂O₄ as showed by abundance peak at *m/z* 327.1595 of [M+H]⁺. ¹H-NMR and ¹³C-NMR indicated that the synthesized compound contains 13 types of proton and 20 types of carbon. Herein we reported that white needle-like crystals of DDIE using FeCl₃ as catalyst had been synthesized, moreover the decreasing of the catalyst reduced the yield of the product.

Keywords: antidiabetic, anti-inflammatory, diabetes, DDIE, *Myristica fragrans* Houtt, nutmeg, PPARγ

1. Introduction

Dehydrodiisoeugenol (DDIE) (Fig.1) is a chemical compound contained in fruit and seed of nutmeg (*Myristica fragrans* Houtt).

Fig. 1. Chemical structure of DDIE (a) and isoeugenol (b)

Built using ChemDrawUltra 8.0.3 of ChemOffice 2004 (www.cambridgesoft.com)

DDIE showed anti-inflammatory (Li and Yang, 2012) and antidiabetic activity on PPARγ receptor (Lestari, 2012). Previous study determined that the level of DDIE, myristicin, and safrole in the ethanol extract of nutmeg seeds was 4.662 %, 17.226 %, and 10.979 %, respectively using RP-HPLC (Saputri, 2014). Isolation of bioactive compounds from plants is a wholesome-work therefore this research was aimed to synthesize DDIE by employing FeCl$_3$ as catalyst.

The coupling of two phenoxy radicals leads to new asymmetric stereo-centers. The reaction can lead to pure enantiomers or mixtures if stereo-control exists due to a catalyst and/or matrix and/or chiral auxiliarities in the starting compound. The coupling will form a very reactive quinone methide intermediate, which could react quickly with a suitable nucleophile and leads finally to a stable dimeric structure called lignin or dilignol (Setälä, 2008).

Leopold categorized eight groups of lignin compound with different elements, whereas DDIE was categorized in group VII. Leopold method in which 50 g of isoeugenol, as the starting material, was reacted with a few crystals of DDIE, had successfully yielded 30 % of the product. This reaction was catalyzed by FeCl$_3$ with ratio of isoeugenol-FeCl$_3$ 1.4:1 (Leopold, 1950). In his paper, Leopold did not provide detailed information, either of the amount of DDIE that he had used for the reaction or the characterization analysis of the product.

FeCl$_3$ was chosen in this reaction, due to its efficiency and green catalyst property in modern organic synthesis (Diaz et al, 2006). The use of FeCl$_3$ as catalyst were reported for arylation of benzyl alcohols and benzyl carboxylates (Zhan and Liu, 2006), benzylation of 1,3-dicarbonyl compounds (Komeyama et al, 2007) and the synthesis of 1-substituted-1H-1,2,3,4-tetrazoles. The latter mentioned that an excess of FeCl$_3$, as catalyst, did not lead to a substantial improvement in the yield while decreasing the catalyst reduced it (Darvish and Khazraee, 2015).

2. Materials and Methods

2.1 Materials

Isoeugenol 99.0 % (Sigma Aldrich-USA) was purchased in Maebashi-Japan, while the other materials were purchased in Jakarta-Indonesia: DDIE standard 500 µg/mL (Kimia Farma-Indonesia), methanol HPLC grade 99.9% (J.T. Baker-USA), ethanol analytical grade 99 % (Merck Millipore-Germany), double-distilled water (IPHA-Indonesia), FeCl$_3$ analytical grade (Merck Millipore-Germany), n-hexane analytical grade 96% (Merck Millipore-Germany), ethyl acetate analytical grade 99.5 % (Merck Millipore-Germany), KBr analytical grade (Merck-Germany).

2.2 Synthesis of DDIE

This compound was synthesized by modifying a method recommended by Leopold (Leopold, 1950):

1 mL of isoeugenol 99.0 % oily solution was slowly added by stirring, into a mixture of 12.06 mL ethanol 95% and 5.36 mL of double-distilled water, until the two immiscible liquids dissolved. Into this mixture, 5 mL of FeCl$_3$ 35% (isoeugenol: FeCl$_3$ = 1.9 : 1) was added and stirred until a yellowish-green precipitate formed (Fig.2a). Then, 230 µL of 500 µg/mL DDIE standard was poured into the solution. The mixture was kept at 5°C for 24 hours. The white crystalline formed (Fig.2b) was vacuum-filtered and washed with 45% ethanol.

2.3 Thin Layer Chromatography

A few mg of the white crystalline product was dissolved in methanol and eluted on silica gel F$_{254}$ (Merck) using a mixture of n-hexane-ethyl acetate (8:2) as eluent. Both isoeugenol 99.0 % and DDIE standard were used as comparison.

2.4 UV Spectroscopy

2 mg of the white crystalline product was dissolved in 100 ml of double-distilled water to get 20µg/ml. This solution was measured its absorbance against water in UV-1700 Pharma (Shimadzu) spectrophotometer.

2.5 IR Spectroscopy

2 mg of the white crystalline product was dispersed in 198 mg of previously dried KBr. The disc was measured its %T using FTIR IRAffinity-1 (Shimadzu).

2.6 HPLC

20 µl of 20 µg/ml solution was injected to C18 column, 250 mm x 4 mm of HPLC 1525 Binary HPLC Pump (Waters) instrument, using a mixture of methanol and double-distilled water (73:27) as mobile phase. Flow rate was set at 1 ml/min. The UV detector was set at 282 nm.

2.7 HRMS

A few mg of the white crystalline product was dissolved in a mixture of acetone and 0.1 % formic acid in acetonitrile-water (1:1). The solution was ionized by heating it at 300 °C after it was injected to a capillary column of HRMS (Waters LCT Premier XE). TOF was used as detector.

2.8 NMR

10 mg of the white crystalline product was dissolved in CDCl$_3$ and injected in ^{1}H-NMR (Agilent), ^{13}C-NMR (Agilent). Spectrum was recorded at 500 MHz and 125 MHz Agilent, respectively, and analyzed using VnmrJ 3.2 software.

3. Result and Discussion

Leopold, who performed a synthesis of DDIE, had modified a method described by Erdtman. He used a ratio of isoeugenol-FeCl$_3$ 1.4:1, and yielded 30 % of product (Leopold, 1950). No detailed reaction or spectra were provided in this paper.

In our project, we modified Leopold method by using lesser amount of FeCl$_3$ (ratio of isoeugenol-FeCl$_3$ is 1.9:1). The yellowish-green precipitate (Fig.2) that was slowly formed, was kept at 5°C for 24 hours after it had been reacted with 230 µl of DDIE standard priorly.

(a)   (b)

Fig. 2. Yellowish-green precipitate of isoeugenol-FeCl$_3$ (a) and the white crystalline product (b)

The phenolic group of isoeugenol when it was reacted with FeCl$_3$, formed a coloured ferric-phenolate intermediate (Fig.3).

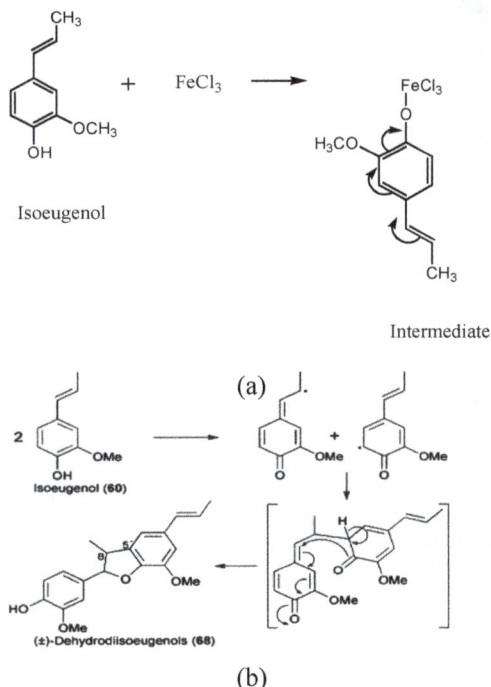

Fig. 3. Proposed scheme for isoeugenol reaction with FeCl$_3$ (built using ChemDrawUltra 8.0.3) (a); Erdtman's scheme for DDIE synthesis (b). Reaction 3b was copied from Davin et al (2008)

Furthermore Davin and his colleagues (2008), reviewed a scheme reaction on FeCl$_3$-catalyzed coupling of isoeugenol to give 8-5'-linked-dehydrodiisoeugenol (Fig.3b) proposed by Erdtman. In his review, Davin traced back the work of Hogar Erdtman who investigated the coupling of allylphenol and isoeugenol, which in the presence of FeCl$_3$, could afford the racemic dehydrogenated products, DDIE (Fig.3b) linked through the 8-5' positions. This coupling initially generated the intermediate quinone methide, whose subsequent intramolecular ring closure afforded (±)-DDIE (Davin et al, 2008).

TLC assay showed that the synthesis product was eluted at Rf 0.32 (Fig.4).

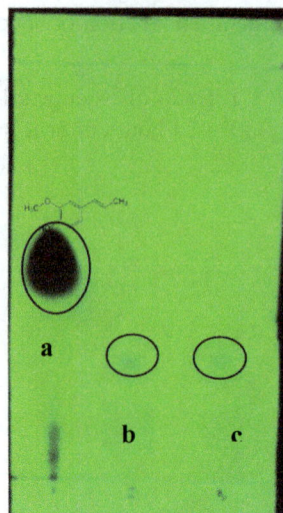

Fig. 4. TLC spots of (a) isoeugenol; (b) DDIE; and (c) synthesis product

The synthesis product's polarity character resembled to that of DDIE standard (Fig.4b and 4c), while the oily solution of isoeugenol (Fig.4a) eluted with the nonpolar solvent faster than the other two compounds, hence it resulted a higher Rf value.

The synthesis product melted in the range between 138-139°C, compared to Leopold's 132-133°C.

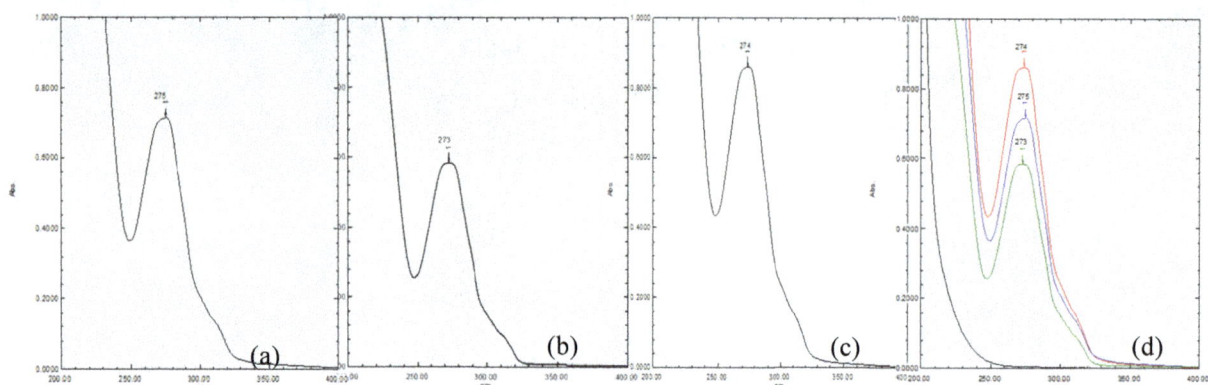

Fig. 5. UV spectrum of (a) DDIE standard; (b) synthesis product; (c) a mixture of DDIE standard-synthesis product (1:1); (d) overlay spectra of (a), (b), and (c)

Both DDIE (Fig.5a) and the synthesis product (Fig.5b) showed maxima at 275 and 273 nm, respectively. This maximum confirmed the λ_{max} prediction of Woodward-Fieser rules, that the chromophore structure of DDIE (Fig.1a) is the aromatic ring with –OH substituent which base value is 270 nm (Pavia et al, 2009; Silverstein et al, 1999).

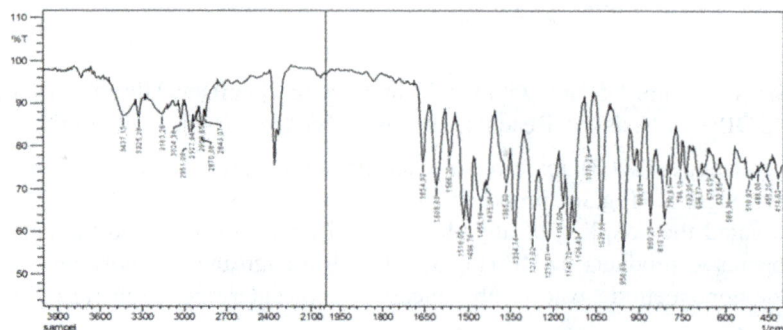

Fig. 6. IR spectrum of synthesis product

IR spectrum of the product (Fig.6) showed a specific broad O-H stretch at 3437.15 cm^{-1}, C-H aromatic and C-H alkene at 3163.26 and 3024.38 cm^{-1}, C-H alkyl stretch at 2951.09 and 2927.94 cm^{-1}. An overtone peak of aromatic was detected at 2100 to 1700 cm^{-1}. C-O peak was detected at 1126.43 cm^{-1}. These peaks resembled those of DDIE standard.

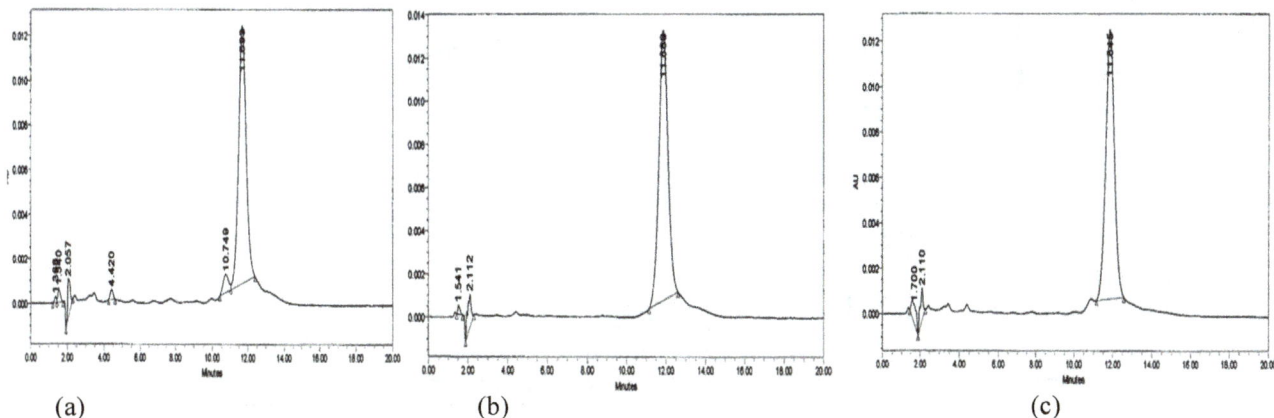

(a) (b) (c)

Fig.7. HPLC chromatogram of (a) DDIE standard; (b) synthesis product; (c) a mixture of DDIE standard-synthesis product (1:1)

Furthermore, HPLC chromatogram indicated a similar character of DDIE (R_t = 11.693 min, Fig.7a) and the synthesis product (R_t = 11.886 min, Fig.7b). Both compounds were eluted slowly due to their hydrophobic character, e.g. aromatics and methyls in the molecule (Fig.1a).

The synthesis product was further analyzed using HRMS with TOF detector, compared to DDIE, $C_{20}H_{22}O_4$ (MW = 326.15186 g/mol). A strong intensity (> 70 %) base molecular peak which indicated (M+H)$^+$ was detected at m/z 327.1595 (Fig.8). This peak was predicted belonged to DDIE.

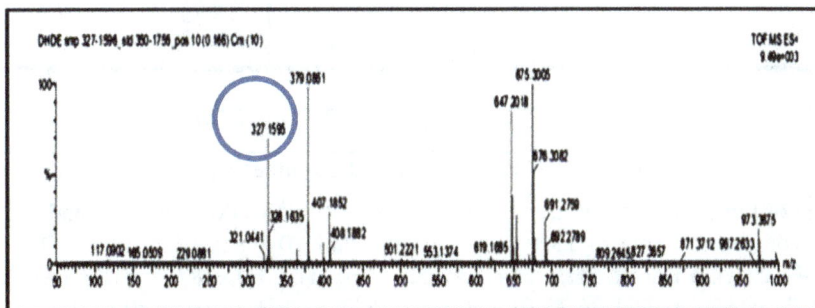

Fig. 8. HRMS spectrum of the synthesis product

The synthesis product was examined under a polarization microscope (Olympus BX53) with 400x magnification, and the solid needle-like crystals could be seen in Fig.9.

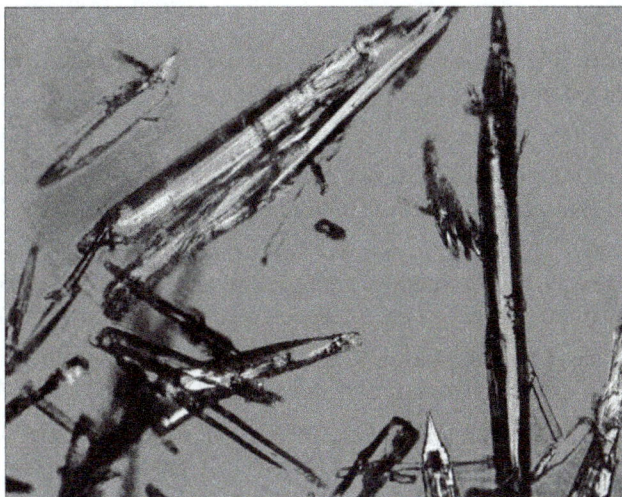

Fig. 9. Needle-like crystals of the synthesis product in 400x magnification

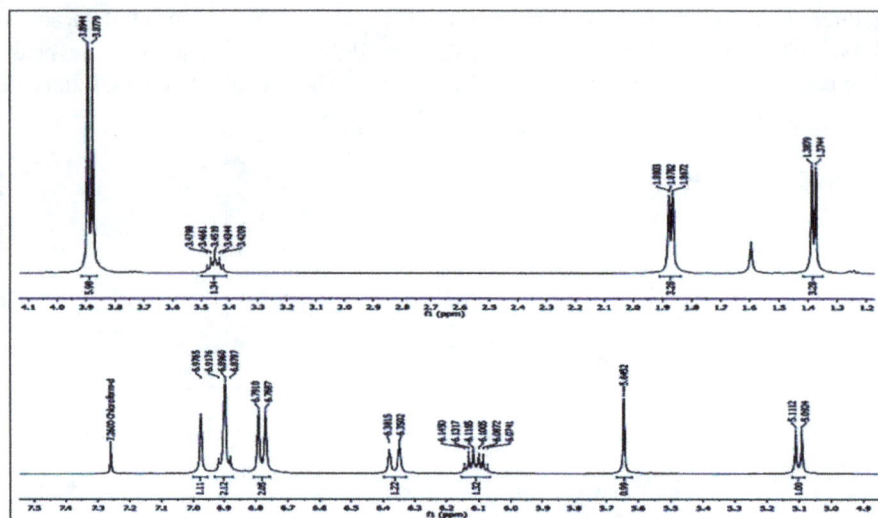

Fig. 10. ^{1}H-NMR spectrum of the synthesis product

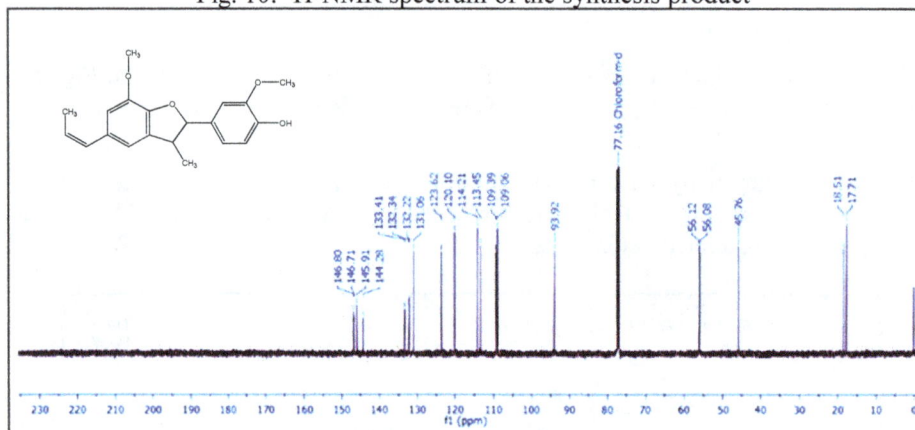

Fig. 11. ^{13}C-NMR spectrum of the synthesis product

The last characterization of the synthesis product was performed using ^{1}H-NMR (Fig.10) and ^{13}C-NMR (Fig.11) which shows the synthesized compound contains 13 types of proton and 20 types of carbon. 2D ^{1}H-NMR (CDCl$_3$/TMS) indicated: δ 1.37 (d, 3H, CH$_3$), δ 1.86 (t, 3H, CH$_3$), 3.42-3.47 (m, 1H, CH), 3.87 (s, 3H, CH$_3$-O), 3.89 (s, 3H, CH$_3$-O), 5.09 (d, 1H, CH), 5.64 (br s, 1H, Ar-OH), 6.35 (d, 1H, C17), 6.76 (s, 1H, C16), 6.79 (s, 1H, C14), 6.87-6.91 (m, 2H, CH$_3$, CH), 6.97 (s, 1H, CH$_3$), while 2D ^{13}C-NMR (CDCl$_3$) indicated: δ 17.7 (C9), 18.5 (C19), 45.7 (C8), 56.0 (CH$_3$-O 56.1 (CH$_3$-O), 93.9 (C7), 109.0 (C2), 109.3 (C12), 113.4 (C16), 114.2 (C5), 120.1 (C-6), 123.6 (C18), 131.0 (C17). 132.2 (C11), 132.1 (C1), 133.4 (C15), 144.2 (C13), 145.9 (C4), 146.7 (C14), 146.8 (C3).

4. Conclusion

Herein we reported that white needle-like crystals of DDIE using FeCl$_3$ as catalyst had been synthesized; moreover the decreasing of the catalyst reduced the yield of the product.

References

Darvish, F., & Khazraee, S. (2015). FeCl$_3$ Catalyzed One Pot Synthesis of 1-Substituted 1H-1,2,3,4-Tetrazoles under Solvent-Free Conditions. *International Journal of Organic Chemistry, 5,* 75-80.

Davin, L. B., Jourdes, M., Patten, A. M., Kim, K. W., Vassão, D. G., & Lewis, N. G. (2008). Dissection of Lignin Macromolecular Configuration and Assembly: Comparison to Related Biochemical Processes in Allyl/Propenyl Phenol and Lignan Biosynthesis. *Natural Product Reports, 25,* 1015-1090. http://dx.doi.org/10.1039/b510386j

Diaz, D. D., Miranda, P. O., Padron, J. I., & Martin, V. S. (2006). Recent Uses of Iron(III) Chloride in Organic Synthesis. *Current Organic Chemistry, 10,* 457-476. http://dx.doi.org/10.2174/138527206776055330_

Komeyama, K., Morimoto, T., Nakayama, Y., & Takaki, K. (2007). Cationic Iron-Catalyzed Intramolecular Hydroalkoxylation of Unactivated Olefins. *Tetrahedron Letters, 48,* 3259-3261. http://dx.doi.org/10.1016/j.tetlet.2007.03.004

Leopold, B. (1950). Aromatic Keto- and Hydroxy-polyethers as Lignin Models. *Acta Chemica Scandinavica, 4,* 1523-1537.

Lestari, K., Hwang, J. K., Kariadi, S. H., Wijaya, A., Ahmad, T., Subarnas, A., & Supriyatna, Muchtaridi, M. (2012). Screening for PPAR-gamma Agonist from *Myristica fragrans* Houtt Seeds for the Treatment of Type 2 Diabetes by *In Vitro* and *In Vivo*. *Medical and Health Science Journal, 12,* 7-15. http://dx.doi.org/10.15208/mhsj.2012.37

Li, F., & Yang, X. W. (2012). Analysis of Anti-Inflammatory Dehydrodiisoeugenol and Metabolites Excreted in Rat Feces and Urine using HPLC-UV. *Biomedical Chromatography, 26*(6), 703-707. http://dx.doi.org/10.1002/bmc.1717/pdf

Pavia, D. L., Lampman, G. M., Kris, G. S., & Vyvyan, J. R. (2009). *Introduction to Spectroscopy*. Brooks/Cole. Cengage Learning. p. 98

Saputri, F. A., Mutakin, M., Lestari, K., & Levita, J. (2014). Determination of Safrole in Ethanol Extract of Nutmeg (*Myristica fragrans* Houtt) Using Reversed-Phase High Performance Liquid Chromatography. *International Journal of Chemistry, 6*(3), 14-20. http://dx.doi.org/10.5539/ijc.v6n3p14

Setälä, H. (2008). Regio- and stereoselectivity of oxidative coupling reactions of phenols: Spirodienones as construction units in lignin. *Academic Dissertation*. http://www.vtt.fi/publications/index.jsp

Silverstein, R. M., Francis, X. W., & David, J. K. (1999). *Spectrometric Identification of Organic Compounds*. John Wiley & Sons, Inc. New York. 2-36.

Zhan, Z. P., & Liu, H. J. (2006). FeCl$_3$-Catalyzed Coupling of Propargylic Acetates with Alcohols. *Synlett, 14,* 2278-2280. http://dx.doi.org/10.1055/s-2006-949645

Carbonyl Chalcogenides Clusters Existing as Disguised Forms of Hydrocarbon Isomers

Enos Masheija Rwantale Kiremire

Correspondence: Enos Masheija Rwantale Kiremire, Department of Chemistry and Biochemistry, University of Namibia, Private Bag 13301, Windhoek, Namibia. E-mail: kiremire15@yahoo.com

Abstract

The 4n Series Method has been utilized to categorize, analyze and predict structures for transition metal carbonyl, borane, hydrocarbon and Zintl ion clusters. The method is being extended to study carbonyl chalcogenide clusters. Adequate examples have been given to demonstrate the application of the 4n series method to categorize clusters and where possible predict their possible skeletal structures. In this paper, the method is being applied to the study of carbonyl chalcogenide cluster complexes. What has been found is the striking structural similarity of a wide range of carbonyl chalcogenide clusters to those of corresponding hydrocarbon clusters. It was observed that when a derived hydrocarbon from a cluster, $F_{CH} = C_nH_q$, is such that n<q, the cluster portrays structural similarity with an equivalent hydrocarbon. On the other hand when n>q, the 'hydrocarbon character' becomes reduced and the typical cluster tendencies increase. When n = q, the situation becomes more or less a borderline case. When q=0, then $F_{CH} = C_n$. When the series becomes bi-capped or more, then the equivalent carbon cations are obtained.

Keywords: clusters, hydrocarbons, fragments, 4n- series, isolobal, capping -clusters, hydrocarbon- character, cations

1. Introduction

Although boron and carbon atoms are next to each other in the periodic table, the boranes and their relatives are dominated by polyhedral shapes unlike the hydrocarbon analogues (Cotton, and Wilkinson, 1980). Arising from the unique polyhedral structures and potential applications, the clusters have continued to fascinate many scientists (Lipscomb, 1963; Wade, 1971, 1976; Pauling, 1977; Mingos, 1972, 1984; Hawthorne, et al, 1999; Jemmis, 2005; Jemmis, et al, 2006; Jemmis, et al, 2008; Welch, 2013). Wade-Mingos rules have been exceedingly useful in categorizing and predicting the shapes of clusters (Wade, 1971, 1976; Mingos, 1972, 1984; Welch, 2013). During the study of the metal carbonyl clusters to determine the nature of their constitution, it was discovered that they generally obey the 14n rule while the main group clusters obey the 4n rule and that the two systems are interrelated via the isolobal relationship (Hoffmann, 1982; Kiremire, 2014, 2015, 2016). Furthermore, it was discerned that clusters could readily be categorized and their shapes predicted using the 4n series method (Kiremire, 2015, 2016).

2. Results and Discussion

2.1 The role of 4n Series Method and Its Unification of Clusters

In order to understand the application of series approach to categorize clusters and how they are interlinked, it is important to give some examples as illustrations. The application of series may be regarded as 'wearing a new type of spectacles' that makes it possible to see more details in a chemical fragment. The series method utilizes a simple algebra of valence electrons of fragments and ligands. The series formula which is in the form of S = 4n+q contains useful information. It provides the number valence electrons contained within a cluster. It can be used to categorize a cluster or generate another type of cluster as required. On closer analysis of series as they are applied to transition metal carbonyls, boranes, heteroboranes, metalloboranes, hydrocarbons and Zintl ions (Greenwood & Earnshaw, 1998), it is readily discerned that clusters are all interrelated. This is summarized in Scheme 1.

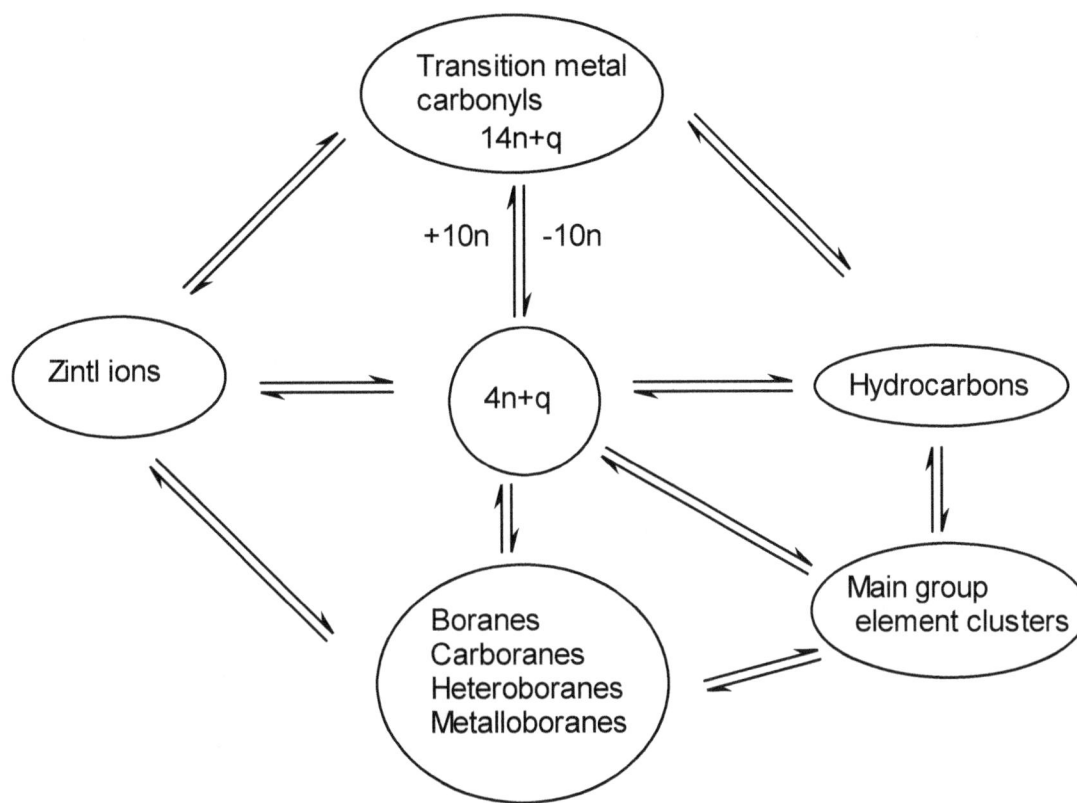

Scheme 1. Interrelationship among clusters

Selected examples of different types of clusters have been taken and transformed into equivalent hydrocarbon clusters for illustrations. These are given in 2.2 to 2.7. They demonstrate how a cluster formula may be decomposed into a series formula from which other clusters may be generated. The details of the method have been covered elsewhere (Kiremire, 2015, 2016). Furthermore, sufficient explanation is provided in each of the examples given.

2.2 B_2H_6

$2[BH] \rightarrow 2[3+1] = 2[4+0] \rightarrow 4n+0(n=2); F_{CH} = 4n+4$

$\quad\quad\quad 4(H) \rightarrow 0+4(n=0)$

$\quad\quad\quad\quad S = 4n+4(n=2)$

$F_{CH} = 4n+4 = C_2H_4$

$[C]$

2

H

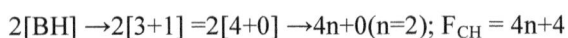

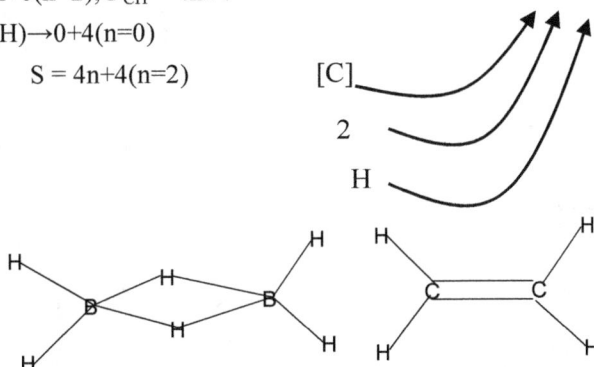

Figure F-1. Shape of B_2H_6 Figure F-2. Shape of C_2H_4

The B_2H_6 molecule belongs to the nido clan series. Its shape is shown in F-1. Its hydrocarbon isomer, C_2H_4 is shown in F-2. The cluster k value is given by $k = 2n-2 = 2(2)-2 = 2$. That means the two skeletal atoms are linked by 2 bonds. In case the of borane (F-1), the bridging H atoms supply the additional electrons to complete the double bonds.

2.3 Os$_3$(CO)$_{12}$

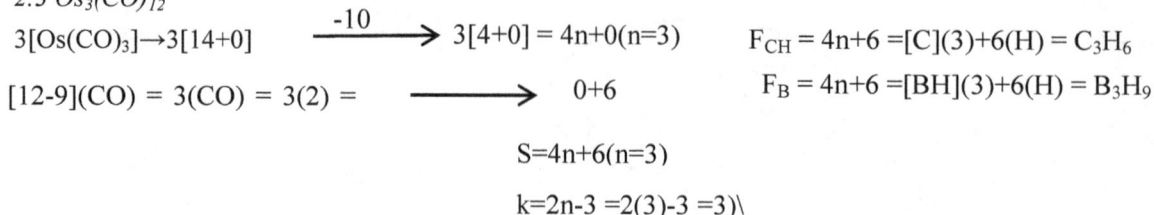

$$3[Os(CO)_3] \rightarrow 3[14+0] \xrightarrow{\quad -10 \quad} 3[4+0] = 4n+0(n=3) \qquad F_{CH} = 4n+6 = [C](3)+6(H) = C_3H_6$$

$$[12-9](CO) = 3(CO) = 3(2) = \longrightarrow 0+6 \qquad F_B = 4n+6 = [BH](3)+6(H) = B_3H_9$$

$$S = 4n+6 (n=3)$$

$$k = 2n-3 = 2(3)-3 = 3)\backslash$$

The skeletal shape of the Os$_3$(CO)$_{12}$ is a triangle. This is shown in Figure F-3. The corresponding 'mapped' isomeric hydrocarbon isomer is given in Figure F-4. When the hydrogen atoms are included in the hydrocarbon skeletal structure, we get figure F-5.

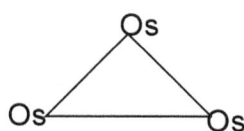

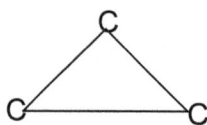

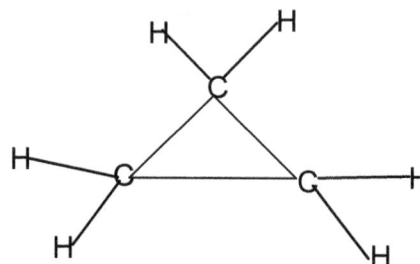

k = 3

k = 3

F-3

F-4

F-5

The reason for mapping the hydrocarbon isomer or designing the sketch to be similar to that of the parent cluster is that the hydrocarbon may have several other possible isomers.

2.4 B$_4$H$_{10}$

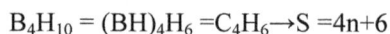

$$B_4H_{10} = (BH)_4H_6 = C_4H_6 \rightarrow S = 4n+6$$

The borane cluster can directly be transformed into a hydrocarbon cluster by factoring the formula into [BH] fragments as illustrated in the case of B$_4$H$_{10}$ without going through the series formula route. This is because [BH] $\leftrightsquigarrow$ [C] in terms valence electron content. The derivation of the series formula from a hydrocarbon cluster is also easy as a cluster C$_n$ is simply S = 4n+0. The hydrogen atoms in the hydrocarbon simply give us the variable figure after 4n. The cluster k value = 2n-2 = 2(4)-3 = 5. A skeletal sketch of B$_4$H$_{10}$ is given in F-6 and that of the C$_4$H$_6$ isomer in F-7 including the H atoms.

k = 5

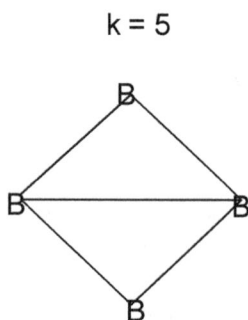

k = 5

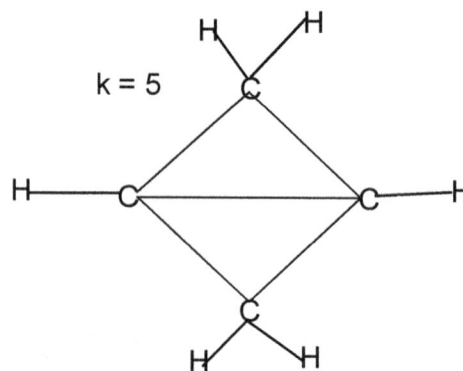

F-7

F-6

2.5 $Re_5(C)(H)(CO)_{16}^{2-}$

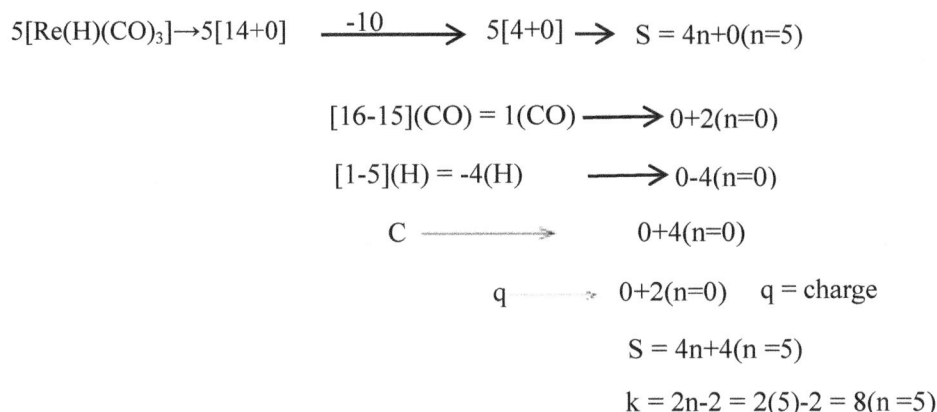

$$5[Re(H)(CO)_3] \rightarrow 5[14+0] \quad \xrightarrow{-10} \quad 5[4+0] \rightarrow S = 4n+0(n=5)$$

$$[16-15](CO) = 1(CO) \longrightarrow 0+2(n=0)$$

$$[1-5](H) = -4(H) \longrightarrow 0-4(n=0)$$

$$C \longrightarrow 0+4(n=0)$$

$$q \quad \longrightarrow \quad 0+2(n=0) \quad q = charge$$

$$S = 4n+4(n=5)$$

$$k = 2n-2 = 2(5)-2 = 8(n=5)$$

Since S =4n+4, the cluster belongs to the nido clan of clusters.The corresponding hydrocarbon and borane clusters can readily be derived from the series formula above. Thus, $F_{CH} = 4n+4 = [C](5)+4(H) = C_5H_4$ and $F_B = [BH](5)+4H = B_5H_9$. The sketches of the skeletal shapes of the rhenium cluster, $Re_5(C)(H)(CO)_{16}^{2-}$, the corresponding borane cluster, B_5H_9 and the hydrocarbon, C_5H_4 clusters are given in F-8, F-9 and F-10 respectively. The hydrocarbon isomer has to be mapped according to the rhenium cluster as was done in F-7 above.

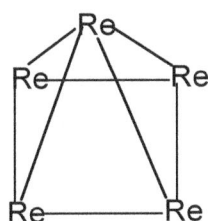
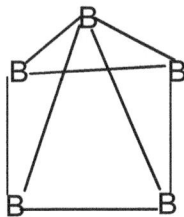
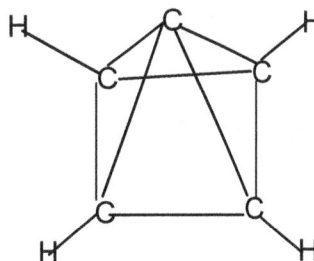

F-8   F-9   F-10

k =8   k =8   k =8

As can be seen from F-10, the C_5H_4 isomer is very much strained especially the pyramical-apex carbon atom which has no hydrogen attachment. Some interesting work on strained hydrocarbons has been going on for sometime (Stohrer and Hoffmann, 1972; Carnes, et al, 2008).

2.6 $Rh_5(CO)_{15}^-$ versus $Os_5(CO)_{15}^{2-}$

Using the same procure as in 2.2 to 2.5, $Rh_5(CO)_{15}^-$ belongs to S =4n+6 (arachno)while $Os_5(CO)_{15}^{2-}$ belongs to S =4n+2(closo) series. The 4n+6(n=5) can be transformed into B_5H_{11} (C_5H_6, k =7)while S = 4n+2 converts into $B_5H_5^{2-}$ (C_5H_2, k =9). $Rh_5(CO)_{15}^-$ has been described as having a trigonal bipyramid shape with elongated axial lengths while the $Os_5(CO)_{15}^{2-}$ cluster was described as having a regular trigonal bipyramid shape (Teo, et al, 1984). This is not surprising as they belong to two different series with different k values. Proposed skeletal shapes modeled by their hydrocarbons are given in F-11 and F-12.

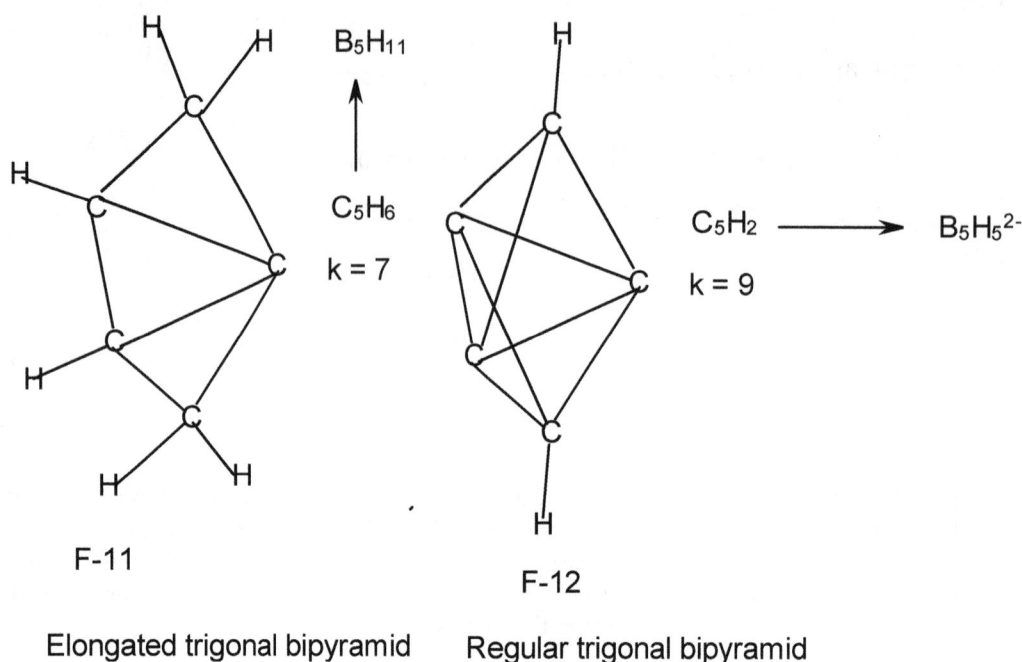

F-11	F-12
Elongated trigonal bipyramid	Regular trigonal bipyramid

The hydrocarbon isomers F-11 and F-12 are very much strained.

2.7 Octahedral Clusters: $Re_6(C)(CO)_{19}^{2-}$, $Os_6(CO)_{18}^{2-}$, $Rh_6(CO)_{16}$, $B_6H_6^{2-}$

The range of octahedral clusters is large. The above examples have been selected to be used as demonstrations of the power of series.

2.7.1 $Re_6(C)(CO)_{19}^{2-}$

$6[Re(H)(CO)_3]$

$\rightarrow 6[14+0] = 6[14+0-10] = 6[4+0] \rightarrow 4n+0(n=6)$.

$$[0-6](H) \rightarrow 0-6(n=0)$$

$$[19-18](CO) = 1(CO) \rightarrow 0+2(n=0)$$

$$(C) \rightarrow 0+4(n=0)$$

$$q \rightarrow 0+2(n=0), q = \text{charge}$$

$$S = 4n+2(n=6), \text{member of a closo clan of series.}$$

In deriving a series formula, it is found easier to use fragments whose electron contents which are 14(transition metals) or 4(for main group elements). In creating suitable fragments, balancing ligands such as H or CO which are not in the original formula of the cluster may be added and then subtracted later as in this example. The ligands which are not defined as part of the skeletal elements are assigned the value of 0. The transition metal clusters $Os_6(CO)_{18}^{2-}$ and $Rh_6(CO)_{16}$ are dealt with in the same way as $Re_6(C)(CO)_{19}^{2-}$ cluster and all of them belong to the 4n+2 series.

2.7.2 $B_6H_6^{2-}$

$6[BH] \rightarrow 6[3+1] = 6[4+0] \rightarrow 4n+0(n=6)$

$$q \rightarrow 0+2(n=0)$$

$$S = 4n+2(n=6)$$

This means that the clusters $Re_6(C)(CO)_{19}^{2-}$, $Os_6(CO)_{18}^{2-}$, $Rh_6(CO)_{16}$ and $B_6H_6^{2-}$ belong to the same family of the clan of series $S = 4n+2$ (n = 6, closo). They all portray a skeletal shape of an octahedral geometry (G-1).

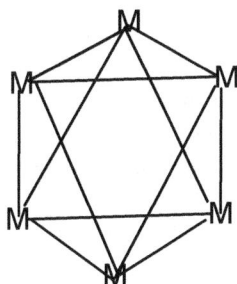

G-1. Octahedral geometry, O_h

Using the above series formula we can derive a hydrocarbon analogue of the clusters. This is given by the formula F_{CH} = 4n+2 = [C](6)+2(H) = C_6H_2. It will be interesting to map C_6H_2 cluster into a skeletal sketch of an octahedral geometry. One of the possible isomers is shown in F-13.

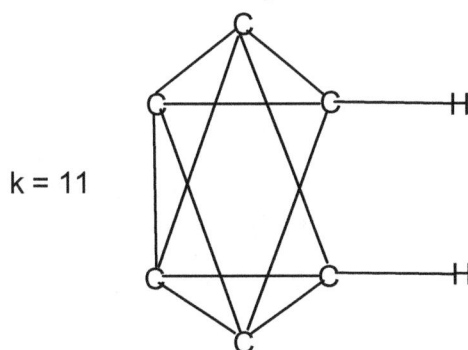

k = 11

F-13

According to the cluster series S = 4n+2 and k =2n-1. If n = 6, then k = 2(6)-1 = 11. This means that C_6H_2 has 11 skeletal linkages.   When the two hydrogen atoms are included, we get the sketch of the isomer shown in F-13. The hydrocarbon isomer is clearly very strained. Another possible hypothetical skeletal isomer of C_6H_2 is given in F-14 and its hydrogen counterpart in F-15.

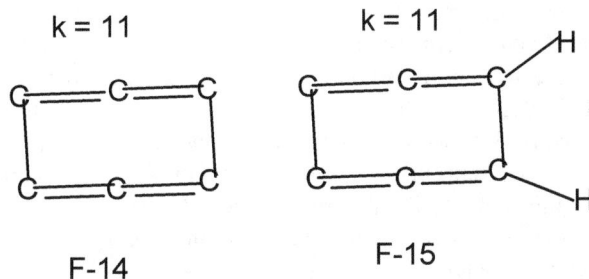

k = 11   k = 11

F-14   F-15

The skeletal linkages exclude those bonds which link the carbon atoms to the H atoms. What is also interesting for the hydrocarbon models such as F-11, F-12, F-13 and F-15 is that each of the carbon atoms obeys the 8-electron rule despite the severe strain of the configuration. The general series equation of the clusters may be represented as S = 4n+q. It appears that as q becomes smaller and smaller, the cluster becomes more and more strained as demonstrated by the corresponding hydrocarbon analogues. With this background, we can examine the carbonyl chalcogenide complexes using the 4n series method.

2.8 Categorization and Structural Prediction of Carbonyl Chalcogenides Using the 4n Series Method

The series method is being utilized to examine chemical systems including among others, atoms, molecules, fragments and clusters. In this paper, the method is being applied to analyze carbonyl chalcogenide complexes. The examples have been selected from various literature sources (Shieh, et al, 2003; Shieh, et al, 2012; Harlan, et al, 1996; Das, et al, 1997). The results of the samples analyzed are given in Table 1. Their series fall in the range of S= 4n-2 (bi-capped) to

S=4n+20. This can be compared with most common boranes which fall in the range S = 4n+2 (closo) to S =4n+8 (hypho). When the digit after 4n is large, this implies that some of the chalcogenide carbonyl clusters behave like hydrocarbons. For instance $C_{10}H_{22}$, belongs to the series S = 4n+22. What is amazing is that the structures of chalcogenide carbonyl clusters with q values such as 6, 8, 10, and 12 resemble some of the isomers of the analogous hydrocarbons. Selected examples are given in CCG-1 to CCG-5 below. The cluster $TeCr_2(Cp)_2(CO)_4$, S = 4n+4(n=3) and its equivalent hydrocarbon is C_3H_4. The cluster belongs to the nido clan series with k value of 4. The $TeCr_2$ skeletal fragment is found to have two double bonds of Cr atoms linked to Te element Cr=Te=Cr. This is similar to the allene isomer of C_3H_4. This is illustrated in CCG-1. The CCG-2 cluster, $Se_2Cr(Cp)(CO)_2^-$, belongs to the arachno series S = 4n+6 with k = 3. In this case, the Se_2Cr skeletal elements form a triangle. The series correspond to C_3H_6 hydrocarbon. The triangular shape of Se_2Cr fragment corresponds to cyclopropane which is one of the isomers of C_3H_6. These shapes are shown in CCG-2. The examples in CCG-3 to CCG-5 have been treated in the same manner. The striking similarity between some carbonyl clusters with hydrocarbons can further be demonstrated with the rhenium carbonyl cluster $Re_6H_5(CO)_{24}^-$ (Housecroft, 2005). This cluster belongs to S =4n+12 series with k value of 6. The corresponding hydrocarbon is C_6H_{12}. The carbonyl cluster has been found to have a cyclohexane shape which is similar to one of the isomers of C_6H_{12} hydrocarbon. Due to the striking structural similarity of derived hydrocarbons with those of the corresponding carbonyl chalcogenides Table 1 comprising of 19 clusters was constructed.

2.9 Extractions from Table 1

2.9.1 *Origin of Conventional Hydrocarbon Series*

There is some useful information that can be derived from Table 1. If we keep the number of skeletal atoms (carbon elements) constant and vary the number of hydrogen atoms (ligands), we can get ranges of hydrocarbon series. For example, C_6H_{14}, C_6H_{12}, C_6H_{10}, C_6H_8, up to C_6. Such series of hydrocarbon clusters are shown in Table 2. From Table 1, one of the carbonyl complex has been found to belong to the series S = 4n-2. Since n=8, this gives us the hydrocarbon analogue of the form C_8-2H. This type of hydrocarbon can be represented by C_8^{2+} cation. The series 4n-2 represents a clan of bi-capped series, Cp = C^2C[M-6]. The capping symbol means that there are 2 skeletal atoms capping onto 6 atoms which belong to 4n+2 closo series with an octahedral geometry. The first three columns on the left of Table 2 represent the well-known alkane, alkene and alkyne series respectively.

2.9.2 Hydrocarbons Arranged according to 4n Series

When Table 2 so constructed is scrutinized, we can see another type of series along the diagonal. These include among others, the series C, C_2, C_3, C_4, C_5, C_6 and so on and others CH_2, C_2H_2, C_3H_2, C_4H_2, C_5H_2, C_6H_2 and so on. These two examples correspond to mono-capped and closo series respectively. A sample of these series are given in Table 3.

2.9.3 Skeletal Linkages (k values) Arranged according to 4n Series

When Table 3 is also analyzed, it is observed that the k values also form series which represent the hydrocarbons in the table. This is given in Table 4. If we take M-2 in Table 4 and move from left to right we get k =4(4n+0), 3(4n+2), 2(4n+4) and 1 for (4n+6). These values correspond to C≡C, C≡C, C=C and C—C bonds respectively. The hydrocarbons having the triple, double and single carbon-carbon bonds are well known. However, the quadruple bond is still controversial (Shaik, et al, 2012) despite the fact that it was also predicted much earlier for C_2 and BeO(Teo, et al, 1984). Certain shapes may be recognized by looking at the series and their corresponding k values. For example, for M-3, k=3(4n+6) may represent cyclopropanes, M-4, k=6(4n+4) can correspond to tetrahedranes, k =4(4n+8),cyclobutanes, M-5; k =9(4n+2), trigonal bipyramid clusters, k=8(4n+4), square pyramid clusters, k =5(4n+10), cyclopentanes and for M-6; k =11(4n+2),octahedral clusters, k =10(4n+10), pentagonal pyramid, k =9(4n+6), benzenes and prismanes, and k =6(4n+12), cyclohexanes.

2.9.4 Rudolph System of Series (Upward Diagonal Relationship)

The Rudolph series of clusters are detected by looking at the diagonal of Table 4 upwards. Let us consider the following series: M-5; k =9 (4n+2), M-4; k =6(4n+4), M-3; and k =3(4n+6). The k values in this case, represent trigonal bipyramid, tetrahedral and trigonal planar clusters respectively. Another set to consider is M-6; k =11(4n+2), M-5; k = 8(4n+4), M-4; k=5(4n+6) and M-3; k=2(4n+8) represent octahedral, square planar and open M-3 cluster systems respectively. These series coincide with those of Rudolph system widely used to predict shapes of clusters together with Wade-Mingos rules (Rudolph, 1976; Wade, 1971).

2.9.5 General Considerations of Hydrocarbon Clusters $F_{CH} = C_nH_q$

Transition metal carbonyl, boranes, Zintl ion and main group element clusters can readily be transformed into hydrocarbon analogue clusters. The hydrocarbon clusters so produced can be mapped onto the corresponding shapes of the clusters from which they were produced. This means that the parent cluster adopts a skeletal shape of one of the corresponding hydrocarbon isomers. Since hydrocarbon clusters are versatile and less bulky, they may be useful in analyzing structural characteristics of clusters. A hydrocarbon cluster may be represented by the formula $F_{CH} = C_nH_q$. Hence, the corresponding series formula is $S = 4n+q$. As q becomes smaller and smaller, the k value increases more and more. This is clearly shown by the k values in Table 4. If we take some of the M-2 k values, for instance, $C_2(k = 4) > C_2H_2(k = 3) > C_2H_4(k=2) > C_2H_6(k=1) > C_2H_8(k=0)$. Thus, the k values decrease as the H ligands are added to the bi-skeletal cluster. This makes sense since the addition of 2H atoms to the C_2 fragment to form C_2H_2 molecule, two electrons from the C_2 are removed to participate in formation of the 2 C-H bonds resulting in the decrease of the k value from 4 to 3. The process continues until $C_2H_8(k = 0)$ cluster is formed. The k = 0 value means that the fragment C_2H_8 is unstable and decomposes to generate the more stable molecules $2CH_4$ which obey the octet rule. The variation of k values horizontally is well illustrated in Table 4. By looking at the hydrocarbon clusters, it appears that if n is kept constant and q is increased, the 'hydrocarbon character' of the cluster increases. That means the k value of the cluster decreases as well. Clearly, the k value is sensitive to the electron density of the valence electrons around the skeletal atoms. As the electron density of the skeletal atoms increases, the k values decrease and vice-versa. As q increases relative to n, a stage is reached where k value of the cluster represents same linkages as the hydrocarbon fragment and both the carbonyl cluster and hydrocarbon molecule portray similar shapes. For instance a cluster with a formula $F_{CH} = C_5H_2$ and k =9 and a cluster with a formula $F_{CH} = C_5H_{10}$ and k= 5, the later system may portray a shape similar to a cyclopentane isomer. Hypothetically, the C_5H_2 is electronically equivalent to the unknown C_5^{2-} ion. However, the heavier tetrel elements M_5^{2-}(M = Sn, Pb) in the form of Zintl ion clusters (Scharfe, et, al, 2010) are well known. Some of the possible skeletal shapes of M-2, M-3, M-4 and M-5 clusters are shown in Schemes 2 and 3. When clusters are analyzed more closely, and the ratio of number of valence skeletal electrons (V, based on 4n series) to the number of skeletal elements(n) is calculated for the clusters in Tables 1-3, it is found that , r =V/n > 5. Therefore, it appears that the "hydrocarbon character" starts when the ratio r =V/n > 5. This observation will be investigated further in the continued work. In a general hydrocarbon formula $F_{CH} = C_nH_q$, 'cluster character' tends to increase when n>>q. The 4n series can be viewed as a unifier that links up clusters such as boranes, carbonyls, hydrocarbons and Zintl ions (Scharfe, et al, 2010).

CCG-1: $TeCr_2(Cp)_2(CO)_4$

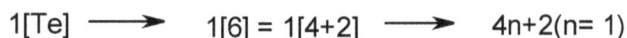

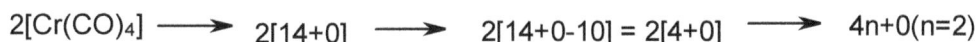

1[Te] $\longrightarrow$ 1[6] = 1[4+2] $\longrightarrow$ 4n+2(n= 1)

2[Cr(CO)$_4$] $\longrightarrow$ 2[14+0] $\longrightarrow$ 2[14+0-10] = 2[4+0] $\longrightarrow$ 4n+0(n=2)

[4-8](CO) $\longrightarrow$ -4(CO) $\longrightarrow$ 0-8(n=0)

2(Cp) $\longrightarrow$ 0+10(n=0)

S = 4n+4(n= 3)   Nido cluster

k = 2n-2 = 2(3)-2 = 4

$F_{CH} = 4n+4 = [C](3)+4(H) = C_3H_4$

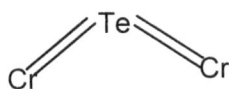

CCG-1

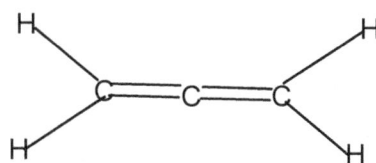

CCG-1-Hydrocarbon

CCG-2: $Se_2Cr(Cp)(CO)_2{}^{-}$

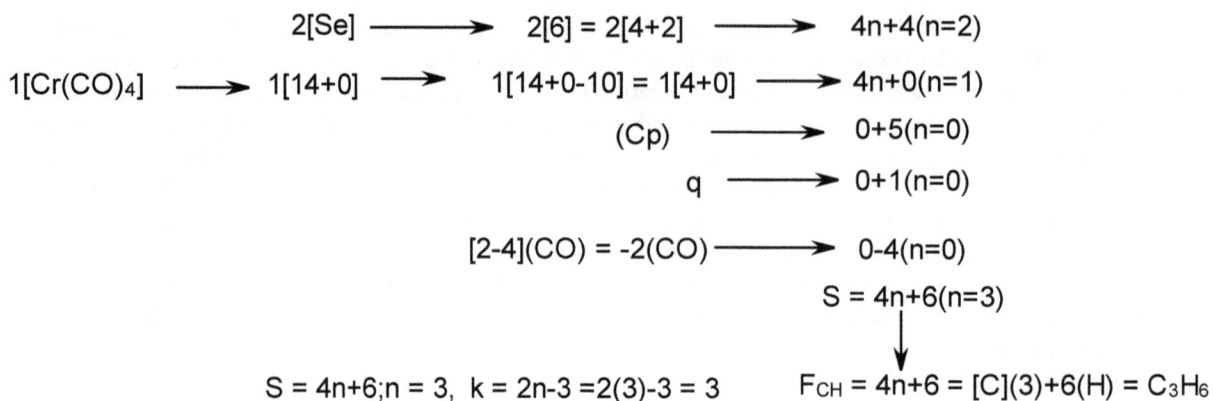

$$2[Se] \longrightarrow 2[6] = 2[4+2] \longrightarrow 4n+4(n=2)$$

$$1[Cr(CO)_4] \longrightarrow 1[14+0] \longrightarrow 1[14+0-10] = 1[4+0] \longrightarrow 4n+0(n=1)$$

$$(Cp) \longrightarrow 0+5(n=0)$$

$$q \longrightarrow 0+1(n=0)$$

$$[2-4](CO) = -2(CO) \longrightarrow 0-4(n=0)$$

$$S = 4n+6(n=3)$$

$$S = 4n+6; n = 3, \ k = 2n-3 = 2(3)-3 = 3 \qquad F_{CH} = 4n+6 = [C](3)+6(H) = C_3H_6$$

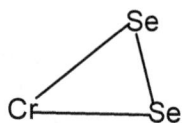

k = 3

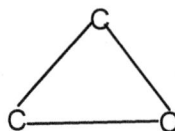

k = 3

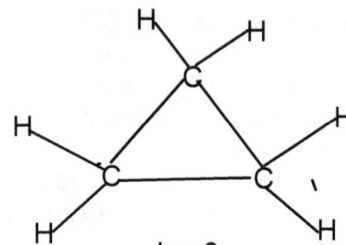

k = 3

CCG-2-Skeletal shape   CCG-2- Hydrocarbon

M-2: k = 1,2,3, 4

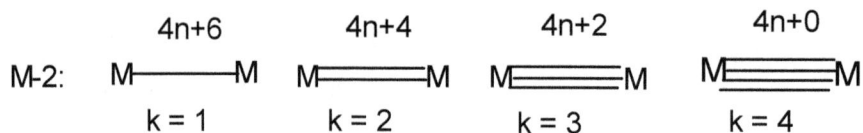

M-2:

4n+6	4n+4	4n+2	4n+0
k = 1	k = 2	k = 3	k = 4

M-3: k = 2, 3, 4

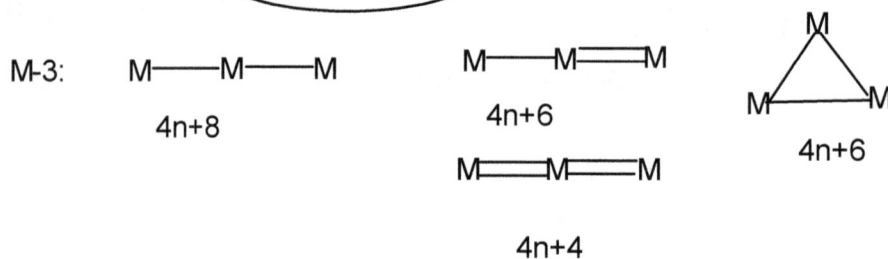

M-3:

4n+8

4n+6

4n+6

4n+4

M-4: k = 3, 4, 5, 6

k = 3   4n+10   k = 4   4n+8

M——M——M——M   M——M——M═══M

M-4   4n+8   4n+6   4n+4

k = 4   k = 5

k = 6

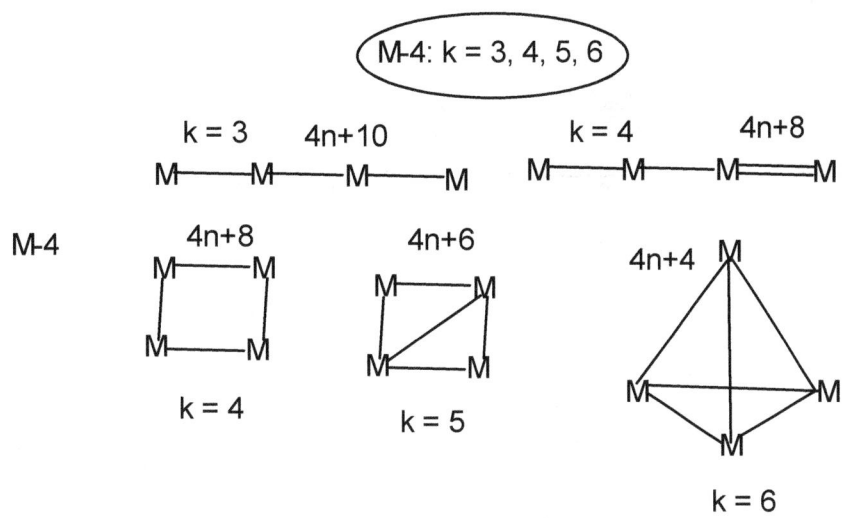

Scheme 2. Possible skeletal shapes M-2, M-3 and M-4 clusters

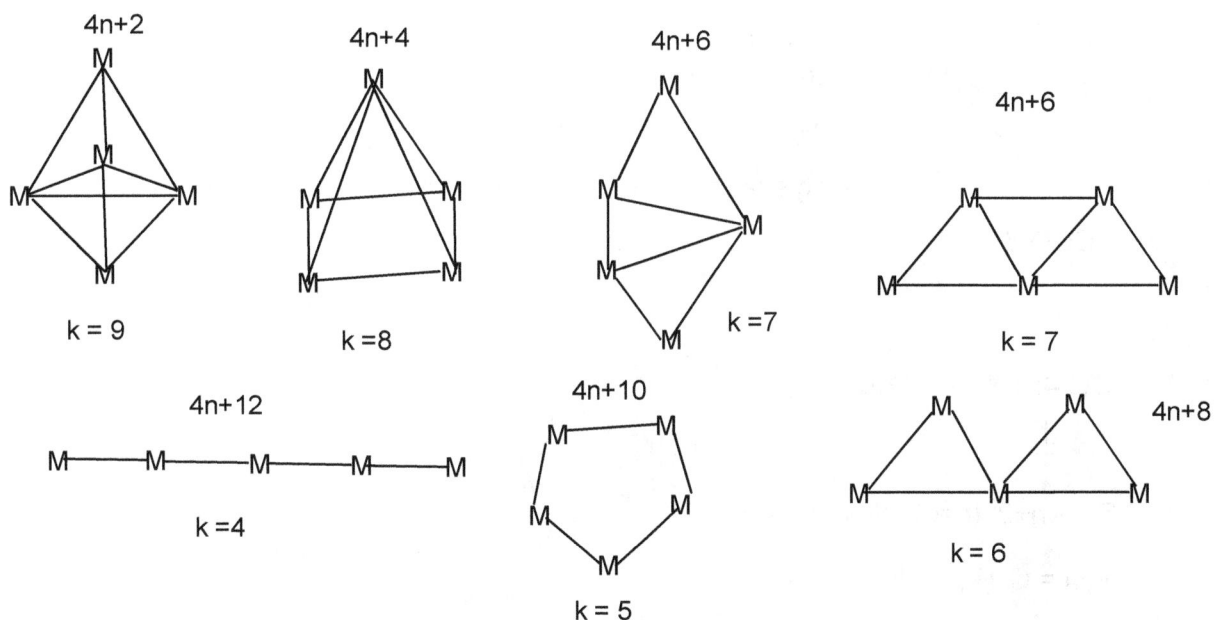

4n+2   4n+4   4n+6   4n+6

k = 9   k = 8   k = 7   k = 7

4n+12   4n+10   4n+8

k = 4   k = 5   k = 6

Scheme 3. Possible skeletal shapes of M-5 cluster systems

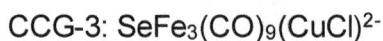

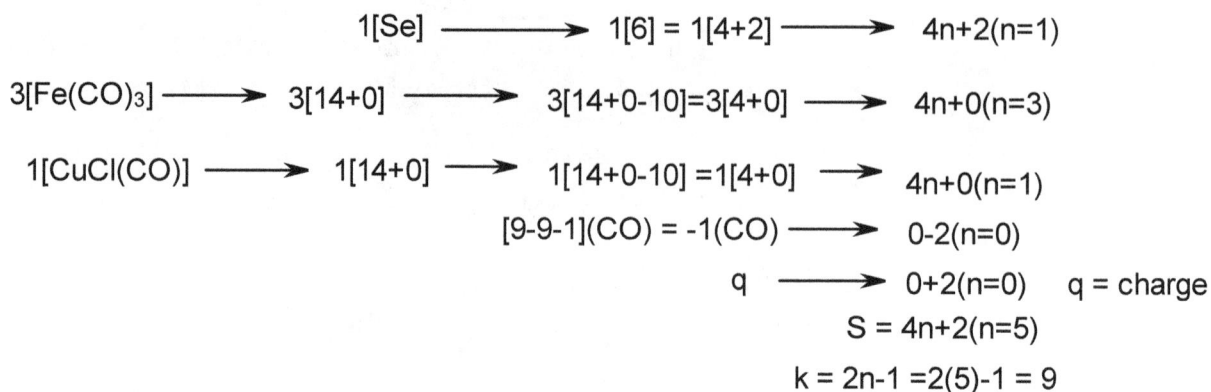

CCG-3: $SeFe_3(CO)_9(CuCl)^{2-}$

$1[Se] \longrightarrow 1[6] = 1[4+2] \longrightarrow 4n+2(n=1)$

$3[Fe(CO)_3] \longrightarrow 3[14+0] \longrightarrow 3[14+0-10]=3[4+0] \longrightarrow 4n+0(n=3)$

$1[CuCl(CO)] \longrightarrow 1[14+0] \longrightarrow 1[14+0-10] =1[4+0] \longrightarrow 4n+0(n=1)$

$[9-9-1](CO) = -1(CO) \longrightarrow 0-2(n=0)$

$q \longrightarrow 0+2(n=0)$ q = charge

$S = 4n+2(n=5)$

$k = 2n-1 =2(5)-1 = 9$

$F_{CH} = 4n+2 = [C](5)+2(H) = C_5H_2 = B_5H_5^{2-}$ This is a closo cluster and is expected to have a shape similar to that of $B_5H_5^{2-}$, trigonal bipyramid.

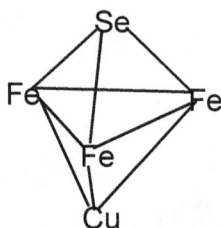

$S = 4n+2$

$k =9$

C_5H_2

CCG-3

Mapping skeletal structure of C_5H_2 will be the same as in F-12.

CCG-4: $TeRu_5(CO)_{14}^{2-}$

$S = 4n+2$, (n=6), closo cluster

$F_{CH} = C_6H_2,$ $F_B = B_6H_6^{2-}$

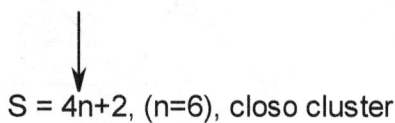

Octahedral shape expected.   CCG-4. Observed skeletal shape

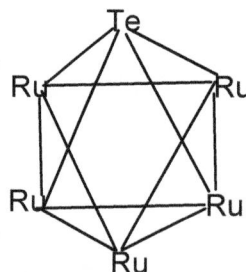

The C_6H_2 hydrocarbon can be mapped onto an octahedral geometry as in F-13.

CCG-5: $TeRu_5(CO)_{14}(CuCl)_2{}^{2-}$

$$1[Te] \longrightarrow 1[4+2] \longrightarrow 4n+2(n=1)$$

$$5[Ru(CO)_3] \longrightarrow 5[14+0] \longrightarrow 5[14+0-10]=5[4+0] \longrightarrow 4n+0(n=5)$$

$$2[CuCl(CO)] \longrightarrow 2[14+0] \longrightarrow 2[14+0-10] = 2[4+0] \longrightarrow 4n+0(n=2)$$

$$[14-15-2](CO) \longrightarrow -3(CO) \longrightarrow 0-6(n=0)$$

$$q \longrightarrow 0+2(n=0)$$

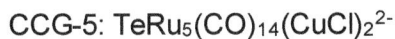

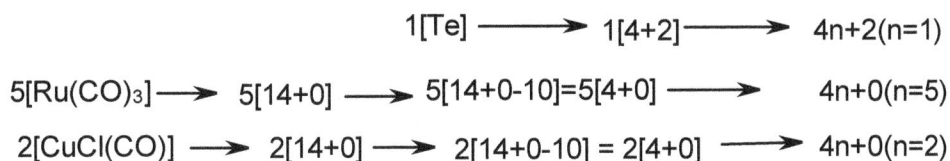

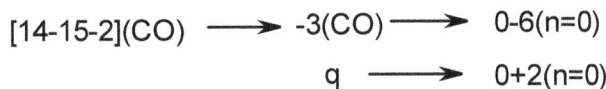

Bi-capped closo cluster $S = 4n-2(n=8)$ $F_C = C_8{}^{2+}$

$Cp = C^2C[M-6]$

This means 2 skeletal atoms are capped onto an octahedral geometry.

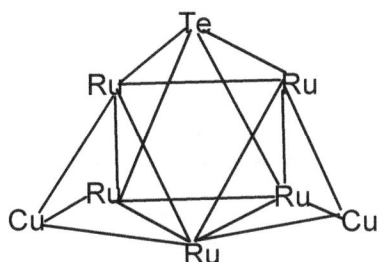

CCG-5. Observed skeletal shape

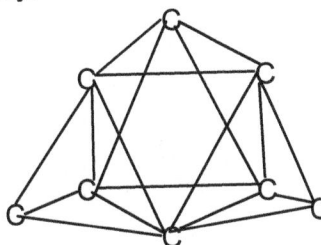

Possible hypothetical skeletal sketch of $C_8{}^{2+}$ ion

Table 1. Carbonyl chalcogenide clusters and their hydrocarbon analogues

Cluster	Series	n	Hydrocarbon	k value
$Cr_2(Cp)_2(CO)_6$	4n+6	2	C_2H_6	1
$SeCr_2(Cp)_2(CO)_4$	4n+4	3	C_3H_4	4
$Se_2Cr(Cp)(CO)_2{}^-$	4n+6	3	C_3H_6	3
$TeCr_2(CO)_{10}{}^{2-}$	4n+8	3	C_3H_8	2
$SeFe_3(CO)_9{}^{2-}$	4n+4	4	C_4H_4	6
$SeCr_3(CO)_6(Cp)_3{}^-$	4n+6	4	C_4H_6	5
$TeCr_3(CO)_{15}{}^{2-}$	4n+10	4	C_4H_{10}	3
$Se_2Cr_3(CO)_{10}{}^{2-}$	4n+2	5	C_5H_2	9
$Se_2Cr_3(CO)_{16}{}^{2-}$	4n+2	5	C_5H_2	9
$Se_2Mn_3(CO)_9$	4n+2	5	C_5H_2	9
$SeFe_3(CO)_9(CuCl)^{2-}$	4n+4	5	C_5H_4	8
$Te_4Cr(CO)_4{}^{2-}$	4n+10	5	C_5H_{10}	5
$SeFe_3(CO)_9(CuCl)_2{}^{2-}$	4n+0	6	C_6	12
$TeRu_5(CO)_{14}{}^{2-}$	4n+2	6	C_6H_2	11
$Te_2Cr_4(CO)_{18}{}^{2-}$	4n+10	6	C_6H_{10}	7
$Re_6H_5(CO)_2{}^{4-}$	4n+12	6	C_6H_{12}	6
$Te_2Cr_4(CO)_{20}{}^{2-}$	4n+14	6	C_6H_{14}	5
$TeRu_5(CO)_{14}(CuCl)_2{}^{2-}$	4n-2	8	$C_8{}^{2+}$	17
$CH_2Te_2Cr_6(CO)_{30}{}^{2-}$	4n+20	9	C_9H_{20}	8

Table 2. Hydrocarbons and carbon clusters arranged according to the number of skeletal carbon atoms

M-1	CH_4	CH_2	C	C^{2+}											
M-2	C_2H_6	C_2H_4	C_2H_2	C_2	C_2^{2+}										
M-3	C_3H_8	C_3H_6	C_3H_4	C_3H_2	C_3	C_3^{2+}									
M-4	C_4H_{10}	C_4H_8	C_4H_6	C_4H_4	C_4H_2	C_4	C_4^{2+}								
M-5	C_5H_{12}	C_5H_{10}	C_5H_8	C_5H_6	C_5H_4	C_5H_2	C_5	C_5^{2+}							
M-6	C_6H_{14}	C_6H_{12}	C_6H_{10}	C_6H_8	C_6H_6	C_6H_4	C_6H_2	C_6	C_6^{2+}						
M-7	C_7H_{16}	C_7H_{14}	C_7H_{12}	C_7H_{10}	C_7H_8	C_7H_6	C_7H_4	C_7H_2	C_7	C_7^{2+}					
M-8	C_8H_{18}	C_8H_{16}	C_8H_{14}	C_8H_{12}	C_8H_{10}	C_8H_8	C_8H_6	C_8H_4	C_8H_2	C_8	C_8^{2}				
M-9	C_9H_{20}	C_9H_{18}	C_9H_{16}	C_9H_{14}	C_9H_{12}	C_9H_{10}	C_9H_8	C_9H_6	C_9H_4	C_9H_2	C_9	C_9^{2+}			
M-10	$C_{10}H_{22}$	$C_{10}H_{20}$	$C_{10}H_{18}$	$C_{10}H_{16}$	$C_{10}H_{14}$	$C_{10}H_{12}$	$C_{10}H_{10}$	$C_{10}H_8$	$C_{10}H_6$	$C_{10}H_4$	$C_{10}H_2$	C_{10}	C_{10}^{2+}		
M-11	$C_{11}H_{24}$	$C_{11}H_{22}$	$C_{11}H_{20}$	$C_{11}H_{18}$	$C_{11}H_{16}$	$C_{11}H_{14}$	$C_{11}H_{12}$	$C_{11}H_{10}$	$C_{11}H_8$	$C_{11}H_6$	$C_{11}H_4$	$C_{11}H_2$	C_{11}	C_{11}^{2+}	
M-12	$C_{12}H_{26}$	$C_{12}H_{24}$	$C_{12}H_{22}$	$C_{12}H_{20}$	$C_{12}H_{18}$	$C_{12}H_{16}$	$C_{12}H_{14}$	$C_{12}H_{12}$	$C_{12}H_{10}$	$C_{12}H_8$	$C_{12}H_6$	$C_{12}H_4$	$C_{12}H_2$	C_{12}	C_{12}^{2+}

Table 3. Hydrocarbon clusters arranged according to series

	Bicp	Monocp	Closo	Nido	Arachno	Hypho	Klapo				
	4n-2	4n+0	4n+2	4n+4	4n+6	4n+8	4n+10	4n+12	4n+14	4n+16	4n+18
M-1	C^{2+}	C	CH_2	CH_4							
M-2	C_2^{2+}	C_2	C_2H_2	C_2H_4	C_2H_6						
M-3	C_3^{2+}	C_3	C_3H_2	C_3H_4	C_3H_6	C_3H_8					
M-4	C_4^{2+}	C_4	C_4H_2	C_4H_4	C_4H_6	C_4H_8	C_4H_{10}				
M-5	C_5^{2+}	C_5	C_5H_2	C_5H_4	C_5H_6	C_5H_8	C_5H_{10}	C_5H_{12}			
M-6	C_6^{2+}	C_6	C_6H_2	C_6H_4	C_6H_6	C_6H_8	C_6H_{10}	C_6H_{12}	C_6H_{14}		
M-7	C_7^{2+}	C_7	C_7H_2	C_7H_4	C_7H_6	C_7H_8	C_7H_{10}	C_7H_{12}	C_7H_{14}	C_7H_{16}	
M-8	C_8^{2}	C_8	C_8H_2	C_8H_4	C_8H_6	C_8H_8	C_8H_{10}	C_8H_{12}	C_8H_{14}	C_8H_{16}	C_8H_{18}
M-9	C_9^{2+}	C_9	C_9H_2	C_9H_4	C_9H_6	C_9H_8	C_9H_{10}	C_9H_{12}	C_9H_{14}	C_9H_{16}	C_9H_{18}
M-10	C_{10}^{2+}	C_{10}	$C_{10}H_2$	$C_{10}H_4$	$C_{10}H_6$	$C_{10}H_8$	$C_{10}H_{10}$	$C_{10}H_{12}$	$C_{10}H_{14}$	$C_{10}H_{16}$	$C_{10}H_{18}$
M-11	C_{11}^{2+}	C_{11}	$C_{11}H_2$	$C_{11}H_4$	$C_{11}H_6$	$C_{11}H_8$	$C_{11}H_{10}$	$C_{11}H_{12}$	$C_{11}H_{14}$	$C_{11}H_{16}$	$C_{11}H_{18}$
M-12	C_{12}^{2+}	C_{12}	$C_{12}H_2$	$C_{12}H_4$	$C_{12}H_6$	$C_{12}H_8$	$C_{12}H_{10}$	$C_{12}H_{12}$	$C_{12}H_{14}$	$C_{12}H_{16}$	$C_{12}H_{18}$

Table 4. The k values of hydrocarbon clusters

	Bicp	k_{bicp}	Monocp	Closo	Nido	Arachno	Hypho	Klapo				
	4n-2	2n+1	4n+0	4n+2	4n+4	4n+6	4n+8	4n+10	4n+12	4n+14	4n+16	4n+18
M-1	C^{2+}	3	2	1	0							
M-2	C_2^{2+}	5	4	3	2	1						
M-3	C_3^{2+}	7	6	5	4	3	2					
M-4	C_4^{2+}	9	8	7	6	5	4	3				
M-5	C_5^{2+}	11	10	9	8	7	6	5	4			
M-6	C_6^{2+}	13	12	11	10	9	8	7	6	5		
M-7	C_7^{2+}	15	14	13	12	11	10	9	8	7	6	
M-8	C_8^{2}	17	16	15	14	13	12	11	10	9	8	7
M-9	C_9^{2+}	19	18	17	16	15	14	13	12	11	10	9
M-10	C_{10}^{2+}	21	20	19	18	17	16	15	14	13	12	11
M-11	C_{11}^{2+}	23	22	21	20	19	18	17	16	15	14	13
M-12	C_{12}^{2+}	25	24	23	21	20	19	18	17	16	15	14

3. Conclusion

The sample of carbonyl chalcogenide complexes that were investigated portrayed a tendency of being dominated by clusters that structurally behave like hydrocarbons. The introduction of copper (group 11) seems to induce capping into the clusters as deduced by 4n series method. If n<q, in the hydrocarbon C_nH_q there is a tendency for a cluster fragment to structurally behave as its equivalent hydrocarbon. If n = q, then a borderline case appears to be attained. On the other hand if n>q, the clustering tendency appears to increase. When q=0, then S = 4n+0, we get carbon series C_n(n =1, 2,3,..) which belong to the clan of mono-capped series. Different clusters can be transformed into hydrocarbons. Such hydrocarbons may be used as model clusters. The 4n series may be viewed as a unifier of many different clusters and as a mirror for seeing special features in atoms, molecules, fragments and clusters.

Acknowledgements

The author wishes to express gratitude to the University of Namibia for the provision of facilities and NAMSOV, Namibia for the financial support, Mrs Merab Kambamu Kiremire for her continued encouragement to write the articles.

References

Carnes, M., Buccella, D., Siegrist, T., Steigerwald, M. L., & Nuckolls, C. (2008). Reactions of Strained Hydrocarbons with Alkenes and Alkyne Metathesis Catalysts. *J. Am. Chem. Soc., 130*(43), 14078-14079. http://dx.doi.org/10.1021/ja806351m

Cotton, F. A., & Wilkinson, G. (1980). *Advanced Inorganic Chemistry, 4th Ed.*, John Wiley and Sons, New York.

Das, B. K., & Kanatzidis, M. G. (1997). Methanothermal synthesis of polynuclear ruthenium telluride carbonyl clusters. *Polyhedron, 16*(17), 3061-3066. http://dx.doi.org/10.1016/S0277-5387(97)00018-1

Greenwood, N. N., & Earnshaw, A. (1998). *Chemistry of the Elements, 2nd Ed.* Butterworth, Oxford.

Harlan, C. J., Gillan, E. G., Bott, A. R., & Barron, A. R. (1996). Tert-Amyl compounds of Aluminium and Gallium: Halides, Hydroxides and Chalcogenides. *Organometallics, 15*, 5479-5488. http://dx.doi.org/10.1021/om9605185

Hawthorne, M. F., & Maderma, A. (1999). Applications of Radiolabelled Boron Clusters to the Diagnosis and Treatment of Cancer. *Chem. Rev., 99*, 3421-3434. http://dx.doi.org/10.1021/cr980442h

Hoffmann, R. (1982). Building Bridges between Inorganic and Organic Chemistry. *Angew. Chem. Int. Ed. Engl., 21*, 711-724. http://dx.doi.org/10.1002/anie.198207113

Housecroft, C. E., & Sharpe, A. G. (2005). *Inorganic Chemistry, 2nd Ed.*, Pearson, Prentice Hall, Harlow, England.

Jemmis, E. D. (2005). Building relationships between polyhedral boranes and elemental boron. *Inorg. Chem., 18*, 620-628.

Jemmis, E. D., & Prasad, D. L. V. K. (2008). Unknowns in the chemistry of Boron. *Current Science, 95*(10), 1277-1283.

Jemmis, E. D., Jayasree, E. G., & Parameswaran, P. (2006). Hypercarbons in polyhedral structures. *Chem. Soc. Rev., 35*, 157-168. http://dx.doi.org/10.1039/B310618G

Kiremire, E. M. (2014). Validation and verification of the Expanded Table for Transition Metal Carbonyl and Main Group Element Cluster Series which obey the 18-Electron and the 8-Electron Rules Respectively. *Orient. J. Chem., 30*(4), 1475-1495. http://dx.doi.org/10.13005/ojc/300404

Kiremire, E. M. (2015b). Classification of Transition Metal Carbonyl Clusters Using the 14n Rule Derived from Number Theory. *Orient. J. Chem., 31*(2), 605-618. http://dx.doi.org/10.13005/ojc/310201

Kiremire, E. M. (2015c). A Unique Bypass to the Carbonyl Cluster Nucleus Using the 14n Rule. *Orient. J. Chem., 31*(3), 1469-1476. http://dx.doi.org/10.13005/ojc/310326

Kiremire, E. M. (2015d). Unification and Expansion of Wade-Mingos Rules with Elementary Number Theory. *Orient. J. Chem., 31*(1), 387-392. http://dx.doi.org/10.13005/ojc/310146

Kiremire, E. M. (2015e). Boranes Hiding Inside Carbonyl Clusters. *Orient. J. Chem., 31(Spl, Edn)*, 121-127.

Kiremire, E. M. (2015f). Categorization and Structural Determination of Simple and More Complex Carbonyl Clusters of Rhenium and Osmium Using K-values and the Cluster Table. *Orient. J. Chem., 31*(1), 293-302. http://dx.doi.org/10.13005/ojc/310133

Kiremire, E. M. (2015g). Isolobal Series of Chemical Fragments. *Orient. J. Chem., 31(spl. Edn)*, 59-70.

Kiremire, E. M. R. (2015a). Capping and Decapping Series of Boranes. *Int. J. Chem. 7*(2), 186-197. http://dx.doi.org/10.5539/ijc.v7n2p186

Kiremire, E. M. R. (2016a). The categorization and Structural Prediction of Transition Metal Carbonyl Clusters Using the 14n Series Numerical Matrix. *Int. J. Chem., 8*(1), 109-125. http://dx.doi.org/10.5539/ijc.v8n1p109

Kiremire, E. M. R. (2016b). Generating Formulas of Transition Metal Carbonyl Clusters of Osmium, Rhodium and Rhenium. *Int. J. Chem., 8*(1), 126-144. http://dx.doi.org/10.5539/ijc.v8n1p126

Kiremire, E. M. R. (2016c). Unusual underground capping carbonyl clusters of palladium. *International J. Chem., 8*(1), 145-158. http://dx.doi.org/10.5539/ijc.v8n1p145

Lipscomb, W. N. (1963). *Boron Hydrides*. W. A. Bejamin, Inc., New York.

Miessler, G., Fischer, P., & Tarr, D. (2014). *Inorganic Chemistry, 5th Edition*, Pearson Education, Inc., Upper Saddle River.

Mingos, D. M. P. (1984). *Polyhedral Skeletal Electron Pair Approach. Acc. Chem. Res., 17*(9), 311-319. http://dx.doi.org/10.1021/ar00105a003

Mingos, D. M. P. (1972). A General Theory for Cluster and Ring Compounds of the Main Group and Transition Elements. *Nature (London), Phys. Sci., 236*, 99-102. http://dx.doi.org/10.1038/physci236099a0

Mingos, D. M. P. (1991). Theoretical aspects of metal cluster chemistry. *Pure and Appl. Chem., 83*(6), 807-812.

http://dx.doi.org/10.1351/pac199163060807

Pauling, L. (1977). Structure of Transition Metal Cluster Compounds: Use of an additional orbital resulting from f, g character of spd bond orbitals. *Proc. Natl. Acad. Sci., USA, 74*(12), 5235-5238. http://dx.doi.org/10.1073/pnas.74.12.5235

Rudolph, R. W. (1976). Boranes and heteroboranes: a paradigm for the electron requirements of clusters? *Acc. Chem. Res., 9*(12), 446-452. http://dx.doi.org/10.1021/ar50108a004

Scharfe, S., & Fässler, T. F. (2010). Polyhedral nine-atom clusters of tetrel elements and intermetalloid derivatives. *Phil. R. Soc. A,* 1265-1284. http://dx.doi.org/10.1098/rsta.2009.0270

Shaik, S., Danovich, D., Wu, W., Su, P., Rzepa, H. S., & Hiberty, P. C. (2012). Quadruple bonding in C_2 and analogous eight valence electron species. *Nature Chemistry, 4*, 195-200. http://dx.doi.org/10.1038/nchem.1263

Shieh, M., Ho, L. F., Guo, Y. W., Lin, S. F., Lin, Y. C., Peng, S. M., & Liu, Y. H. (2003). Carbonylchromium monotelluride complexes. *Organometallics, 22*, 5020-5026. http://dx.doi.org/10.1021/om034033+

Shieh, M., Miu, C. Y., Chu, Y. Y., & Lin, C. N. (2012). Recent progress in the chemistry of anionic groups 6-8 carbonyl chacogenide clusters. *Coord. Chem. Rev., 256*, 637-694. http://dx.doi.org/10.1016/j.ccr.2011.11.010

Stohrer, W. D., & Hoffmann, R. (1972). Bond-Stretch Isomerism and Polytopol Rearrangements in $(CH)_5^+,(CH)_5^-$ and $(CH)_4(CO)$. *J. Am. Chem. Soc., 94*(5), 1661-1668. http://dx.doi.org/10.1021/ja00760a039

Stohrer, W. D., & Hoffmann, R. (1972). Electronic and Reactivity of Strained Tricyclic hydrocarbons. *J. Am. Chem. Soc., 94*(2), 779-786. http://dx.doi.org/10.1021/ja00758a017

Teo, B. K., Longoni, G., & Chung, F. R. K. (1984). Applications of Topological Electron-Counting Theory to Polyhedral Metal Clusters. *Inorg. Chem., 23*(9), 1257-1266. http://dx.doi.org/10.1021/ic00177a018

Wade, A. (1976). Structural and Bonding Patterns in Cluster Chemistry. *Adv. Inorg. Chem. Radiochem., 18*, 1-66. http://dx.doi.org/10.1016/s0065-2792(08)60027-8

Wade, K. (1971). The structural significance of the number of skeletal bonding electron-pairs in carboranes, the higher boranes and borane ions and various transition metal carbonyl cluster compounds. *Chem. Commun.,* 792-793. http://dx.doi.org/10.1039/c29710000792

Welch, A. J. (2013). The significance of Wade's rules. *Chem. Commun., 49*, 3615-3616. http://dx.doi.org/10.1039/c3cc00069a

3

Generating Formulas of Transition Metal Carbonyl Clusters of Osmium, Rhodium and Rhenium

Enos Masheija Rwantale Kiremire

Correspondence: Enos Masheija Rwantale Kiremire, Department of Chemistry and Biochemistry, University of Namibia, Private Bag 13301, Windhoek, Namibia. E-mail:kiremire15@yahoo.com

Abstract

The series for carbonyl clusters of transition metals have been developed. They may be considered to be formed with a fragment centered around the 14 electron valence content. The capping series are based on the fragment of 12 electron valence content. The formulas of clusters can be decomposed into series from which the shapes of clusters may be predicted. The electron counting numbers of carbonyl clusters can be identified.

Keywords: Carbonyl and borane clusters, fragments, isolobal principle, numerical sequence of fragments, building blocks, cluster series, 14n and 4n rules, valence electron content, cluster number

1. Introduction

The discovery of the first mono-skeletal transition metal carbonyl $Ni(CO)_4$ one hundred and twenty five years ago (Mond, et al, 1890), opened a wide gate for the synthesis of hundreds of metal carbonyl complexes. The structure of the nickel carbonyl $Ni(CO)_4$ at the time was proposed to be as shown in Figure 1(F-1) in which the nickel atom exerted a valence of 2 while the valence of carbon and oxygen were 4 and 2 respectively. But the correct shape as we know it today is tetrahedral, Figure 1(F-2).

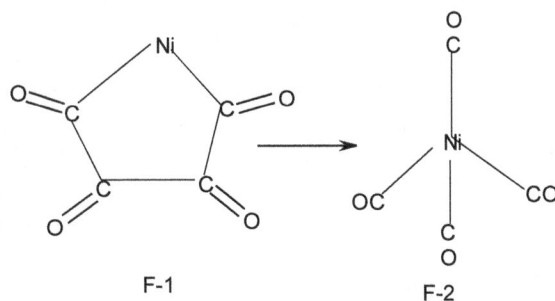

Figure 1. Historical Shapes of $Ni(CO)_4$

Other carbonyls such as $Fe(CO)_5$(Mond, et al, 1891), $Fe_2(CO)_9$(1905), $Re_2(CO)_{10}$(Hieber and Fuchs, 1941) and many others were discovered later. A good number of metal carbonyls and other related metal complexes have been applied in a good number of major industrial processes such as oxo-process, Monsanto process, Wacker oxidation process and Wilkinson catalyst in hydrogenation processes (Roth,1975;Ojima, et al, 2000;Grubbs, et al,2012; Wilkinson, et al, 1965; Tolman,1972). Many more carbonyl complexes of higher nuclearity skeletal metal atoms were synthesized (Dahl, et al, 2010). With emergency of borane chemistry (Stock, 1933; Longuet-Higgins , 1943; Lipscomb, 1963; Mingos, 1972; Lipscomb, 1976; Wade, 1976, Rudolph, 1976; Hoffmann, 1982; Mingos, 1984; King, 1986;Greenwood and Earnshaw,1998; Shriver and Atkins, 1999;Jemmis et al, 2001; 2002; 2003; Balakrishnarajan and Hoffmann, 2004; Jemmis, 2005; Housecroft and Sharpe, 2005; Jemmis, et al., 2008; Welch, 2013), it was found that the structures of carbonyl clusters could be explained using Polyhedral Skeletal Electron Pair Theory (PSEPT) or Wade-Mingos rules (Wade, 1971;Mingos, 1972).). It was also discovered that the structures of NIDO, ARACHNO, HYPHO and KLAPO clusters could be regarded as being successive derived from respective CLOSO clusters (Rudolph,1976). Transition metal carbonyl clusters and Zintyl ion clusters and others were found to be related to the borane clusters and their analogues (Crabtree, 2005; Miessler, et al, 2014). It became a great challenge to scientists to explain the intriguing geometrical shapes and bonding of the clusters let alone their chemical reactivity (Rudolph, 1976;King,2002). Due to their unique chemical reactivity, special synthetic techniques and handling them were invented (Stock, 1933). New clusters comprising of borane, main group elements and transition metal fragments were synthesized and these provoked the creation of ISOLOBAL PRINCIPLE which linked up the various fields of chemistry (Hoffmann, et all, 1976; Hoffmann,1982; Mingos, 1991, Shriver, et al, 1999).

It has taken the author some time in searching for simple formulas to teach undergraduate chemistry the Lewis shapes of molecules and clusters of boranes and transition metal carbonyls (Kiremire, 2008; Kiremire, et al, 2014; Kiremire, 2015a). This has resulted in the creation of series discussed in this paper, the introduction of the proposed cluster number (k), the capping symbol (C^nC) and generating functions of cluster fragments and molecules of the series(Kiremire,2014a; 2014b;Kiremire,2015b) also to be discussed in this paper. The ISOLOBAL SERIES of chemical fragments unifies all the clusters such as boranes, heteroboranes, metalloboranes, metallacarboranes, metallocenes, transition metal carbonyl clusters, Zintyl ion clusters as well as other clusters of main group elements'(Hoffman, 1982;Kiremire, 2015c).

A close scrutiny of the formulas of the clusters revealed that the transition metal carbonyl clusters can be categorized by 14n rule (Kiremire,2015c). In addition, it has recently been found that not only can they be categorized according to 14n but can also be generated using the same. The Wade-Mingos rules(Wade, 1971, 1976, Mingos, 1984, 1991, Welch, 2013) point to the existence of series through the n+1, n+2, n+3 or electron counting 4n+0(14n+0), 4n+2(14n+2) and so on based on their Molecular Orbital Theory calculations (Mingos, 1972; Mingos, 1984) . This paper presents a simple technique of the deriving the cluster formulas using the 14n(4n) series.

2. Results and Discussion

2.1 Formation of Cluster Series Using the 14n+0 as a Base

During the study of osmium carbonyl clusters(Kiremire,2015c, 2015d) it was found that $Os(CO)_3$ fragment which carries 14 electron valence content can be utilized as a fulcrum to generate as many hypothetical osmium carbonyl clusters as possible. With this fragment a range of multiples of the fragment can be produced. For instance, doubling the fragment, $2[Os(CO)_3]$ we produce $Os_2(CO)_6$ fragment. Tripling the fragment, we get $Os_3(CO)_9$. Other multiples are $Os_4(CO)_{12}$, $Os_5(CO)_{15}$, $Os_6(CO)_{18}$, $Os_7(CO)_{21}$, $Os_8(CO)_{24}$, $Os_9(CO)_{27}$ and $Os_{10}(CO)_{30}$ and so on. These are shown in one of the columns of Table 1. If we divide each formula by the coefficient of osmium in that formula (or simply dividing the formula by the multiplier of the initial fragment) we just obtain a constant of $Os(CO)_3$ in each case. For instance, $\frac{1}{2}$ $[Os_2(CO)_6] = Os(CO)_3$, $\frac{1}{3}$ $[Os_3(CO)_9] = Os(CO)_3$, $\frac{1}{5}$ $[Os_5(CO)_{15}] = Os(CO)_3$, $\frac{1}{10}$ $[Os_{10}(CO)_{30}] = Os(CO)_3$ and so on. Since the fragment $Os(CO)_3$ is the starting point of the column and its valence content is 14, it was deemed fit to label this column 14n or 14n+0 series. It is important to note that this result was previously obtained from molecular orbital calculations(Mingos,1991).When a CO ligand is added to $Os(CO)_3$,that is, $Os(CO)_3$+CO, we get a new fragment $Os(CO)_4$. The addition of the CO increases the valence electron content by 2. Therefore, we can regard $Os(CO)_4$ fragment as being the first member of the series S = 14n+2 since $Os(CO)_3$ fragment was considered to be the first member of the series S = 14n+0. Other fragments can be transformed in the same way. For example, $Os_2(CO)_6$+CO $\rightarrow$ $Os_2(CO)_7$ and $Os_9(CO)_{27}$ +CO $\rightarrow$ $Os_9(CO)_{28}$. Indeed if n =1 is substituted into the series S = 14n+2 of the fragment $Os(CO)_4$ we get S = 14x1+2 = 16 which is the same as the valence content of $Os(CO)_4$. If we continue the process and add another CO to $Os(CO)_4$ we get $Os(CO)_5$. This means that we have added a total of 4 electrons to $Os(CO)_3$ fragment. Therefore the fragment $Os(CO)_5$ can be considered to be the first member of the series S =14n+4. If we substitute n = 1 into S =14n+4, we get 18. This means that $Os(CO)_5$ complex obeys the 18 electron rule. Thus, by adding a CO to each member of the S = 14n+0 series in Table 1, we generate members of 14n+2 series. Similarly, by adding a CO to each member of the 14n+2 we get the clusters of S = 14n+4 series. Some of these are given in Table 2 which is simply an extension of Table 1. Again starting with the 14 electron valence fragment of osmium, $Os(CO)_3$ and removing one CO ligand, we get a 12 valence electron fragment $Os(CO)_2$. The removal of the CO ligand means the loss of 2 electrons. Therefore the fragment $Os(CO)_2$ can be considered to be the first member of 14n-2 series of osmium carbonyl clusters. If we remove a CO ligand from every member of the 14n+0 members, then the members of the series S = 14n-2 are produced. Some of these are shown in Table 2. Further removal of a CO ligand from $Os(CO)_2$ we get a 10 electron valence fragment $Os(CO)$ which can be regarded as the first member of the S = 14n-4 series. The removal of CO from the members of S = 14n-2 can be extended to generate the clusters or fragments of 14n-4 series. The carbonyl clusters have been categorized as 14n+0(4n+0) mono-capped closo, 14n+2(4n+2) as closo, 14n+4(4n+4) as nido, 14n-2(4n-2) as bi-capped closo, 14n-4(4n-4) tri-capped closo and so on. This classification has been expanded and is given in Table 3. What this means is that if we take the 14n+0 as the middle column, moving to the right successively adding CO ligands we create series with positive integers after 14n(4n) whereas moving to the left of the 14n+0 series successively removing the CO ligand we create negative integers after (14n(4n). A good number of the known osmium carbonyl clusters** are given in Table 4. A good number of the clusters in the table are the same or electronically equivalent to those generated in Table 3.These include, $Os_2(CO)_9$, $Os_3(CO)_{12}$, $Os_4(CO)_{14}$, $Os_5(CO)_{16}$, $OS_6(CO)_{18}$, $Os_6(CO)_{19}$ which equivalent to $Os_6(CO)_{18}^{2-}$, and $Os_7(CO)_{21}$, $Os_7(CO)_{22}$, $Os_8(CO)_{23}$,$Os_9(CO)_{25}$(disguised as $Os_9(CO)_{24}^{2-}$), and $Os_{10}(C)(CO)_{24}^{2-}$.The derivation of the known carbonyl clusters using series underpins the great significance of the same. Table 2 can be extended as needed. A sketch of how the osmium carbonyl clusters can readily be generated and is shown in Figure 1 for 14n+0. It is quite clear that the movement along a column from one fragment to the next the difference is equivalent to $Os(CO)_3$ fragment which equivalent to 14 valence electrons. A sketch showing the derivation of uncapping series (those with 14n followed by positive even integers)is given such as S = 14n+2, 14n+4, 14n+6, and so on which are obtained by successive addition of CO on the right of 4n+0 series and the capping series (those with negative even integers after 14n) obtained by successive removing the CO from the 14n+0 series is given in Figure 2. The 14n series represent mono-capped series. The formation of carbonyl clusters of a given type of series is similar is similar to the formation of compounds between two elements one of which is exerting a fixed valence while the other portrays variable valence or oxidation states. For instance chromium(Cr) and fluorine(F) form the compounds CrF_2, CrF_3, CrF_4, CrF_5 and CrF_6. In these compounds, fluorine exerts a fixed valence of one while the chromium exerts the oxidation states (valencies) of 2,3, 4, 5 and 6. Likewise, for a fixed cluster series say 4n+2, the variation of n = 1, 2, 3, 4, 5 and so on to generate different clusters is similar to the variation of the chromium oxidation state. Now let us consider the osmium carbonyl clusters of 14n+2 (CLOSO) series. As we know, the building bloc fragment for osmium carbonyl clusters is $[Os(CO)_3]$. The building block has14 valence electrons. When n = 1, we generate F = 14n+2 = $[Os(CO)_3](1)+CO = Os(CO)_4$ fragment. While n = 2, we produce, F =$[Os(CO)_3](2)+CO = Os_2(CO)_6+CO = Os_2(CO)_7$ and for n values 3,4,5,and 6 we get $Os_3(CO)_{10}$,$Os_4(CO)_{13}$,$Os_5(CO)_{16}$,and $Os_6(CO)_{19}$ (disguised as $Os_6(CO)_{18}^{2-}$). Thus, n behaves as a variable valence similar to that in transition metals or main group elements. For example, in the case of compounds such as ClF, ClF_3, ClF_5 the fluorine atom can be regarded as exerting a valence of = 1, while Cl is portraying the valences 1, 3, and 5 or in SF_4 and SF_6 compounds the sulphur atom is using the valences V= 4 and 6. But when sulphur combines with oxygen which exerts a valence of 2, it can form compounds SO_2 and SO_3. The change of combination of sulphur with F to combination with O to produce different compounds is similar to changing of n values when generating cluster fragments when the series changes from CLOSO 4n+2 to NIDO 4n+4. In order to elaborate this point, let us consider the closo series S = 14n+2 and nido series S = 14n+4 for rhodium carbonyl clusters. The buiding bloc for rhodium may be taken as $[Rh(H)(CO)_2]$. This is a fragment loaded with 14 valence electrons. When n = 6, the corresponding cluster carbonyl formulas are F= 14n+2 = $[Rh(H)(CO)_2](6)+CO = Rh_6H_6(CO)_{12}+CO = Rh_6(CO)_3(CO)_{12}+CO = Rh_6(CO)_{15}+CO =Rh_6(CO)_{16}$. Therefore the corresponding nido (S = 14n+4 is when n = 6 is simply obtained by adding a CO ligand to rhodium cluster $Rh_6(CO)_{16}$. That is, $Rh_6(CO)_{16} +CO = Rh_6(CO)_{17}$. Let us compare the complexes generated when n = 4 for the same transition metal element rhodium. The closo complex F = 14n+2 = $[Rh(H)(CO)_2](4)+CO = Rh_4H_4(CO)_8+CO = Rh_4(CO)_2(CO)_8+CO= Rh_4(CO)_{10} +CO = Rh_4(CO)_{11}$. The corresponding nido cluster will be obtained by just adding a CO to the closo cluster $Rh_4(CO)_{11} +CO = Rh_4(CO)_{12}$. In this way, both $Rh_6(CO)_{16}$ and $Rh_4(CO)_{11}$ belong to the same closo series S = 14n+2 while $Rh_6(CO)_{17}$ and $Rh_4(CO)_{12}$ clusters belong to the nido series S = 14n+4.

Although we are discussing carbonyl clusters in this article, the boranes which obey the 4n rule, behave in the same manner. For instance S = 4n+2 closo series, we know that no neutral boranes are known and the building fragment in this case is [BH]. Hence to generate the charged closo boranes, we must leave the integer 2 to represent the charge. Thus, for n =4, F = 4n+2 = $[BH](4)+2 = B_4H_4^{2-}$,n= 5, F =$[BH](5)+2 = B_5H_5^{2-}$,n=6, F=$[BH](6)+2 = B_6H_6^{2-}$, and n =12, F=$[BH](12)+2 = B_{12}H_{12}^{2-}$. The exception appears to be n = 1, F =$[BH](1)+2H = BH_3$. Let us generate more boranes with S = 4n+4(nido) series. When n = 1, F = $[BH](1) +4H = BH_5$, n= 2, F = $[BH](2)+4H = B_2H_6$, n= 3, F =$[BH](3)+4H = B_3H_7$,n= 4, F = $[BH](4)+4H =B_4H_8$,n = 5, F =$[BH](5)+4H = B_5H_9$, and n= 6, F= $[BH](6)+4H = B_6H_{10}$.

In the discussion above, the rhodium fragment $[Rh(H)(CO)_2]$ was utilized to produce four hypothetical clusters. The fragment can be used to generate many more clusters. This has been done and the results are given in Tables 5. A good number of known rhodium carbonyls or their equivalents(Greenwood and Earnshaw,1998) can be identified in the Table 6. The examples include $Rh_4(CO)_{12}$ disguised as $Rh_4H_4(CO)_{10}$, $Rh_6(CO)_{16}$ which is disguised as $Rh_6(H)_6(CO)_{13}$, $Rh_7(CO)_{16}^{3-}$disguised as $Rh_7H_7(CO)_{14}$, $Rh_{10}(CO)_{21}^{2-}$ disguised as $Rh_{10}H_{10}(CO)_{17}$and $Rh_{11}(CO)_{23}^{3-}$disguised as $Rh_{11}H_{11}(CO)_{19}$. Clearly, the hypothesis of a 14 electron valence fragment with plus or minus CO to generate cluster series is still valid for rhodium carbonyls. Many other known carbonyls of rhodium(Greenwood and Earnshaw,1998) can be identified in the hypothetically derived Table 5. A diagram for deriving rhodium carbonyl clusters are shown in Figure 3.

The utilization of 14 valence electron mono-skeletal carbonyl fragment was also tested on rhenium (Re) element. Since rhenium has 7 valence electrons, we need to add some suitable number of carbonyl ligands to constitute 14 electrons. In this regard, an appropriate fragment $[Re(H)(CO)_3]$ was selected. The presence of (H) in the fragment is to accompany Re which has an odd number of valence electrons and CO ligand has an even number of valence electrons and the required fragment is an even number of 14 electrons. Thus the selected fragment has $[7+1+3\times2 = 14]$electrons. This became the first member of the $S = 14+0$ series of rhenium. The mono-capped series of rhenium were derived starting with $[Re(H)(CO)_3]$ fragment and ending at $Re_{20}H_{20}(CO)_{60}$. The hypothetically derived rhenium carbonyl clusters are given in Table 7. A good number of known rhenium carbonyl clusters or their equivalents (Greenwood and Earnshaw,1998)were identified in the Table 8. Among these include, $Re_2H_2(CO)_8$, $Re_3H_3(CO)_{10}^{2-}$,$Re_4H_4(CO)_{12}$, $Re_4(CO)_{16}^{2-}$, $Re_5(C)(CO)_{16}(H)^{2-}$,$Re_6(C)(CO)_{19}^{2-}$, and $Re_7(C)(CO)_{21}^{3-}$. The principle of utilizing the 14n+0 mono-skeletal fragment and CO ligand for generating the members of 14n series appears to be applicable to a number of transition metal elements. A sketch for deriving rhenium carbonyl clusters is shown in Figure 4.

2.2 The Capping Series based on the Closo Series $S = 4n+2$

Using the example of osmium carbonyl series, we can identify the capping series which are lie on the diagonal fragments in the Table 2. For example, the $Os_6(CO)_{18}^{2-}[Os_6(CO)_{19}]$is a member of the closo series $S = 14n+2$ has the capping clusters with formulas $Os_7(CO)_{21}$[mono-capped O_h](Mingos,1991), $Os_8(CO)_{23}$ [bi-capped O_h](Hughes and Wade, 2000), $Os_9(CO)_{25}$[tri-capped O_h], and $Os_{10}(CO)_{27}$ [tetra-capped O_h]. But $Os_8(CO)_{23}$ occurs as $Os_8(CO)_{22}^{2-}$,while $Os_9(CO)_{25}$ and $Os_{10}(CO)_{27}$ occur as $Os_9(CO)_{24}^{2-}$ and $Os_{10}(CO)_{26}^{2-}$ respectively. $Os_5(CO)_{16}$ has the capping clusters with formulas $Os_6(CO)_{18}$(mono-capped)(, $Os_7(CO)_{20}$(bi-capped), $Os_8(CO)_{22}$(tri-capped), $Os_9(CO)_{24}$(tetra-capped) and $Os_{10}(CO)_{26}$(penta-capped). The $Os_6(CO)_{18}$ has been reported earlier as a mono-capped trigonal bipyramid (D_{3h}) complex(Hughes and Wade, 2000).Thus, the capping clusters that have been encountered are based upon the CLOSO systems $S = 14n+0(4n+0)$. If we use the generated osmium carbonyl clusters in Table 2 as examples, apart from the capping series of closo $Os_5(CO)_{16}$ and $Os_6(CO)_{19}$ already mentioned, the other capping series based on the closo clusters $S = 14n+0$ are as follows: for $Os(CO)_4{\rightarrow}Os_2(CO)_6{\rightarrow}Os_3(CO)_8{\rightarrow}Os_4(CO)_{10}{\rightarrow}Os_5(CO)_{12}{\rightarrow}Os_6(CO)_{14}$; for $Os_2(CO)_7{\rightarrow}Os_3(CO)_9{\rightarrow}Os_4(CO)_{11}{\rightarrow}Os_5(CO)_{13}{\rightarrow}Os_6(CO)_{15}{\rightarrow}Os_7(CO)_{17}$; for $Os_3(CO)_{10}{\rightarrow}Os_4(CO)_{12}{\rightarrow}Os_5(CO)_{14}{\rightarrow}Os_6(CO)_{16}{\rightarrow}Os_7(CO)_{18}{\rightarrow}Os_8(CO)_{20}$; for $Os_7(CO)_{22}{\rightarrow}Os_8(CO)_{24}{\rightarrow}Os_9(CO)_{26}{\rightarrow}Os_{10}(CO)_{28}$; for $Os_8(CO)_{25}{\rightarrow}Os_9(CO)_{27}{\rightarrow}Os_{10}(CO)_{29}$;and for $Os_9(CO)_{28}{\rightarrow}Os_{10}(CO)_{30}$. In all these capping series examples, we can identify a very important capping fragment of formula $Os(CO)_2$. This fragment carries 12 valence electrons and is the first member of bi-capping series 14n-2 or 4n-2.

2.3 Symbol for capping $Cp = C^nC$

In our earlier work, a symbol for capping was introduced (Kiremire,2015a). The proposed symbol is $Cp = C^nC$. Since for the series $S = 4n+0$ $(14n+0)$ represents mono-capped series, a symbol $Cp = C^1C$ was introduced. The series $S = 4n-2$ represented bi-capped series and $S = 4n-4$ tri-capped series, and $S = 4n-6$ tetra-capped series, and so on. This means that for every addition of (-2) integer after 4n or 14n represents an additional capping, then for $S= 4n-2 = S = 4n+(-2)$, and $Cp = C^1C+C^1C = C^2C$. This is because 14n implies a single capping C^1C and +(-2) after it is another cap C^1C and so we get a total of two caps C^2C. Also, $S = 4n-4 = 4n+2(-2)$, and $Cp = C^1C+C^2C = C^3C$. Therefore, for $S = 4n-6 = 4n+3(-2)$, $Cp = C^1C+C^3C = C^4C$. Let us take the case of osmium again, $Os_6(CO)_{18}^{2-}$, $S = 4n+2$ $(14n+2)$ this is an octahedral complex [M-6] where the symbol [M-6](F-1) represents a closo cluster comprising of 6 atoms. The cluster $Os_7(CO)_{21}$, $S = 4n+0(14n+0)$ is referred to as a mono-capped octahedral cluster(F-2). The capping symbol for this will be $Cp = C^1C$[M-6], a mono-capped closo octahedral complex. The complex $Os_8(CO)_{23}$, $S = 4n-2$ will be $Cp = C^2C$[M-6], that is, bi-capped closo octahedral complex(F-3). The capping on the octahedral geometry (O_h) can go on to form giant clusters. For example, $Ni_{38}Pt_6(CO)_{48}(H)^{5-}$, $S= 4n-74$; $Cp = C^{38}C$[M-6]. This means the cluster of 44 skeletal atoms has six of them which constitute an octahedral closo cluster as a nucleus. This has been found to be the case(Zanello, P.,2011).

2.4 Derivation of Carbonyl Cluster Formulas Using Series

It has been found that the building bloc of the carbonyl series has a valence electron content of 14. This number then forms the backbone for generating series. Let us create a closo $[14n+2]$ cluster of osmium carbonyl comprising of six skeletal atoms. This means that the value of n = 6 and the osmium building bloc will be $Os(CO)_3$, $V = 8+3\times2 = 14$. Therefore the cluster formula $F = 14n+2 = [Os(CO)_3](6)+CO = Os_6(CO)_{18}+CO = Os_6(CO)_{19}$. But since we know that in case of osmium the cluster a negative charge of 2 is present. Hence we can derive the formula as $F =[Os(CO)_3](6)+2 = Os_6(CO)_{18}^{2-}$. Let us take the example of rhenium. A possible backbone fragment for this element would be $[Re(H)(CO)_3]$,$V = 7+1+3\times2 = 14$. Therefore the hypothetical closo cluster when n =6, will be $[Re(H)(CO)_3](6)+ CO = [Re_6(H)_6(CO)_{18}]+CO = [Re_6(CO)_3(CO)_{18}]+CO = [Re_6(CO)_{22}]$. The known octahedral closo clusters include $Re_6(C)(CO)_{19}^{2-}$ which is electronically equivalent to $Re_6(CO)_2(CO)_{19}(CO) = Re_6(CO)_{22}$ in terms of valence electrons. In principle, we can derive hypothetical carbonyl cluster for a metal such as nickel. In this case, the backbone fragment will be $[Ni(CO)_2]$, $V = 10+2\times2 = 14$. Hence, the hypothetical closo cluster when n = 6 will have the formula $F= 14n+2 = [Ni(CO)_2](6)+2 = [Ni_6(CO)_{12}]^{2-}$ or $[Ni_6(CO)_{12}]+CO = Ni_6(CO)_{13}$. The complex$[Ni_6(CO)_{12}]^{2-}$ is known(Greenwood and Earnshaw,1998).

$$Os = Os(CO)_3$$

Os
Os-Os
Os-Os-Os
Os-Os-Os-Os
Os-Os-Os-Os-Os
Os-Os-Os-Os-Os-Os
Os-Os-Os-Os-Os-Os-Os
Os-Os-Os-Os-Os-Os-Os-Os

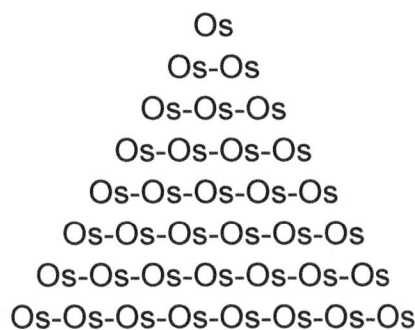

Figure 2a. Sketch representing the process of forming members of 14n+0 series

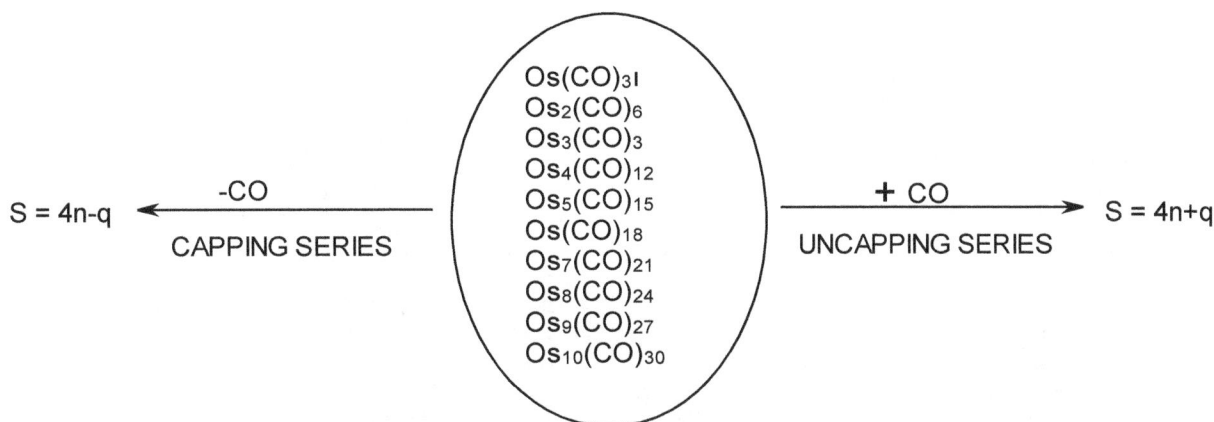

$$S = 4n-q \xleftarrow[\text{CAPPING SERIES}]{-CO}$$

$Os(CO)_{31}$
$Os_2(CO)_6$
$Os_3(CO)_3$
$Os_4(CO)_{12}$
$Os_5(CO)_{15}$
$Os(CO)_{18}$
$Os_7(CO)_{21}$
$Os_8(CO)_{24}$
$Os_9(CO)_{27}$
$Os_{10}(CO)_{30}$

$$\xrightarrow[\text{UNCAPPING SERIES}]{+CO} S = 4n+q$$

Figure 2b. Generating Capping and Decapping Series of Osmium carbonyls

Table 1. Derivation of 14n+0 series

14n	n	V	V/n	
$Os(CO)_3$	1	14	14	$Os(CO)_3$
$Os_2(CO)_6$	2	28	14	$Os(CO)_3$
$Os_3(CO)_9$	3	42	14	$Os(CO)_3$
$Os_4(CO)_{12}$	4	56	14	$Os(CO)_3$
$Os_5(CO)_{15}$	5	70	14	$Os(CO)_3$
$Os_6(CO)_{18}$	6	84	14	$Os(CO)_3$
$Os_7(CO)_{21}$	7	98	14	$Os(CO)_3$
$Os_8(CO)_{24}$	8	112	14	$Os(CO)_3$
$Os_9(CO)_{27}$	9	126	14	$Os(CO)_3$
$Os_{10}(CO)_{30}$	10	140	14	$Os(CO)_3$

Table 2. Hypothetically derived osmium carbonyl clusters

C^5C 14n-8	C^4C 14n-6	C^3C 14n-4	C^2C 14n-2	C^1C 14n+0	CLOSO 14n+2	NIDO 14n+4	ARACHNO 14n+6
	Os	$Os(CO)_1$	$Os(CO)_2$	$Os(CO)_3$	$Os(CO)_4$	$Os(CO)_5$	
$Os_2(CO)_2$	$Os_2(CO)_3$	$Os_2(CO)_4$	$Os_2(CO)_5$	$Os_2(CO)_6$	$Os_2(CO)_7$	$Os_2(CO)_8$	$Os_2(CO)_9$
$Os_3(CO)_5$	$Os_3(CO)_6$	$Os_3(CO)_7$	$Os_3(CO)_8$	$Os_3(CO)_9$	$Os_3(CO)_{10}$	$Os_3(CO)_{11}$	$Os_3(CO)_{12}$
$Os_4(CO)_8$	$Os_4(CO)_9$	$Os_4(CO)_{10}$	$Os_4(CO)_{11}$	$Os_4(CO)_{12}$	$Os_4(CO)_{13}$	$Os_4(CO)_{14}$	$Os_4(CO)_{15}$
$Os_5(CO)_{11}$	$Os_5(CO)_{12}$	$Os_5(CO)_{13}$	$Os_5(CO)_{14}$	$Os_5(CO)_{15}$	$Os_5(CO)_{16}$	$Os_5(CO)_{17}$	$Os_5(CO)_{18}$
$Os_6(CO)_{14}$	$Os_6(CO)_{15}$	$Os_6(CO)_{16}$	$Os_6(CO)_{17}$	$Os_6(CO)_{18}$	$Os_6(CO)_{19}$	$Os_6(CO)_{20}$	$Os_6(CO)_{21}$
$Os_7(CO)_{17}$	$Os_7(CO)_{18}$	$Os_7(CO)_{19}$	$Os_7(CO)_{20}$	$Os_7(CO)_{21}$	$Os_7(CO)_{22}$	$Os_7(CO)_{23}$	$Os_7(CO)_{24}$
$Os_8(CO)_{20}$	$Os_8(CO)_{21}$	$Os_8(CO)_{22}$	$Os_8(CO)_{23}$	$Os_8(CO)_{24}$	$Os_8(CO)_{25}$	$Os_8(CO)_{26}$	$Os_8(CO)_{27}$
$Os_9(CO)_{23}$	$Os_9(CO)_{24}$	$Os_9(CO)_{25}$	$Os_9(CO)_{26}$	$Os_9(CO)_{27}$	$Os_9(CO)_{28}$	$Os_9(CO)_{29}$	$Os_9(CO)_{30}$
$Os_{10}(CO)_{26}$	$Os_{10}(CO)_{27}$	$Os_{10}(CO)_{28}$	$Os_{10}(CO)_{29}$	$Os_{10}(CO)_{30}$	$Os_{10}(CO)_{31}$	$Os_{10}(CO)_{32}$	$Os_{10}(CO)_{33}$

Table 3. Classication of Clusters

	PROPOSED		
	SYMBOL	NAME	
14n+10	4n+10		KLAPO
14+8	4n+8		HYPHO
14+6	4n+6		ARACHNO
14n+4	4n+4		NIDO
14n+2	4n+2		CLOSO
14n+0	4n+0	C^1C	MONOCAP
14n-2	4n-2	C^2C	BICAP
14n-4	4n-4	C^3C	TRICAP
14n-6	4n-6	C^4C	TETRACAP
14n-8	4n-8	C^5C	PENTACAP
14n-10	4n-10	C^6C	HEXACAP
14n-12	4n-12	C^7C	HEPTACAP
14n-14	4n-14	C^8C	OCTACAP
14n-16	4n-16	C^9C	
14n-18	4n-18	$C^{10}C$	
14n-20	14n-20	$C^{11}C$	

Table 4. Some known osmium carbonyl clusters

CLUSTER	SERIES	CLUSTER	SERIES	CLUSTER	SERIES
$Os_{20}(CO)_{40}^{2-}$	4n-38	$Os_7(CO)_{21}$	4n+0	$Os_5(H)(CO)_{15}^{-}$	4n+2
$Os_{17}(CO)_{36}^{2-}$	4n-20	$Os_7(H)(CO)_{21}^{-}$	4n+2	$Os_4(H)_4(CO)_{12}$	4n+4
$Os_{10}(C)(CO)_{24}$	4n-8	$Os_7(H)_2(C)(CO)_{19}$	4n+2	$Os_4(CO)_{15}$	4n+6
$Os_{10}(CO)_{26}^{2-}$	4n-14	$Os_6(CO)_{18}^{2-}$	4n+2	$Os_4(CO)_{16}$	4n+8
$Os_{10}(C)(CO)_{24}^{2-}$	4n-14	$Os_6(H)(CO)_{18}^{-}$	4n+2		
$Os_{10}(C)(CO)_{24}^{4-}$	4n-12	$Os_6(P)(CO)_{18}Cl$	4n+6		
$Os_{10}(H)(C)(CO)_{24}^{-}$	4n-6	$Os_6(H)_2(CO)_{19}$	4n+4		
$Os_{10}(H)_4(CO)_{24}^{2-}$	4n-6	$Os_6(CO)_{18}$	4n+0	$Os_4(H)_2(CO)_{12}^{2-}$	4n+4
$Os_{10}(H)_2(C)(CO)_{24}$	4n-6	$Os_6(CO)_{17}L_4$	4n+6	$Os_4(H)_2(CO)_{13}$	4n+4
$Os_9(CO)_{24}^{2-}$	4n-4	$L=P(OMe)_3$		$Os_4(CO)_{14}$	4n+4
$Os_9(H)(CO)_{24}^{-}$	4n-4	$Os_5(CO)_{19}$	4n+8	$Os_3(CO)_{12}$	4n+6
$Os_8(CO)_{23}$	4n-2	$Os_5(CO)_{16}$	4n+2	$Os_2(CO)_9$	4n+6
$Os_8(CO)_{22}^{2-}$	4n-2	$Os_5(H)_2(CO)_{16}$	4n+4		
$Os_8(CO)_{12}$	4n-24	$Os_5(CO)_{18}$	4n+6		

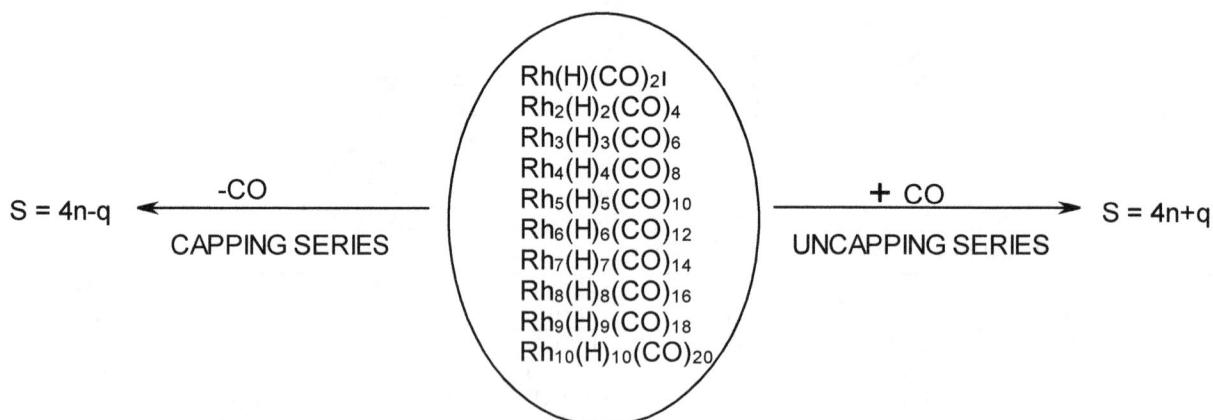

Figure 3. Generating Capping and Decapping Series of Rhodium carbonyls

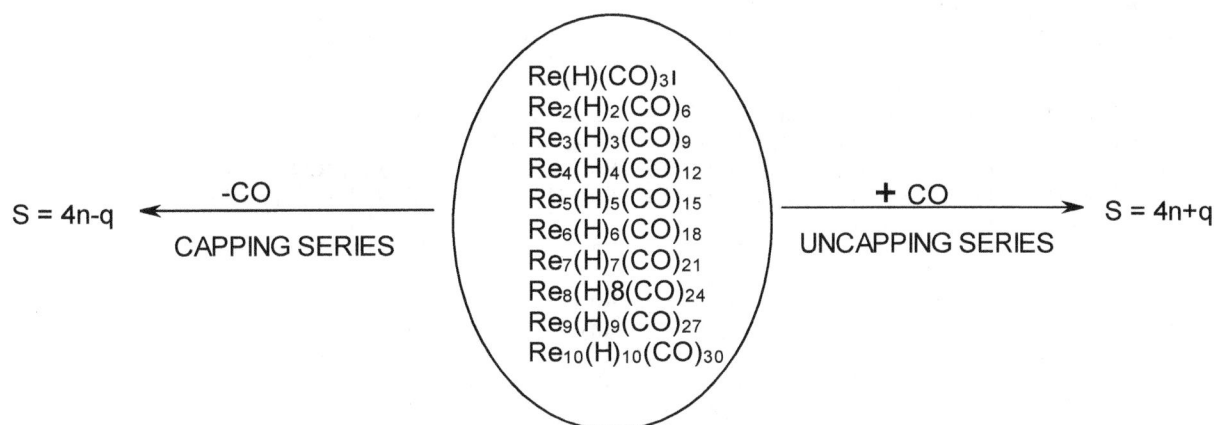

$$S = 4n-q$$

-CO

CAPPING SERIES

$Re(H)(CO)_{31}$
$Re_2(H)_2(CO)_6$
$Re_3(H)_3(CO)_9$
$Re_4(H)_4(CO)_{12}$
$Re_5(H)_5(CO)_{15}$
$Re_6(H)_6(CO)_{18}$
$Re_7(H)_7(CO)_{21}$
$Re_8(H)8(CO)_{24}$
$Re_9(H)_9(CO)_{27}$
$Re_{10}(H)_{10}(CO)_{30}$

+ CO

UNCAPPING SERIES

$$S = 4n+q$$

Figure 4. Generating Capping and Decapping Series of Rhenium carbonyls

Table 5. Hypothetically derived rhodium carbonyl clusters

14n-6	14n-4 Rh(H)	14n-2 Rh(H)(CO)_1	14n+0 Rh(H)(CO)_2	14n+2 Rh(H)(CO)_3	14n+4 Rh(H)(CO)_4	14n+6 Rh(H)(CO)_5
$Rh_2(H)_2(CO)_1$	$Rh_2(H)_2(CO)_2$	$Rh_2(H)_2(CO)_3$	$Rh_2(H)_2(CO)_4$	$Rh_2(H)_2(CO)_5$	$Rh_2(H)_2(CO)_6$	$Rh_2(H)_2(CO)_7$
$Rh_3(H)_3(CO)_3$	$Rh_3(H)_3(CO)_4$	$Rh_3(H)_3(CO)_5$	$Rh_3(H)_3(CO)_6$	$Rh_3(H)_3(CO)_7$	$Rh_3(H)_3(CO)_8$	$Rh_3(H)_3(CO)_9$
$Rh_4(H)_4(CO)_5$	$Rh_4(H)_4(CO)_6$	$Rh_4(H)_4(CO)_7$	$Rh_4(H)_4(CO)_8$	$Rh_4(H)_4(CO)_9$	$Rh_4(H)_4(CO)_{10}$	$Rh_4(H)_4(CO)_{11}$
$Rh_5(H)_5(CO)_7$	$Rh_5(H)_5(CO)_8$	$Rh_5(H)_5(CO)_9$	$Rh_5(H)_5(CO)_{10}$	$Rh_5(H)_5(CO)_{11}$	$Rh_5(H)_5(CO)_{12}$	$Rh_5(H)_5(CO)_{13}$
$Rh_6(H)_6(CO)_9$	$Rh_6(H)_6(CO)_{10}$	$Rh_6(H)_6(CO)_{11}$	$Rh_6(H)_6(CO)_{12}$	$Rh_6(H)_6(CO)_{13}$	$Rh_6(H)_6(CO)_{14}$	$Rh_6(H)_6(CO)_{15}$
$Rh_7(H)_7(CO)_{11}$	$Rh_7(H)_7(CO)_{12}$	$Rh_7(H)_7(CO)_{13}$	$Rh_7(H)_7(CO)_{14}$	$Rh_7(H)_7(CO)_{15}$	$Rh_7(H)_7(CO)_{16}$	$Rh_7(H)_7(CO)_{17}$
$Rh_8(H)_8(CO)_{13}$	$Rh_8(H)_8(CO)_{14}$	$Rh_8(H)_8(CO)_{15}$	$Rh_8(H)_8(CO)_{16}$	$Rh_8(H)_8(CO)_{17}$	$Rh_8(H)_8(CO)_{18}$	$Rh_8(H)_8(CO)_{19}$
$Rh_9(H)_9(CO)_{15}$	$Rh_9(H)_9(CO)_{16}$	$Rh_9(H)_9(CO)_{17}$	$Rh_9(H)_9(CO)_{18}$	$Rh_9(H)_9(CO)_{19}$	$Rh_9(H)_9(CO)_{20}$	$Rh_9(H)_9(CO)_{21}$
$Rh_{10}(H)_{10}(CO)_{17}$	$Rh_{10}(H)_{10}(CO)_{18}$	$Rh_{10}(H)_{10}(CO)_{19}$	$Rh_{10}(H)_{10}(CO)_{20}$	$Rh_{10}(H)_{10}(CO)_{21}$	$Rh_{10}(H)_{10}(CO)_{22}$	$Rh_{10}(H)_{10}(CO)_{23}$
$Rh_{11}(H)_{11}(CO)_{19}$	$Rh_{11}(H)_{11}(CO)_{20}$	$Rh_{11}(H)_{11}(CO)_{21}$	$Rh_{11}(H)_{11}(CO)_{22}$	$Rh_{11}(H)_{11}(CO)_{23}$	$Rh_{11}(H)_{11}(CO)_{24}$	$Rh_{11}(H)_{11}(CO)_{25}$
$Rh_{12}(H)_{12}(CO)_{21}$	$Rh_{12}(H)_{12}(CO)_{22}$	$Rh_{12}(H)_{12}(CO)_{23}$	$Rh_{12}(H)_{12}(CO)_{24}$	$Rh_{12}(H)_{12}(CO)_{25}$	$Rh_{12}(H)_{12}(CO)_{26}$	$Rh_{12}(H)_{12}(CO)_{27}$
$Rh_{13}(H)_{13}(CO)_{23}$	$Rh_{13}(H)_{13}(CO)_{24}$	$Rh_{13}(H)_{13}(CO)_{25}$	$Rh_{13}(H)_{13}(CO)_{26}$	$Rh_{13}(H)_{13}(CO)_{27}$	$Rh_{13}(H)_{13}(CO)_{28}$	$Rh_{13}(H)_{13}(CO)_{29}$
$Rh_{14}(H)_{14}(CO)_{25}$	$Rh_{14}(H)_{14}(CO)_{26}$	$Rh_{14}(H)_{14}(CO)_{27}$	$Rh_{14}(H)_{14}(CO)_{28}$	$Rh_{14}(H)_{14}(CO)_{29}$	$Rh_{14}(H)_{14}(CO)_{30}$	$Rh_{14}(H)_{14}(CO)_{31}$

4n-6	4n-4	4n-2	4n	4n+2	4n+4
$Rh_{14}(H)_{14}(CO)_{25}$	$Rh_{14}(H)_{14}(CO)_{26}$	$Rh_{14}(H)_{14}(CO)_{27}$	$Rh_{14}(H)_{14}(CO)_{28}$	$Rh_{14}(H)_{14}(CO)_{29}$	$Rh_{14}(H)_{14}(CO)_{30}$
$Rh_{15}(H)_{15}(CO)_{27}$	$Rh_{15}(H)_{15}(CO)_{28}$	$Rh_{15}(H)_{15}(CO)_{29}$	$Rh_{15}(H)_{15}(CO)_{30}$	$Rh_{15}(H)_{15}(CO)_{31}$	$Rh_{15}(H)_{15}(CO)_{32}$
$Rh_{16}(H)_{16}(CO)_{29}$	$Rh_{16}(H)_{16}(CO)_{30}$	$Rh_{16}(H)_{16}(CO)_{31}$	$Rh_{16}(H)_{16}(CO)_{32}$	$Rh_{16}(H)_{16}(CO)_{33}$	$Rh_{16}(H)_{16}(CO)_{34}$
$Rh_{17}(H)_{17}(CO)_{31}$	$Rh_{17}(H)_{17}(CO)_{32}$	$Rh_{17}(H)_{17}(CO)_{33}$	$Rh_{17}(H)_{17}(CO)_{34}$	$Rh_{17}(H)_{17}(CO)_{35}$	$Rh_{17}(H)_{17}(CO)_{36}$
$Rh_{18}(H)_{18}(CO)_{33}$	$Rh_{18}(H)_{18}(CO)_{34}$	$Rh_{18}(H)_{18}(CO)_{35}$	$Rh_{18}(H)_{18}(CO)_{36}$	$Rh_{18}(H)_{18}(CO)_{37}$	$Rh_{18}(H)_{18}(CO)_{38}$
$Rh_{19}(H)_{19}(CO)_{35}$	$Rh_{19}(H)_{19}(CO)_{36}$	$Rh_{19}(H)_{19}(CO)_{37}$	$Rh_{19}(H)_{19}(CO)_{38}$	$Rh_{19}(H)_{19}(CO)_{39}$	$Rh_{19}(H)_{19}(CO)_{40}$
$Rh_{20}(H)_{20}(CO)_{37}$	$Rh_{20}(H)_{20}(CO)_{38}$	$Rh_{20}(H)_{20}(CO)_{39}$	$Rh_{20}(H)_{20}(CO)_{40}$	$Rh_{20}(H)_{20}(CO)_{41}$	$Rh_{20}(H)_{20}(CO)_{42}$
$Rh_{21}(H)_{21}(CO)_{39}$	$Rh_{21}(H)_{21}(CO)_{40}$	$Rh_{21}(H)_{21}(CO)_{41}$	$Rh_{21}(H)_{21}(CO)_{42}$	$Rh_{21}(H)_{21}(CO)_{43}$	$Rh_{21}(H)_{21}(CO)_{44}$
$Rh_{22}(H)_{22}(CO)_{41}$	$Rh_{22}(H)_{22}(CO)_{42}$	$Rh_{22}(H)_{22}(CO)_{43}$	$Rh_{22}(H)_{22}(CO)_{44}$	$Rh_{22}(H)_{22}(CO)_{45}$	$Rh_{22}(H)_{22}(CO)_{46}$
$Rh_{23}(H)_{23}(CO)_{43}$	$Rh_{23}(H)_{23}(CO)_{44}$	$Rh_{23}(H)_{23}(CO)_{45}$	$Rh_{23}(H)_{23}(CO)_{46}$	$Rh_{23}(H)_{23}(CO)_{47}$	$Rh_{23}(H)_{23}(CO)_{48}$
$Rh_{24}(H)_{24}(CO)_{45}$	$Rh_{24}(H)_{24}(CO)_{46}$	$Rh_{24}(H)_{24}(CO)_{47}$	$Rh_{24}(H)_{24}(CO)_{48}$	$Rh_{24}(H)_{24}(CO)_{49}$	$Rh_{24}(H)_{24}(CO)_{50}$
$Rh_{25}(H)_{25}(CO)_{47}$	$Rh_{25}(H)_{25}(CO)_{48}$	$Rh_{25}(H)_{25}(CO)_{49}$	$Rh_{25}(H)_{25}(CO)_{50}$	$Rh_{25}(H)_{25}(CO)_{51}$	$Rh_{25}(H)_{25}(CO)_{52}$
$Rh_{26}(H)_{26}(CO)_{49}$	$Rh_{26}(H)_{26}(CO)_{50}$	$Rh_{26}(H)_{26}(CO)_{51}$	$Rh_{26}(H)_{26}(CO)_{52}$	$Rh_{26}(H)_{26}(CO)_{53}$	$Rh_{26}(H)_{26}(CO)_{54}$
$Rh_{27}(H)_{27}(CO)_{51}$	$Rh_{27}(H)_{27}(CO)_{52}$	$Rh_{27}(H)_{27}(CO)_{53}$	$Rh_{27}(H)_{27}(CO)_{54}$	$Rh_{27}(H)_{27}(CO)_{55}$	$Rh_{27}(H)_{27}(CO)_{56}$
$Rh_{28}(H)_{28}(CO)_{53}$	$Rh_{28}(H)_{28}(CO)_{54}$	$Rh_{28}(H)_{28}(CO)_{55}$	$Rh_{28}(H)_{28}(CO)_{56}$	$Rh_{28}(H)_{28}(CO)_{57}$	$Rh_{28}(H)_{28}(CO)_{58}$
$Rh_{29}(H_{29}(CO)_{55}$	$Rh_{29}(H_{29}(CO)_{56}$	$Rh_{29}(H_{29}(CO)_{57}$	$Rh_{29}(H_{29}(CO)_{58}$	$Rh_{29}(H_{29}(CO)_{59}$	$Rh_{29}(H_{29}(CO)_{60}$
$Rh_{30}(H)_{30}(CO)_{57}$	$Rh_{30}(H)_{30}(CO)_{58}$	$Rh_{30}(H)_{30}(CO)_{59}$	$Rh_{30}(H)_{30}(CO)_{60}$	$Rh_{30}(H)_{30}(CO)_{61}$	$Rh_{30}(H)_{30}(CO)_{62}$

4n-16	4n-14	4n-12	4n-10	4n-8	4n-6
$Rh_{14}(H)_{14}(CO)_{20}$	$Rh_{14}(H)_{14}(CO)_{21}$	$Rh_{14}(H)_{14}(CO)_{22}$	$Rh_{14}(H)_{14}(CO)_{23}$	$Rh_{14}(H)_{14}(CO)_{24}$	$Rh_{14}(H)_{14}(CO)_{25}$
$Rh_{15}(H)_{15}(CO)_{22}$	$Rh_{15}(H)_{15}(CO)_{23}$	$Rh_{15}(H)_{15}(CO)_{24}$	$Rh_{15}(H)_{15}(CO)_{25}$	$Rh_{15}(H)_{15}(CO)_{26}$	$Rh_{15}(H)_{15}(CO)_{27}$
$Rh_{16}(H)_{16}(CO)_{24}$	$Rh_{16}(H)_{16}(CO)_{25}$	$Rh_{16}(H)_{16}(CO)_{26}$	$Rh_{16}(H)_{16}(CO)_{27}$	$Rh_{16}(H)_{16}(CO)_{28}$	$Rh_{16}(H)_{16}(CO)_{29}$
$Rh_{17}(H)_{17}(CO)_{26}$	$Rh_{17}(H)_{17}(CO)_{27}$	$Rh_{17}(H)_{17}(CO)_{28}$	$Rh_{17}(H)_{17}(CO)_{29}$	$Rh_{17}(H)_{17}(CO)_{30}$	$Rh_{17}(H)_{17}(CO)_{31}$
$Rh_{18}(H)_{18}(CO)_{28}$	$Rh_{18}(H)_{18}(CO)_{29}$	$Rh_{18}(H)_{18}(CO)_{30}$	$Rh_{18}(H)_{18}(CO)_{31}$	$Rh_{18}(H)_{18}(CO)_{32}$	$Rh_{18}(H)_{18}(CO)_{33}$
$Rh_{19}(H)_{19}(CO)_{30}$	$Rh_{19}(H)_{19}(CO)_{31}$	$Rh_{19}(H)_{19}(CO)_{32}$	$Rh_{19}(H)_{19}(CO)_{33}$	$Rh_{19}(H)_{19}(CO)_{34}$	$Rh_{19}(H)_{19}(CO)_{35}$
$Rh_{20}(H)_{20}(CO)_{32}$	$Rh_{20}(H)_{20}(CO)_{33}$	$Rh_{20}(H)_{20}(CO)_{34}$	$Rh_{20}(H)_{20}(CO)_{35}$	$Rh_{20}(H)_{20}(CO)_{36}$	$Rh_{20}(H)_{20}(CO)_{37}$
$Rh_{21}(H)_{21}(CO)_{34}$	$Rh_{21}(H)_{21}(CO)_{35}$	$Rh_{21}(H)_{21}(CO)_{36}$	$Rh_{21}(H)_{21}(CO)_{37}$	$Rh_{21}(H)_{21}(CO)_{38}$	$Rh_{21}(H)_{21}(CO)_{39}$
$Rh_{22}(H)_{22}(CO)_{36}$	$Rh_{22}(H)_{22}(CO)_{37}$	$Rh_{22}(H)_{22}(CO)_{38}$	$Rh_{22}(H)_{22}(CO)_{39}$	$Rh_{22}(H)_{22}(CO)_{40}$	$Rh_{22}(H)_{22}(CO)_{41}$
$Rh_{23}(H)_{23}(CO)_{38}$	$Rh_{23}(H)_{23}(CO)_{39}$	$Rh_{23}(H)_{23}(CO)_{40}$	$Rh_{23}(H)_{23}(CO)_{41}$	$Rh_{23}(H)_{23}(CO)_{42}$	$Rh_{23}(H)_{23}(CO)_{43}$
$Rh_{24}(H)_{24}(CO)_{40}$	$Rh_{24}(H)_{24}(CO)_{41}$	$Rh_{24}(H)_{24}(CO)_{42}$	$Rh_{24}(H)_{24}(CO)_{43}$	$Rh_{24}(H)_{24}(CO)_{44}$	$Rh_{24}(H)_{24}(CO)_{45}$
$Rh_{25}(H)_{25}(CO)_{42}$	$Rh_{25}(H)_{25}(CO)_{43}$	$Rh_{25}(H)_{25}(CO)_{44}$	$Rh_{25}(H)_{25}(CO)_{45}$	$Rh_{25}(H)_{25}(CO)_{46}$	$Rh_{25}(H)_{25}(CO)_{47}$
$Rh_{26}(H)_{26}(CO)_{44}$	$Rh_{26}(H)_{26}(CO)_{45}$	$Rh_{26}(H)_{26}(CO)_{46}$	$Rh_{26}(H)_{26}(CO)_{47}$	$Rh_{26}(H)_{26}(CO)_{48}$	$Rh_{26}(H)_{26}(CO)_{49}$
$Rh_{27}(H)_{27}(CO)_{46}$	$Rh_{27}(H)_{27}(CO)_{47}$	$Rh_{27}(H)_{27}(CO)_{48}$	$Rh_{27}(H)_{27}(CO)_{49}$	$Rh_{27}(H)_{27}(CO)_{50}$	$Rh_{27}(H)_{27}(CO)_{51}$
$Rh_{28}(H)_{28}(CO)_{48}$	$Rh_{28}(H)_{28}(CO)_{49}$	$Rh_{28}(H)_{28}(CO)_{50}$	$Rh_{28}(H)_{28}(CO)_{51}$	$Rh_{28}(H)_{28}(CO)_{52}$	$Rh_{28}(H)_{28}(CO)_{53}$
$Rh_{29}(H_{29})(CO)_{50}$	$Rh_{29}(H_{29})(CO)_{51}$	$Rh_{29}(H_{29})(CO)_{52}$	$Rh_{29}(H_{29})(CO)_{53}$	$Rh_{29}(H_{29})(CO)_{54}$	$Rh_{29}(H_{29})(CO)_{55}$
$Rh_{30}(H)_{30}(CO)_{52}$	$Rh_{30}(H)_{30}(CO)_{53}$	$Rh_{30}(H)_{30}(CO)_{54}$	$Rh_{30}(H)_{30}(CO)_{55}$	$Rh_{30}(H)_{30}(CO)_{56}$	$Rh_{30}(H)_{30}(CO)_{57}$

Table 6. Some known rhodium carbonyl clusters

$Rh_4(CO)_{12}$	4n+4	$Rh_{12}(CO)_{25}(H)_2$	4n-8
$Rh_4(CO)_{11}{}^-$	4n+4	$Rh_{13}(CO)_{12}(H)_2{}^{3-}$	4n-36
$Rh_5(CO)_{15}{}^-$	4n+6	$Rh_{13}(CO)_{24}(H)_2{}^{3-}$	4n-12
$Rh_5Pt(CO)_{15}{}^-$	4n+2	$Rh_{14}(CO)_{25}{}^{4-}$	4n-16
$Rh_6(CO)_{13}(C)^{2-}$	4n-2	$Rh_{14}(CO)_{25}(N)_2{}^{2-}$	4n-8
$Rh_6(CO)_{15}(N)^-$	4n+6	$Rh_{14}(CO)_{33}(C)_2{}^{2-}$	4n+6
$Rh_6(CO)_{15}(C)^{2-}$	4n+6	$Rh_{14}(CO)_{25}(H)^{3-}$	4n-16
$Rh_6(CO)_{14}(N)(H)^{2-}$	4n+6	$Rh_{14}(CO)_{25}{}^{4-}$	4n-16
$Rh_6(CO)_{16}$	4n+2	$Rh_{14}(CO)_{26}{}^{2-}$	4n-16
$Rh_7(CO)_{16}{}^{3-}$	4n+0	$Rh_{15}(CO)_{28}(C)_2{}^-$	4n-10
$Rh_7(CO)_{15}(N)^{2-}$	4n+2	$Rh_{15}(CO)_{25}{}^{3-}$	4n-22
$Rh_8(CO)_{19}(C)$	4n+2	$Rh_{15}(CO)_{28}(C)^{2-}$	4n-10
$Rh_9(CO)_{19}{}^{3-}$	4n-4	$Rh_{15}(CO)_{27}{}^{3-}$	4n-18
$Rh_9(CO)_{21}(P)^{2-}$	4n+4	$Rh_{17}(CO)_{30}{}^{3-}$	4n-22
$Rh_{10}(CO)_{22}(P)^{3-}$	4n+2	$Rh_{17}(CO)_{32}(S)_2{}^{3-}$	4n-6
$Rh_{10}(CO)_{21}{}^{2-}$	4n-6	$Rh_{22}(CO)_{35}(H)^{5-}$	4n-34
$Rh_{11}(CO)_{23}{}^{3-}$	4n-6	$Rh_{22}(CO)_{37}{}^{4-}$	4n-32
$Rh_{12}(CO)_{23}(N)_2(H)^{3-}$	4n+0	$Rh_{23}(CO)_{38}(N)_4{}^-$	4n-18
$Rh_{12}(CO)_{24}(C)_2{}^{2-}$	4n-2	$Rh_{28}(CO)_{41}(N)_4(H)_2{}^{4-}$	4n-32
$Rh_{12}(CO)_{23}(C)_2{}^{4-}$	4n-6		

Table 7. Hypothetically derived rhenium carbonyl clusters

4n-6	4n-4	4n-2	4n	4n+2	4n+4	4n+6
ReH	$ReH(CO)$	$ReH(CO)_2$	$ReH(CO)_3$	$ReH(CO)_4$	$ReH(CO)_5$	$ReH(CO)_6$
$Re_2H_2(CO)_3$	$Re_2H_2(CO)_4$	$Re_2H_2(CO)_5$	$Re_2H_2(CO)_6$	$Re_2H_2(CO)_7$	$Re_2H_2(CO)_8$	$Re_2H_2(CO)_9$
$Re_3H_3(CO)_6$	$Re_3H_3(CO)_7$	$Re_3H_3(CO)_8$	$Re_3H_3(CO)_9$	$Re_3H_3(CO)_{10}$	$Re_3H_3(CO)_{11}$	$Re_3H_3(CO)_{12}$
$Re_4H_4(CO)_9$	$Re_4H_4(CO)_{10}$	$Re_4H_4(CO)_{11}$	$Re_4H_4(CO)_{12}$	$Re_4H_4(CO)_{13}$	$Re_4H_4(CO)_{14}$	$Re_4H_4(CO)_{15}$
$Re_5H_5(CO)_{14}$	$Re_5H_5(CO)_{13}$	$Re_5H_5(CO)_{14}$	$Re_5H_5(CO)_{15}$	$Re_5H_5(CO)_{16}$	$Re_5H_5(CO)_{17}$	$Re_5H_5(CO)_{18}$
$Re_6H_6(CO)_{15}$	$Re_6H_6(CO)_{16}$	$Re_6H_6(CO)_{17}$	$Re_6H_6(CO)_{18}$	$Re_6H_6(CO)_{19}$	$Re_6H_6(CO)_{20}$	$Re_6H_6(CO)_{21}$
$Re_7H_7(CO)_{18}$	$Re_7H_7(CO)_{19}$	$Re_7H_7(CO)_{20}$	$Re_7H_7(CO)_{21}$	$Re_7H_7(CO)_{21}$	$Re_7H_7(CO)_{21}$	$Re_7H_7(CO)_{21}$
$Re_8H_8(CO)_{21}$	$Re_8H_8(CO)_{22}$	$Re_8H_8(CO)_{23}$	$Re_8H_8(CO)_{24}$	$Re_8H_8(CO)_{25}$	$Re_8H_8(CO)_{26}$	$Re_8H_8(CO)_{27}$
$Re_9H_9(CO)_{24}$	$Re_9H_9(CO)_{25}$	$Re_9H_9(CO)_{26}$	$Re_9H_9(CO)_{27}$	$Re_9H_9(CO)_{28}$	$Re_9H_9(CO)_{29}$	$Re_9H_9(CO)_{30}$
$Re_{10}H_{10}(CO)_{27}$	$Re_{10}H_{10}(CO)_{28}$	$Re_{10}H_{10}(CO)_{29}$	$Re_{10}H_{10}(CO)_{30}$	$Re_{10}H_{10}(CO)_{31}$	$Re_{10}H_{10}(CO)_{32}$	$Re_{10}H_{10}(CO)_{33}$
$Re_{11}H_{11}(CO)_{30}$	$Re_{11}H_{11}(CO)_{31}$	$Re_{11}H_{11}(CO)_{32}$	$Re_{11}H_{11}(CO)_{33}$	$Re_{11}H_{11}(CO)_{34}$	$Re_{11}H_{11}(CO)_{35}$	$Re_{11}H_{11}(CO)_{36}$
$Re_{12}H_{12}(CO)_{33}$	$Re_{12}H_{12}(CO)_{34}$	$Re_{12}H_{12}(CO)_{35}$	$Re_{12}H_{12}(CO)_{36}$	$Re_{12}H_{12}(CO)_{37}$	$Re_{12}H_{12}(CO)_{38}$	$Re_{12}H_{12}(CO)_{39}$
$Re_{13}H_{13}(CO)_{36}$	$Re_{13}H_{13}(CO)_{37}$	$Re_{13}H_{13}(CO)_{38}$	$Re_{13}H_{13}(CO)_{39}$	$Re_{13}H_{13}(CO)_{40}$	$Re_{13}H_{13}(CO)_{41}$	$Re_{13}H_{13}(CO)_{42}$
$Re_{14}H_{14}(CO)_{39}$	$Re_{14}H_{14}(CO)_{40}$	$Re_{14}H_{14}(CO)_{41}$	$Re_{14}H_{14}(CO)_{42}$	$Re_{14}H_{14}(CO)_{43}$	$Re_{14}H_{14}(CO)_{44}$	$Re_{14}H_{14}(CO)_{45}$
$Re_{15}H_{15}(CO)_{42}$	$Re_{15}H_{15}(CO)_{43}$	$Re_{15}H_{15}(CO)_{44}$	$Re_{15}H_{15}(CO)_{45}$	$Re_{15}H_{15}(CO)_{46}$	$Re_{15}H_{15}(CO)_{47}$	$Re_{15}H_{15}(CO)_{48}$
$Re_{16}H_{16}(CO)_{45}$	$Re_{16}H_{16}(CO)_{46}$	$Re_{16}H_{16}(CO)_{47}$	$Re_{16}H_{16}(CO)_{48}$	$Re_{15}H_{15}(CO)_{49}$	$Re_{15}H_{15}(CO)_{50}$	$Re_{15}H_{15}(CO)_{51}$
$Re_{17}H_{17}(CO)_{51}$	$Re_{17}H_{17}(CO)_{51}$	$Re_{17}H_{17}(CO)_{51}$	$Re_{17}H_{17}(CO)_{51}$	$Re_{15}H_{15}(CO)_{52}$	$Re_{15}H_{15}(CO)_{53}$	$Re_{15}H_{15}(CO)_{54}$
$Re_{18}H_{18}(CO)_{51}$	$Re_{18}H_{18}(CO)_{52}$	$Re_{18}H_{18}(CO)_{53}$	$Re_{18}H_{18}(CO)_{54}$	$Re_{15}H_{15}(CO)_{55}$	$Re_{15}H_{15}(CO)_{56}$	$Re_{15}H_{15}(CO)_{57}$
$Re_{19}H_{19}(CO)_{54}$	$Re_{19}H_{19}(CO)_{55}$	$Re_{19}H_{19}(CO)_{56}$	$Re_{19}H_{19}(CO)_{57}$	$Re_{15}H_{15}(CO)_{58}$	$Re_{15}H_{15}(CO)_{59}$	$Re_{15}H_{15}(CO)_{60}$
$Re_{20}H_{20}(CO)_{57}$	$Re_{20}H_{20}(CO)_{58}$	$Re_{20}H_{20}(CO)_{59}$	$Re_{20}H_{20}(CO)_{60}$	$Re_{15}H_{15}(CO)_{61}$	$Re_{15}H_{15}(CO)_{62}$	$Re_{15}H_{15}(CO)_{63}$

Table 8. Some known rhenium carbonyl clusters

	$S=4n+q$	$k=2n-q/2$	Category	Proposed Symbol
$Re_2(H)_2(CO)_8$	$4n+4$	2	nido	
$Re_2(H)_3(CO)_8^-$	$4n+6$	1	arachno	
$Re_2(CO)_{10}$	$4n+6$	1	arachno	
$Re_3(H)_3(CO)_{10}^{2-}$	$4n+4$	4	nido	
$Re_3(H)_4(CO)_{10}^-$	$4n+4$	4	nido	
$Re_3(H)(CO)_{12}^{2-}$	$4n+6$	3	arachno	
$Re_3(H)_2(CO)_{12}^-$	$4n+6$	3	arachno	
$Re_3(H)_3(CO)_{12}$	$4n+6$	3	arachno	
$Re_3(H)_2(CO)_{13}^-$	$4n+8$	2	hypho	
$Re_3(H)(CO)_{14}$	$4n+8$	2	hypho	
$Re_4(H)_4(CO)_{12}$	$4n+0$	8	monocapped	$C^1C[M-3]$
$Re_4(H)_5(CO)_{12}^-$	$4n+2$	7	closo	
$Re_4(H)_4(CO)_{13}^{2-}$	$4n+4$	6	nido	
$Re_4(H)_6(CO)_{12}^{2-}$	$4n+4$	6	nido	
$Re_4(CO)_{16}^{2-}$	$4n+6$	5	arachno	
$Re_4(H)_5(CO)_{14}^-$	$4n+6$	5	arachno	
$Re_4(H)_4(CO)_{15}^{2-}$	$4n+8$	4	hypho	
$Re_4(H)_4(CO)_{16}$	$4n+8$	4	hypho	
$Re_5(C)(H)(CO)_{12}^{2-}$	$4n+4$	8	nido	Square pyramid
$Re_6(C)(CO)_{19}^{2-}$	$4n+2$	11	closo	Octahedral, O_h
$Re_6(C)(H)(CO)_{19}^-$	$4n+2$	11	closo	Octahedral, O_h
$Re_6(H)(CO)_{18}^{3-}$	$4n+2$	11	closo	Octahedral, O_h
$Re_6(H)_2(CO)_{18}^{2-}$	$4n+2$	11	closo	Octahedral, O_h
$Re_6(H)_8(CO)_{18}^{2-}$	$4n+4$	10	nido	
$Re_6(H)_7(CO)_{18}^-$	$4n+2$	11	closo	Octahedral, O_h
$Re_7(C)(H)(CO)_{21}^{3-}$	$4n+0$	14	monocapped O_h	$C^1C[M-6]$
$Re_7(C)(H)(CO)_{21}^{2-}$	$4n+0$	14	monocapped O_h	$C^1C[M-6]$
$Re_7(C)(H)_2(CO)_{21}^-$	$4n+4$	12	nido	
$Re_7(C)(CO)_{22}^-$	$4n+0$	14	monocapped O_h	$C^1C[M-6]$
$Re_8(C)(CO)_{24}^{2-}$	$4n-2$	17	Bi-capped O_h	$C^2C[M-6]$

2.5 Applications of 4n Series to Classify Clusters and Predict Their Possible Shapes

We have been demonstrated that the formation of transition metal carbonyl clusters obeys the law of natural series. Also in earlier work, it was shown that chemical fragments and clusters can be categorized into series (Kiremire, 2015c).In addition, using series, we can readily concert cluster formulas into series from which their geometrical shapes may be deduced.

2.5.1 Procedure for Deriving Series and Predicting Symmetries of Clusters

According to this approach, suitable skeletal elements are identified in the cluster formula. The cluster formula is then fragmented into suitable mono-skeletal fragments. If a skeletal element is from the main group elements of the periodic table, then it is best to convert it into a fragment which has a content of four valence electrons say,[C], [BH], [BeH_2], [B^-], and [N^+]. Each of these fragments is regarded as being a member of $S = 4n+0$ series where n = 1 since we are focusing on one skeletal fragment. Since main group elements which obey the octet rule also tend to obey the 4n rule of the cluster series, if a mono-skeletal fragment has more than four valence electrons, the extra become important in categorization of the series. For example, [CH] fragment has five electrons which can be expressed as [4+1]. In such case, the series will be $S = 4n+1$ where n =1. For the fragment [N] which also has five valence electrons it will belong to the series $S = 4n+1$ while [O] with six electrons[4+2] will belong to $S = 4n+2$. Similarly, [CH_2] and [NH] fragments are also members of $S= 4n+2$ series. The [F] atom with seven valence electrons belongs to $S = 4n+3$. The 4n function may be considered to constitute the backbone of the main group element cluster series while the additional digits after 4n determine the type of series. For example, if $S= 4n+(0) \rightarrow$ mono-capped series, $4n+(2) \rightarrow$ closo series, $4n+(4) \rightarrow$ nido series, $4n+(6) \rightarrow$ arachno series, $4n+(8) \rightarrow$ hypho series and $4n+(10) \rightarrow$ klapo series. The bicapped and higher series carry negative integers after 4n as in the case of 4n-2 or 4n-4. If a skeletal fragment is short of 4 backbone valence electrons, its series is still expressed according to 4n rule but reflecting the shortage of valence electrons. For instance, the fragment [B] has three valence electrons. So in terms of the series it is expressed as $S =[4-1] = 4n-1$ where n = 1, and [Be] = [4-2] $\rightarrow 4n-2$ series. In case of transition metal carbonyl clusters, it is assumed that the 18-electron rule is obeyed by the skeletal atoms. This means that the skeletal elements obey the 14n rule of the series. The same procedure as in the case of 4n for the main group elements is followed. That is, the carbonyl cluster formula is first decomposed into mono-skeletal fragments. The fragments are chosen in such a way that it conforms to 14 valence backbone configuration. Examples that fit this category are [Os(CO)_3],[Rh(H)(CO)_2], [ReH(CO)_3] ,[Ni(CO)_2], [C_5H_5Rh], [C_5H_5(CO)Re] and [C_5H_5Fe^-]. If a mono-skeletal fragment has a shortfall of 14, it is reflected in its series. For instance,

[C_5H_5Os] has 13 valence electrons. Its series is expressed as [14-1] = 14n-1. Since 14n is isolobal to 4n, we can also write and use 14n-1 as 4n-1. A transition metal atom if taken alone, its valence electrons may be expressed in the form of 14n series. For instance, [Fe]→[8] = [14-6] = S =14n-6 or 4n-6. In clusters where a transition metal fragment is combined with hydrocarbon fragments such as $C_4H_4Fe(CO)_3$ or borane fragments such as $B_4H_8Ir_2(Cp^*)_2$, it is best to work with 4n series only. In these two examples, we can decompose the formulas as follows $C_4H_4Fe(CO)_3$→4[C]+4H+1[Fe(CO)_3] = 4[4]+4+1[8+3x2] = 4[4n]+4+1[14n] =(4n)+4+(4n) = 4n+4. That is, S = 4n+4 →nido cluster. The 14n has be utilized as 4n due to isolobal relationship. The 4 hydrogen atoms have contributed valence electrons to the cluster and are not regarded as skeletal elements. The value of n =4+1 = 5.The cluster is regarded as having 5 skeletal elements. Using the series formula, we can derive a corresponding or an equivalent borane cluster as follows F = 4n+4 = [BH](5)+4H = B_5H_5+4H = B_5H_9. This borane cluster has a square pyramid shape missing one atom to have an octahedral closo shape of $B_6H_6^{2-}$. Let us consider the cluster $B_4H_8Ir_2(Cp^*)_2$. This formula can be decomposed into series as follows: $B_4H_8Ir_2(Cp^*)_2$→S = 4[BH]+4H+ 2[IrCp^*]=4[4]+4+2[9+5] = 4[4n]+4+2[14n] = 4[4n]+4+2[4n] = (4n)+4+(4n) = 4n+4. This means the cluster belongs to the series S =4n+4→Nido cluster with n = 4+2 = 6. It is interesting to derive the equivalent borane cluster. This is given by the formula F =[BH](6)+4H = B_6H_6+4H = B_6H_{10}. This cluster has a pentagonal pyramid shape which a derivative of $B_7H_7^{2-}$ with a missing atom. Therefore the borane iridium cluster will have a shape similar to that of B_6H_{10}.

Summing up: Cluster Formula →Mono-skeletal fragments + other support ligands if needed →Series→ Possible geometry.

The following section gives selected examples to illustrate the application of series to categorize clusters and predict their possible shapes. Since the use of series to categorize clusters is not well known to many readers, it is best to give some highlights of the background information. One important discovery regarding series is that the series S = 14n+q for the transition metal clusters is isolobal to S = 4n+q of the main group elements. For instance, 14n ⊸ 4n, 14n+1 ⊸ 4n+1, 14n+2 ⊸ 4n+2, 14n+3⊸4n+3, 14n+4⊸4n+4, 14n-2⊸4n-2, and so on. This means that we can use 14n series or 4n series whichever is more convenient. In this paper, the 4n series will be used throughout. Some examples of isolobal series are shown below.

ISOLOBAL SERIES

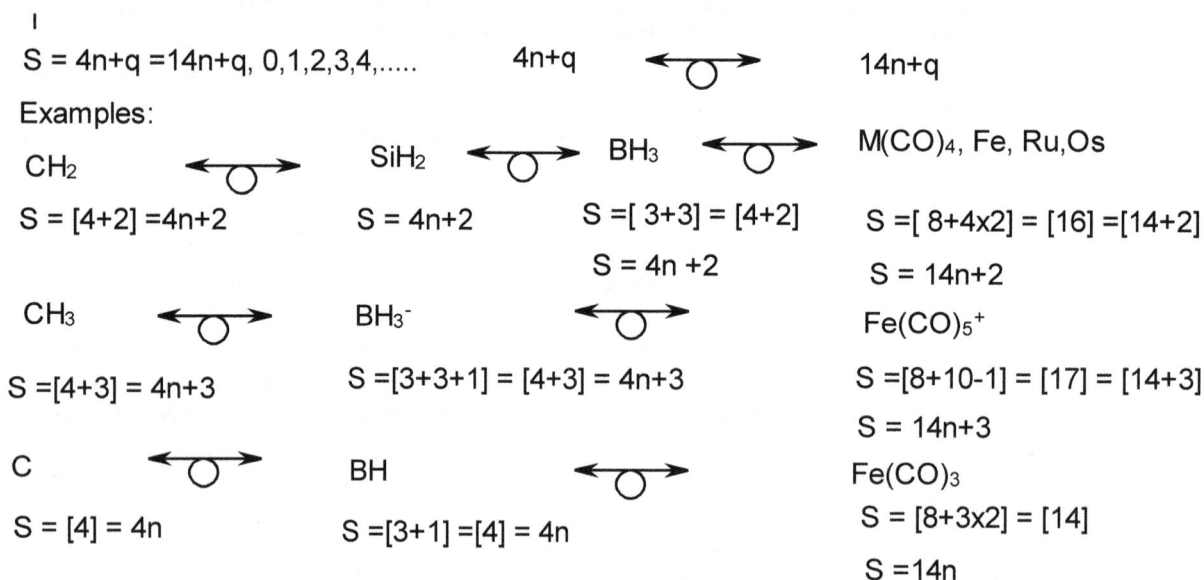

S = 4n+q =14n+q, 0,1,2,3,4,.....   4n+q ⟷ 14n+q

Examples:

CH_2 ⟷   SiH_2 ⟷   BH_3 ⟷   M(CO)_4, Fe, Ru,Os

S = [4+2] =4n+2   S = 4n+2   S =[3+3] = [4+2]   S =[8+4x2] = [16] =[14+2]

                                                 S = 4n +2   S = 14n+2

CH_3 ⟷   BH_3^- ⟷   Fe(CO)_5^+

S =[4+3] = 4n+3   S =[3+3+1] = [4+3] = 4n+3   S =[8+10-1] = [17] = [14+3]

                                                                     S = 14n+3

C ⟷   BH ⟷   Fe(CO)_3

S = [4] = 4n   S =[3+1] =[4] = 4n   S = [8+3x2] = [14]

                                                                     S =14n

Scheme 1. Some examples of isolobal series

2.6 Cluster Number k = 2n-q/2 for Series S = 4n+q

Let us consider C_2H_6. The carbon atom [C] fragment like the [BH] fragment belongs to 4n+0 series since it has 4 valence electrons. Therefore the fragment [CH] of valence electron content of 5 will belong to 4n+1 series. The other fragments [CH_2] and [CH_3] will be members of the series 4n+2 and 4n+3 respectively. When two [CH_3] fragments are combined we get the C_2H_6 molecule. The corresponding series of the molecule will be given by S = [4n+3] + [4n+3] = 2[4n+3] = 4n+6. We can also derive the series of C_2H_6 by decomposing the molecule as follows; C_2H_6 →C_2 + 6H, C_2 → 4n+0 and 6H → 6 valence electrons. Hence, the C_2H_6 series is S = (4n+0)+6 = 4n+6. The number of carbon-carbon bonds in C_2H_6 are given by k = 2n-6/2 = 2n-3. Since there are two skeletal atoms, n = 2. Hence k = 2(2)-3 = 1. This means, there is only one bond joining the two carbon atoms together in C_2H_6. Let us consider another example, C_2H_4.

The ethylene molecule belongs to S = 4n+4 series which can easily be derived as was done for C_2H_6. The corresponding k value is given by k =2n-4/2 , n = 2 and hence k = 2n-2 = 2(2) -2 = 2. This means that there is a double bond C=C in C_2H_4. For the C_2H_2 molecule, it belongs to the series S = 4n+2 and k = 2n-1 = 2(2)-1 = 3. This corresponds to the C≡C triple bond. For C_2 molecule, it belongs to S = 4n+0, k =2n = 2(2) = 4. This means that the C_2 molecule is linked by four bonds. These are shown in Figures 4 and 5. The structures of C_2H_6, C_2H_4 and C_2H_2 structures are well known. However, the case of C_2 having a quadruple bond caused shock waves in scientific community as it was always believed to have a double bond as deduced from molecular orbital theory(Douglas, McDaniel, Alexander, 1994). In actual fact, the isoelectronic species of C_2 including BN, CN^+, and CB^- were found to have quadruple bonds obtained from high level computations (Shaik, et al, 2012).

Empirical analysis of series and bonds linking skeletal elements which collectively obey the octet or 18 electron rule have the same formula for k value. Thus, when S = 4n+q (main group elements) or 14n+q for transition metal carbonyl clusters, then k = 2n+x where x = q/2 and q = 0, ±1, ±2, ±3, ±4, ...For example, if S = 4n+4 or S = 14n+4, k = 2n-2 in each case. For many hydrocarbons and inorganic molecules and ions, the k-value is quite precise. For instance, N_2, S = 4n+2 and k = 2n-2 = 3. Thus N_2 has a triple bond, N≡N. In the case of S_8, S = 4n+16, k = 2n-8 = 8 whereas S_8^{2+}, S = 4n+14, k =2n-7 = 9. Hence, 8 and 9 linkages are observed for these species respectively(Greenwood and Earnshaw,1998) The k value is so precise in predicting the number of carbon-carbon bonds especially in hydrocarbons. This is well demonstrated by the k-values in prismanes, C_6H_6, C_9H_6 and $C_{12}H_6$ and the sketches of the molecules are shown in

Figure 3.

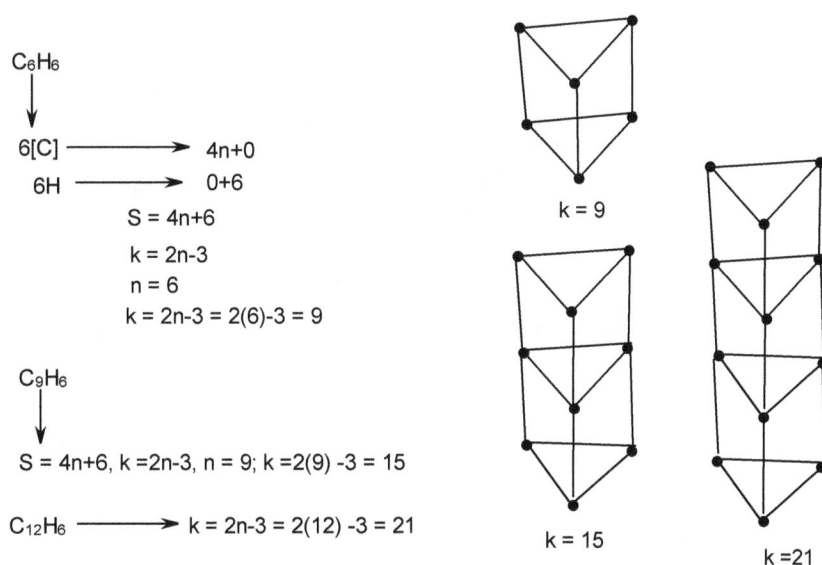

C_6H_6

6[C] ⟶ 4n+0
6H ⟶ 0+6

S = 4n+6

k = 2n-3

n = 6

k = 2n-3 = 2(6)-3 = 9

C_9H_6

S = 4n+6, k =2n-3, n = 9; k =2(9) -3 = 15

$C_{12}H_6$ ⟶ k = 2n-3 = 2(12) -3 = 21

k = 9

k = 15

k =21

Figure 3 Sketches of prismanes C_6H_6, C_9H_6, $C_{12}H_6$

2.7 Rigid Polyhedral Shapes

In general, the shapes of borane clusters were found to be rigid polyhedral three diamensional shapes. The common closo boranes (4n+2) have naturally standardized shapes which are well documented especially for $B_5H_5^{2-}$ to $B_{12}H_{12}^{2-}$(Cotton and Wilkinson, 1980). For instance, closo $B_5H_5^{2-}$ has a trigonal bipyramid (D_{3h}) while the closo $B_6H_6^{2-}$ has an octahedral shape (O_h). The $B_7H_7^{2-}$ has pentagonal bipyramid (D_{5h}) shape. The other borane and related heteroboranes have shapes which are linked to the respective parent closo through the Rudolph decapping relationship. Selected examples to illustrate the use of series to predict the possible shape of a cluster are given below.

$Mn_2(CO)_{10} \longrightarrow 2[Mn(CO)_5] \longrightarrow S = 2[7+5\times2] = 2[17] = 2[14+3] = 2[14n+3] = 14n+6 = 4n+6$

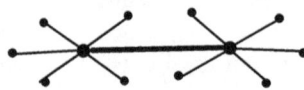

$$n = 2, \ k = 2n-3 = 2(2)-3 = 1$$

$\bullet$ = Mn

k = 1 $\cdot$ = CO

$Rh_2(CO)_2(Cp)_2 \longrightarrow 2[RhCp]+2CO \longrightarrow 2[9+5] +2\times2 = 2[14]+4 = 2[14n]+4 = 14n+4 = 4n+4$

$$n = 2, \ k = 2n-2 = 2(2)-2 = 2$$

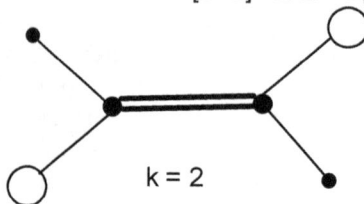

$\bigcirc$ = Cp

k = 2 $\bullet$ = CO

$\bullet$ = Rh

$Mo_2(CO)_4(Cp)_2 \longrightarrow 2[Mo(CO)_2(Cp)] = 2[6+2\times2+5] = 2[15] = 2[14+1] = 2[14n+1] = 14n+2] = 4n+2$

$$n = = 2, \ k = 2n-1 = 2(2)-1 = 3$$

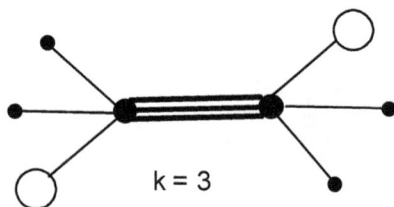

$\bigcirc$ = Cp

$\bullet$ = CO

k = 3 $\bullet$ = Mo

$Os_3(CO)_{12} \longrightarrow 3[Os(CO)_3] +3CO \longrightarrow 3[8+3\times2]+3\times2 = 3[14]+6 = 3[14n] +6 = 14n+6 = 4n+6$

$$n = 3, \ k = 2n-3 = 2(3)-3 = 3$$

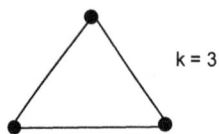

k = 3

$\bullet$ = Os

Skeletal shape of $Os_3(CO)_{12}$

$Ir_4(CO)_{12} \longrightarrow 4[Ir(CO)_3] \longrightarrow 4[9+3\times2] = 4[15] = 4[14+1] = 4[14n+1] = 14n+4 = 4n+4$

$$n = 4, \ k = 2n-2 = 2(4)-2 = 6$$

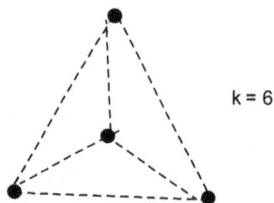

k = 6

Projection map of Skeletal atoms of $Ir_4(CO)_{12}$

$Os_5(C)(CO)_{15}$ $\longrightarrow$ $5[Os(CO)_3]$ $\longrightarrow$ $5[8+6] = 5[14] = 5[14n] = 14n = 4n+0$

$+ C$ $\longrightarrow$ $0+4$

$S = 4n+4$

$n = 5, \ k = 2n-2 \ = 2(5)-2 = 8$

$k = 8$

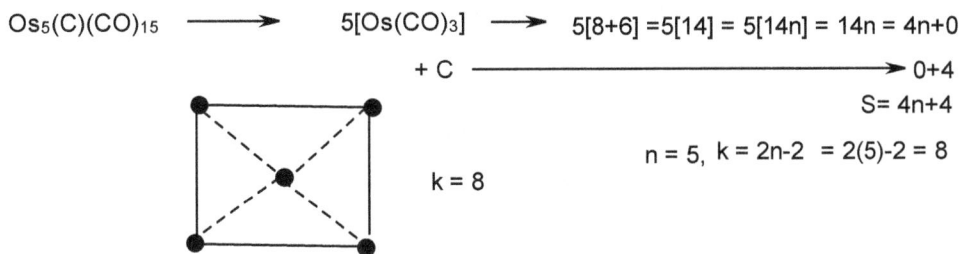

Projection map of skeletal elements of $Os_5(C)(CO)_{15}$

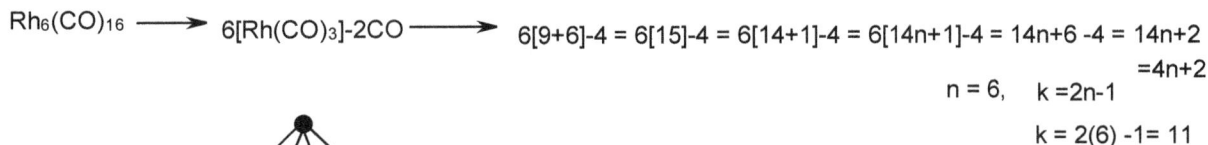

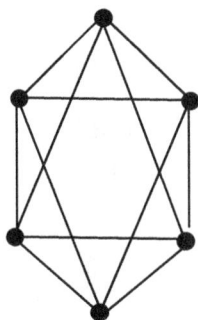

$Rh_6(CO)_{16}$ $\longrightarrow$ $6[Rh(CO)_3]-2CO$ $\longrightarrow$ $6[9+6]-4 = 6[15]-4 = 6[14+1]-4 = 6[14n+1]-4 = 14n+6 -4 = 14n+2$

$= 4n+2$

$n = 6, \quad k = 2n-1$

$k = 2(6) -1 = 11$

$k = 11$

Sketch of Octahedral skeletal geometry of $Rh_6(CO)_{16}$

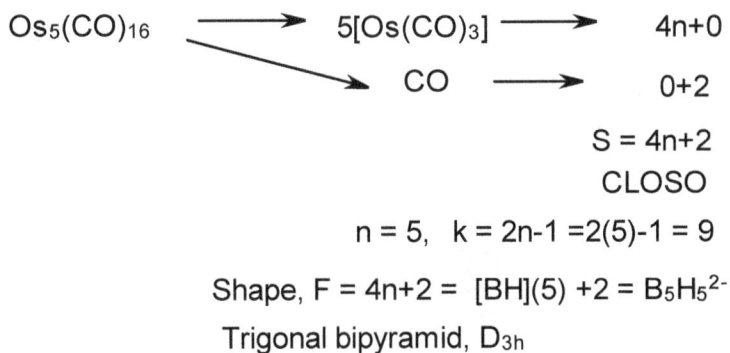

$Os_5(CO)_{16}$ $\longrightarrow$ $5[Os(CO)_3]$ $\longrightarrow$ $4n+0$

$\longrightarrow$ CO $\longrightarrow$ $0+2$

$S = 4n+2$

CLOSO

$n = 5, \ k = 2n-1 = 2(5)-1 = 9$

Shape, $F = 4n+2 = [BH](5) +2 = B_5H_5^{2-}$

Trigonal bipyramid, D_{3h}

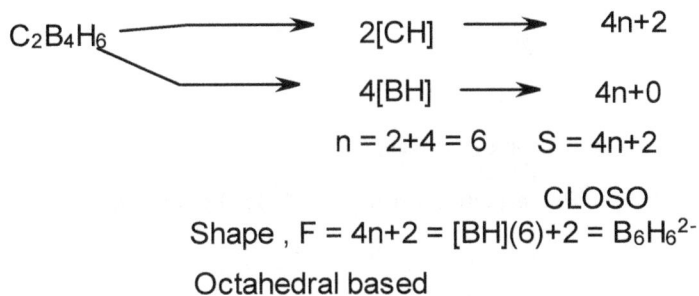

$k = 9$

Trigonal bipyramid, D_{3h}

$C_2B_4H_6$ $\longrightarrow$ $2[CH]$ $\longrightarrow$ $4n+2$

$\longrightarrow$ $4[BH]$ $\longrightarrow$ $4n+0$

$n = 2+4 = 6 \quad S = 4n+2$

CLOSO

Shape , $F = 4n+2 = [BH](6)+2 = B_6H_6^{2-}$

Octahedral based

$TlSn_9^{3-}$ $\longrightarrow$ $1[Tl^{3-}]$ $\longrightarrow$ $1[3+3] = 1[4+2] = 4n+2$

$9[Sn]$ $\longrightarrow$ $9[4] = 9[4n+0] = 4n+0$

$n = 1+9 = 10$, $S = 4n+2$ CLOSO

Shape $= F = 4n+2 = [BH](10)+2 = B_{10}H_{10}^{2-}$

Shape similar to $B_{10}H_{10}^{2-}$

$B_4H_6\{CoCp)_2$ $\nearrow$ $4[BH]$ $\longrightarrow$ $4[4] = 4[4n] = 4n+0$

$\rightarrow$ $2H$ $\longrightarrow$ $0+2$

$\searrow$ $2[CoCp]$ $\longrightarrow$ $2[9+5] = 2[14] = 2[4n] = 4n+0$

$n = 4+2 = 6$ $S = 4n+2$ CLOSO

Shape, $F = 4n+2 = [BH](6)+2 = B_6H_6^{2-}$

Shape similar to $B_6H_6^{2-}$

$CB_9H_{10}CoCp^-$ $\longrightarrow$ $1[CH]$ $\longrightarrow$ $1[4+1+1] = 4n+2$

$\searrow$ $9[BH]$ $\longrightarrow$ $9[4] = 4n+0$

$\searrow$ $1[CoCp]$ $\longrightarrow$ $1[9+5] = 1[14] = 4n+0$

$n = 1+9+1 = 11$, $S = 4n+2$, CLOSO

Shape, $F = 4n+2 = [BH](11)+2 = B_{11}H_{11}^{2-}$

Shape similar to $B_{11}H_{11}^{2-}$

$C_2B_3H_7Fe(CO)_3$ $\longrightarrow$ $2[CH]$ $\longrightarrow$ $2[4+1] = 2[4n+1] = 4n+2$

$\searrow$ $3[BH]$ $\longrightarrow$ $3[3+1] = 3[4] = 4n+0$

$\searrow$ $2H$ $\longrightarrow$ $0+2$

$\searrow$ $1[Fe(CO)_3]$ $\longrightarrow$ $1[8+3\times2] = 1[14] = 1[4] = 4n+0$

$n = 2+3+1 = 6$ $S = 4n+4$,   NIDO

Shape, $F = 4n+4 = [BH](6) +4H = B_6H_6+4H = B_6H_{10}$

Shape similar to B_6H_{10} Pentagonal pyramid. Its CLOSO derivative is $B_7H_7^{2-}$

$Pd_6Ru_6(CO)_{24}{}^{2-}$

↓

$6[Pd(CO)]$ ⟶ $6[10+2] = 6[14-2] = 6[4-2] = 6[4n-2] = 4n-12$

$6[Ru(CO)_3]$ ⟶ $6[8+6] = 6[14] = 6[4] = 6[4n] = 4n+0$

q ⟶ $0+2$

$$S = (4n-12) + (4n+0) + (0+2) = 4n-10$$

$$S = 4n-10$$

$$Cp = C^1C + C^5C = C^6C[M-6]$$

Cluster is a hexa-capped cluster.

$Os_{17}(CO)_{36}{}^{2-}$

↓

$17[Os(CO)_2]$ ⟶ $17[8+4] = 17[14-2] = 17[4-2] = 17[4n-2] = 4n-34$

$2CO$ ⟶ $0+4$

q ⟶ $0+2$

$$S = 4n-28$$

$$Cp = C^1C + C^{14}C = C^{15}C[M-2]$$

This means the cluster has two atoms at the nucleus being a closo member of $4n+2$ series.

$Os_{20}(CO)_{40}{}^{2-}$

↓

$20[Os(CO)_2]$ ⟶ $20[8+4] = 20[14-2] = 20[4-2] = 20[4n-2] = 4n-40$

q ⟶ $0+2$

$$S = 4n-38$$

$$Cp = C^1C + C^{19}C = C^{20}C[M-0]$$

$$[M-0] = 4n+2 = [Os(CO)_3](0) + CO$$

$$[M-0] = 0 + CO = CO$$

$$Cp = C^1C + C^{19}C = C^{20}C[CO]$$

This means that the cluster series starts capping with a CO fragment.

$Cp*IrB_3H_5Co_2(CO)_5$

$\downarrow$

$1[Cp*Ir]$ $= 1[5+9] = 1[14]$ $\longrightarrow$ $4n+0$

$3[BH] = 3[3+1] = 3[4]$ $\longrightarrow$ $4n+0$

$2H$ $\longrightarrow$ $0+2$

$2[Co(CO)_3 = 2[9+6] = 2[15] = 2[14+1] = 2[4+1]$ $\longrightarrow$ $4n+2$

$F_B = $ $4n+2 = [BH](6)+2 = B_6H_6^{2-}$
$-CO$ $\longrightarrow$ $0-2$

$S = 4n+2 \longrightarrow$ CLOSO

$n = 1+3+2 = 6$

Shape similar to $B_6H_6^{2-}$ $\longrightarrow$ Octahedral, O_h

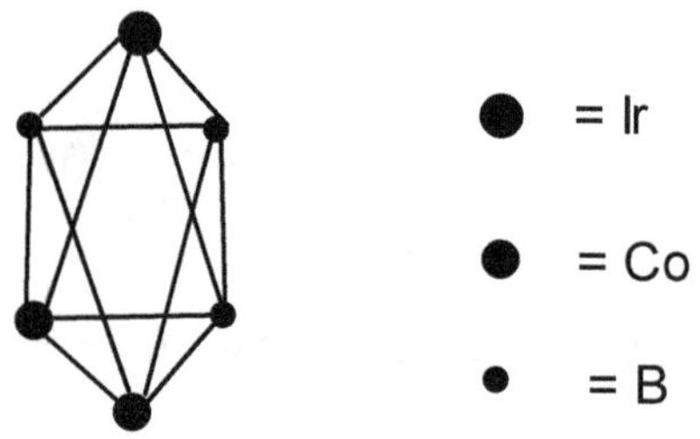

● $= Ir$

● $= Co$

• $= B$

$C_5H_5Mn(CO)_3$

$\downarrow$

• $= CO$

● $= Mn$

$5[C]$ $\longrightarrow$ $4n+0$

$5H$ $\longrightarrow$ $0+5$

$1[Mn(CO)_3]$ $\longrightarrow$ $4n-1$

• $= C$

$=1[7+6] = 1[14-1] = 1[4-1]$

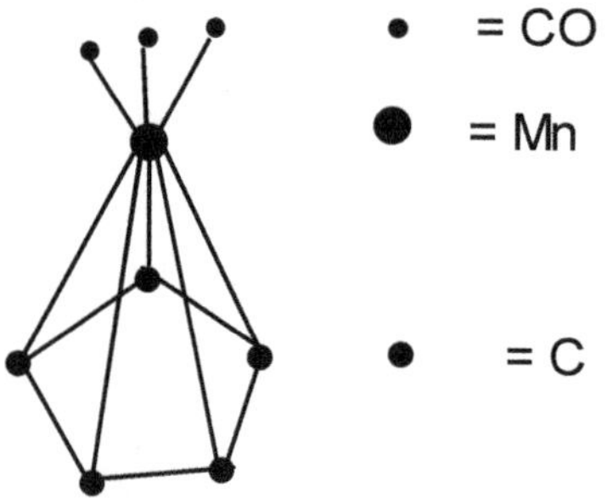

$n = 5+1 = 6$ $S = 4n+4 \longrightarrow$ NIDO $\longrightarrow$ Shape Derived ftom $B_7H_7^{2-}$

$k = 2n-2 = 2(6)-2 = 10$

$F_B = 4n+4 = [BH](6)+4H = B_6H_6+4H = B_6H_{10}$

Shape similar to that of B_6H_{10}

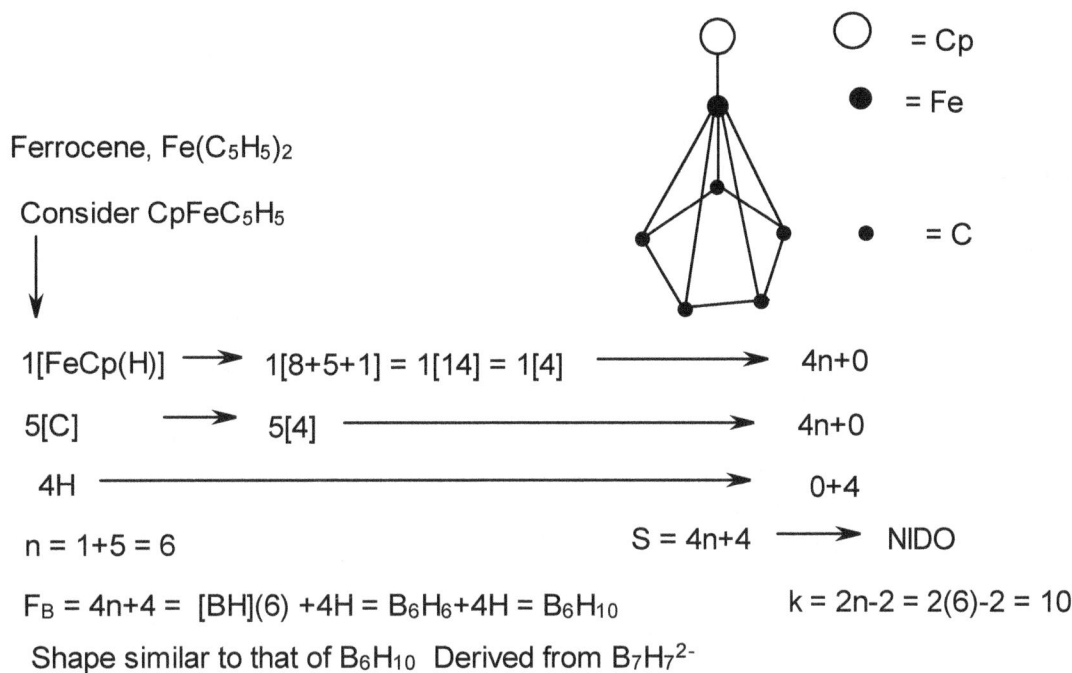

Ferrocene, $Fe(C_5H_5)_2$

Consider $CpFeC_5H_5$

$1[FeCp(H)] \longrightarrow 1[8+5+1] = 1[14] = 1[4] \longrightarrow 4n+0$

$5[C] \longrightarrow 5[4] \longrightarrow 4n+0$

$4H \longrightarrow 0+4$

$n = 1+5 = 6$ $S = 4n+4 \longrightarrow$ NIDO

$F_B = 4n+4 = [BH](6) +4H = B_6H_6+4H = B_6H_{10}$ $k = 2n-2 = 2(6)-2 = 10$

Shape similar to that of B_6H_{10} Derived from $B_7H_7^{2-}$

Ferrocene, $Fe(C_5H_5)_2$

Consider $C_5H_5FeC_5H_5$

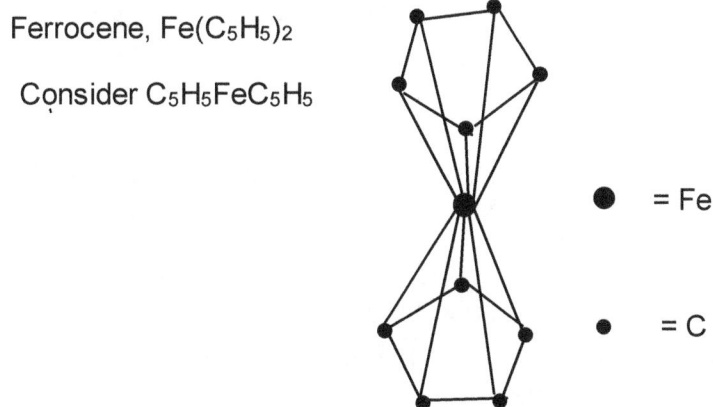

Shape will be two pentagonal pyramids joined through Fe atom

3. Conclusions

The transition metal carbonyl clusters are formed according to a numerical sequence centered on number 14. The 14n series can be generated by multiples of 14. The clusters that belong to $S = 14n+0$ constitute mono-capped closo series. The addition of two valence electrons on members of mono-capped series produces closo series $S = 14n+2$. Other series can be produced by successive addition of 2 extra valence electrons. The two electrons could be supplied by a CO ligand or in some cases two hydrogens or two negative charges. Successive subtraction of 2 valence electrons from the 14n series creates more capping series. Cluster fragments of a particular type of series say 14n+2 can be produced by changing the value of n and adding a CO for integer 2 and substituting a building block fragment with a valence electron content of 14. In case of osmium, the building bloc is $Os(CO)_3$, rhodium, $RhH(CO)_2$, rhenium, $ReH(CO)_3$, and palladium, $Pd(CO)_2$. A cluster formula can be decomposed into chemical fragments centered on mono-skeletal elements and other free atoms or molecules. From the decomposition products series can be derived. From the series a possible geometrical shape of the cluster may be predicted. The series are useful in predicting shapes of hydrocarbons, boranes, carboranes, and their complexes with metals, carbonyl clusters, zintyl ions and other clusters from simple to more complex ones.

Acknowledgements

The author wishes to acknowledge the University of Namibia for financial support and an enabling environment and NAMSOV, Namibia for the offer of a research grant and my wife Merab Kambamu Kiremire for her continued encouragement to write this paper.

References

Balakrishnarajan, M. M., & Hoffmann, R. (2004). Electron Deficient Bonding in Rhomboid Rings. *J. Am. Chem. Soc.,* *126*, 13119-13131. http://dx.doi.org/10.1021/ja0467420

Cotton, F. A., & Wilkinson, G. (1980). *Advanced Inorganic Chemistry, 4th Ed.*, John Wiley and Sons, New York.

Crabtree, R. H. (2005). *The Organometallic Chemistry of the Transition Metals,4th Ed.*, John-Wiley& Sons, New Jersey. http://dx.doi.org/10.1002/0471718769

Douglas, B., McDaniel, D., & Alexander, J. (1994). *Advanced Concepts and Models of Inorganic Chemistry, 3rd Ed.*, John-Wiley& Sons, New York.

Greenwood, N. N., & Earnshaw, A. (1998). *Chemistry of the Elements, 2nd Ed.* Butterworth, Oxford.

Hoffmann, R. (1982). Building Bridges Between Inorganic and Organic Chemistry. *Angew. Chem. Int. Ed. Engl., 21*, 711-724. http://dx.doi.org/10.1002/anie.198207113

Housecroft, C. E., & Sharpe, A. G. (2005). *Inorganic Chemistry, 2nd Ed.*, Pearson, Prentice Hall, Harlow, England.

Hughes, H. K., & Wade, K. (2000). Metal-metal and metal-ligand bond strengths in metal carbonyl clusters. Coord. *Chem. Rev., 197*, 191-229. http://dx.doi.org/10.1016/S0010-8545(99)00208-8

Jardine, F. H., Young, J. F., & Wilkinson, G. (1965). Wilkinson's Catalyst, Chlorotris(triphenyphoshine) rhodium(I), [$RhCl(PPh_3)_3$], 1711-1732.

Jemmis, E. D. (2005). Building relationships between polyhedral boranes and elemental boron. *Inorg. Chem. 18*, 620-628.

Jemmis, E. D., & Balakrishnarajan, M. M. (2001). Polyhedral boranes and elemental boron. Direct structural relations and diverse electronic requirements. *J. Am. Chem. Soc., 123*, 4324-4330. http://dx.doi.org/10.1021/ja0026962

Jemmis, E. D., & Jayasree, E. G. (2003). Analogies between boron and carbon. *Acc. Chem. Res., 36*, 816-824. http://dx.doi.org/10.1021/ar0300266

Jemmis, E. D., & Prasad, D. L. V. K. (2008). Unknowns in the chemistry of Boron. *Current Science, 95(10)*, 1277-1283.

Jemmis, E. D., Balakrishnarajan, M. M., & Pancharatna, P. D. (2001). Unifying electron counting rule for Macropolyhedral Boranes, Metallaboranes, and Metallocenes. *J. Am. Chem. Soc., 123*(18), 4313-4323. http://dx.doi.org/10.1021/ja003233z

Jemmis, E. D., Balakrishnarajan, M. M., & Pancharatna, P. D. (2002). Electronic Requirements for Macropolyhedral Boranes. *Chem. Rev. 102(1),* 93-144. http://dx.doi.org/10.1021/cr990356x

Jemmis, E. D., Jayasree, E. G., & Parameswaran, P. (2006). Hypercarbons in polyhedral structures. *Chem. Soc. Rev., 35*, 157-168. http://dx.doi.org/10.1039/B310618G

King, R. B. (1986). Metal Cluster Topology. *Inorg. Chimica Acta, 116*, 99-107. http://dx.doi.org/10.1016/S0020-1693(00)82162-3

Kiremire, E. M. (2014). Numerical Sequence of Borane Series. *Orient. J. Chem., 30(3)*, 1055-1060. http://dx.doi.org/10.13005/ojc/300317

Kiremire, E. M. (2015b). Classification of Transition Metal Carbonyl Clusters Using the 14n Rule Derived from Number Theory. *Orient. J. Chem., 31(2),* 605-618. http://dx.doi.org/10.13005/ojc/310201

Kiremire, E. M. (2015c). Isolobal Series of Chemical Fragments. Orient. J. Chem., 31(Spl. Edn), 59-70.

Kiremire, E. M. R. (2008). The K-value enumeration of hydrocarbons, alkanes, alkenes and alkynes. *Orient. J. Chem., 24(3),* 795-800.

Kiremire, E. M. (2015a). Categorization and Structural Determination of Simple and More Complex Carbonyl Clusters of Rhenium and Osmium Using k values and the Cluster Table. *Orient. J. Chem., 31(1),* 293-302. http://dx.doi.org/10.13005/ojc/310133

Kiremire, E. M. (2015d). A Unique Bypass into the Cluster Nucleus Using 14n Rule. *Orient. J. Chem. 31(3)*, 1469-1476. http://dx.doi.org/10.13005/ojc/310326

Lipscomb, W. N. (1963). *Boron Hydrides*, Benjamin, New York.

Lipscomb, W. N. (1976). Boranes and their Relatives. *Chemistry*, 224-245.

Longuet-Higgins, H. C., & Bell, R. P. (1943). The Structure of the Boron Hydrides. *J. Chem. Soc.*, 250-255. http://dx.doi.org/10.1039/jr9430000250

Mednikov, E., & Dahl, L. F. (2010). Syntheses, structures and properties of primarily nanosized homo/heterometallic palladium CO/PR$_3$-ligated clusters. *Phil. Trans. R. Soc., 368*, 1301-1331. http://dx.doi.org/10.1098/rsta.2009.0272

Miessler, G, Fischer, P., & Tarr, D. (2014). *Inorganic Chemistry, 5th Edition*, Pearson Education, Inc., Upper Saddle River.

Mingos, D. M. P. (1972). A General Theory for Cluster and Ring Compounds of the Main Group and Transition Elements. *Nature(London), Phys. Sci., 236*, 99-102. http://dx.doi.org/10.1038/physci236099a0

Mingos, D. M. P. (1984). *Polyhedral Skeletal Electron Pair Approach. Acc. Chem. Res., 17(9)*, 311-319. http://dx.doi.org/10.1021/ar00105a003

Mingos, D. M. P. (1991). Theoretical Aspects of metal cluster chemistry. *Pure & Appl. Chem., 63(6)*, 807-812. http://dx.doi.org/10.1351/pac199163060807

Mondo, L., & Langer, C. (1891). On Iron Carbonyls. *J. Chem. Soc. Trans., 59*, 1090-1093. http://dx.doi.org/10.1039/ct8915901090

Mondo, L., Langer, C., & Quinke, F. (1890). Action of Carbon monoxide on Nickel. *J. Chem. Soc. Trans., 57*, 749-753. http://dx.doi.org/10.1039/ct8905700749

Ojima, I., Tsai, C. Y, Tzamarioudaki, M., & Bonafoux, D. (2000). Oxo-Process. *Org. React.,* 56, 1-354.

Rossi, F., & Zanello, P. (2011). Electron Reservoir Activity of High-Nuclearity Transition Metal Carbonyl Clusters. *Portugaliae Electrochimica Acta. 29(5)*, 309-327. http://dx.doi.org/10.4152/pea.201105309

Roth, J. F. (1975). The Production of Acetic Acid-Rhodium Carbonylation of Methanol. *Platinum Metals Rev., 19(1)*, 12-14.

Rudolph, R. W. (1976). Boranes and heteroboranes: a paradigm for the electron requirements of clusters? *Acc. Chem. Res., 9(12)*, 446-452. http://dx.doi.org/10.1021/ar50108a004

Shaik, S., Danovich, D., Wu, W., Su, P., Rzepa, H. S., & Hiberty, P. C. (2012). Quadruple bonding in C$_2$ and analogous eight valence electron species. *Nature Chemistry, 4*, 195-200. http://dx.doi.org/10.1038/nchem.1263

Shriver, D. F., & Atkins, P. W. (1999). *Inorganic Chemistry, 3rd Ed.* Oxford University Press.

Stock, A. (1933). The Hydrides of Boron and Silicon. New York, Cornell University Press. http://dx.doi.org/10.1021/j150356a019

Teo, P., Wickens, Z., K., Dong, G., & Grubbs, R. H. (2012). Efficient and Highly Aldehyde Selective Wacker Oxidation. *Org. Letters, 14(13)*, 3237-3239. http://dx.doi.org/10.1021/ol301240g

Tolman, C. A. (1972). The 16 and 18 Electron Rule in Organometallic Chemistry and Homogeneous catalysis. *Chem. Soc. Rev.*, 337-353. http://dx.doi.org/10.1039/cs9720100337

Wade, K. (1971). The structural significance of the number of skeletal bonding electron-pairs in carboranes, the higher boranes and borane ions and various transition metal carbonyl cluster compounds. *Chem. Commun.*, 792-793. http://dx.doi.org/10.1039/c29710000792

Wade, K. (1976). Structural and Bonding Patterns in Cluster Chemistry. *Adv. Inorg. Chem. Radiochem.. 18*, 1-16. http://dx.doi.org/10.1016/s0065-2792(08)60027-8

Welch, A. J. (2013). The Significance of Wade's Rules. *Chem. Commun., 49*, 3615-3616. http://dx.doi.org/10.1039/c3cc00069a

Williams, R. E. (1971). Carboranes, and boranes, polyhedral and polyhedral fragments. *Inorg. Chem., 10*, 210-214. http://dx.doi.org/10.1021/ic50095a046

4

Mechanism of Alkaline Lignin Oxidation Using Laccase-methyl Syringate Mediator System

Bin Yao[1], Praveen Kolla[2], Ranjit Koodali[3], Chia-Ming Wu[3], Alevtina Smirnova[1,2]

[1]Materials Engineering and Science Program, South Dakota School of Mines and Technology, Rapid City, SD, 57701

[2]Chemistry and Applied Biological Sciences Department, South Dakota School of Mines and Technology, Rapid City, SD, 57701

[3]Chemistry Department, University of South Dakota, Vermillion, SD, 57065

Correspondence: Alevtina Smirnova, Chemistry and Applied Biological Sciences Department, South Dakota School of Mines and Technology, Rapid City, SD, 57701. E-mail: Alevtina.Smirnova@sdsmt.edu.

Abstract

The mechanism of alkaline lignin oxidation in presence of laccase-methyl syringate (MS) mediator is discussed in terms of morphological changes that take place during exposure of the lignin to the phosphate buffer solution (pH=6.5) for 72hr at 70°C. The SEM analysis of lignin before and after enzymatic treatment reveals the morphological changes explained by the interaction of the lignin surface groups with laccase-methyl syringate system. The BET analysis confirms that this interaction causes the change in the surface area from 2.75 to 5.50cm^2/g. The corresponding pore-size distribution in lignin sample treated with laccase-methyl syringate is much broader in comparison to the untreated lignin and the pores within 25-150nm range are detected as a result of the BJH analysis. The electrochemical study of lignin, lignin with laccase, and lignin with laccase in presence of the mediator in the buffer solution has been performed in the potential range from -0.3 to +1.0V vs. Ag, AgCl, Cl$^-$ reference electrode. The cyclic voltammetry confirms reversible oxidation-reduction behavior of the methyl syringate natural mediator in anaerobic and aerobic environment. Specifically in anaerobic conditions three oxidation anodic peaks (0.265V, 0.474V and 0.884V) and two reduction peaks (0.421V and 0.103V) were detected out of which the oxidation peak at 0.474V was assigned to the formation of MS· radical. In aerobic conditions the methyl syringate demonstrates two oxidation peaks (0.473V and 0.812V) and two reduction peaks (0.410V and 0.135V). The mechanism of MS radical stability in oxygen vs. anaerobic environment is proposed based on formation of MS· radicals.

Key words: methyl syringate, laccase, lignin electrooxidation, laccase-mediator system

1. Introduction

Lignin is the second most abundant natural polymer following cellulose, but the commercial use of lignin which is mostly produced by paper industry is only 2%. The rest of the available lignin is usually burned for providing the heat (Chapple et al., 2007) that does not satisfy the need for effective power generation. Different ways of lignin recycling by using catalysts and ionic liquids are reflected in numerous publications (Deepa & Dhepe, 2015., Nanayakkara et al., 2014) however these approaches are often expensive and do not meet the requirements for sustainable lignin recycling. One of the environmentally friendly approaches is to use naturally existing enzymes that are capable of degrading lignin and biomass (Christopher et al., 2014).

Lignin is recalcitrant to degradation because of the complex cross-linked network structure. The white-rot fungi have the efficiency and selectivity to biodegrade the lignin due to production of the lignolytic enzymes (Niladevi, 2009) which are categorized to peroxidases and laccases (Hori et al., 2014). Laccase (oxygen oxidoreductase) is a multicopper oxidase that has an ability to participate in the biodegradation of lignin (Higuchi, 2004; Hoopes & Dean, 2004). Due to the low electrochemical reduction potential, laccase can only oxidize the phenolic lignin moiety (<20% of total lignin) and not the non-phenolic aromatic structure (80% of total lignin) (Camarero et al., 1994). Moreover, lignin's microporous network is not readily accessible for laccase molecules due to their large size (~5-6 nm) (Gascón et al., 2014) and further decreases the overall oxidation efficiency.

The mediator is a small molecule which could be oxidized by laccase and reduced by the substrate. Mediators with low molecular weight have a redox potential higher than 0.9V and can serve as the electron carriers. On the other hand, mediators, as small organic molecules, can enter the pores of lignin and perform a role of electron carriers between the laccase and the lignin. The mediator in an oxidized state has high redox potential (Li et al., 1999) and the ability to oxidize the non-phenolic moiety of lignin. To facilitate the oxidation of lignin by an enzyme, the natural mediators (e.g. 3,5-dimethoxy-4-hydroxybenzaldehyde and methyl syringate (Díaz-González et al., 2011)) and synthetically produced mediators (e.g. ABTS (Bourbonnais & Paice, 1990) or HBT (Call, 1994)) have been proposed. Compared to artificial mediators, the natural mediators are less toxic and more economically feasible. Mediators can be oxidized by the laccase and reduced by the substrate. The optimal mediator is able to keep the cyclic and reversible redox reaction going by reducing to the initial form. However, the mediator should not inhibit the laccase activity during the process of alkali lignin degradation.

Natural phenolic mediators, such as methyl syringate (Wells, 2006) from *Myceliophthora thermophile* (Alejandro Rico1, 2013) can be oxidized by laccase and enhances the oxidization of non-phenolic moiety of lignin and lignin model compounds. As an example, methyl syringate (Díaz-González et al., 2011) and 4-hydroxy-3, 5 dimethoxyacetophenone act as enhancers with laccase from *Trametes villosa* in electrochemical oxidation of kraft and flax lignin, respectively (González Arzola et al., 2009).

The goal of the present study is to investigate the mechanism of alkali lignin oxidation in presence of methyl syringate in the laccase mediator system (LMS) in terms of morphological changes and electrochemical behavior. The electrochemical comparison of the laccase, mediator, and the laccase-mediator system behavior toward degradation of lignin directly deposited on the surface of the rotating glassy carbon electrode in aerobic and anaerobic environment is presented for the first time.

2. Materials and Methods

2.1 Materials

Methyl syringate (MS) mediator and laccase were provided by Novozyme Corp. (Denmark). Disodium hydrogen phosphate (Na_2HPO_4), monosodium phosphate (NaH_2PO_4), 2, 2'-azino-bis (3-ethylbenzothiazoline-6-sulphonic acid) (ABTS), and alkali lignin were purchased from Sigma-Aldrich (US). The laccase activity was measured as an initial velocity during oxidation of 1.6 mM ABTS from Roche to its cation radical (ε_{420} =36000 $M^{-1}.cm^{-1}$) in 0.2M Na_2HPO_4-NaH_2PO_4 buffer solution (pH 6.5) at 30°C. The laccase activity of the enzyme was 326 U/mL. One activity unit (U) was defined as the amount of enzyme transforming 1 μmol of ABTS per min.

2.2 Morphological Study

The lignin samples before and after LMS treatment by LMS were tested with a Zeiss Supra 40VP variable-pressure field emission SEM. The scanning electron micrographs were recorded under an electron beam acceleration of 1 keV at a 5 mm working distance using an in-lens detector. The surface morphology of the materials was characterized using multi point N_2-physisorption isotherms at 77K by means of an automated Quantachrome Nova2200e. The degassing of the catalysts was conducted at 80°C in N_2 overnight prior to the analysis.

2.3 Electrochemical Evaluation

Cyclic voltammetry (CV) - experiments were performed with a Pine bi-potentiostat (AFCBP1, Pine Research Instrumentation, USA) attached to an analytical three-electrode configuration (RDE0018) from Princeton Applied Research. All measurements were carried out in a 100-mL cell at room temperature. A glassy carbon electrode (GCE) with deposited lignin film was used as a working electrode. A platinum wire was used as a counter-electrode and a silver-silver chloride electrode (Ag, AgCl, Cl⁻) served as a reference electrode. Before each experiment, the surface of the glassy carbon electrode was polished on a diamond-polishing pad followed by washing with distilled water. To prepare a working electrode, a 20 μL of alkali lignin dispersed in phosphate buffer solution (0.1 mg/mL) was dropped onto the polished surface of the GC working electrode and allowed to dry for 15 min at room temperature. All CVs were recorded in 0.2 M phosphate buffer solution Na_2HPO_4-NaH_2PO_4 (pH=6.5) in absence or presence of 1.48 mM methyl syringate. The electrode potential was scanned from -0.3 to 1.0 V at different scan rates of 50 mV/s, 100 mV/s and 200 mV/s. Glassy carbon electrode (GCE) in aqueous solutions is considered to be an inert electrode for hydronium ion reduction (Gattrell & Kirk, 1990; Calderon et al., 2013).

3. Results and Discussion

3.1 Morphological Study of Lignin Oxidation in Presence of the Laccase -methyl Syringate System

The SEM analysis reveals the changes in the surface morphology associated with alkali lignin before and after the LMS treatment (Figure 1). The surface morphology of the lignin (Figure 1a-d) improves after the LMS treatment (Figure 1e-h). Specifically, there are more pores present on the surface of the lignin compared to the non-treated lignin (Hamed, 2013). This observation is confirmed by the BET analysis (Figure 2).

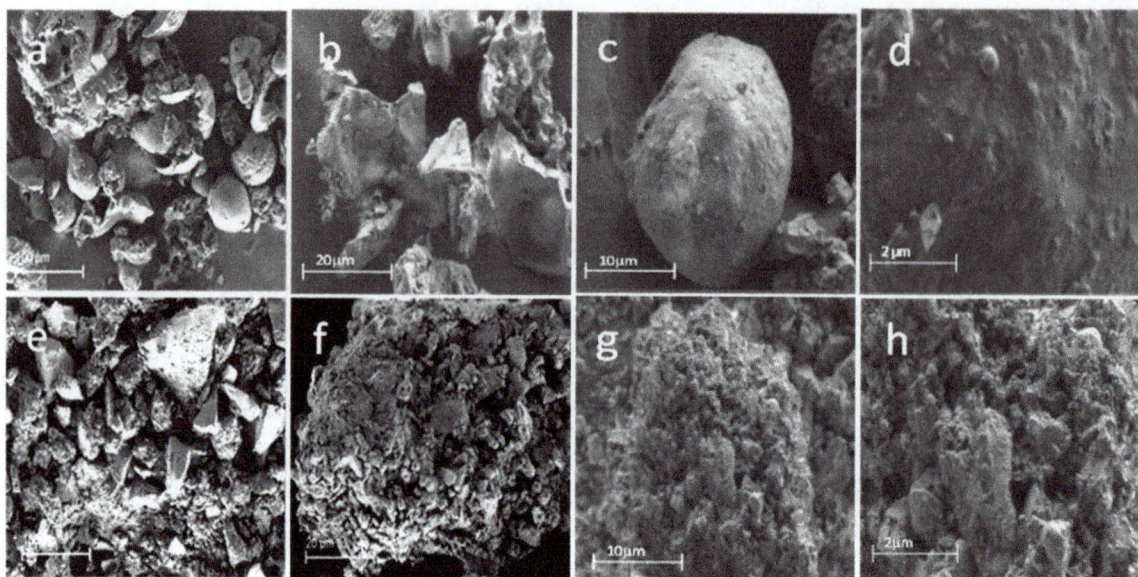

Figure 1. SEM of alkali lignin before (a-d) and after (e-h) treatment with LMS.

The Brunauer–Emmett–Teller (BET) analysis of the alkali lignin has been conducted before and after contact with the laccase-mediator system (5% *Myceliophthora Thermophila* laccase and 5% methyl syringate mediator) in oxygen-rich environment at T=70°C for 72 hrs in Na_2HPO_4-NaH_2PO_4, pH=6.5. The buffer solution contained 5% *Myceliophthora Thermophila* laccase and 5% methyl syringate mediator. Multi point BET isotherms and Barrett–Joyner–Halenda (BJH) pore size distribution in the range between 2-200 nm were presented (Figure 2 a, b). The N_2 adsorption-desorption isotherms (Figure 2a) show a hysteresis due to nitrogen capillary-condensation phenomenon which is typical to mesoporous materials. The Specific Surface Areas (SSAs) and pore volumes corresponding to partial pressures of p/p_o=0.30 and 0.99, respectively were improved up to 2 times after the LMS treatment. Pore-Size-Distribution (PSD) comparison of the alkali lignin (Figure 2b) indicates that the pore-volume was improved due to formation of new mesopores after the LMS degradation. The results indicate that the chosen LMS is active and participates in lignin degradation. This observation is in good agreement with SEM analysis and confirms that at the chosen conditions the LMS is actively involved in the lignin degradation process.

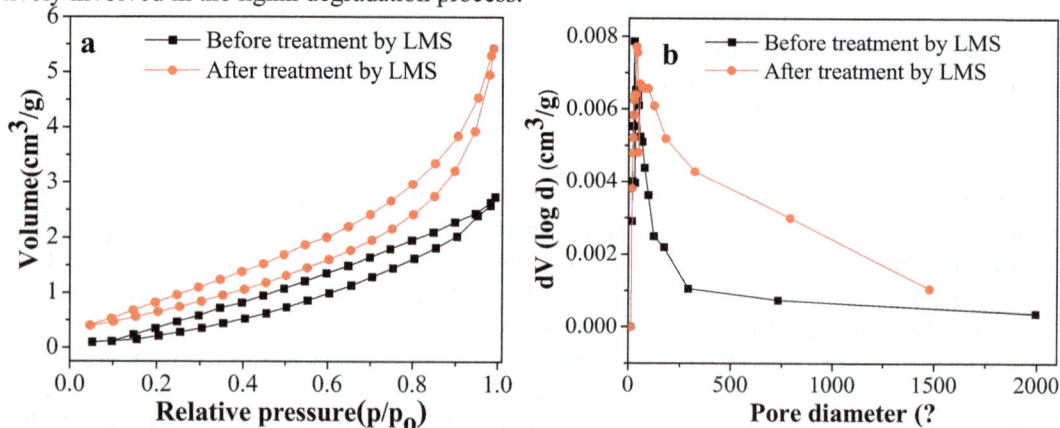

Figure 2. Comparison of the nitrogen adsorption-desorption isotherms (77K) of the alkali lignin (a) and the pore size distribution before and after LMS treatment (b).

3.2 Electrochemical Study of the Lignin-LMS System

3.2.1 Alkali Lignin in Presence of Laccase

The lignin oxidation in presence of LMS demonstrates three different redox electrochemical reactions coupled in the process of lignin decomposition (Figure 3). The presented mechanism indicates that the enzyme alone (Lac^+) is not capable of complete lignin oxidation. As expected, in presence of atmospheric oxygen, an oxygen reduction reaction (ORR) takes place and the atmospheric oxygen participates in the oxidation of laccase (Figure 3, step 1) (Wong, 2009). The laccase has an ability to participate in a redox process on the surface of the reducing substrate involving the four-electron transfer from substrate to each molecular oxygen (Wong, 2009). The laccase can oxidize the mediator to the mediator radical which has high redox potential. (Figure 3, step 2). However, the mediator at a high redox potential (>0.9 V) is capable of oxidizing the lignin non-phenolic groups by itself (Figure 3, step 3). Therefore the laccase can oxidize the mediator as an electron transfer agent (Figure 3, step 2) with a redox potential normally higher than 0.9 V. Then mediator, in turn, can oxidize lignin (Figure 3, step 3). It is known that in the laccase- mediator systems the mediators form stable free ion radicals performing as oxidizing compounds (Morozova et al., 2007). The charged mediator species (Med$^+$) can diffuse away from the enzyme easily penetrating the lignocellulose matrix and initiating the process of lignin oxidation and depolymerization.

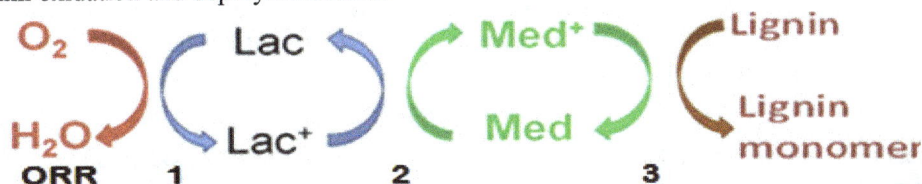

To evaluate the electrochemical behavior of the alkali lignin in presence of the laccase, the CV plots of the laccase in nitrogen (imitating anaerobic conditions) and oxygen have been measured (Figure 4a-b). Fifteen cycles were performed for each test to ensure stabilization of the cyclic voltammetry data. For the laccase alone, the redox transformation in the potential range between -0.3 and 1.0 V vs. Ag/AgCl, Cl$^-$ electrode has not been detected at 50 mV/s scan rate indicating that the laccase without lignin does not have a noticeable redox activity. In case of lignin, the CV at 50mV/s shows an oxidation peak E_{pa} at 0.249V and the corresponding reduction peak E_{pc} at 0.035V in nitrogen. One of the explanations for the observed lignin electrochemical activity detected in anaerobic environment could be related to the molecular oxygen within the pores of lignin. However, considering that this peak remains constant after fifteen cycles an alternative explanation can be given. Specifically, this activity can be due to the electrochemically active phenolic groups on the lignin surface. After addition of laccase to the phosphate buffer solution in presence of lignin on the RDE surface, the oxidation and reduction peaks of lignin shift to the lower potentials of 0.245V and 0.023V, respectively (Figure 4a). Furthermore, the corresponding increase in current density indicates that there is an interaction between the active groups on the surface of lignin and the laccase. The capacitive behavior of the alkali lignin on the surface of glassy carbon in presence of laccase increases, which is in correlation with the SEM data (Figure 1).

The laccase oxidation by molecular oxygen is the first step of lignin decomposition process (Figure 3) (Morozova et al., 2007). Thus, considering our assumption that the electrochemically active groups on the lignin surface are responsible for the redox reversible behavior, similar experiments were performed in oxygen atmosphere. The CVs in anaerobic conditions were compared to the results acquired in oxygen. One anodic and one cathodic faradaic peak at 0.190V to 0.250V, respectively, corresponding to a reversible redox behavior of the phenolic groups on the lignin surface is visible in Figure 4b. The anodic peak is a result of lignin oxidation of by the laccase in the oxidized state (Lac$^+$) which is involved in the oxidation of phenolic groups. However, there is almost no difference between the CV plots in N$_2$ and O$_2$ for laccase that proves our concept that in phosphate buffer solution in the range of the applied voltages the peaks are attributed to the phenolic groups of lignin. As expected, there is no difference in the electrochemical behavior of the laccase alone in N$_2$ or O$_2$ (Figure 4a, b) since the laccase does not have electrochemically active groups in this potential range. Considering the lignin alone there is a shift in the oxidation peak voltage of about 59mV in N$_2$ (0.249V) vs. O$_2$ (0.190V) due to the presence of phenolic groups on the lignin surface. When the laccase is added to lignin, the difference between N$_2$ and O$_2$ becomes more visible: the corresponding shift increases to 85mV in N$_2$. Furthermore, in the case of N$_2$ the redox currents of the system including both lignin and laccase in comparison to the pure lignin in absence of laccase are higher which is in correlation with a positive effect of laccase in the lignin oxidation.

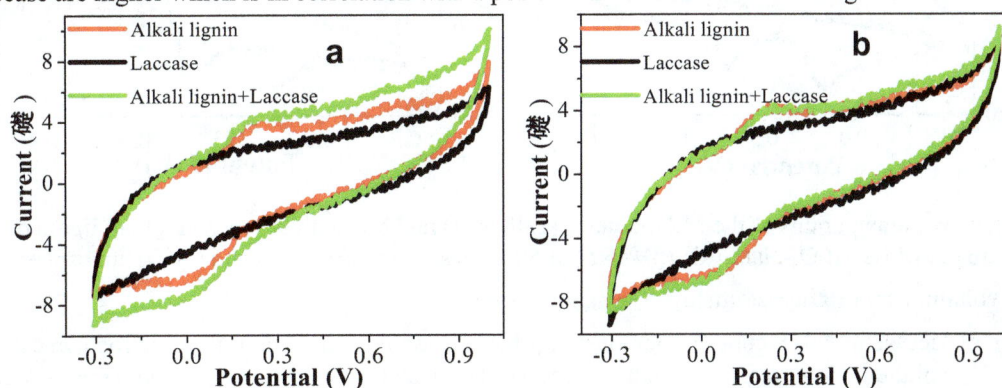

Figure 4. Cyclic voltammograms of alkali lignin, laccase (326 U/mL), and alkali lignin-laccase system in 0.2 M Na$_2$HPO$_4$-NaH$_2$PO$_4$ (pH=6.5) buffer solution at 50 mV/s in N$_2$ (a) and O$_2$ (b) atmosphere.

3.2.2 Alkali Lignin in Presence of the Methyl Syringate Mediator

To determine the reactivity of lignin surface groups with methyl syringate in the oxidized state, (Figure 5) the cyclic voltammograms of MS alone and with alkali lignin in nitrogen atmosphere were performed. The MS alone demonstrates reversible redox behavior (Figure 5a) with three oxidation anodic peaks (E_{pa}) at 0.265V, 0.474V and 0.884V and only two reduction peaks (E_{pc}) at 0.421V and 0.103V. The oxidation peak at 0.474V can be assigned to the formation of (MS·) radical involved in the redox reversible process.

The oxidation peak at 0.265V is irreversible and can be assumed as resulting from the byproduct (MS·+) formed during the mediator oxidization. During MS oxidation, the radical cation (MS·+) formed from methyl syringate molecule is produced and can further dissociate forming the radical MS·:

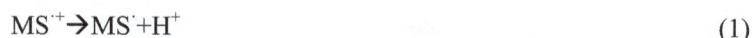

$$MS^{·+} \rightarrow MS^{·} + H^{+} \tag{1}$$

The electrochemical oxidation of MS and alkali lignin results in decrease of the MS oxidation peak. The peak at 0.884V is the result of the oxidation of the methyl syringate radical (Díaz González et al., 2009).

The difference between the oxidation and the reduction peaks (E_{pa}-E_{pc}) is 0.473 V and the ratio of the peak currents equals to i_{pa}/i_{pc} =3.40. This result indicates that the MS· radicals produced in the process of the electrochemical oxidation of MS are not stable and decay to a non-reducible compounds in the nitrogen atmosphere.

Addition of lignin to the MS solution in N_2 (Figure 5a) significantly changes the current density and the shape of the CV plot. In comparison to pure lignin the current increases at high potentials (> 0.8V vs. Ag, AgCl, Cl⁻). The decrease of the current peak at 0.506 V for a mixture of lignin with MS in comparison to MS alone can be attributed to the partial loss of the redox activity of the MS in presence of lignin and the MS interaction with the lignin surface groups. Furthermore, the polymerization of the phenolic oxidation products on the electrode surface can take place due to the formation of a polymeric passivating layer (Aracri et al., 2013). After the methyl syringate is added, the oxidation of the non-phenolic groups of lignin corresponding to a weak cathodic peak appears far away from the oxidation peak (Aracri et al., 2013). This effect indicates partial blocking of MS electron transfer to and from the electrode, produced by lignin deposition on the electrode surface. Possible explanation could be related to the relatively high Lignin: MS ratio that was 350:1 significantly exceeding the previous values of 10:1 (Aracri et al., 2013). Furthermore the ratio of 10:1 considered earlier for lignin model compound (that was present in solution, rather than on the GC electrode surface) can impose mass-transport limitations and decrease the catalytic efficiency (CE) (Aracri et al., 2013).

Similar to nitrogen atmosphere, in the oxygen environment the cyclic voltammetry of the MS shows a reversible redox behavior. The values of the two oxidation MS anodic peaks (E_{pa}) at 0.473V and 0.812V, and the two reduction cathodic peaks (E_{pc}) at 0.410V and 0.135V (Figure 5b) are close to those obtained in N_2. The current density for each of the oxidation anodic currents increases twice due to the presence of oxygen compared to the anodic currents in nitrogen.

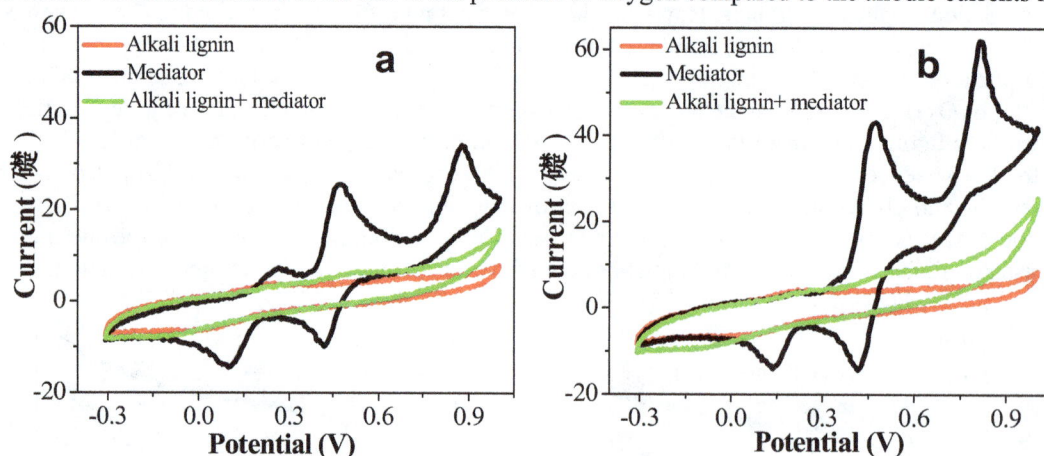

Figure 5. Cyclic voltammograms of the MS mediator (1.48 mM) and MS with 0.1 mg/mL alkali lignin on RDE surface in 0.2 M Na_2HPO_4-NaH_2PO_4 (pH=6.5) at 50mV/s scan rate in N_2 (a) and O_2 (b) atmosphere.

3.2.3 Cyclic Voltammetry of Laccase-methyl Syringate System

The CV plots of laccase in N_2 (Figure 6a) demonstrate that laccase does not show the electrochemical activity in the phosphate buffer solution. However, in absence of laccase the mediator demonstrates a reversible oxidation-reduction behavior. After the laccase is added to the MS mediator, the two oxidation peaks (0.474V, 0.884V) of mediator shift to higher potentials and two reduction peaks (0.421V, 0.103V) shift to lower potentials. Moreover, the currents of the two oxidation peaks increase, which indicates that the laccase has ability to oxidize the mediator even in nitrogen atmosphere.

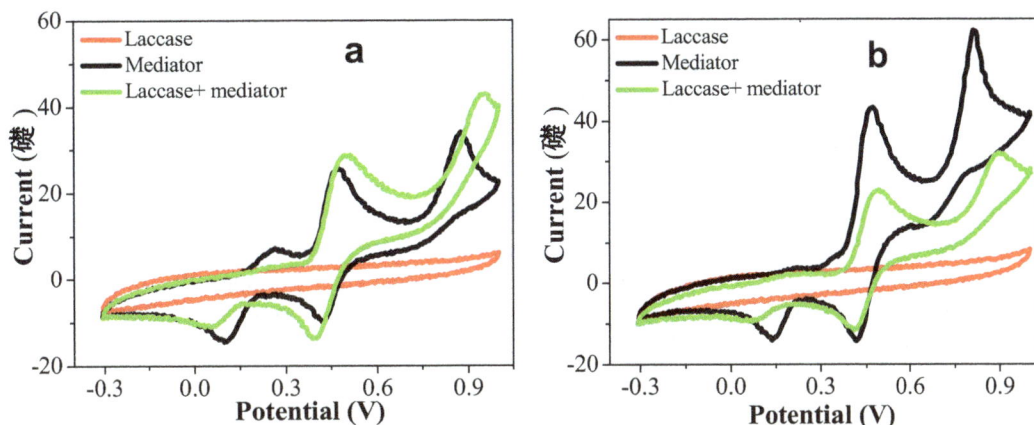

Figure 6. Cyclic voltammograms of laccase (326 U/mL), mediator (1.48mM) and laccase -mediator (1.48mM) in 0.2 M Na$_2$HPO$_4$-NaH$_2$PO$_4$ (pH=6.5) at 50mV/s scan rate in N$_2$ (a) and O$_2$ (b) atmosphere.

However, there is a significant difference between the CV plots in N$_2$ and O$_2$ indicating that the peak current densities are higher in O$_2$ than in N$_2$. Furthermore, the voltage shifts take place. Specifically, in O$_2$ the CV plot for the laccase-mediator system (Figure 6b) shows the first peak at lower potential (0.491V). However, the second oxidation peak for laccase-mediator is shifted to the lower voltage (0.898V) due to the oxidation of the mediator promoted by the ORR in presence of oxygen.

Addition of the laccase to the mediator (Figure 6b) causes significant current drop due to low stability of MS radicals in oxygen atmosphere and deposition of the laccase-oxidized species on the surface of the rotating disk electrode (Coote & Henry, 2005). These species decrease the mass transport limitations for the mediator molecules, thus minimizing the synergistic effect between the laccase and the mediator. Additional CV data for the mediator, laccase, and the mediator with laccase at different rotation speeds (0-900rpm) and scan rates (50-200mV/sec) generated in this study indicate that in all the cases the mediator redox currents in comparison to the laccase with mediator system are higher in oxygen vs. nitrogen.

3.2.4 Alkali Lignin in Presence of Laccase-methyl Syringate System

The comparison of the CV plots of lignin-mediator system and lignin-laccase-mediator system in N$_2$ atmosphere show that there is a little shift between two of the oxidation peaks of lignin-mediator (0.536V) and lignin-laccase-mediator (0.586V). The first oxidation peak of lignin-laccase system at 0.249V has the highest current as a result of oxidation of phenolic group of lignin by laccase (Figure 7a). The oxidation of the non-phenolic groups appears at 0.506V. The oxidation peak currents at 0.506V increase after the laccase is added into the lignin-mediator system. This can be explained by the high stability of the MS· radical that has enough time to react with the non-phenolic groups on the surface of lignin.

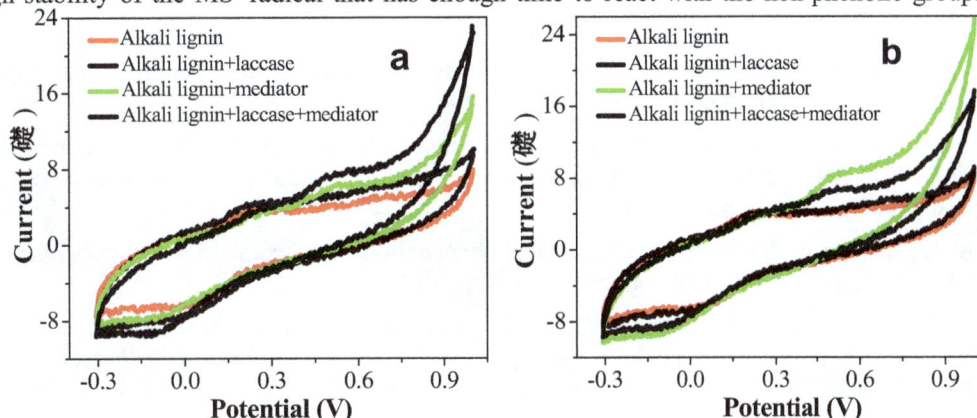

Figure 7. Cyclic voltammograms of alkali lignin (0.1 mg/mL), alkali lignin-laccase (326 U/mL), alkali lignin-mediator (1.48mM) and alkali lignin -laccase -mediator in 0.2 M Na$_2$HPO$_4$-NaH$_2$PO$_4$ (pH=6.5) at 50mV/s scan rate in N$_2$ (a) and O$_2$ (b) atmosphere.

In the oxygen atmosphere, the oxidation peak of lignin-mediator system at 0.525V shifts to 0.521V when the laccase is added. According to the mechanism (Figure 3), more mediator radicals could be produced in oxygen atmosphere than in the nitrogen atmosphere. However, depending on CV analysis, the stability of the radicals has higher effect on the CV potential than the number of produced radicals in oxidation of the non-phenolic groups of lignin (Figure 7b). The reason of the current decrease can be explained by the fact that the radicals are not stable in oxygen condition and do not have enough time to oxidize the non-phenolic groups of lignin.

4. Conclusions

The performed morphological studies of the alkali lignin in presence of laccase-mediator system indicate that the surface of alkali lignin is modified resulting in significant surface modification. The BET demonstrates that the specific surface area and pore volume of the alkali lignin increases up to two times after the LMS treatment. The reason for the improvement of the pore volume is in formation of the new mesopores, which indicates that the proposed LMS system can actively participate in the lignin degradation. The cyclic voltammetry has been used as a simple and powerful tool to predict the efficiency of the laccase mediator systems in the lignin oxidative process under the specified experimental conditions. In presence of MS, an increase in lignin oxidative behavior at higher potentials suggests that the oxidation of the non-phenolic structures in lignin takes place. Based on the electrochemical study, an assumption was made that the stability of MS· radicals is more important than the number of the MS· mediator radicals.

Acknowledgements

This work was supported by NSF EPSCOR grant (Award No. 1330842).

References

A. Wells, M. T. A.T. E. (2006). Green oxidations with laccase–mediator systems, *Biochem. Soc. Trans., 34*, 304–308.

Alejandro, R. J. R., Jose, C. D. R., Angel, T. M., & Ana, G. (2013). Pretreatment with laccase and a phenolic mediator degrades lignin and enhances saccharification of Eucalyptus feedstock, *Biotechnol Biofuels., 7*.

Aracri, E., Tzanov, T., & Vidal, T. (2013). Use of Cyclic Voltammetry as an Effective Tool for Selecting Efficient Enhancers for Oxidative Bioprocesses: Importance of pH, *Ind Eng Chem Res., 52*, 1455-1463. http://dx.doi.org/10.1021/ie3027586.

Bourbonnais, R., & Paice, M. G. (1990). Oxidation of non-phenolic substrates: An expanded role for laccase in lignin biodegradation, *FEBS Letters., 267*, 99-102. http://dx.doi.org/10.1016/0014-5793(90)80298-W.

Calderon, E. H., Dangate, M., Manfredi, N., Abbotto, A., Salamone, M. M., Ruffo, R., & Mari, C. M. (2013). Electrochemical and Spectroelectrochemical Properties of a New Donor¨CAcceptor Polymer Containing 3,4-Dialkoxythiophene and 2,1,3-Benzothiadiazole Units, *Polymers.*. http://dx.doi.org/10.3390/polym5031068.

Call, H. P. (1994). Process for modifying, breaking down or bleaching lignin materials containing lignin or like substances, *World patent application* WO 94/29510.

Camarero, S., Galletti, G. C., & Martínez, A. T. (1994). Preferential degradation of phenolic lignin units by two white rot fungi, *Appl. Environ. Microbiol., 60*, 4509-4516.

Chapple, C., Ladisch, M., & Meilan, R. (2007). Loosening lignin's grip on biofuel production, *Nat Biotech., 25*, 746-748.

Christopher, L. P., Yao, B., & Ji, Y. (2014). Lignin biodegradation with laccase-mediator systems, *Front. Chem. Eng.Res., 2*, 1-13. http://dx.doi.org/10.3389/fenrg.2014.00012.

Coote, M. L., & Henry, D. J. (2005). Effect of Substituents on Radical Stability in Reversible Addition Fragmentation Chain Transfer Polymerization: An ab Initio Study, *Macromolecules., 38*, 1415-1433. http://dx.doi.org/10.1021/ma047814a.

Deepa, A. K., & Dhepe, P. L. (2015). Lignin Depolymerization into Aromatic Monomers over Solid Acid Catalysts, *ACS Catal., 5*, 365-379. http://dx.doi.org/10.1021/cs501371q.

Diaz-Gonzalez, M., Vidal, T., & Tzanov, T. (2011). Phenolic compounds as enhancers in enzymatic and electrochemical oxidation of veratryl alcohol and lignins, *Appl Microbiol Biotechnol., 89*, 1693-1700. http://dx.doi.org/10.1007/s00253-010-3007-3.

Diaz Gonzalez, M., Vidal, T., & Tzanov, T. (2009). Electrochemical Study of Phenolic Compounds as Enhancers in Laccase-Catalyzed Oxidative Reactions, *Electroanal., 21*, 2249-2257. http://dx.doi.org/10.1002/elan.200904678.

Gascon, V., Diaz, I., Márquez-Alvarez, C., & Blanco, R. (2014). Mesoporous Silicas with Tunable Morphology for the Immobilization of Laccase, *Molecules., 19*, 7057.

Gattrell, M., & Kirk, D. W. (1990). The electrochemical oxidation of aqueous phenol at a glassy carbon electrode, *Can. J. Chem. Eng., 68*, 997-1003. http://dx.doi.org/10.1002/cjce.5450680615.

González. A. K., Arévalo, M. C., & Falcón, M. A. (2009). Catalytic efficiency of natural and synthetic compounds used as laccase-mediators in oxidising veratryl alcohol and a kraft lignin, estimated by electrochemical analysis, *Electrochim. Acta., 54*, 2621-2629. http://dx.doi.org/10.1016/j.electacta.2008.10.059.

Hamed, S. A. M. (2013). In-vitro studies on wood degradation in soil by soft-rot fungi: Aspergillus niger and Penicillium chrysogenum, *Int. Biodeterior. Biodegradation., 78*, 98-102 http://dx.doi.org/10.1016/j.ibiod.2012.12.013.

Higuchi, T. (2004). Microbial degradation of lignin: Role of lignin peroxidase, manganese peroxidase, and laccase, *Proc. Jpn. Acad., Series. B, 80*, 204-214.

Hoopes, J. T., & Dean, J. F. D. (2004). Ferroxidase activity in a laccase-like multicopper oxidase from Liriodendron tulipifera, *Plant Physiol Biochem., 42*, 27-33. http://dx.doi.org/10.1016/j.plaphy.2003.10.011.

Hori, C., Gaskell, J., Igarashi, K., Kersten, P., Mozuch, M., Samejima, M., & Cullen, D. (2014). Temporal Alterations in the Secretome of the Selective Ligninolytic Fungus Ceriporiopsis subvermispora during Growth on Aspen Wood Reveal This Organism's Strategy for Degrading Lignocellulose, *Appl Environ Microbiol., 80*, 2062-2070. http://dx.doi.org/10.1128/aem.03652-13.

Li, K., Xu, F., & Eriksson, K. E. L. (1999). Comparison of Fungal Laccases and Redox Mediators in Oxidation of a Nonphenolic Lignin Model Compound, *Appl Environ Microbiol., 65*, 2654-2660.

Morozova, O. V., Shumakovich, G. P., Shleev, S. V., & Yaropolov, Y. I. (2007). Laccase-mediator systems and their applications: A review, *Appl. Biochem. Microbiol., 43*, 523-535. http://dx.doi.org/10.1134/S0003683807050055.

Nanayakkara, S., Patti, A. F., & Saito, K. (2014). Lignin Depolymerization with Phenol via Redistribution Mechanism in Ionic Liquids, *ACS Sustain Chem Eng., 2*, 2159-2164. http://dx.doi.org/10.1021/sc5003424.

Niladevi, K. N. (2009). "Ligninolytic Enzymes," in *Biotechnology for Agro-Industrial Residues Utilisation,* eds. P. Singh Nee' Nigam & A. Pandey, *Springer Netherlands.*, 397-414.

Wong, D. S. (2009). Structure and Action Mechanism of Ligninolytic Enzymes, *Appl Biochem Biotechnol., 157*, 174-209. http://dx.doi.org/10.1007/s12010-008-8279-z.

5

Patulin, Deoxynivalenol, Zearalenone and T-2 Toxin Affect Viability and Modulate Cytokine Secretion in J774A.1 Murine Macrophages

Jonathan H. Loftus[1,3], Gregor S. Kijanka[2], Richard O'Kennedy[1,2], Christine E. Loscher[3]

[1]Applied Biochemistry Group, School of Biotechnology, Dublin City University, Dublin 9, Ireland

[2]Biomedical Diagnostics Institute, Dublin City University, Dublin 9, Ireland

[3]Immunomodulation Group, School of Biotechnology, Dublin City University, Dublin 9, Ireland

Correspondence: Christine Loscher, Immunomodulation Group, School of Biotechnology, Dublin City University, Dublin 9, Ireland. E-mail: christine.loscher@dcu.ie

Abstract

Mycotoxins are secondary fungal metabolites, which occur in food and feed. They have detrimental effects on the health of humans and animals, and they are known to cause immunosuppression. In this study the effect of patulin, deoxynivalenol (DON), zearalenone (ZEN) and T-2 toxin exposure on the viability and the secretion of key pro- and anti-inflammatory cytokines from the murine macrophage cell line, J774A.1, was investigated. Exposure of macrophages to high doses of ZEN (100,000 pg/mL) and T-2 toxin (10,000 and 100,000 pg/mL) resulted in a significant decrease ($P < 0.05$ and $P < 0.01$) in cell viability. Exposure of macrophages to these mycotoxins resulted in a dose-dependent modulation of cytokine secretion. Specifically, exposure to low doses of patulin (0.001, 0.1 and 1 pg/mL) resulted in a statistically significant decrease in the secretion of the pro-inflammatory cytokines interleukin (IL) 6 (IL-6) and tumor necrosis factor alpha (TNF-α), following stimulation with lipopolysaccharide (LPS), a component of Gram-negative bacterial cell walls. Treatment with low doses of DON (0.001 pg/mL) and ZEN (0.001 and 0.01 pg/mL) significantly decreased ($P < 0.01$) the secretion of the pro-inflammatory cytokine IL-12p40, while several doses of T-2 toxin (0.001, 0.01, 0.1, 1 and 100 pg/mL) caused a significant decrease the expression of IL-6. Each of the mycotoxins also significantly increased the production of the anti-inflammatory cytokine IL-10, both before and after LPS stimulation. This data provides further insight into the mechanisms by which mycotoxins modulate the host immune response to exert their immunosuppressive activity.

Keywords: mycotoxins, macrophage, inflammation, immunosuppression, immunomodulation

1. Introduction

Mycotoxins are the naturally occurring toxic secondary metabolites of some fungal species. They are food contaminants which are proven to have detrimental effects on human and animal health. Their immunosuppressive and carcinogenic nature is of particular concern. The Food and Agriculture Organization (FAO) has estimated that 25% of the world's food supply is contaminated with mycotoxins (Marroquín-Cardona et al., 2015). Mycotoxin contamination can occur at any time during harvest or storage of food.

Over 400 different mycotoxins have been identified. They are structurally diverse (Figure 1) but are produced primarily by three genera of fungi: *Aspergillus*, *Penicillium*, and *Fusarium*. Patulin (~150 Da), a polyketide lactone is produced by *Penicillium* and *Aspergillus*. The trichothecenes (~250-550 Da) are a group of sesquiterpene epoxides produced mainly by *Fusarium* species while zearalenone (~318 Da) is a resorcyclic acid lactone also produced by *Fusarium*.

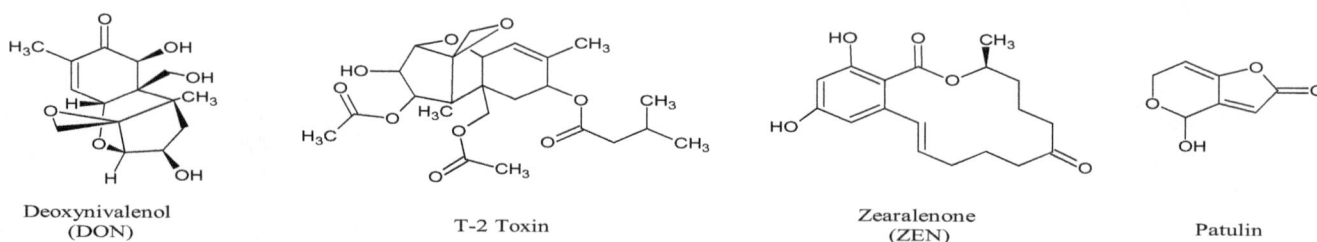

Figure 1. Chemical structure of deoxynivalenol (DON), T-2 toxin, zearalenone (ZEN) and patulin.

Mycotoxin contamination can occur in a variety of plants used as food, including commodities such as cereal grains (barley, corn, rye and wheat), coffee, dairy products, fruits, nuts, peanuts, and spices. Contamination can also arise in animal products, like milk, caused by animals consuming contaminated feeds. Generally, crops that are stored for more than a few days become a potential target for mold growth and mycotoxin formation. The primary way in which humans become exposed to mycotoxins is by eating contaminated food such as grain, corn, and other foodstuff or, alternatively, by consuming animals or animal products that have eaten contaminated feed. Mycotoxins are extremely stable compounds so they can easily avoid damage in the digestive system and reach the human bloodstream (Stoev, 2015).

Mycotoxins have the potential to elicit a wide variety of toxicological effects, including immune-suppression and immune-stimulation. Exposure to sublethal doses of many mycotoxins has the potential to either stimulate or suppress immune functions, such as lymphocyte proliferation, cell-mediated immunity and humoral immunity, depending on the dose and exposure time (Edite Bezerra da Rocha et al., 2014). Immunosuppression is viewed as one of the most significant effects of mycotoxin exposure and it has a major economic impact. Studies have already highlighted the immunosuppressive nature of several mycotoxins such as aflatoxins, trichothecenes, ochratoxin A and zearalenone (Ferrante et al., 2002; Sharma et al., 2004; Ferrante et al., 2008; Ubagai et al., 2008; Marzocco et al., 2009; Bianco et al., 2012; Bruneau et al., 2012; Alassane-Kpembi et al., 2013; Jia et al., 2014). Immunostimulation and immunosuppression can occur with the same toxin depending on the exposure conditions, including dosage and time. Such experimental observations have given rise to the term 'immunomodulation' which accounts for this dual effect. It is important to stress that any deregulation of immune cell homeostasis can result in serious consequences for immune functions, increasing susceptibility to infections and cancer, as well as favouring the development of autoimmune diseases. Therefore, any significant changes in the functionality of immune cells must be considered as a significant hazard (Marin et al., 2013).

Several studies have shown that the group of mycotoxins, aflatoxins, can affect macrophages, both in vitro and in vivo. Liu et al., (2002) showed that fumonisin B and aflatoxin B1 were immunotoxic to swine alveolar macrophages, employing techniques such as DNA laddering, nuclear fragmentation and phagocytosis, and by analysis of apoptosis-related heat shock protein 72 (hsp72) and cytokines IL-1β and TNF-α. Numerous animal studies have demonstrated that aflatoxins have immunosuppressive activity. Poultry (chickens and turkeys), pigs and lambs in particular, are susceptible to aflatoxin-induced immunosuppression (Devegowda and Murthy, 2005). Furthermore, cell-mediated immunity is affected by aflatoxin exposure. Murine macrophages exposed to aflatoxins both in vivo and in vitro have exhibited decreased cytokine secretion. Other macrophage functions, such as release of reactive intermediates and phagocytosis are decreased in macrophages exposed to AFB1 (Liu et al., 2002). Although the immunomodulatory effects of the aflatoxins have been demonstrated previously, the mechanism by which these compounds exert their immunosuppressive effects is still unknown. Previous investigations have shown that AFB1 pretreatment, followed by LPS stimulation, decreases CD14 expression in murine peritoneal macrophages. However, CD14 expression was unaffected in cells that were pre-treated with AFB1, but not challenged with LPS (Moon and Pyo, 2000).

Mycotoxins can produce cellular depletion in lymphoid organs, cause alterations in T-cell and B-cell function, suppress antibody responses, suppress NK cell activity, decrease delayed-type hypersensitivity responses, and increase susceptibility to infectious disease. T-2 toxin was implicated as a developmental immunotoxicant, which targets fetal lymphocyte progenitors (Holladay et al., 2002). The effect of the trichothecenes group of mycotoxins on immune function has shown that the mechanism of impairment is related to inhibition of protein synthesis. High doses of trichothecenes induce lymphocyte apoptosis along with immune suppression. Low doses promote expression of cytokines including IL-1, IL-2, IL-5, and IL-6. Tricothecenes also activate mitogen-activated protein kinases (MAPK's) in vivo and in vitro, via the ribotoxic stress response (Zhou et al., 2003; Pestka et al., 2004). Pestka and Zhou (2006) shows that pre-exposure with LPS sensitises murine RAW264.7 macrophages and peritoneal murine macrophages to the pro-inflammatory effects of DON.

ZEN has been shown to be immunotoxic (Luongo et al., 2008) and has also been observed to be both a suppressor and inductor of the production of inflammatory cytokines (Salah-Abbès et al., 2008). Several alterations of immunological parameters were found associated with ZEN concentrations in vitro (Murata et al., 2003). Additive and synergistic effects of mycotoxins were demonstrated for various combinations of mycotoxins, such as ochrations, aflatoxins and ZEN, (Grenier and Oswald, 2011; Halbin et al., 2013; Lei et al., 2013). In vitro studies have shown that mycotoxin combinations can act additively, synergistically and even in some cases antagonistically (Bruneau et al., 2012; Clarke et al., 2014).

Macrophage activation by LPS results in expression of a number of pro-inflammatory cytokines, including IL-12p40, TNF-α and IL-6. This is then followed by the expression of anti-inflammatory cytokines, such as IL-10, to stop production of the pro-inflammatory cytokines and to ensure regulation of the immune response (Bruneau *et al.*, 2012). Although these cytokines have important roles in host defence, their over- or under-expression can lead to problems such as inflammatory diseases. Since the macrophage is a key cell in the innate immune response, we investigated the effect of mycotoxins exposure on cell viability and cytokine secretion in a murine macrophage cell line, J774A.1. The aim of this study was to gain better understanding of the immunomodulatory effects of these compounds on macrophages, in order to shed light on their immunosuppressive activity.

2. Materials and Methods

2.1 Reagents and Chemicals

Patulin, deoxynivalenol, T-2 toxin, zearalenone and dimethyl sulphoxide (DMSO) were purchased from Sigma Aldrich (St. Louis, MO). Dulbecco's Modified Eagle's Medium (DMEM), Fetal Bovine Serum (FBS) and Penicillin/Streptomycin were purchased from Invitrogen (Carlsbad, CA). Lipopolysaccharide (LPS) isolated from *E. coli*, serotype R515, was purchased from Enzo Life Sciences (Farmingdale, NY). CellTiter 96® Aqueous One Solution was purchased from Thermo Fisher Scientific (Rockford, IL). DuoSet cytokine ELISA kits were purchased from R&D Systems (Minneapolis, MN).

2.2 Cell Culture

The murine macrophage cell line J774A.1 was obtained from the European Collection of Cell Cultures (ECACC; Salisbury, UK). The cells were grown in DMEM supplemented with 10% (w/v) fetal bovine serum, 50U penicillin and 50 µg streptomycin. All cultures were maintained in a 37°C, in 5% CO_2 humidified atmosphere. For cell viability analysis, 100 µL of a J774A.1 cell suspension, at a concentration of 1×10^6 cells/mL, was added to each well of a NUNC™96 well tissue culture plate and allowed to adhere for 1 h. The cells were then incubated with the appropriate concentration of each mycotoxin for 24 h ±100 ng/mL LPS. The negative control consisted of cells incubated alone and with 0.1% (v/v) methanol (vehicle). As a positive control, cells were incubated with 10% v/v DMSO. 24 hours after the addition of LPS, 20 µL of the CellTiter 96® Aqueous One Solution was added to each well of the 96-well plate. The plates were incubated for 4 hours at 37°C in 5 % CO_2 and absorbance read at 490 nm. For cytokine expression analysis, 250 µL of a J774A.1 cell suspension, at a concentration of 1×10^6 cells/mL, was added to each well of a NUNC™96 well tissue culture plate and allowed to adhere for 1 h. The cells were then incubated with the appropriate concentration of each mycotoxin for 24 h ±100 ng/mL LPS to induce an inflammatory response. The negative control was made up of cells that were incubated alone and with 0.1% (v/v) methanol (vehicle). At the end of the incubation time, the supernatant was removed and stored at -20°C until analysis. All experiments were carried out in triplicate.

2.3 Determination of Cell Viability

The CellTiter 96® Aqueous One Solution (Pierce, UK) was employed for spectrophotometric quantification of cell viability. The assay was carried out according to the manufacturer's instructions. Percentage viability was calculated compared to untreated control cells. Each sample was assayed in triplicate.

2.4 Cytokine ELISAs

TNF-α, IL-10, IL-6 and IL-12p40 ELISAs were performed according to the manufacturer's instructions (R&D systems, UK). Each sample was assayed in triplicate for all of the cytokines indicated.

2.5 Statistical Analysis

Data was analyzed by ANOVA using Prism version 5 (GraphPad, La Jolla, CA, USA). If the ANOVA table was significant ($P < 0.05$), Post-hoc analysis was performed using Dunnett's test.

3. Results

3.1 ZEN and T-2 Toxin Affect J774A.1 Macrophage Cell Viability

The viability of the J774A.1 murine macrophage cell line, in the presence of patulin, DON, ZEN and T-2 toxin in concentrations ranging from 10 to 100,000 pg/mL, was assessed using the MTS assay. The inflammatory processes in the macrophage are activated by the endotoxin, LPS. Therefore, LPS was included in the MTS analysis, to determine the effect of toxin exposure on macrophage viability. The results indicate that patulin and DON have no cytotoxic effect on J774A.1 macrophages at any of the chosen concentrations. For ZEN there is a significant ($P < 0.05$) decrease in the percentage viability at a concentration of 100,000 pg/mL, which is cytotoxic without LPS stimulation (Figure 2). T-2 toxin shows a significant cytotoxic effect at 100,000 pg/mL ($P < 0.01$) for unstimulated cells and at 10,000 pg/mL ($P < 0.05$) and 100,000 pg/mL ($P < 0.01$) for LPS-stimulated cells. All experiments show no effect of the vehicle control on cell viability with and without LPS stimulation.

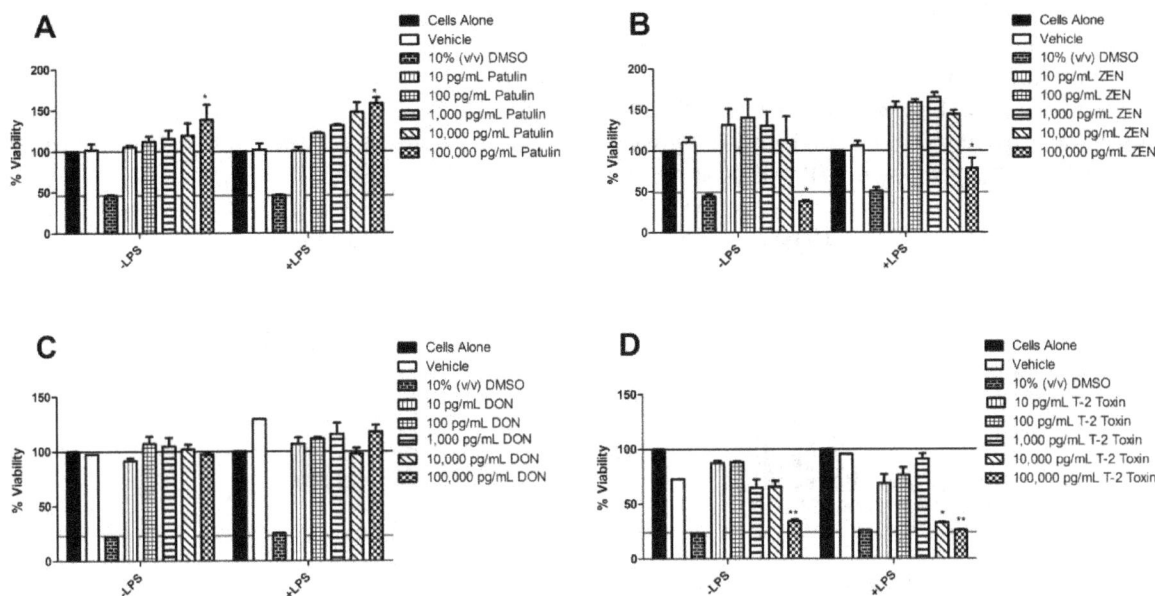

Figure 2. MTS proliferation assay to analyse the effect of patulin (A), ZEN (B), DON (C), and T-2 toxin (D) on the J774A.1 macrophage cell line.

J774A.1 cells were treated for 24 hours with each mycotoxin, ranging in concentration from 10 to 100,000 pg/mL, with and without LPS (100 ng/mL) stimulation. DMSO (10%, v/v) was included as a positive control of cytotoxicity. Results indicate a mean value from three independent experiments and error bars represent mean ± 3SE (standard error of mean) (n=3). Significance: $P < 0.05*$; $P < 0.01**$; $P < 0.001***$ as illustrated, all relative to cells not exposed to toxin.

3.2 Exposure of Macrophages to Single Mycotoxins Modulates Cytokine Secretion

DuoSet ELISA kits (R&D systems) were used to quantify the levels of cytokines (IL-6, IL-10, IL-12p40 and TNF-α) in the supernatants of J774A.1 macrophage cells treated with patulin, DON, ZEN and T-2 toxin in increasing concentrations from 0.001 to 100 pg/mL for 24 hours with and without LPS (100 ng/mL) stimulation.

The results for patulin (Figure 3) (Table 1) show that there is a significant ($P < 0.001$) increase in the expression of IL-6 at 10 pg/mL patulin without LPS stimulation, however, significant suppression is observed at 0.1 pg/mL ($P < 0.01$) and 1 pg/mL ($P < 0.05$). The production of IL-10 is increased significantly at 1 pg/mL ($P < 0.01$), 10 pg/mL ($P < 0.001$) and 100 pg/mL ($P < 0.05$) without LPS stimulation, and at $0.1 - 10$ pg/mL ($P < 0.01$) following LPS treatment. An increase in IL-12p40 is observed at 10 pg/mL ($P < 0.001$) and 100 pg/mL ($P < 0.01$) without LPS stimulation and also 10 pg/mL and 100 pg/mL ($P < 0.001$) with LPS stimulation. TNF-α was increased at 10 pg/mL patulin (-LPS) ($P < 0.001$) and (+LPS) ($P < 0.05$), however, its levels were decreased at 0.001 pg/mL and 0.1 pg/mL ($P < 0.05$).

DON (Figure 4) (Table 2) causes a significant increase in IL-6 production at 1 pg/mL ($P < 0.01$) and 10 pg/mL ($P < 0.001$) without LPS, and 10 pg/mL ($P < 0.05$) with LPS. Levels of IL-10 are increased significantly without LPS stimulation at 0.1 pg/mL ($P < 0.01$), 1 pg/mL ($P < 0.001$), 10 pg/mL ($P < 0.001$) and 100 pg/mL ($P < 0.05$) but only up for 10 pg/mL ($P < 0.001$) with LPS stimulation. IL-12p40 is significantly increased ($P < 0.001$) at 10 pg/mL both + and - LPS, while IL-12p40 is decreased ($P < 0.05$) by 0.001 pg/mL DON following LPS stimulation. TNF-α is significantly increased at 0.1 pg/mL ($P < 0.05$), 1 pg/mL ($P < 0.01$) and 10 pg/mL ($P < 0.001$) without LPS, but only increased for 10 pg/mL ($P < 0.001$) with LPS.

The results for ZEN (Figure 5) (Table 3) show that IL-6 is significantly increased ($P < 0.001$) by $0.1 - 100$ pg/mL ZEN with no LPS stimulation. However, no significant changed are observed for IL-6 following LPS treatment. IL-10 was significantly increased by 10 pg/mL ($P < 0.001$) before LPS treatment and by 0.01 pg/mL ($P < 0.05$), 0.1 pg/mL ($P < 0.01$) and 1 pg/mL ($P < 0.01$) after LPS treatment. here is a significant increase in the levels of IL-12p40 at 10 pg/mL ($P < 0.001$) (-LPS) but a significant decrease in its levels at 0.001 and 0.01 pg/mL ($P < 0.01$) (+LPS). TNF-α was significantly increased at 0.1 pg/mL ($P < 0.05$), 1 pg/mL ($P < 0.05$) and 10 pg/mL ($P < 0.001$) without LPS stimulation and 1 pg/mL ($P < 0.05$) and 10 pg/mL ($P < 0.001$) with LPS treatment.

T-2 toxin (Figure 6) (Table 4) exposure resulted in a significant increase in IL-6 production at 0.1 pg/mL (P < 0.05) 1 pg/mL (P < 0.01) and 10 pg/mL (P < 0.001) (-LPS). However, significant suppression of IL-6 was observed at 0.001 pg/mL (P < 0.001), 0.01 pg/mL (P < 0.001), 0.1 pg/mL (P < 0.01), 1 pg/mL (P < 0.05) and 100 pg/mL (P < 0.01) following LPS stimulation. IL-10 levels were increased (P < 0.001) by 10 pg/mL T-2 without LPS but significant decrease was seen at 0.001 pg/mL (P < 0.05), 0.01 pg/mL (P < 0.001) and 100 pg/mL (P < 0.05). Following LPS stimulation an increase (P < 0.01) of IL-10 was only seen at 10 pg/mL T-2 toxin. IL-12p40 levels were increased at 10 pg/mL (P < 0.01) (-LPS), and 1pg/mL (P < 0.05) and 10 pg/mL (P < 0.001) (+LPS). TNF-α production was significantly increased by 1 pg/mL (P < 0.05) and 10 pg/mL (P < 0.001), both with and without LPS stimulation. However, LPS stimulation also showed a significant decrease in TNF-α at 0.01 pg/mL (P < 0.05) and 1 pg/mL (P < 0.05) T-2 toxin.

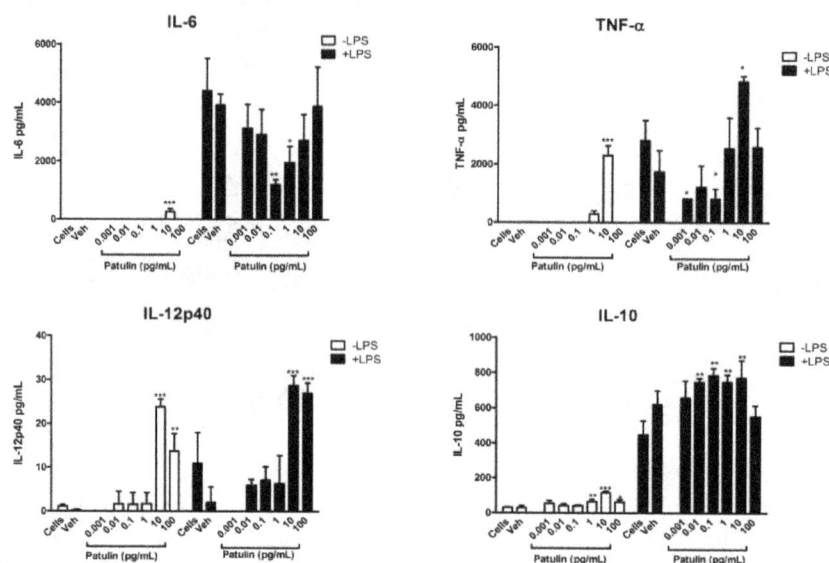

Figure 3. Expression of IL-6, IL-10, IL-12p40 and TNF-α in the supernatants of J774A.1 murine macrophage cells, cultured for 24 hours in the presence of **patulin** ranging in concentration from 0.001 to 100 pg/mL, with and without LPS (100 ng/mL) stimulation.

The toxin treatment alone is represented by the white bars (□) and the toxin treatment with LPS challenge is represented by the black bars (■). Data was expressed as the mean for each experiment (n=3), with error bars indicating ± 3SE. Significance: P < 0.05*; P < 0.01**; P < 0.001*** as illustrated, all relative to cells not exposed to toxin.

Table 1. The effect of increasing concentrations of **patulin** (0.001 – 100 pg/mL) on the expression levels of cytokines from J774A.1 macrophages with and without LPS (100 ng/mL) treatment. Significance represented by: no change –, increase P < 0.05↑; P < 0.01↑↑; P < 0.001↑↑↑, decrease P < 0.05↓; P < 0.01↓↓; P < 0.001↓↓↓, compared to control cells.

Patulin (pg/mL)	IL-6	IL-10	IL-12p40	TNF-α
0.001	–	–	–	–
0.01	–	–	–	–
0.1	–	–	–	–
1	–	↑↑	–	–
10	↑↑↑	↑↑↑	↑↑↑	↑↑↑
100	–	↑	↑↑	–
Patulin (pg/mL) + LPS (100 ng/mL)				
0.001	–	–	–	↓
0.01	–	↑↑	–	–
0.1	↓↓	↑↑	–	↓
1	↓	↑↑	–	–
10	–	↑↑	↑↑↑	↑
100	–	–	↑↑↑	–

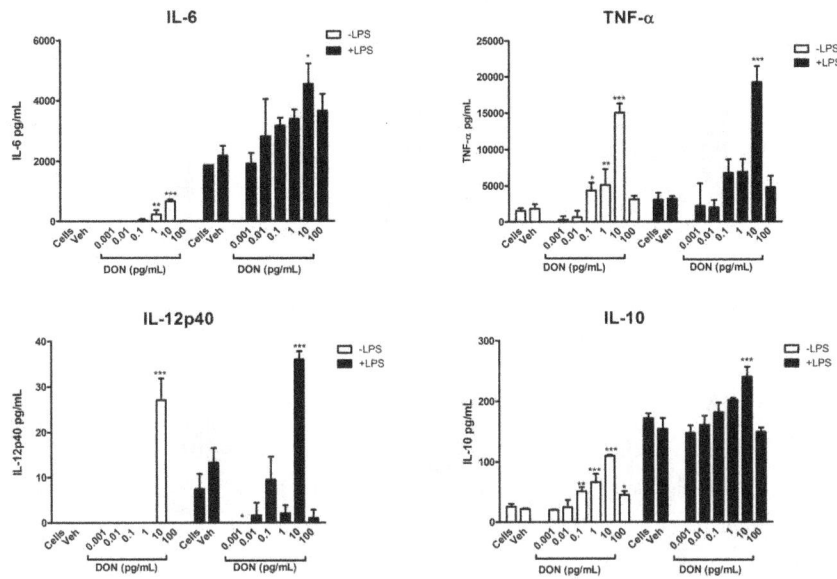

Figure 4. Expression of IL-6, IL-10, IL-12p40 and TNF-α in the supernatants of J774A.1 murine macrophage cells, cultured for 24 hours in the presence of **DON** ranging in concentration from 0.001 to 100 pg/mL, with and without LPS (100 ng/mL) stimulation.

The toxin treatment alone is represented by the white bars (□) and the toxin treatment with LPS challenge is represented by the black bars (■). Data was expressed as the mean for each experiment (n=3), with error bars indicating ± 3SE. Significance: $P < 0.05*$; $P < 0.01**$; $P < 0.001***$ as illustrated, all relative to cells not exposed to toxin.

Table 2. The effect of increasing concentrations of **DON** (0.001 – 100 pg/mL) on the expression levels of cytokine from J774A.1 macrophages with and without LPS (100 ng/mL) treatment. Significance represented by: no change –, increase $P < 0.05$↑; $P < 0.01$↑↑; $P < 0.001$↑↑, decrease $P < 0.05$↓; $P < 0.01$↓↓; $P < 0.001$↓↓↓, compared to control cells.

DON (pg/mL)	IL-6	IL-10	IL-12p40	TNF-α
0.001	–	–	–	–
0.01	–	–	–	–
0.1	–	↑↑	–	↑
1	↑↑	↑↑↑	–	↑↑
10	↑↑↑	↑↑↑	↑↑↑	↑↑↑
100	–	↑	–	–
DON (pg/mL) + LPS (100 ng/mL)				
0.001	–	–	↓	–
0.01	–	–	–	–
0.1	–	–	–	–
1	–	–	–	–
10	↑	↑↑↑	↑↑↑	↑↑↑
100	–	–	–	–

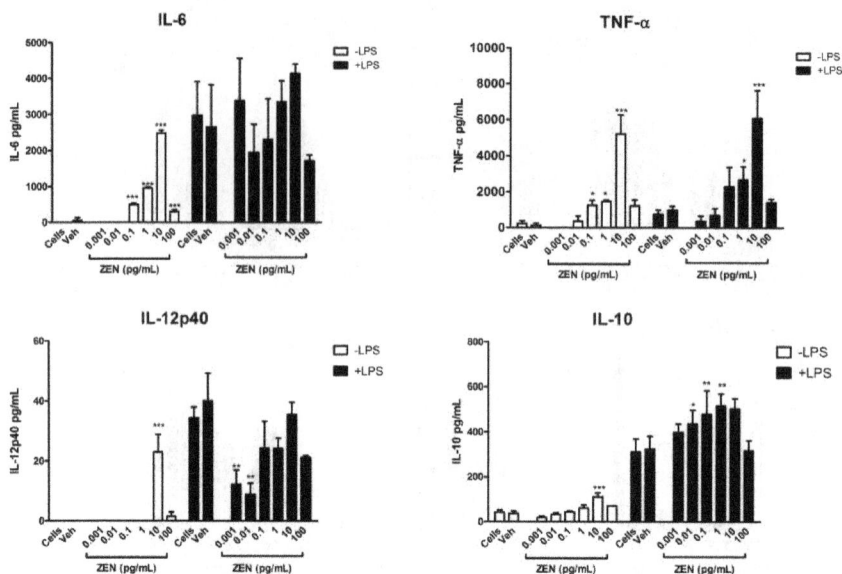

Figure 5. Expression of IL-6, IL-10, IL-12p40 and TNF-α in the supernatants of J774A.1 murine macrophage cells, cultured for 24 hours in the presence of **ZEN** ranging in concentration from 0.001 to 100 pg/mL, with and without LPS (100 ng/mL) stimulation.

The toxin treatment alone is represented by the white bars (□) and the toxin treatment with LPS challenge is represented by the black bars (■). Data was expressed as the mean for each experiment (n=3), with error bars indicating ± 3SE. Significance: $P < 0.05$*; $P < 0.01$**; $P < 0.001$*** as illustrated, all relative to cells not exposed to toxin.

Table 3. The effect of increasing concentrations of ZEN (0.001 – 100 pg/mL) on the expression levels of cytokines from J774A.1 macrophages with and without LPS (100 ng/mL) treatment. Significance represented by: no change –, increase $P < 0.05$↑; $P < 0.01$↑↑; $P < 0.001$↑↑, decrease $P < 0.05$↓; $P < 0.01$↓↓; $P < 0.001$↓↓↓, compared to control cells.

ZEN (pg/mL)	IL-6	IL-10	IL-12p40	TNF-α
0.001	–	–	–	–
0.01	–	–	–	–
0.1	↑↑↑	–	–	↑
1	↑↑↑	–	–	↑
10	↑↑↑	↑↑↑	↑↑↑	↑↑↑
100	↑↑↑	–	–	–
ZEN (pg/mL) + LPS (100 ng/mL)				
0.001	–	–	–	–
0.01	–	↑	↓↓	–
0.1	–	↑↑	↓↓	–
1	–	↑↑	–	↑
10	–	–	–	↑↑↑
100	–	–	–	–

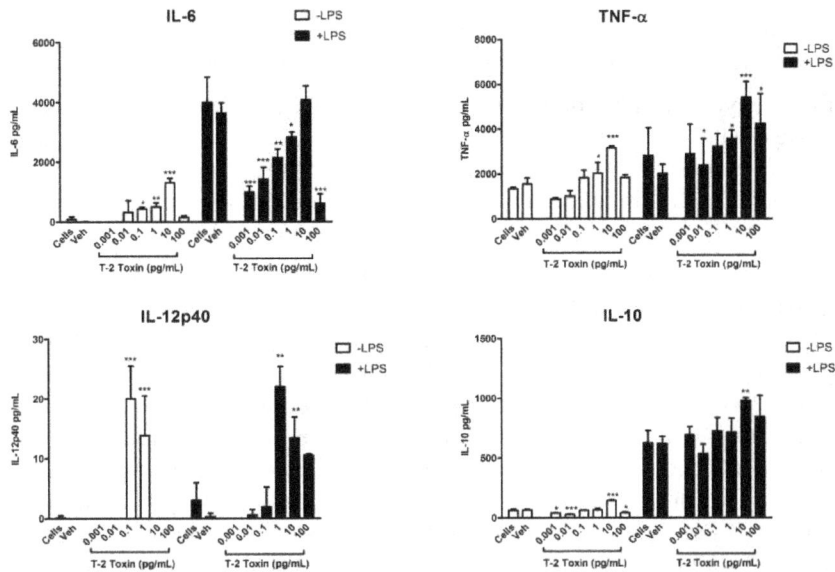

Figure 6. Expression of IL-6, IL-10, IL-12p40, and TNF-α in the supernatants of J774A.1 murine macrophage cells, cultured for 24 hours in the presence of **T-2 toxin** ranging in concentration from 0.001 to 100 pg/mL, with and without LPS (100 ng/mL) stimulation.

The toxin treatment alone is represented by the white bars (□) and the toxin treatment with LPS challenge is represented by the black bars (■). Data was expressed as the mean for each experiment (n=3), with error bars indicating ± 3SE. Significance: $P < 0.05$*; $P < 0.01$**; $P < 0.001$*** as illustrated, all relative to cells not exposed to toxin.

Table 4. The effect of increasing concentrations of T-2 toxin (0.001 – 100 pg/mL) on the expression levels of cytokines from J774A.1 macrophages with and without LPS (100 ng/mL) treatment. Significance represented by: no change –, increase $P < 0.05$↑; $P < 0.01$↑↑; $P < 0.001$↑↑↑, decrease $P < 0.05$↓; $P < 0.01$↓↓; $P < 0.001$↓↓↓, compared to control cells.

T-2 toxin (pg/mL)	IL-6	IL-10	IL-12p40	TNF-α
0.001	–	↓	–	–
0.01	–	↓↓↓	–	–
0.1	↑	–	–	–
1	↑↑	–	–	↑
10	↑↑↑	↑↑↑	↑↑	↑↑
100	–	↓	–	–
T-2 toxin (pg/mL) + LPS (100 ng/mL)				
0.001	↓↓↓	–	–	↓
0.01	↓↓↓	–	–	↓
0.1	↓↓	–	–	–
1	↓	–	↑	↑
10	–	↑↑	↑↑↑	↑↑↑
100	↓↓↓	–	–	↑

4. Discussion

Under normal healthy conditions, the immune system defends the host against pathogenic infections. However, certain environmental factors, including toxins, can alter the development and function of the immune response, leading to autoimmunity, hypersensitivity or immunosuppression. The understanding of the biological processes underlying immune system dysfunction by mycotoxins is incomplete.

Mycotoxin contamination is widespread and exposure to low levels of the toxins is unavoidable, therefore it is imperative to understand the effect of exposure to low doses. The effects of various mycotoxins on the immune system has been widely investigated (Sharma *et al.*, 2004; Ferrante et al., 2008; Ubagai *et al.*, 2008; Marzocco *et al.*, 2009;

Bianco *et al.*, 2012; Alassane-Kpembi *et al.*, 2013; Clarke *et al.*, 2014; Solhaug *et al.*, 2015), but there is a lack of understanding of the exact mechanisms behind the interactions involved. The effects of individual toxins can vary depending on concentration, time and other environmental factors.

The aim of this study was to characterise the individual effects of patulin, DON, ZEN and T-2 toxin in order to improve understanding of the individual molecular mechanisms of mycotoxin immunomodulation.

The effect that each of these mycotoxins has on the viability of J774A.1 murine macrophages was examined. The result of this investigation showed that patulin and DON did not display cytotoxic effects on the macrophage cells at any of the chosen concentrations. In contrast, ZEN has a significant ($P < 0.05$) effect on cell viability at 100,000 pg/mL and is cytotoxic without LPS stimulation at this concentration. T-2 toxin becomes significantly cytotoxic ($P < 0.01$) at 10,000 and 100,000 pg/mL.

Other studies have demonstrated the cytotoxic nature of mycotoxins. Clarke *et al.* (2014) examined the cytotoxicity of AFB1, fumonisin B1 (FB1) and ochratoxin A (OTA) on Caco-2, MDBK and RAW264.7 cell lines. They also found these toxins to be cytotoxic, but at much higher concentrations to that used in this study. Hymery *et al.* (2014) found that the mycotoxin cyclopiazonic acid was toxic to the human cell lines; CD34+, monocytes, THP-1 and Caco-2. Notably, Marzocco *et al.* (2009) showed that both nivalenol (NIV) and DON (10-100µM) significantly stimulate apoptosis in J774A.1 macrophages in a concentration-dependent manner on cultured J774A.1 murine macrophages. Capasso *et al.* (2015) demonstrated that low doses of DON used alone have minor toxic effects, while induce cytotoxicity and inflammation when used in combination with particulate matter. Ferrante *et al.* (2008) showed that OTA (30nM–100µM) induces a time and concentration dependent cytotoxic effect on J774A.1 macrophages, increased when cells were co-stimulated with LPS (100 ng/mL), a concentration that alone did not modify the cellular viability.

Mycotoxins may not be cytotoxic at low concentrations but this does not mean they are not causing dysfunction to the immune system. The cytokine expression analysis presented here, clearly shows that these mycotoxins have varying effects on the expression of IL-6, IL-10, IL-12p40 and TNF-α, which are dependent on toxin dosage and exposure to LPS stimulation. Depending on the dosage, it is possible for mycotoxins to be immunostimulatory as well as immunosuppressive.

Exposure of macrophages to patulin, DON, ZEN and T-2 toxin resulted in a dose-dependent modulation of cytokine secretion. For example, a significant increase in pro-inflammatory cytokines was observed in cells that were exposed to high concentrations (10 pg/mL) of the mycotoxins without LPS treatment. This pro-inflammatory response can also be seen for each toxin at the 10 pg/mL concentration after LPS treatment, with statistically significant increases in the secretion of IL-6, IL-12p40 and TNF-α. DON also demonstrates a potent pro-inflammatory response overall at 10 pg/mL both with and without LPS stimulation. Exposure to low doses of patulin resulted in a statistically significant decrease in the secretion of IL-6 and TNF-α following stimulation with LPS. Treatment with low doses of DON and ZEN significantly decreased the secretion of IL-12p40, while nearly all doses of T-2 toxin caused a significant decrease the expression of IL-6.

There are very few published reports on the effect of mycotoxins on immune regulation, particularly cytokine secretion. Ferrante *et al.* (2008) observed that OTA (3µM) alone induces a significant increase in cyclooxygenase-2 (COX-2) and inducible nitric oxide synthase (iNOS) expression, while at the highest concentration (10µM) a reduced expression of both enzymes was shown. When cells were co-stimulated with LPS, OTA showed a concentration-dependent reduction of COX-2 and iNOS expression. These results confirm the pro-inflammatory role of OTA by itself, and demonstrate the impaired capability of OTA-treated macrophages to respond properly to noxious stimuli, such as LPS, mimicking the environmental co-exposure to both compounds. Bruneau *et al.* (2012) investigated the effect of AFB1, AFB2 and AFG1 exposure, alone and in combination, on the secretion of key pro- and anti-inflammatory cytokines from the murine macrophage cell line, J774A.1. Exposure of macrophages to low doses of aflatoxin (0.01 or 0.1 ng/mL) resulted in a statistically significant change in the secretion of a number of cytokines following stimulation with LPS. Treatment with AFB1 or AFB2 alone significantly decreased the secretion of IL-10, while the secretion of the pro-inflammatory cytokine IL-6 was significantly increased. The study also demonstrated how aflatoxin exposure affects expression levels of key cell surface markers involved in the inflammatory response. TLR2 and CD14 expression levels decreased significantly, TLR4 expression was unaffected. Jia *et al.* (2014) showed that ZEN increased mRNA and protein expression of TLR4 and inflammatory cytokines in kidney in dose-dependent manner. Their results indicated that TLR4-mediated inflammatory reactions signal pathway was one of the mechanisms of ZEN-mediated toxicity in the kidney.

5. Conclusion

In conclusion, the results from this investigation demonstrate that although these compounds come from similar sources, due to the differences in structure, they exert variable effects on cytokine production. Overall, the results indicate that individually patulin, DON, ZEN and T-2 toxin have the ability to modulate the typical inflammatory response of macrophages to LPS stimulus. In particular, these toxins affected the secretion of cytokines that are critical for the normal host responses to infection. This is the first report on the effect of patulin, DON, ZEN and T-2 toxin on the modulation of these important pro- and anti-inflammatory cytokines. From this research, it is clear that these mycotoxins have the ability to cause macrophage dysfunction. Some of the molecular processes of toxicity are still not fully understood. However, this report has provided data on the potential mechanisms by which mycotoxins inhibit macrophage functions and, therefore, host defence functions, through deregulation of cytokine profiles.

Acknowledgments

This project is funded by the Irish Research Council and Science Foundation Ireland.

References

Alassane-Kpembi, I., Kolf-Clauw, M., Gauthier, T., Abrami, R., Abiola, F. A., Oswald, I. P., & Puel, O. (2013). New insights into mycotoxin mixtures: The toxicity of low doses of type B trichothecenes on intestinal epithelial cells is synergistic. *Toxicol. Appl. Pharmacol., 272*(1), 191-198. http://dx.doi.org/10.1016/j.taap.2013.05.023.

Bianco, G., Russo, R., Marzocco, S., Velotto, S., Autore, G., & Severino, L. (2012). Modulation of macrophage activity by aflatoxins B1 and B2 and their metabolites aflatoxins M1 and M2. *Toxicon., 59*(6), 644–650. http://dx.doi.org/10.1016/j.toxicon.2012.02.010.

Bruneau, J. C., Stack, E., O'Kennedy, R., & Loscher, C. E. (2012). Aflatoxins B1, B2 and G1 modulate cytokine secretion and cell surface marker expression in J774A.1 murine macrophages. *Toxicol. In Vitro., 26*(5), 686-693. http://dx.doi.org/10.1016/j.tiv.2012.03.003.

Capasso, L., Longhin, E., Caloni, F., Camatini, M., & Gualtieri, M. (2015). Synergistic inflammatory effect of PM10 with mycotoxin deoxynivalenol on human lung epithelial cells. *Toxicon., 104*, 65-72. http://dx.doi.org/10.1016/j.toxicon.2015.08.008.

Clarke, R., Connolly, L., Frizzell, C., & Elliott, C. T. (2014). Cytotoxic assessment of the regulated, co-existing mycotoxins aflatoxin B1, fumonisin B1 and ochratoxin, in single, binary and tertiary mixtures. *Toxicon., 90*, 70-81. http://dx.doi.org/10.1016/j.toxicon.2014.07.019.

Devegowda, G., & Murthy, T. K. N. (2005). Mycotoxins: their adverse effects in poultry and some practical solutions. In The Mycotoxin Blue Book, 1st Edition. Editors, Diaz, D. E. Nottingham University Press, Nottingham, UK, 25-26.

Edite Bezerra da Rocha, M., Freire, F. D. C. O., Erlan, F. M. F., Izabel, F. G. M., & Rondina, D. (2014). Mycotoxins and their effects on human and animal health. *Food Control., 36*(1), 159-165. http://dx.doi.org/10.1016/j.foodcont.2013.08.021.

Ferrante, M. C., Meli, R., Raso, G. M., Esposito, E., Severino, L., Di Carlo, G., & Lucisano, A. (2002). Effect of fumonisin B on structure and function of macrophage plasma membrane. *Toxicol. Lett., 129*(3), 181–187. http://dx.doi.org/10.1016/S0378-4274(01)00476-3.

Ferrante, M. C., Raso, M. G., Bilancione, M., Esposito, E., Iacono, A., & Meli, R. (2008). Differential modification of inflammatory enzymes in J774A.1 macrophages by ochratoxin a alone or in combination with lipopolysaccharide. *Toxicol. Lett., 181*(1), 40–46. http://dx.doi.org/10.1016/j.toxlet.2008.06.866.

Grenier, B., & Oswald, I. (2011). Mycotoxin co-contamination of food and feed: meta-analysis of publications describing toxicological interactions. *World Mycotoxin J., 4*, 285-313. http://dx.doi.org/10.3920/wmj2011.1281.

Halbin, K. J., Kouakou, B., & Dago, G. (2013). Low level of ochratoxin A enhances aflatoxin B1 induced cytotoxicity and lipid peroxidation in both human intestinal (caco-2) and hepatoma (HepG2) cell lines. J. *Nutr. Food Sci., 2*(6), 294-30. http://dx.doi.org/10.11648/j.ijnfs.20130206.15.

Holladay, S. D., & Blaylock, B. L. (2002). The mouse as a model for developmental immunotoxicology. *Hum. Exp. Toxicol., 9-10*, 525-531. http://dx.doi.org/10.1191/0960327102ht292oa.

Hymery, N., Masson, F., Barbier, G., & Coton, E. (2014). Cytotoxicity and immunotoxicity of cyclopiazonic acid on human cells. *Toxicol. In Vitro., 28*(5), 940-947. http://dx.doi.org/10.1016/j.tiv.2014.04.003.

Jia, Z., Liu, M., Qu, Z., Zhang, Y., Yin, S., & Shan, A. (2014). Toxic effects of zearalenone on oxidative stress, inflammatory cytokines, biochemical and pathological changes induced by this toxin in the kidney of pregnant rats. *Environ. Toxicol. Pharmacol., 37*(2), 580-591. http://dx.doi.org/10.1016/j.etap.2014.01.010.

Lei, M., Zhang, N., & Desheng, Q. (2013). *In vitro* investigation of individual and combined cytotoxic effects of aflatoxin B1 and other selected mycotoxins on the cell line porcine kidney 15. *Exp. Toxicol. Pathol.*, *63*, 1149-1157. http://dx.doi.org/10.1016/j.etp.2013.05.007.

Liu, B., Yu, F., Chan, M., & Yang, Y. (2002). The effects of mycotoxins, fumonisin B1 and aflatoxin B1, on primary swine alveolar macrophages. *Toxicol. Appl. Pharmacol.*, *3*, 197-204. http://dx.doi.org/10.1006/taap.2002.9406.

Luongo, D., Severino, L., Bergamo, P., De Luna, R., Lucisano, A., & Rossi, M. (2006). Interactive effects of fumonisin B1 and alpha-zearalenol on proliferation and cytokine expression in Jurkat T cells. *Toxicol. In. Vitro.*, *8*, 1403-1410. http://dx.doi.org/10.1016/j.tiv.2006.06.006.

Marin, S., Ramos, A. J., Cano-Sancho, G., & Sanchis, V. (2013). Mycotoxins: Occurrence, toxicology, and exposure assessment. *Food Chem. Toxicol.*, *60*, 218-237. http://dx.doi.org/10.1016/j.fct.2013.07.047.

Marroquín-Cardona, A. G., Johnson, N. M., Phillips, T. D., & Hayes, A. W. (2014). Mycotoxins in a changing global environment – A review. *Food and Chem. Toxicol.*, *69*, 220-230. http://dx.doi.org/10.1016/j.fct.2014.04.025.

Marzocco, S., Russo, R., Bianco, G., Autore, G., & Severino, L. (2009). Pro-apoptotic effects of nivalenol and deoxynivalenol trichothecenes in J774A.1 murine macrophages. *Toxicol. Lett.*, *189*(1), 21–26. http://dx.doi.org/10.1016/j.toxlet.2009.04.024.

Moon, E. Y., & Pyo, S. (2000). Aflatoxin B1 inhibits CD14-mediated nitric oxide production in murine peritoneal macrophages. *Int. J. Immunopharmacol.*, *3*, 237-246. http://dx.doi.org/10.1016/S0192-0561(99)00081-8.

Mosser, D. M., & Edwards, J. P. (2008). Exploring the full spectrum of macrophage activation. *Nat. Rev. Immunol.*, *8*, 958–969. http://dx.doi.org/10.1038/nri2788.

Murata, H., Sultana, P., Shimada, N., & Yashioka, M. (2003). Structure–activity relationships among zearalenone and its derivatives based on bovine neutrophil chemiluminescence. *Vet. Hum. Toxicol.*, *45*(1), 18–20.

Pestka, J. J., & Zhou, H. R. (2006). Toll-like receptor priming sensitizes macrophages to proinflammatory cytokine gene induction by deoxynivalenol and other toxicants. *Toxicol. Sci.*, *2*, 445-455. http://dx.doi.org/10.1093/toxsci/kfl012.

Pestka, J. J., Zhou, H. R., Moon, Y., & Chung, Y. J. (2004). Cellular and molecular mechanisms for immune modulation by deoxynivalenol and other trichothecenes: unraveling a paradox. *Toxicol. Lett.*, *1*, 61-73. http://dx.doi.org/10.1016/j.toxlet.2004.04.023.

Salah-Abbès, B., Abbès S., Houas, Z., Abdel-Wahhab, M. A., & Oueslati, R. (2008). Zearalenone induces immunotoxicity in mice: possible protective effects of radish extract (Raphanus sativus). *J. Pharm. Pharmacol.*, *60*, 761-70. http://dx.doi.org/10.1211/jpp.60.6.0012.

Sharma, N., He, Q., & Sharma, R. P. (2004). Augmented fumonisin B toxicity in co-cultures: Evidence for crosstalk between macrophages and non-parenchymatous liver epithelial cells involving proinflammatory cytokines. *Toxicol.*, *203*(1–3), 239–251. http://dx.doi.org/10.1016/j.tox.2004.06.017.

Solhaug, A., Wisbech, C., Christoffersen, T. E., Hult, L. O., Lea, T., Eriksen, G. S., & Holme, J. A. (2015). The mycotoxin alternariol induces DNA damage and modify macrophage phenotype and inflammatory responses. *Toxicol. Lett.*, *239*(1), 9-21. http://dx.doi.org/10.1016/j.toxlet.2015.08.1107.

Stoev, S. D. (2015). Foodborne mycotoxicoses, risk assessment and underestimated hazard of masked mycotoxins and joint mycotoxin effects or interaction. *Environ. Toxicol. Pharmacol.*, *39*(2), 794-809. http://dx.doi.org/10.1016/j.etap.2015.01.022.

Ubagai, T., Tansho, S., Ito, T., & Ono, Y. (2008). Influences of aflatoxin B1 on reactive oxygen species generation and chemotaxis of human polymorphonuclear leukocytes. *Toxicol. In Vitro.*, *4*, 1115-1120. http://dx.doi.org/10.1016/j.tiv.2008.01.007.

Zhou, H. R., Islam, Z., & Pestka, J. J. (2003). Rapid, sequential activation of mitogenactivated protein kinases and transcription factors precedes proinflammatory cytokine mRNA expression in spleens of mice exposed to the trichothecene vomitoxin. *Toxicol. Sci.*, *1*, 130-142. http://dx.doi.org/10.1093/toxsci/kfg006.

6

Reciprocal Potential Oscillations across the Electrolytic Cells Connected in Series

Chetansing K. Rajput

Correspondence: Dr. Chetansing K. Rajput, M.B.B.S., T.N.M.C., Mumbai University, ACST, 306, Vikrikar Bhavan, Nashik-422010, India. E-mail: chetansingkrajput@gmail.com

Abstract

The series arrangement of identical electrolytic cells having copper anodes and chloride electrolyte is found to demonstrate an extremely asymmetric and reciprocally oscillating voltage drop across the series of these coupled cells. The origin of this phenomenon is attributed to the presence of different extents of cuprous oxide phase on surfaces of the anodes of different cells, in the series. Such series arrangement of multiple electrolytic cells introduces a novel phenomenon of the non-linear temporal behaviour of coupled cells. Most importantly, such configuration proves the very existence of adsorbed Cu_2O on surface of copper anode, in acidic chloride media. And hence, it signifies the important role of Cu_2O in the electro-dissolution mechanism of copper anode, even at lower pH values.

Keywords: reciprocal potential oscillations, copper anode, chloride electrolyte, potential oscillations, electrochemical oscillations

1. Introduction

Electro-dissolution of copper anode in acidic chloride media has always been the area of research interest of many investigators. As per general acceptance, the anodic dissolution of copper, in acidic chloride electrolyte, proceeds via formation of CuCl phase and soluble cuprous chloride complexes, like $CuCl_2^-$, $CuCl_3^{2-}$, etc. (Lee et al. 1986; Kear et al. 2004; Crundwell et al. 1992; Nobe et al. 1979). However, there are arguments regarding the formation of Cu_2O phase in acidic environment. The potential–pH diagrams do not permit the presence of copper oxide in acidic chloride electrolyte. Also, according to previous studies, oxide is hard to form in the bulk electrolyte, at lower pH values (Tromans et al. 1991; Sourisseau et al. 2005).

However, this does not exclude the existence of Cu_2O as an adsorbed layer on anodic surface. What happens in the thin surface film at electrode-electrolyte interface, where mass transport and electron transfer is limited to a small space, is a different scenario than what happens in bulk electrolyte.

This paper describes the phenomenon of potential oscillations between the electrolytic cells, connected in series. The rise in voltage across one of the coupled cells, is accompanied by the fall of voltages across other cells in the series. Hence, such kind of electrochemical oscillations can be termed as 'Reciprocal Potential Oscillations' (R.P.O.s), between the connected cells. Also, the amplitude of these R.P.O.s is found to show strong dependence upon Cl⁻ concentration in electrolyte. Hence, this phenomenon can be successfully exploited for the detection of chloride ions and determination of their concentration in the electrolyte. This phenomenon can also be used to assemble an electrochemical clock, to demonstrate the exact aetiology and mechanism of the electrochemical oscillations, etc.

Also, the aim of present study is to reveal the mechanism of this new type of temporal behaviour in electrochemistry. The interpretation of its mechanism is presented on basis of the periodic formation and dissolution of the passivating cuprous oxide layer on anodic surface. The non-linear dynamics of Cu | Chloride system can be resolved with such series configuration of electrolytic cells. Also, such coupled cells enable the detection of adsorbed Cu_2O in acidic environment, with usual characterization techniques.

2. Experimental

2.1 Instrumentation

The observed phenomenon of R.P.O.s can be demonstrated with the series arrangement of a couple of identical electrolytic cells, each cell having a copper anode,1.01millimeter in diameter and 55 millimetre long (minimum purity of copper = 99.9%) and 100 millilitre of 0.5M HCl (Fisher Scientific), as electrolyte. All solutions are prepared in double distilled water.

A digital voltmeter is connected in parallel to each of the two electrolytic cells. A DC supply is used for external voltage application. (The phenomenon of RPOs can be well demonstrated with application of any smaller external voltage across the series of these cells. However, in these experiments, 30 volts DC is applied for getting the potential oscillations of larger amplitudes, which enable the clearer differentiation between the active and passive anodes.) All experiments are performed at 25^0C.

2.2 Characterization

Fourier transform infrared (FT-IR) spectroscopy, Scanning electron microscopy, Energy dispersive X-ray analysis and X-Ray diffractometry are introduced for the characterization of anodic films. The FT-IR measurements are performed with Shimadzu IR Affinity spectrophotometer, over the range of 400-4000 cm^{-1}. Scanning electron microscopy (SEM) measurements are carried out with FESEM ULTRA PLUS instrument by ZEISS Co. SEM – coupled energy dispersive X-ray spectroscopy (EDS) is performed for elemental analysis. The XRD measurements are performed with PANalytical X'Pert Pro X Ray diffractometer, with Cu-Kα radiation over angular range $10^0 \leq 2\theta \leq 90^0$, in 0.0167 step size ($\lambda$=1.5418 Å).

3. Results and Discussion

3.1 Reciprocal Potential Oscillations

Upon applying the voltage of 30 volts across the series of these electrolytic cells, the voltage across each of the two individual cells rises and falls, periodically and reciprocally, from as high as 29 volts to as low as 1 volt (i.e. the average amplitude of oscillations = 28v), with frequency of about 1/240 Hz, which signifies that the anode of each cell attains the similar state of activity / passivity after around every 240 seconds of electrolysis. The reciprocal potential oscillations between the coupled cells can be graphically represented, as shown in figure 1.

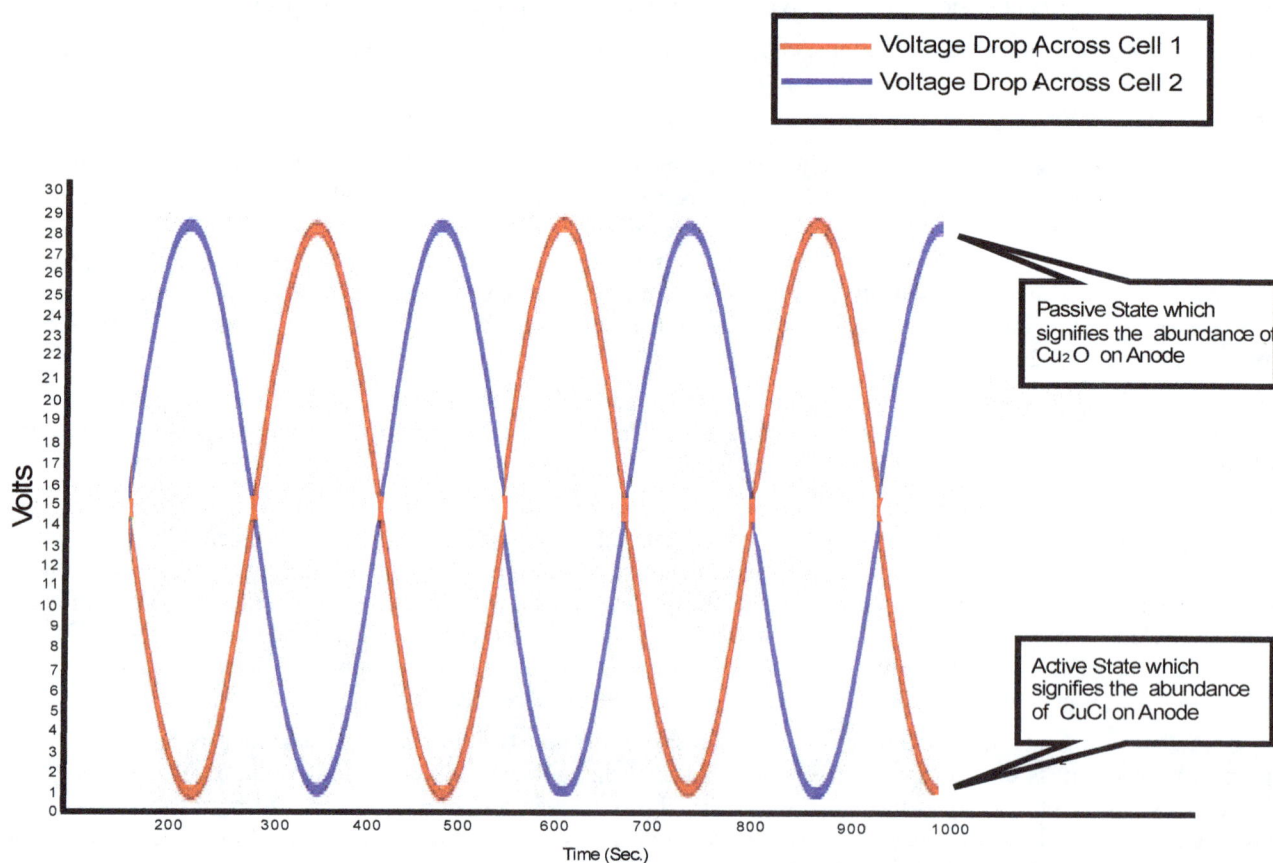

Figure 1.Chrono- Potentiograms of the two described cells, connected in series

3.2 Electro-dissolution Mechanism and the Passivation of Copper Anode

As per general acceptance, the electro-dissolution of copper anode in chloride media proceeds via following two steps reaction mechanism (Lee et al. 1986; Kear et al. 2004; Crundwell et al. 1992; Nobe et al. 1979).

Initial step is the adsorption of Cl⁻ on Cu surface, followed by the formation of adsorbed CuCl by reaction (1)

$$Cu + Cl^- \quad \rightarrow CuCl + e^- \qquad \cdots\cdots\cdots \qquad (1)$$

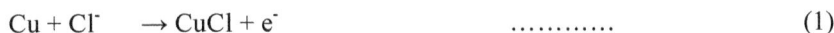

This CuCl film has poor adhesion to copper surface and in presence of Cl⁻ ions, it is transformed into the soluble cuprous chloride complex.

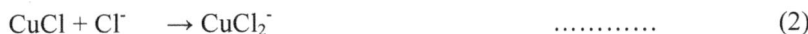

$$CuCl + Cl^- \quad \rightarrow CuCl_2^- \qquad \cdots\cdots\cdots \qquad (2)$$

It is also well known that, at higher concentration of Cl⁻ [~ 1mol L-1], the higher cuprous – chloride complexes are formed.

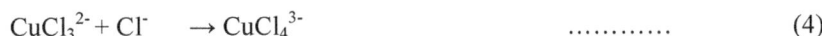

$$CuCl_2^- + Cl^- \quad \rightarrow CuCl_3^{2-} \qquad \cdots\cdots\cdots \qquad (3)$$

$$CuCl_3^{2-} + Cl^- \quad \rightarrow CuCl_4^{3-} \qquad \cdots\cdots\cdots \qquad (4)$$

In neutral or alkaline pH media, at elevated concentration of dissolved CuCl²⁻ species, the equilibrium of following reaction (5) shifts to right, and cuprous oxide is precipitated by hydrolysis of CuCl²⁻ (Kear et al. 2004).

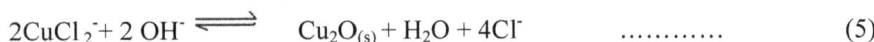

$$2CuCl_2^- + 2\,OH^- \rightleftharpoons \quad Cu_2O_{(s)} + H_2O + 4Cl^- \qquad \cdots\cdots\cdots \qquad (5)$$

Moreover, cuprous oxide is also formed by the oxidation of anodic copper, through following reaction (6)

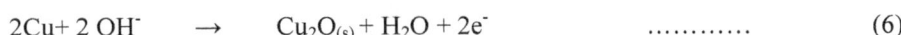

$$2Cu + 2\,OH^- \quad \rightarrow \quad Cu_2O_{(s)} + H_2O + 2e^- \qquad \cdots\cdots\cdots \qquad (6)$$

There is consensus among the investigators, about the existence of Cu_2O at neutral or alkaline pH values (Tromans et al. 1991). Cuprous oxide, formed by reaction (5) & (6), is responsible for the passivation of copper anode. However, under continuous attack of aggressive Cl⁻ ions, this passivating Cu_2O layer is converted into CuCl phase, by reaction 7.

$$Cu_2O + 2\,Cl^- \quad + 2\,H^+ \rightleftharpoons \quad 2\,CuCl + H_2O \qquad \cdots\cdots\cdots \qquad (7)$$

It is well known that, a single electrolytic cell, having copper anode in chloride media, exhibits very small current oscillations. Under potentiostatic conditions, a typical three electrodes cell, with copper as working electrode in chloride electrolyte, is long known to produce these tiny current fluctuations, which are generally attributed to the formation of the CuCl film on surface of anode (Lee et al. 1985; Basset et al. 1990).

However, such single cell is not found to produce obvious potential oscillations. Also, the role of Cu_2O in the non-linear behaviour of Cu│acidic chloride system has not been recognized yet, since the Pourbaix diagrams of Cu-Cl-H₂O system do not permit the existence of Cu_2O in acidic chloride media (Tromans et al. 1991), as shown in figure 2. And hence, the oxide formation by reactions (5), (6) and (7) is considered to be implausible, in acidic environment.

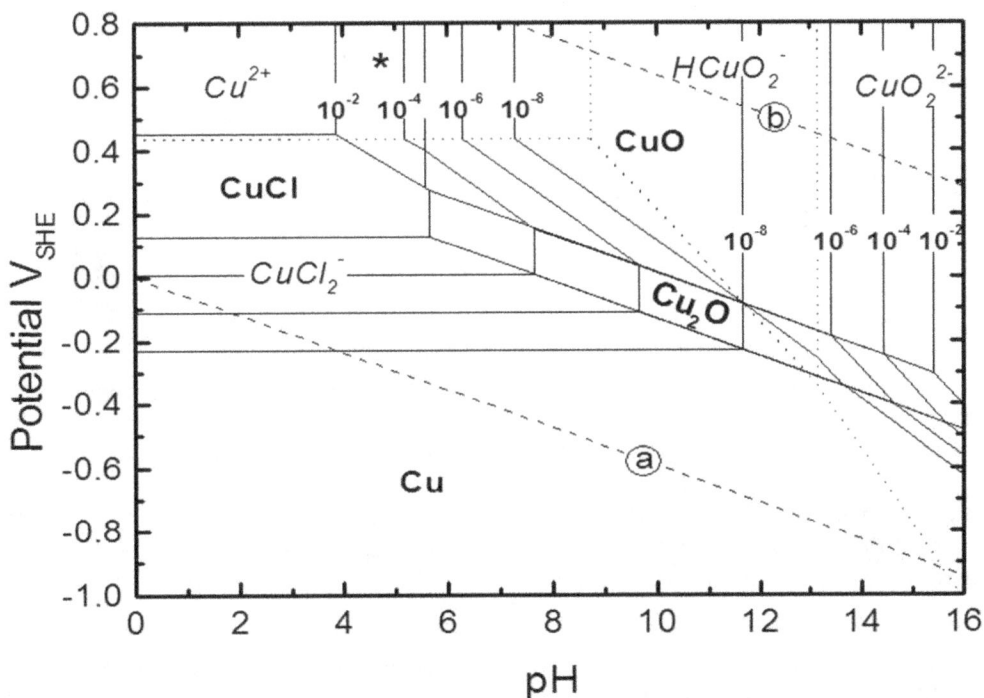

Figure 2. Potential-pH diagram for Cu-Cl-H₂O system at 25⁰C

3.3 Origin and Mechanism of Reciprocal Potential Oscillations

Now it has been observed that, if such multiple electrolytic cells having copper anode and acidic chloride electrolyte, are connected in series, the competitive adsorption of Cl⁻ and OH⁻ species on copper anodes, and subsequent formation of cuprous chloride (CuCl) and cuprous oxide (Cu_2O) accordingly, produce the interesting phenomenon of reciprocally oscillating voltage drop, across the series of these cells. Modestov et al had proposed that, in 0.5M chloride solution, at pH 8.5 or higher, Cu_2O is formed first, whereas, at pH 5.7 and lower, CuCl is formed initially on copper surface, which is followed by Cu_2O formation under the CuCl layer. (Modestov et al.1995).

Hence, the activity/passivity status of anode can be attributed to the relative extents of two anodic partial reactions viz. formation of porous CuCl film and formation of passivating Cu_2O layer under the CuCl film. The formation of Cu_2O passivates the copper anode and hinders copper dissolution as CuCl and $CuCl_2^-$, which results in the rise of potential difference across the cell. Cu_2O layer is clearly protective and passivating in nature, since the current falls significantly following Cu_2O film formation. Hence, it commences the trans-passive dissolution of copper anode, possibly as Cu^{++} species. However, this Cu_2O layer is under continuous attack of aggressive Cl^- anions and the stability of Cu_2O is inversely proportional to the concentration of chloride ions (Bianchi et al 1978). The aggressiveness of Cl^- causes the localized breakdown of this oxide film and pitting commences by nucleation of precipitated CuCl at week points in this underlying Cu_2O film. This causes the thinning of Cu_2O layer, by formation of the outer porous layer rich in chloride species and finally the dissolution of Cu_2O layer from copper anode, in form of copper chloride complex (reaction 5)/cuprous chloride salt film (reaction 7). The attack of aggressive chloride ions replaces Cu_2O in the passivating layer, forming CuCl phase by reaction (7). This poorly adherent, porous CuCl film permits the diffusion of Cl^- ions, and hence, the generalized active dissolution of copper anode, by reactions (1) & (2), commences. It leads to the transition of copper anode to active state i.e. low voltage state, which is indicated by the fall of voltage drop across that cell.

This mechanism of R.P.O.s can be schematically represented, as shown here in figure 3.

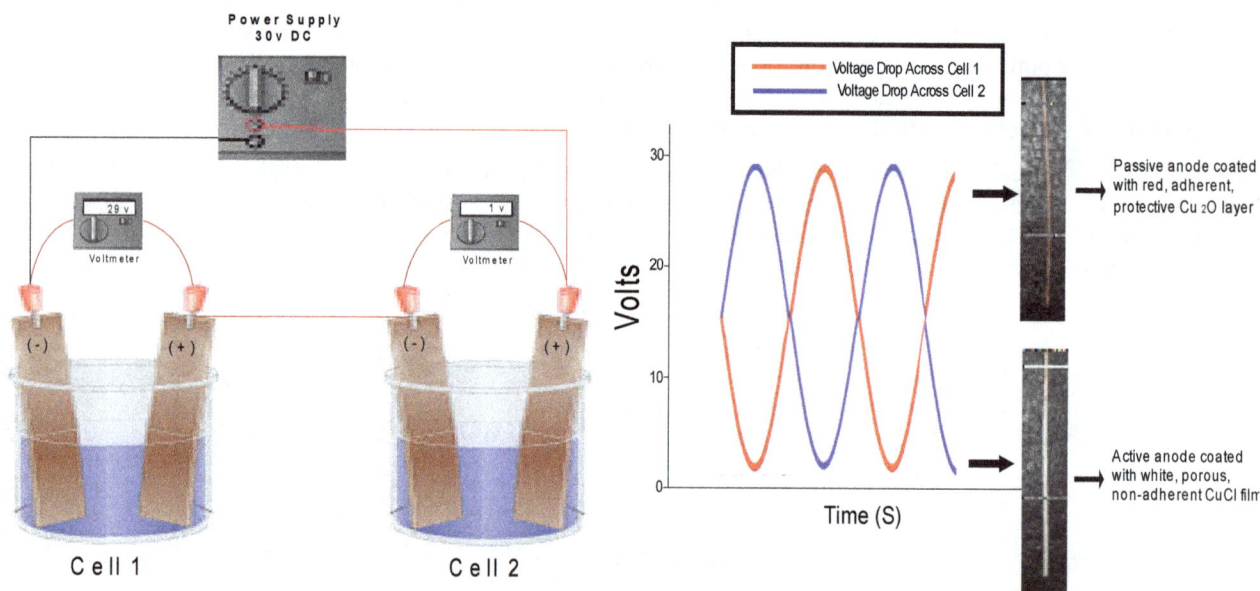

Figure 3. Schematic representation of the mechanism of Reciprocal Potential Oscillations

Thus, at any given time, the relative extent of Cu_2O phase, adsorbed on anodic surface, decides the passivity / activity status of that anode, which in turn, determines the voltage drop across that particular electrolytic cell. The potential oscillations across the coupled cells accompany the dynamic film processes and the changing Cu_2O:CuCl ratios on the anodes. The active-passive transitions of anodes in the different cells in series are synchronized to produce an integrated and harmonious pattern, which manifests as the R.P.O.s between the coupled cells. Hence, such series arrangement of electrolytic cells clearly differentiates between the anodes in active and passive states, and also enables the detection of Cu2O on passive anodic surface, in acidic chloride media.

3.4 Characterization of Anodic Films

FT-IR, SEM, EDS and XRD measurements are introduced for characterization of the anodic films, during active and passive states. The SEM images of anodic surfaces, and the chemical composition of anodic films revealed by EDS spectra signify the presence of cuprous oxide on anodic surface during passive state, while only cuprous chloride is detected during the active state of anode, as shown here, in figure 4.

Figure 4. SEM micrographs and EDS spectra of the anodic films during (a) passive, and (b) active state

FT-IR spectrum is recorded over the range of 400-4000cm^{-1}. The FT-IR spectrum of the passivating layer on anodic surface, during high voltage state, indicates the presence of Cu_2O. The peak at 654.86 cm^{-1} corresponds to the vibrational mode of Cu—O functional group in Cu_2O phase.

Also, the XRD measurements, during the passive state of anode, reveal the presence of Cu_2O on anodic surface. The characteristic Cu_2O peak centred at 2θ of 36.49^0 is observed and it corresponds to the lattice planes (111) of Cu_2O. This peak for Cu2O is consistent with the JCPDS – PDF File No. 05-0667.

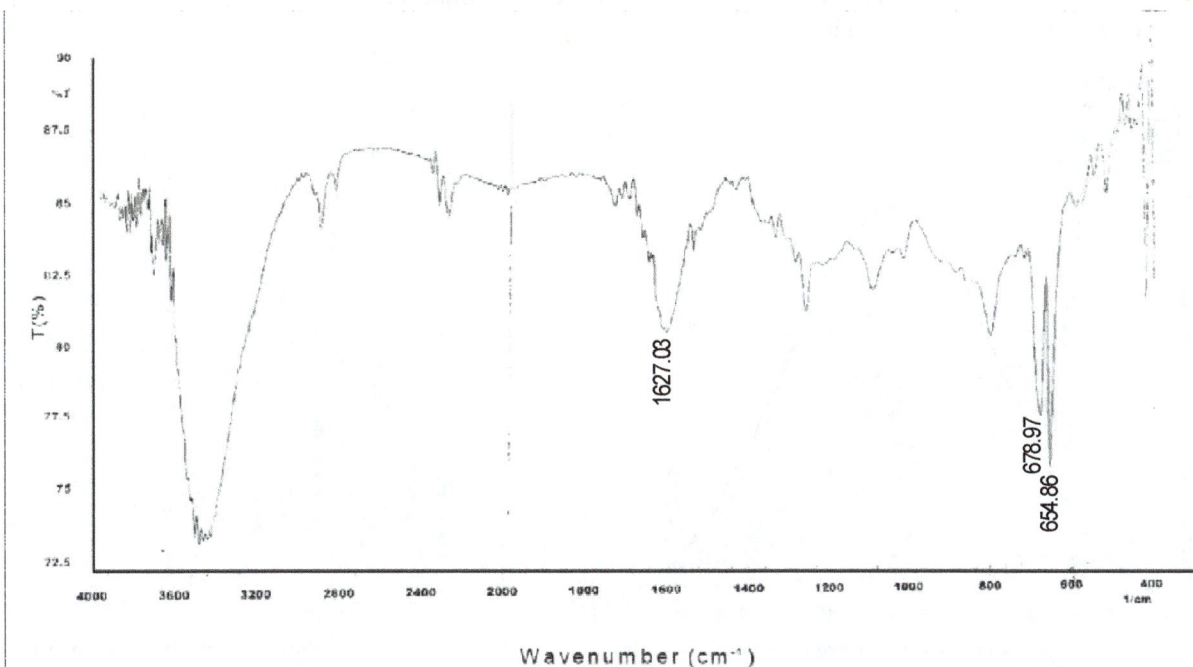

Figure 5.The FT-IR Spectrum of the passivating layer on anode, during high voltage state

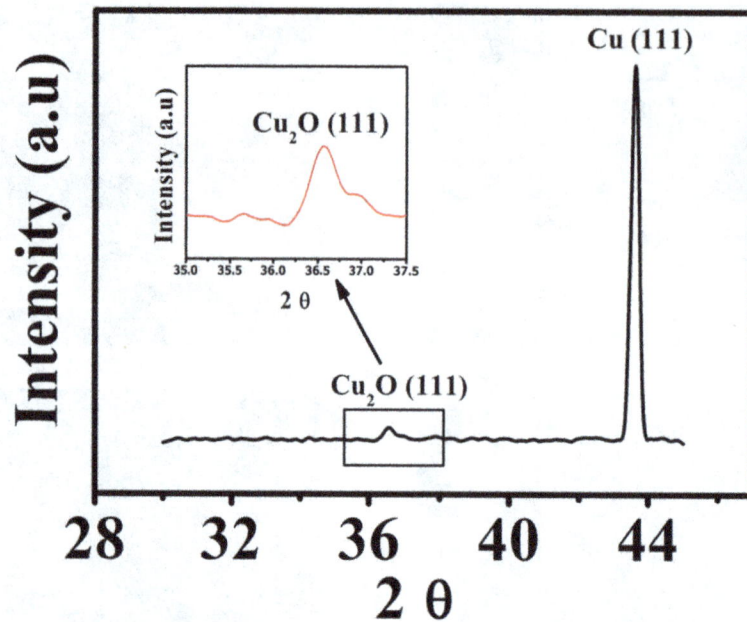

Figure 6. The XRD patterns of the anodic surfaces during the passive state

These results indicate the role of Cu_2O in passivation of copper anode, even in acidic electrolyte. And hence, it signifies that, the periodic formation and dissolution of cuprous oxide layer on anodic surface is responsible for the non-linear temporal behaviour of Cu | chloride system.

3.5 Influence of Chloride Ions Concentration on the Amplitude of R.P.O.s

The amplitude of these RPOs is found to show strong dependence upon the Cl⁻ concentration in electrolyte. At about 0.5 M concentration of HCl electrolyte, the coupled cells demonstrate R.P.O.s of maximum range (1v to 29v) and maximum amplitude (28v). The range and amplitude of R.P.O.s are found to be sequentially decreasing with higher as well as lower concentrations of HCl, as shown in figure 7.

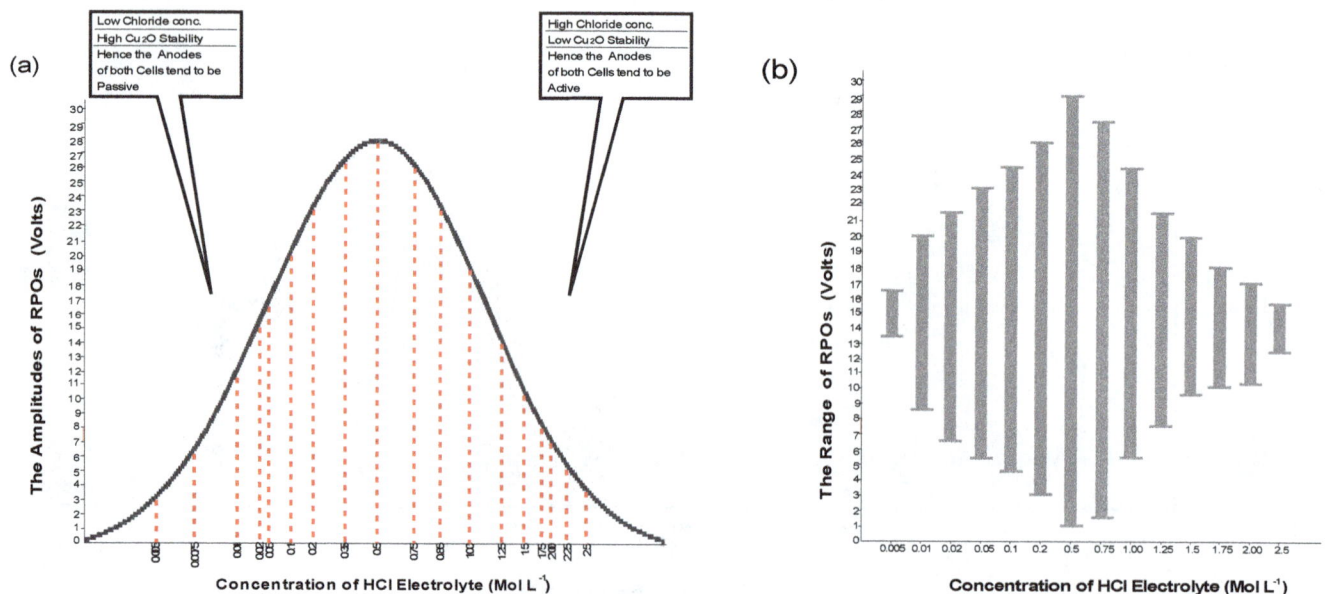

Figure 7. The Normal Distribution Curve showing variations in (a) The Average Amplitude and (b) The Average Range

of R.P.O.s, in the described cells, during 600 seconds of electrolysis, with various concentrations of HCl electrolyte

The dependence of the amplitude of RPOs on Cl⁻ concentration can be explained on basis of the varying stability of Cu_2O at different Cl⁻ concentrations. At very low concentration of aggressive chloride anions, the stability of passivating cuprous oxide layer is very high. The deficient chloride concentration is incapable of the substantial dissolution of oxide from anode surface. Hence, the anodes of both the coupled cells tend to be passive. With increasing Cl⁻ concentration in electrolyte, an equilibrium is established between the formation of passivating Cu_2O layer and its leaching from the anodic surface, in form of copper (I) chloride salt film / soluble cuprous chloride complexes. At such optimum concentration (≈ 0.5MHCl), the cells connected in series demonstrate the voltage asymmetry and R.P.O.s at maximum amplitude. However, with HCl concentration higher than 0.5M and so, the anodes cannot sustain passivity, since the abundance of aggressive chloride anions diminish the stability of Cu_2O layer, which is now incapable of passivating the anodes up to optimum level. Both the anodes tend to be active and hence, amplitude of R.P.O.s is found to be decreasing again, with increasing concentration of Cl⁻. The dependence of amplitude of R.P.O.s on Cl⁻ concentration gives a typical bell shaped curve, as shown in figure 7(a). Hence, this phenomenon can be successfully exploited for the detection of chloride ions and determination of their concentration in the electrolyte.

Aqueous solutions of all chloride electrolytes, including KCl, NaCl, $CuCl_2$, etc. demonstrate this effect, very well. Most importantly, this effect is well demonstrated by HCl electrolyte, with the concentrations ranging from ~ 0.005 M to ~ 2.5 M. Thus, contrary to common belief and potential–pH diagrams for Cu–acidic chloride electrolytic system, this phenomenon of R.P.O.s, observed with HCl electrolyte signifies the very existence of adsorbed Cu_2O on copper anode, in acidic chloride media.)Cu_2O formation in acidic environment can be explained on basis of local pH changes. Copper cation abundance causes the electro-migration of H^+ ions from vicinity of anode to bulk solution, resulting in the rise of local pH, which facilitates the formation of Cu_2O on anodic surface.

Lastly, it has also been observed that the occurrence of R.P.O.s is independent of the composition of cathode and catholyte. The cathode composed of any metal is not found to affect the occurrence of this phenomenon. Also, if salt-bridges are introduced to separate the anolyte and the catholyte, it is observed that the occurrence of R.P.O.s is not affected by the constitution of catholyte.

4. Conclusions

Two identical electrolytic cells with copper anodes and certain concentrations of HCl electrolyte were connected in series and the voltages across the individual cells were plotted against time. This work demonstrated an extremely asymmetric voltage drop and reciprocal potential oscillations between the two coupled cells. SEM-EDS, FT-IR spectroscopy and XRD measurements are introduced for characterization of the anodic films. The characterization data revealed the presence of cuprous oxide on anodic surface, remarkably only during the passive state of anode. Hence, the origin of these R.P.O.s is attributed to the presence of different extents Cu_2O phase, on surfaces of the anodes of different cells in series.   Also, the amplitude of R.P.O.s is found to show strong dependence upon Cl⁻ concentration in electrolyte. Most importantly, such series configuration of electrolytic cells enabled to prove the very existence of Cu_2O, adsorbed on the surface of the passive copper anode, in acidic chloride electrolyte. And hence, this work signified the important role of Cu_2O, in both, the electro-dissolution of copper anode in acidic chloride media, and the non-linear temporal behaviour of the Cu | chloride system.

References

Basset M. R., & Hudson J. L. (1990). The Oscillatory Electro dissolution of Copper in Acidic Chloride Solution. *Journal of the Electrochemical Society, 137*(6), 1815-1826. http://cat.inist.fr/?aModele=afficheN&cpsidt=19260778

Bianchi, G., Fiori, G., Longhi, P., & Mazza, F. (1978). Horse Shoe Corrosion of copper alloys in flowing sea water: mechanism and possibility of cathodic of condenser tubes in power stations. *Corrosion–NACE, 34*, 396-406.

Crundwell, F. K. (1992). The anodic dissolution of copper in hydrochloric acid solutions. *Electrochimica Acta, 37*(15), 2707-2714. http://www.sciencedirect.com/science/article/pii/001346869285197S

Kear, G., Barker, B. D., & Wlash, F. C. (2004). Electrochemical corrosion of unalloyed copper in chloride media—A critical review. *Corrosion Science, 46*, 109. http://www.sciencedirect.com/science/article/pii/S0010938X02002573

Lee, H. P., & Nobe, K. (1986). Kinetics and Mechanisms of Cu Electro dissolution in Chloride Media. *Journal of The Electrochemical Society, 133*(10), 2035-2043. http://m.jes.ecsdl.org/content/133/10/2035.abstract

Lee, H. P., Nobe, K., & Pearlstein, A. J. (1985). Film Formation and Current Oscillations in the Electro dissolution of Cu in Acidic Chloride Media. *Journal of the Electrochemical Society, 132*(5), 1031. http://m.jes.ecsdl.org/

Modestov, A. D., Zhou, G. D., Ge, H. H., & Loo, B. H. (1995). A study by voltammetry and photocurrent response method of copper electrode behaviour in acidic and alkaline solutions containing chloride ions. *Journal of Electroanalytical Chemistry, 380*, 63. http://www.sciencedirect.com/science/article/pii/002207289403577P

Nobe, K., & Braun, M. (1979). Electro dissolution Kinetics of Copper in Acidic Chloride Solutions. *Journal of the Electrochemical Society, 126*(10), 1666. http://m.jes.ecsdl.org/

Sourisseau, T., Chauveau, E., & Baroux, B. (2005). Mechanism of copper action on pitting phenomena observed on stainless steels in chloride media. *Corrosion Science, 147*(5), 1097-1117. http://www.sciencedirect.com/science/article/pii/S0010938X04001179

Tromans, D., & Sun, R. H. (1991). Anodic Polarization Behaviour of Copper in Aqueous Chloride/Benzotriazole Solutions. *Journal of the Electrochemical Society, 138*(11), 3235-3244. http://m.jes.ecsdl.org/content/138/11/3235.abstract

7

A Hypothetical Model for the Formation of Transition Metal Carbonyl Clusters Based upon 4n Series Skeletal Numbers

Enos Masheija Rwantale Kiremire

Correspondence: Enos Masheija Rwantale Kiremire, Department of Chemistry and Biochemistry, University of Namibia, Private Bag 13301, Windhoek, Namibia. E-mail: kiremire15@yahoo.com

Abstract

Skeletal numbers of elements have been introduced as derivatives of the 4n series method. They are based on the number of valence electrons present in the skeletal element. They are extremely useful in deducing possible shapes of skeletal elements in molecules or clusters especially the small to medium ones. For large skeletal clusters, the skeletal numbers may simply be regarded as identity numbers. In carbonyl clusters, they can be used as a guide to facilitate the distribution of the ligands such as CO, H and charges onto the skeletal atoms. A naked skeletal cluster may be viewed as a reservoir for skeletal linkages which get utilized when ligands or electrons get bound to it. The sum of linkages used up by the ligands bound to a skeletal fragment and the remaining cluster skeletal numbers is equal to the number of the skeletal linkages present in the original 'naked parent' skeletal cluster. The skeletal numbers can be used as a quick way of testing whether or not a skeletal atom obeys the 8-or 18-electron rules.

Keywords: Linkages, skeletal, numbers, clusters, fragments, 18-electron rule, carbonyls, series

1. Introduction

The recently developed 4n series method has been found to analyze and categorize atoms, molecules, fragments and clusters (Kiremire, 2016a, 2016b, 2016c). The method highly complements Wade-Mingos rules which have been in existence for more than forty years (Wade, 1991; Mingos, 1972, 1984, 1991).Other methods for dealing with electron counting in clusters have been devised (Lipscomb, 1976; King, 1986a, 1986b; Jensen, 1978; Teo, et al, 1984; Wales, 2005; Wheeler and Hoffmann, 1986; Jemmis, et al, 2000, 2001a, 2001b). However on closer scrutiny of the 4n series method, it has become apparent that we could go further and assign skeletal linkage numbers to elements and ligands which greatly simplifies the prediction of structures of molecules and clusters. The skeletal numbers are especially very helpful in assigning a specific number of carbonyl ligands to skeletal metal elements and hence generating carbonyl cluster isomers. Furthermore, it makes it easier to deduce whether or not clusters obey the 18-electron rule. The observation of behavior of k values in the hydrocarbons involving the addition of hydrogen atoms to skeletal carbon fragments (Kiremire, 2016e) have triggered the need to observe the behavior of k values by adding the carbonyl (CO) ligands to transition metal skeletal fragments. The impact of this work has been to introduce the concept of assigning skeletal numbers to the atoms of the main group elements and the transition metals.

2. Results and Discussion

2.1 Assignment of Skeletal Linkages (K Values) to Elements

The procedure for categorization and structural prediction of fragments, molecules and clusters using the 4n series method is now well established (Kiremire, 2016a, 2016b, 2016c). What is more interesting and exciting is that on closer scrutiny of the 4n series method is that skeletal elements and ligands can actually be assigned skeletal k values. For instance, the single carbon atom 1[C], with valence electron content of four [4], $S=1[4+0]=4n+0(n=1)$, and $k=2n(n=1)=2$. Hence, a carbon atom is assigned a k value of 2. For the nitrogen atom, N with valence electron content of 5, $S=4n+1$ $(n=1)$ and $k=2n-0.5=1.5$ while boron, B with valence electrons 3, $S=4n-1(n=1)$ and $k=2n+0.5=2.5$. The assigned k values of the main group elements are given in Table 1. In the case of transition metals, the $S = 14+q$ is taken as equivalent (Kiremire, 2015a; Hoffmann, 1982) to $S = 4n+q$.The k values of transition metals are given in Table 2. Furthermore, the k values of naked metal clusters from 1 to 10 for first row transition metals are given in Table 3. The addition of a hydrogen atom to a carbon atom produces the fragment [CH] which has 5 valence electrons like a nitrogen atom [N]. Hence its series is given by $S = 4n+1$ and the k value will also be given by $k=2n-0.5=2(1)-0.5=1.5$. But the carbon atom [C] belongs to $S =4n+0$ with $k=2n+0=2(1)+0=2$. This means that a simple operation of C(k =2) + H $\rightarrow$ CH (k=1.5) results in the decrease of k value by 0.5. Hence, it makes sense if we could assign a value of k=-0.5 to a hydrogen atom (H) ligand.

Table 1. Skeletal Values of the Main Group Elements

Group	Series, $S = 4n+q$	$k = 2n-q/2$					
1	4n-3	3.5	Li	Na	K	Rb	Cs
2	4n-2	3	Be	Mg	Ca	Sr	Ba
3	4n-1	2.5	B	Al	Ga	In	Tl
4	4n+0	2	C	Si	Ge	Sn	Pb
5	4n+1	1.5	N	P	As	Sb	Bi
6	4n+2	1	O	S	Se	Te	Po
7	4n+3	0.5	F	Cl	Br	I	At
8	4n+4	0	Ne	Ar	Kr	Xe	Rn

Table 2. Skeletal Values of Transition Metals

3d-TM*	4d-TM*	5d-TM*	Series, $S = 4n+q$	$k = 2n-q/2$	Comments	Possible cluster
Sc	Y	La	4n-11	7.5	Possible Dimerization	M_2L_{14} or ML_7H, ML_7X
Ti	Zr	Hf	4n-10	7	Mono-skeletal cluster	ML_7
V	Nb	Ta	4n-9	6.5	Possible Dimerization	M_2L_{12} ML_6H, ML_6X
Cr	Mo	W	4n-8	6	Mono-skeletal cluster	ML_6
Mn	Tc	Re	4n-7	5.5	Possible Dimerization	M_2L_{10} or ML_5H, ML_5X
Fe	Ru	Os	4n-6	5	Mono-skeletal cluster	ML_5
Co	Rh	Ir	4n-5	4.5	Possible Dimerization	M_2L_8 or ML_4H, ML_4X
Ni	Pd	Pt	4n-4	4	Mono-skeletal cluster	ML_4
Cu	Ag	Au	4n-3	3.5	Possible Dimerization	M_2L_6, or M_3H, ML_3X
Zn	Cd	Hg	4n-2	3	Mono-skeletal cluster	ML_3

* Transition Metals, L = 2-Electron donor, e.g, CO, PPh_3, X = Cl, Br, I

Table 3. Skeletal Values of Selected Naked Skeletal Clusters of First Row Transition Metals

n	1	2	3	4	5	6	7	8	9	10
Sc_n	7.5	15	22.5	30	37.5	45	52.5	60	67.5	75
Ti_n	7	14	21	28	35	42	49	56	63	70
V_n	6.5	13	19.5	26	32.5	39	45.5	52	58.5	65
Cr_n	6	12	18	24	30	36	42	48	54	60
Mn_n	5.5	11	16.5	22	27.5	33	38.5	44	49.5	55
Fe_n	5	10	15	20	25	30	35	40	45	50
Co_n	4.5	9	13.5	18	22.5	27	31.5	36	40.5	45
Ni_n	4	8	12	16	20	24	28	32	36	40
Cu_n	3.5	7	10.5	14	17.5	21	24.5	28	31.5	35
Zn_n	3	6	9	12	15	18	21	24	27	30

2.2 The Power of Skeletal Numbers

2.2.1 The Skeletal Numbers and Main Group Elements

When skeletal atoms are combined alone or with ligands such as hydrogen atoms to form fragments or molecules, the corresponding skeletal numbers are handled in the same manner. Let us take the following simple illustrations.

$C(k=2)+C(k=2)\rightarrow C_2(k=4)\rightarrow C\equiv C$; $B(k=2.5)+N(k=1.5)\rightarrow BN(k=4)\rightarrow B\equiv N$; $CN^+(k=2+1.5+0.5=4)\rightarrow[(C\equiv N)^+]$; $CB^-(k=2+2.5-0.5=4)\rightarrow[(C\equiv B)^-]$; $CH(k=1.5)+CH(k=1.5)\rightarrow C_2H_2(k=3)\rightarrow HC\equiv CH$; $C(k=2)+C(k=2)+2H(k=-1)\rightarrow C_2H_2(k=4-1=3)\rightarrow HC\equiv CH$; $CO(k=2+1=3)\rightarrow C\equiv O$; $NO^+(k=1.5+1+0.5=3)\rightarrow[(N\equiv O)^+]$, $CN^-(k=2+1.5-0.5=3)\rightarrow[(C\equiv N)^-]$; $C_2H_4[k=2(2)-4(0.5)=2]\rightarrow H_2C=CH_2$; $O_2[k=2(1)=2]\rightarrow O=O$; $C_2H_6[2(2)-6(0.5)=1]\rightarrow H_3C-CH_3$; $F_2(k=0.5+0.5=1)\rightarrow F-F$; $C_4H_4(k=8-2=6)$, $C_6H_6[k=12-3=9$, benzene], benzene hexacarbonitrile$[C_{12}N_6$, $k=12(2)+6(1.5)=33$, see BC-1]. The fact that the number of bonds of benzenehexacarbonitrile as counted from the structure BC-1 is the same as that calculated from the skeletal numbers of the atoms in the formula underpins the power of the skeletal numbers. It should also be noted that a negative charge is taken the same as an electron from a donor atom by the 4n series while a positive charge does the opposite.

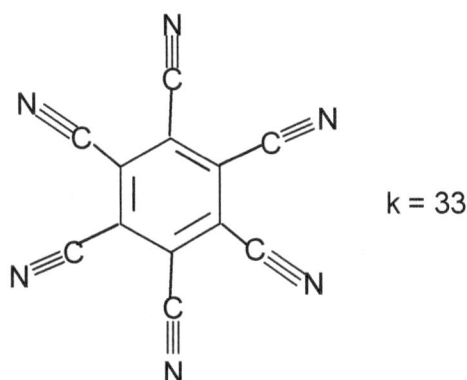

$k = 33$

BC-1   Structure of benzenehexacarbonitrile

The chemical fragments C_2, CN^+, BN, and CB^- have been shown to possess quadruple bonds (Shaik, et al, 2012) by high level computations although the concept is still controversial. However the skeletal numbers derived from the 4n series approach agree with their results. Also most of the bond orders of chemical fragments obtained from molecular orbital energy level diagrams (Housecroft, et al, 2005) agree with the k values obtained from skeletal numbers of atoms. For the diatomic fragments, the k value obtained is simply the same as the bond order.

2.3 The Effect of Adding CO Ligands on Skeletal Linkages of Clusters

2.3.1 Single Metal Skeletal Element (M_1)

2.3.1.1 Fe Skeletal Element

Let us illustrate this by successive addition of CO ligands to Fe skeletal element. This summed up in Scheme 1. What happens to the k value when the CO ligands are step-by-step added to a single metal atom? Let us use Fe (S=4n-6) atom again as our illustration. The Fe atom belongs to the series S=4n-6, k=2n+3=2(1)+3=5. Addition of the first : CO ligand, we get the Fe(CO) fragment. This can be expressed by a simple equation Fe + CO→ Fe(CO). Since in 4n series, we are dealing with valence electron content, the : CO ligand contributes two more electrons to the series S =4n-6 +2→S=4n-4. This means that we get the fragment Fe(CO) which belongs to the series S = 4n-4 and k=2n+2 =2(1)+2=4. Thus, the k value of Fe (k=5) has decreased to Fe(CO)(k=4). Further addition of: CO ligand, we get another fragment Fe(CO)$_2$ (S=4n-2, k=2n+1=3). The next fragment becomes Fe(CO)$_3$ (S=4n+0, k=2n=2). This will be followed by Fe(CO)$_4$(S=4n+2, k=2n-1=2(1)-1=1). The last fragment will be Fe(CO)$_5$(S=4n+4, k=2n-2=2(1)-2=0). Clearly the k value of 5 for the Fe skeletal atom implies the number of the skeletal coordinate bonds or electron pairs to be received from the ligands to form coordinate bonds so as to enable the Fe skeletal atom attain the 18-electron rule. Since the Fe atom has 8 valence electrons, it makes sense that it requires additional 5 pairs (10 electrons) electrons so as to obey the 18-electron rule. We know that the addition of one CO ligand decreases the skeletal k value by 1, therefore we could assign a value of k=-1 to the CO ligand. In this way, the CO ligand or any other ligands may be regarded as 'neutralizing agents' of the cluster skeletal bonds or linkages of the original naked parent skeletal fragment. Each electron provided by a ligand or a charge, neutralizes the skeletal linkages by k value of 0.5 as deduced from the 4n series. That is, a single electron reduces the skeletal value of a cluster by 0.5. This implies that we could as well assign a hydrogen atom (H) ligand, a value of k=-0.5.

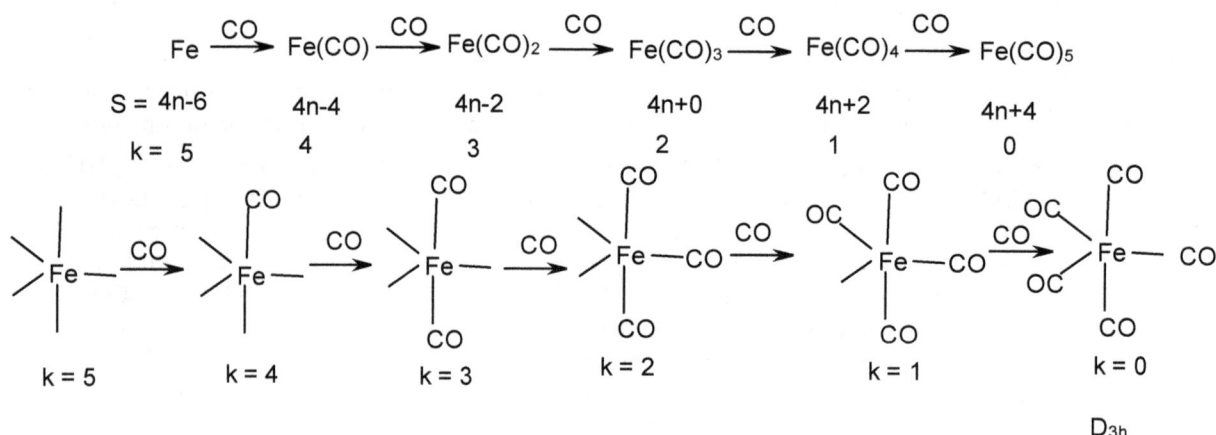

Scheme 1. Successive addition of CO ligands on Fe skeletal element

2.3.1.2 Mn Skeletal Element

Another good example is Mn skeletal element with $S=4n-7$ and a k value of $2n+3.5=2(1)+3.5=5.5$. The successive addition of CO ligands to Mn is summarized in Scheme 2. As in the case of Fe element, the k value decreases by 1 until it reaches the value of 0.5. If another CO ligand were to be added, we would get $Mn(CO)_6$ with $S=4n+5$, and k value of $2n-2.5=2(1)-2.5=-0.5$. The $Mn(CO)_5$ fragment produced may be regarded as a form of free radical. It can receive an electron or a hydrogen atom in order for it to obey the 18-electron rule. These possibilities are shown in Scheme 3. What is also observed is that for the cluster number to reach zero, $k=0$ (see Table 4), the number of ligands must have the same numerical value as the metallic skeletal element. The changes in k values for first row mono-skeletal transition metal elements when CO ligands are added are indicated in Table 4. The mono-skeletal elements Sc(k=7.5), Ti(k=7), V(6.6), Cr(k=6), Mn(k=5.5) and Fe(k=5) can take up carbonyl ligands including hydrogen atoms and or negative charges to "neutralize" the appropriate k values. Hence, the chemical fragments and clusters such as $HFe(CO)_4^-$, $Fe(CO)_4^{2-}$, $Co(CO)_4^-$, $Ti(CO)_6^{2-}$, $V(CO)_6^-$ are known(Cotton, et al, 1980;Housecroft, et al, 2005;Meissler, et al, 2014). The complex $CpTi(CO)_4^-$(Ti, k=7) is also known (Gardner, et al, 1987).

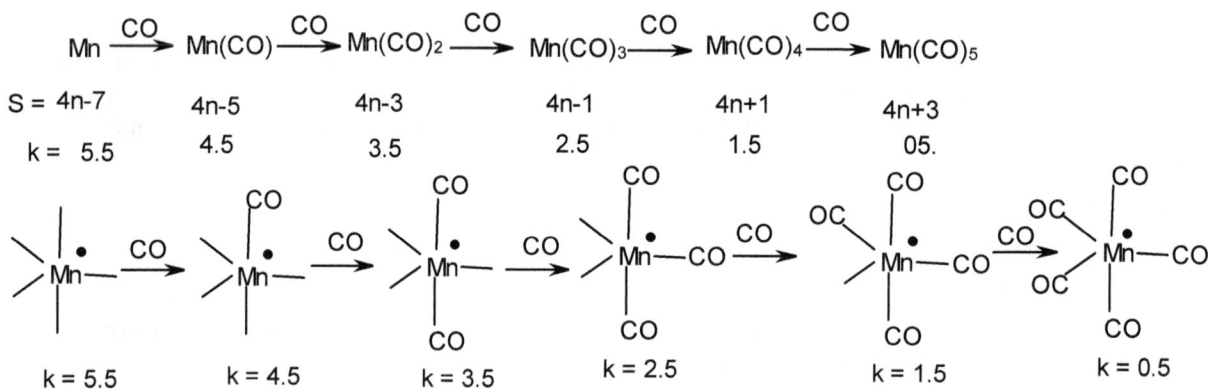

Scheme 2. Successive addition of CO ligands on Mn skeletal element

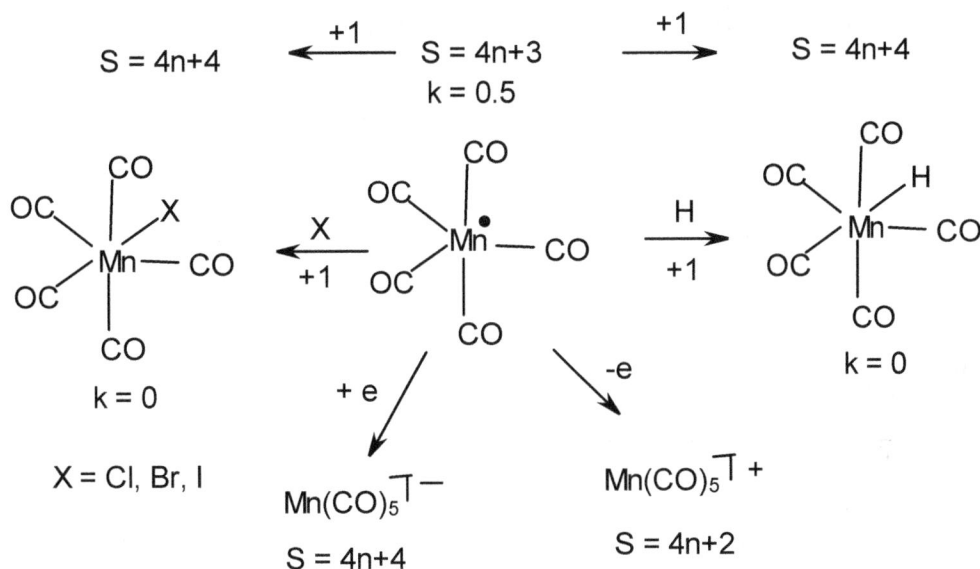

Scheme 3. Possible reactions of $Mn(CO)_5$ fragment

2.3.1.3 A Simple Test for the 18-Electron Rule

In principle, the carbonyl complexes to be formed by mono-skeletal atoms of the first row transition elements are expected to have hypothetical formulas as follows: $Sc \rightarrow k=7.5 \rightarrow Sc(CO)_7(H)$; $Ti \rightarrow k=7 \rightarrow Ti(CO)_7$; $V \rightarrow k=6.5 \rightarrow V(CO)_6(H)$; $Cr \rightarrow k=6 \rightarrow Cr(CO)_6$; $Mn \rightarrow k=5.5 \rightarrow Mn(CO)_5(H)$; $Fe \rightarrow k=5 \rightarrow Fe(CO)_5$; $Co \rightarrow k=4.5 \rightarrow Co(CO)_4(H)$; $Ni \rightarrow k=4 \rightarrow Ni(CO)_4$; $Cu \rightarrow k=3.5 \rightarrow Cu(CO)_3(H)$; and $Zn \rightarrow k=3 \rightarrow Zn(CO)_3$. The homologous transition metals of the corresponding elements in the second and third rows are expected to form similar carbonyl complexes. For instance, $Fe(CO)_5$ is expected to have sister clusters $Ru(CO)_5$ and $Os(CO)_5$. All these three carbonyls are known(Cotton and Wilkinson, 1980; Douglas, et al, 1994; Housecroft, et al, 2005;Miessler, et al, 2014).The changes in k values for the first row transition metals as CO ligands are added is shown in Table 4. When all the skeletal linkages of an element are fully filled up (saturated, see Scheme 1), the mono-skeletal element then obeys the 18-electron rule. Other examples include, $(\eta^5\text{-}C_5H_5)_2Fe(k=5\text{-}2x2.5=0)$, $(\eta^6\text{-}C_6H_6)_2Cr(k=6\text{-}2x3=0)$, $(\eta^6\text{-}C_6H_6)Cr(CO)_3(k=6\text{-}3\text{-}3=0)$, $(\eta^5\text{-}C_5H_5)Fe(CO)_2Cl(k=5\text{-}2.5\text{-}2\text{-}0.5=0)$, $Fe(CO)_4^{2-}(k=5\text{-}4\text{-}1=0)$, $(\eta^5\text{-}C_5H_5)_2Co^+(k=4.5\text{-}2x2.5+0.5=0)$, $(\eta^5\text{-}C_5H_5)Ni(NO)(k=4\text{-}2.5\text{-}1.5=0)$, $(\eta^5\text{-}C_5H_5)_2Ti(CO)_2(k=7\text{-}2x2.5\text{-}2=0)$, $(\eta^5\text{-}C_5H_5)V(CO)_4(k=6.5\text{-}2.5\text{-}4=0)$, $(\eta^5\text{-}C_5H_5)_2Mo(CO)(k = 6\text{-}2.5x2\text{-}1=0)$, $(\eta^5\text{-}C_5H_5)Mn(CO)_3(k=5.5\text{-}2.5\text{-}3=0)$. When a mono-skeletal element cluster attains a value of $k = 0$, then that element should be regarded as having obeyed the 18-electron rule.

Table 4. The k Values of Cluster Fragments Generated by adding Carbonyl Ligands to Selected   Mono-skeletal Transition Metal Elements

CLUSTER	SERIES =4n+q	k VALUE =2n-q/2	CLUSTER	SERIES =4n+q	k VALUE =2n-q/2	CLUSTER	SERIES =4n+q	k VALUE =2n-q/2
Sc	4n-11	7.5	Ti	4n-10	7.0	V	4n-9	6.5
Sc(CO)	4n-9	6.5	Ti(CO)	4n-8	6.0	V(CO)	4n-7	5.5
Sc(CO)$_2$	4n-7	5.5	Ti(CO)$_2$	4n-6	5.0	V(CO)$_2$	4n-5	4.5
Sc(CO)$_3$	4n-5	4.5	Ti(CO)$_3$	4n-4	4.0	V(CO)$_3$	4n-3	3.5
Sc(CO)$_4$	4n-3	3.5	Ti(CO)$_4$	4n-2	3.0	V(CO)$_4$	4n-1	2.5
Sc(CO)$_5$	4n-1	2.5	Ti(CO)$_5$	4n+0	2.0	V(CO)$_5$	4n+1	1.5
Sc(CO)$_6$	4n+1	1.5	Ti(CO)$_6$	4n+2	1.0	V(CO)$_6$	4n+3	0.5
Sc(CO)$_7$	4n+3	0.5	Ti(CO)$_7$	4n+4	0	V(CO)$_7$	4n+5	-0.5
Sc(CO)$_8$	4n+5	-0.5						

Table 4 continued

CLUSTER	SERIES =4n+q	k VALUE =2n-q/2	CLUSTER	SERIES =4n+q	k VALUE =2n-q/2	CLUSTER	SERIES =4n+q	k VALUE =2n-q/2
Cr	4n-8	6	Mn	4n-7	5.5	Fe	4n-6	5
$Cr(CO)$	4n-6	5	$Mn(CO)$	4n-5	4.5	$Fe(CO)$	4n-4	4
$Cr(CO)_2$	4n-4	4	$Mn(CO)_2$	4n-3	3.5	$Fe(CO)_2$	4n-2	3
$Cr(CO)_3$	4n-2	3	$Mn(CO)_3$	4n-1	2.5	$Fe(CO)_3$	4n+0	2
$Cr(CO)_4$	4n+0	2	$Mn(CO)_4$	4n+1	1.5	$Fe(CO)_4$	4n+2	1
$Cr(CO)_5$	4n+2	1	$Mn(CO)_5$	4n+3	0.5	$Fe(CO)_5$	4n+4	0
$Cr(CO)_6$	4n+4	0	$Mn(CO)_6$	4n+5	-0.5			

Table 4 continued

CLUSTER	SERIES =4n+q	k VALUE =2n-q/2	CLUSTER	SERIES =4n+q	k VALUE =2n-q/2	CLUSTER	SERIES =4n+q	k VALUE =2n-q/2
Co	4n-5	4.5	Ni	4n-4	4	Cu	4n-3	3.5
$Co(CO)$	4n-3	3.5	$Ni(CO)$	4n-2	3	$Cu(CO)$	4n-1	2.5
$Co(CO)_2$	4n-1	2.5	$Ni(CO)_2$	4n+0	2	$Cu(CO)_2$	4n+1	1.5
$Co(CO)_3$	4n+1	1.5	$Ni(CO)_3$	4n+2	1	$Cu(CO)_3$	4n+3	0.5
$Co(CO)_4$	4n+3	0.5	$Ni(CO)_4$	4n+4	0	$Cu(CO)_4$	4n+5	-0.5
$Co(CO)_5$	4n+5	-0.5						

2.3.2 M_2 Skeletal Systems

Let us consider what happens to the k values when the CO ligands are added to two skeletal transition metal atoms. In the case of chromium, Cr, S =4n-8 and k=2n+4=2(1)+4=6. Therefore Cr_2 will have k values of 12 arising from S=4n-16, and k=2n+8=2(2)+8=12. The variation in k values for selected bi-skeletal fragments is indicated in Table 5. This means

Cr_2 (k=12)+12CO(-12)→$Cr_2(CO)_{12}$, k=0. This clearly means that there is no linkage between the two Cr skeletal elements and hence the $Cr_2(CO)_{12}$ cluster decomposes into two fragments $2Cr(CO)_6$ each of which obeys the 18-electron rule. On the other hand,

Cr_2 (k=12)+ 11CO(k=-11)→$Cr_2(CO)_{11}$(k=1). This implies, if the complex $Cr_2(CO)_{11}$ were to be formed, in principle it will expected to have one metal-metal bond linking up the two chromium skeletal atoms, Cr-Cr. Other M_2 systems can be interpreted in the same way. For instance, Mn_2(k=11)+ 11C0(k=-11)→$M_2(CO)11$(S=4n+8,k=2n-4=2(2)-4=0) while Mn_2(k=11)+10CO(k=-10)→$M_2(CO)_{10}$(k=1). Thus, it is expected that a single Mn-Mn bond will be observed in $Mn_2(CO)_{10}$ complex. This is what is reported in literature (Cotton and Wilkinson, 1980; Douglas, et al, 1994; Housecroft, et al, 2005; Miessler, et al, 2014). The series showing the stepwise addition of the CO ligands are given in Table 5 for Cr_2, Mn_2 and Fe_2 fragments. Similarly, single bonds   are expected for $Fe_2(CO)_9$(k=5x2-9=1), $Co_2(CO)_8$(k =4.5x2-8=1), $Ni_2(CO)_7$(k=4x2-7=1, $Cu_2(CO)_6$ (k=3.5x2-6=1)   and $Zn_2(CO)_5$(k3x2-5)=1, $(C_p)_2Zn_2$ (k =3x2-2x2.5=1),$Re_2(\eta^5-C_5H_5)_2(CO)_5$(k=2x5.5-2x2.5-5=1), $Mo_2(\eta^5-C_5H_5)_2(CO)_6$(k=2x6-2x2.5-6=1), $Mo_2(\eta^5-C_5H_5)_2(CO)_4$(k=2x6-2x2.5-4=3), $Cr_2(\eta^5-C_5H_5)_2(CO)_6$(k=2x6-2x2.5-6=1), $Cr_2(\eta^5-C_5H_5)_2(CO)_6$(k=2x6-2x2.5-6=1), and $Fe_2(\eta^5-C_5H_5)_2(CO)_4$(k=2x5-2x2.5-4=1). Theoretical studies have been carried out on some bimetallic skeletal carbonyl systems (Ignatyev, et al, 2000; Schaefer III, et al, 2001) as well as the complexes (Freund, et al, 1979) $Cr_2(CO)_{10}^{2-}$ (Cr, k=6) and $W_2(CO)_{10}^{2-}$ (W, k=6). On the basis of k values, these complexes are expected to have single, Cr-Cr and W-W bonds. The existence of complexes such as $HCo(CO)_4$ (Co, k =4.5)and ReH_9^{2-}(Re, k =5.5) can easily be appreciated in terms of k values of the skeletal elements(Pauling, 1977).

Table 5. The k Values of Cluster Fragments Generated by adding Carbonyl Ligands to Selected Bi-skeletal Transition Metal Elements

CLUSTER	SERIES	k VALUE	CLUSTER	SERIES	k VALUE	CLUSTER	SERIES	k VALUE
Cr_2	4n-16	12	Mn_2	4n-14	11	Fe_2	4n-12	10
$Cr_2(CO)$	4n-14	11	$Mn_2(CO)$	4n-12	10	$Fe_2(CO)$	4n-10	9
$Cr_2(CO)_2$	4n-12	10	$Mn_2(CO)_2$	4n-10	9	$Fe_2(CO)_2$	4n-8	8
$Cr_2(CO)_3$	4n-10	9	$Mn_2(CO)_3$	4n-8	8	$Fe_2(CO)_3$	4n-6	7
$Cr_2(CO)_4$	4n-8	8	$Mn_2(CO)_4$	4n-6	7	$Fe_2(CO)_4$	4n-4	6
$Cr_2(CO)_5$	4n-6	7	$Mn_2(CO)_5$	4n-4	6	$Fe_2(CO)_5$	4n-2	5
$Cr_2(CO)_6$	4n-4	6	$Mn_2(CO)_6$	4n-2	5	$Fe_2(CO)_6$	4n+0	4
$Cr_2(CO)_7$	4n-2	5	$Mn_2(CO)_7$	4n+0	4	$Fe_2(CO)_7$	4n+2	3
$Cr_2(CO)_8$	4n+0	4	$Mn_2(CO)_8$	4n+2	3	$Fe_2(CO)_8$	4n+4	2
$Cr_2(CO)_9$	4n+2	3	$Mn_2(CO)_9$	4n+4	2	$Fe_2(CO)_9$	4n+6	1
$Cr_2(CO)_{10}$	4n+4	2	$Mn_2(CO)_{10}$	4n+6	1	$Fe_2(CO)_{10}$	4n+8	0
$Cr_2(CO)_{11}$	4n+6	1	$Mn_2(CO)_{11}$	4n+8	0			
$Cr_2(CO)_{12}$	4n+8	0						

2.4 M_x (x>2) Systems

M_3 Systems are treated in the same way as in M_2 systems.

Let us consider the Cr_3 naked skeletal cluster. The chromium atom has 6 valence electrons. Therefore according to the series it belongs to S = 14n-8(n=1). When the series formula is converted to the main group series type, it becomes S= 4n-8(n=1) and k=2n+4(n=1)=2(1)+4 = 6. For Cr_3, S=4n-24(n=3) and k=2n+12(n=3)=2(3)+12=18. Table 6 shows the effect on k values of adding the: CO ligands to the small naked skeletal clusters Cr_3(k = 18), Fe_3(k =15), Co_4(k=18), Ru_5(k=25) and Rh_6(k =27). As can be seen from Table 6, the addition of 18CO ligands, all the 18 skeletal linkages will be neutralized to zero, k=0. Thus, Cr_3 (k=18) +18CO(-18)→$Cr_3(CO)_{18}$(k=0). It is clear that the Cr_3 cluster will evenly be fragmented into 3[$Cr(CO)_6$, k=0]. It has been discovered that for a cluster (M_n), as the CO ligands are added in a step-wise manner, a saturation point is reached when k=n-1. Taking the Cr_3 cluster (n=3) as an example, the saturation point occurs when the k_2 value fragment cluster is reached. At that point, the cluster becomes $Cr_3(CO)_{16}$, with S=4n+8 and its hydrocarbon equivalent is C_3H_8. As we very much know, the C_3H_8 hydrocarbon belongs to the alkane series C_nH_{2n+2}. In the case of Fe_3 (k = 15), we need to add on 15CO ligands to neutralize all the skeletal linkages. Thus,

Fe_3(k=15)+15CO(-15)→$Fe_3(CO)_{15}$(k=0)→3[$Fe(CO)_5$, k=0]. A saturation point is reached when k=2 and the skeletal cluster becomes $Fe_3(CO)_{13}$. Some known M_3 clusters include, $M_3(CO)_{12}$(M=Fe, Ru, Os) and these clusters do possess three metal-metal bonds and with the k value of k=3. The variation of k value was also studied for a large naked metallic fragment Pd_{23}. This is shown in Table 7.

2.5 General Observations

2.5.1 The Following Observations Are Noted Regarding the Hypothetical Model

- For every addition of 1 [:CO] ligand to a fixed naked metallic fragment, the skeletal k value decreases by 1.

- The cluster series last digit (determinant) increases by 2 due to the 2 electrons donated by the CO ligand.

- The capping of the series decreases more and more.

- The capping ends at S = 4n+0 which represents mono-capped series. The series also represents the carbon clusters, C_n; n=1→C_1, n=2→C_2, n=3→ C_3 and so on.

- At the series formula becomes S=4n+2(CLOSO series), the hydrocarbon analogues of the carbonyl clusters commence. For instance, if n=1→CH_2, n=2→C_2H_2, n=3, C_3H_2, n=4→C_4H_2 and so forth.

- The addition of the CO ligands reaches a SATURATION POINT when k=x-1 for a given cluster M_x. This means for x=1, k=0, x=2, k=1, x=3, k=2. For instance, Cr_2 the saturation point is reached at $Cr_2(CO)_{11}$ (S=4n+6, k=1, $F_{CH}=C_2H_6$), and for Mn_2, we get $Mn_2(CO)_{10}$(S=4n+6, k=1, $F_{CH}=C_2H_6$).

- Complete fragmentation of the naked cluster takes place when k = 0 and all the initial skeletal linkages have been consumed by the CO ligands. For example in the case of the Cr_2(K=12) skeletal fragment, if we add 12CO ligands we get $Cr_2(CO)_{12}$,then k value of the cluster attains the value of k=0 and $Cr_2(CO)_{12}$ and hence the cluster decomposes into two fragments as follows, $Cr_2(CO)_{12}$ → 2[$Cr(CO)_6$].

- The initial capping process of the naked skeletal fragment increases with the increase in the number of the skeletal metal atoms. For instance, 1[Cr], S=14n-8(n=1)→4n-8(n=1) and k =2n+4=2(1)+4=6, the capping is given by the formula (Kiremire,2015,2016e) $C_p = C^1 + C^4 = C^5 C[M^{-4}]$. For the Cr_2 fragment, $S = 4n-16(n=2)$, $k = 2n+8 = 2(2)+8 = 12$ and $C_p = C^1 + C^8 = C^9 C[M^{-7}]$; Cr_3, $S = 4n-24$, $k = 18$ and $C_p = C^1 + C^{12} = C^{13} C[M^{-10}]$. In the case of a particular naked metallic cluster, M_x, $S = 4n-r$, the capping stops at $S = 4n+0$ when the number of CO ligands $z = r/2$. In the case of Cr_1, $S = 4n-8$, the number of CO ligands needed to produce $S = 4n+0$, are $8/2=4$, for Cr_2, $S = 4n-16$, the number of CO ligands needed are $16/2=8$ and for Cr_3, $S=4n-24$, the number of CO ligands are $24/2=12$. This means that keeping the starting naked metallic cluster constant and adding CO ligands the capping decreases until $S = 4n+0$ is attained. Addition of more CO ligands, generates clusters which have hydrocarbon analogues.

- Also a very important observation is made, that is, the sum of the ligands on the cluster (corresponding to the utilized k values) and the cluster linkages (corresponding to the unutilized k values) present is equal to the original k value of the parent naked metallic cluster. This point is discussed in more details under the heading Fundamental Principle.

2.5.2 Limits of the Carbonyl Cluster Series

The carbonyl cluster complexes may be regarded as being formed by adding the carbonyl ligands stepwise to a metallic cluster fragment. This may be expressed by a simple equation below.

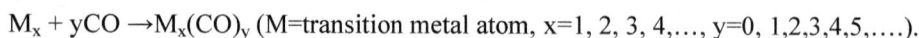

$M_x + yCO \rightarrow M_x(CO)_y$ (M=transition metal atom, x=1, 2, 3, 4,…, y=0, 1,2,3,4,5,….).

The naked skeletal fragment, M_x possesses a fixed number of skeletal linkages which can readily be determined by 4n series based on its valence electrons. These linkages are neutralized one by one for every addition of the CO ligand. If we represent the series as $S = 4n+q$, then when $q \leq 0$ the 'metallic character' increases as the CO ligands are removed, the 'metallic character' of the fragment decreases with the addition of the CO ligands until the series becomes $4n+0$. This series represents the carbon cluster family, $F=C_n$ (n=1, 2,3,4,5, 6,…). The large carbon clusters such as C_{60} (fullerenes) and C_{70} are members of the carbon clusters which belong to the series $S=4n+0$. The $4n+0$ series is the borderline which may be regarded as the highest level of the 'hydrocarbon' series but also as the beginning of the 'metallic' series. It is interesting to note that carbon has vast industrial applications due its unique properties. The 'hydrocarbon' type series can be expressed by the series $S=4n+q$ ($q \geq 0$). We can also regard the series range such as $S=4n+2, 4n+4, 4n+6$, and so on as indicating the increase in the 'hydrocarbon character' of the carbonyl clusters despite the inclusion of the metallic skeletal elements. Let us consider the following changes in metallic fragments of chromium.

$Cr \rightarrow S=4n-8$, $k=2n+4$ $(n=1)=6$, Cr $[S=4n-8, k=6]+CO(k=-1) \rightarrow Cr(CO)[S=4n-6,k=5] \rightarrow Cr(CO)_2[S=4n-4, k=4] \rightarrow Cr(CO)_3$ $[S=4n-2, k=3] \rightarrow Cr(CO)_4[S =4n+0; C, k=2] \rightarrow Cr(CO)_5[S=4n+2; CH_2, k=1] \rightarrow Cr(CO)_6$ $[S=4n+4, CH_4, k=0]$. $Cr_2 \rightarrow S=4n-16$, $k=2n+8(n=2)=12$, $Cr_2(S=4n-16, k=12)+CO(k=-1) \rightarrow Cr_2(CO)(S=4n-14, k=11) \rightarrow Cr_2(CO)_2(S=4n-12, k=10) \rightarrow Cr_2(CO)_3(S=4n-10, k=9) \rightarrow Cr_2(CO)_4(S=4n-8, k=8) \rightarrow Cr_2(CO)_5(S=4n-6, k=7) \rightarrow Cr_2(CO)_6(S=4n-4, k=6) \rightarrow Cr_2(CO)_7(S=4n-2, k=5) \rightarrow Cr_2(CO)_8(S=4n+0;C_2 ,k=4) \rightarrow Cr_2(CO)_9(S=4n+2;C_2H_2, k=3) \rightarrow Cr_2(CO)_{10}(S=4n+4;C_2H_4, k=2) \rightarrow Cr_2(CO)_{11}(S=4n+6;C_2H_6, k=1) \rightarrow Cr_2(CO)_{12}(S=4n+8;C_2H_8, k=0) \rightarrow Cr_2(CO)_{13}(S=4n+10;C_2H_{10} k=-1)$.

The interpretation of $k = 0$ is quite interesting.

$Cr(CO)_6$, $S=4n+4$, $k=2n-2=2(1)-2=0$; CH_4, $S=4n+4$, $k=2n-2=2(1)-2=0$. $Cr_2(CO)_{12}$, $S=4n+8$, $k=2n-4=2(2)-4=0$; C_2H_8, $S=4n+8$, $k =2n-4=2(2)-4=0$.

This implies that $Cr_2(CO)_{12}$ decomposes into two fragments: $Cr_2(CO)_{12} \rightarrow 2[Cr(CO)_6]$ and $C_2H_8 \rightarrow 2CH_4$. Also, C_2H_{10} (k =-1) $\rightarrow 2CH_4(k=0)+2H[k = 2(-0.5)=-1]$; $Cr_2(CO)_{13}$ (k=-1) $\rightarrow 2[Cr(CO)_6, k=0]+CO(k=-1)$. Thus, the cluster series $S=4n-8$, $4n-6, 4n-4, 4n-2$ and $4n+0$ associated with Cr fragment may be regarded as having some type of 'metallic character'. On the other hand the series $S = 4n+0, 4n+2$, and $4n+4$ which correspond to the fragments C, CH_2 and CH_4 may be regarded as having some type of 'hydrocarbon character'. Hence, the $S= 4n+0$ series is the borderline between metallic type and hydrocarbon type of series. In fact, $S = 4n+0$ is refers to the clusters or fragments which are referred to as being mono-capped (Kiremire, 2015a, 2016e). Other series of fragments may be interpreted in the same way. A good example to illustrate the hydrocarbon character of series is $[H_5Re_6(CO)_{24}]^-$ complex. Using the skeletal numbers, the k value of the cluster is given by $k=6(5.5)+5(-0.5)+24(-1)+1(-0.5)=33-2.5-24-0.5=6$. Since for the series $S=4n+q$, $k=2n-q/2$, then $k=6=2n-q/2(n=6)$; $6=2(6)-q/2$, $q=12$. Hence, $S=4n+12(n=6) \rightarrow C_6H_{12}$. This implies that the shape of $[H_5Re_6(CO)_{24}]^-$ complex will be similar to one of the isomers of C_6H_{12}. This is found to be the case (Housecroft and Sharpe, 2005 and the ideal shapes are sketched in Figure 1. The reverse of the hypothetical formation of carbonyl cluster may be represented as: $M_x(CO)_y \rightarrow M_x + yCO$. This involves the removal of CO ligands from the original cluster. In this manner, the cluster may be viewed as going from a hydrocarbon-type to a more metallic type of cluster fragments. Ideally, this should give rise to the recovery of naked parent metallic fragment M_x. The corresponding cluster k values of the fragment should be on the increase. Very interesting extensive work on the metal carbonyls that involves stripping off the CO ligands has been done by several research groups (Butcher, et al, 2002, 2003; Crawford, et al, 2006; Critchley, et al, 1999; Dyson, et al, 2001; Henderson, et al, 1998, 2009). When the structure of the series is carefully analyzed, the hydrocarbon chain type terminates or reaches a saturation point or limit that corresponds to the corresponding hydrocarbon alkane series $F = C_nH_{2n+2}$. Hence, for $M_1 \rightarrow CH_4 \rightarrow 4n+4$, and if M=Fe, then the complex will be $Fe(CO)_5$. For $M_2 \rightarrow C_2H_6 \rightarrow 4n+6$; $M_2 = Cr_2 \rightarrow Cr_2(CO)_{11}$; $M_2 = Mn_2 \rightarrow Mn_2(CO)_{10}$; $M_2 = Fe_2 \rightarrow Fe_2(CO)_9$; and for $M_{23} = Pd_{23} \rightarrow C_{23}H_{48} \rightarrow S = 4n+48 \rightarrow Pd_{23}(CO)_{70}$ (see Table 7).

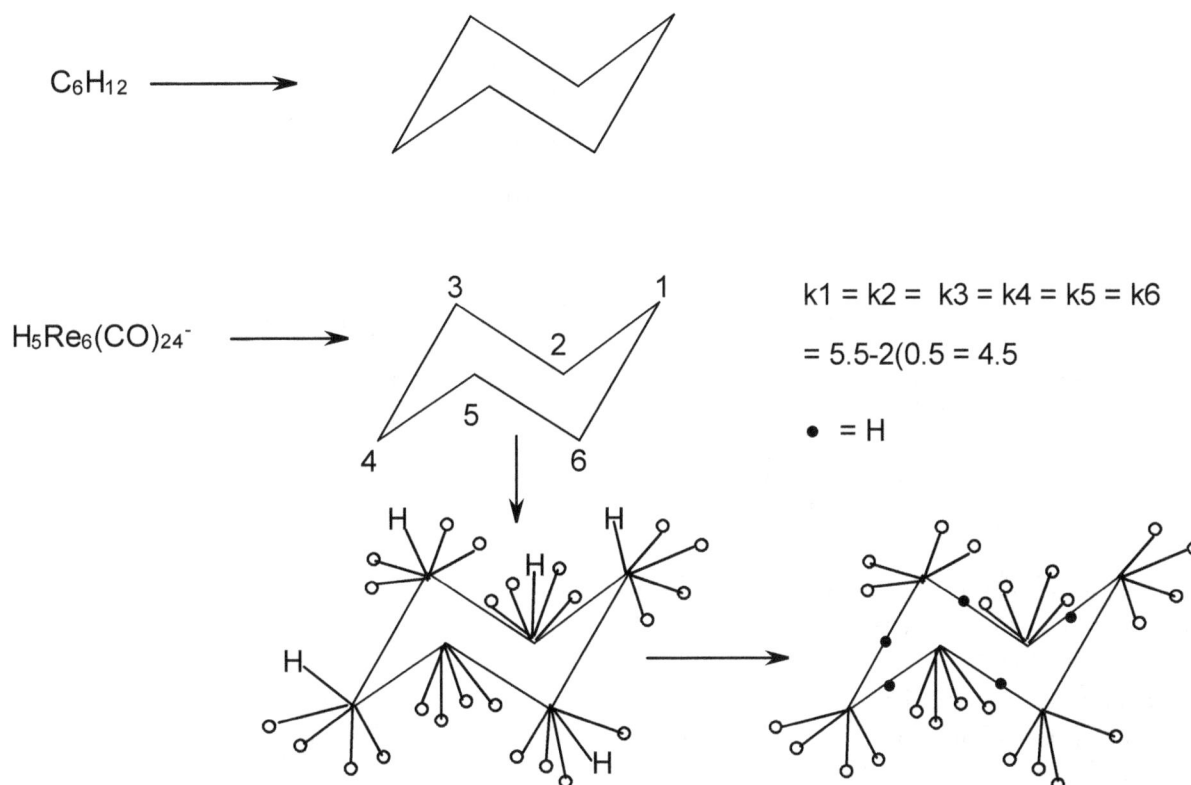

Figure 1. Structural similarity between $H_5Re_6(CO)_{24}^-$ and C_6H_6

Table 6. The k Values of Selected Transition Metal Carbonyl Fragments (M_x, x = 3-6)

CLUSTER	SERIES	k VALUE	CLUSTER	SERIES	k VALUE	CLUSTER	SERIES	k VALUE
Cr_3	4n-24	18	Fe_3	4n-18	15	Co_4	4n-20	18
$Cr_3(CO)$	4n-22	17	$Fe_3(CO)$	4n-16	14	$Co_4(CO)$	4n-18	17
$Cr_3(CO)_2$	4n-20	16	$Fe_3(CO)_2$	4n-14	13	$Co_4(CO)_2$	4n-16	16
$Cr_3(CO)_3$	4n-18	15	$Fe_3(CO)_3$	4n-12	12	$Co_4(CO)_3$	4n-14	15
$Cr_3(CO)_4$	4n-16	14	$Fe_3(CO)_4$	4n-10	11	$Co_4(CO)_4$	4n-12	14
$Cr_3(CO)_5$	4n-14	13	$Fe_3(CO)_5$	4n-8	10	$Co_4(CO)_5$	4n-10	13
$Cr_3(CO)_6$	4n-12	12	$Fe_3(CO)_6$	4n-6	9	$Co_4(CO)_6$	4n-8	12
$Cr_3(CO)_7$	4n-10	11	$Fe_3(CO)_7$	4n-4	8	$Co_4(CO)_7$	4n-6	11
$Cr_3(CO)_8$	4n-8	10	$Fe_3(CO)_8$	4n-2	7	$Co_4(CO)_8$	4n-4	10
$Cr_3(CO)_9$	4n-6	9	$Fe_3(CO)_9$	4n+0	6	$Co_4(CO)_9$	4n-2	9
$Cr_3(CO)_{10}$	4n-4	8	$Fe_3(CO)_{10}$	4n+2	5	$Co_4(CO)_{10}$	4n+0	8
$Cr_3(CO)_{11}$	4n-2	7	$Fe_3(CO)_{11}$	4n+4	4	$Co_4(CO)_{11}$	4n+2	7
$Cr_3(CO)_{12}$	4n+0	6	$Fe_3(CO)_{12}$	4n+6	3	$Co_4(CO)_{12}$	4n+4	6
$Cr_3(CO)_{13}$	4n+2	5	$Fe_3(CO)_{13}$	4n+8	2	$Co_4(CO)_{13}$	4n+6	5
$Cr_3(CO)_{14}$	4n+4	4	$Fe_3(CO)_{14}$	4n+10	1	$Co_4(CO)_{14}$	4n+8	4
$Cr_3(CO)_{15}$	4n+6	3	$Fe_3(CO)_{15}$	4n+12	0	$Co_4(CO)_{15}$	4n+10	3
$Cr_3(CO)_{16}$	4n+8	2				$Co_4(CO)_{16}$	4n+12	2
$Cr_3(CO)_{17}$	4n+10	1				$Co_4(CO)_{17}$	4n+14	1
$Cr_3(CO)_{18}$	4n+12	0				$Co_4(CO)_{18}$	4n+16	0

Table 6 continued

CLUSTER	SERIES	k VALUE	CLUSTER	SERIES	k VALUE
Ru_5	4n-30	25	Rh_6	4n-30	27
$Ru_5(CO)$	4n-28	24	$Rh_6(CO)$	4n-28	26
$Ru_5(CO)_2$	4n-26	23	$Rh_6(CO)_2$	4n-26	25
$Ru_5(CO)_3$	4n-24	22	$Rh_6(CO)_3$	4n-24	24
$Ru_5(CO)_4$	4n-22	21	$Rh_6(CO)_4$	4n-22	23
$Ru_5(CO)_5$	4n-20	20	$Rh_6(CO)_5$	4n-20	22
$Ru_5(CO)_6$	4n-18	19	$Rh_6(CO)_6$	4n-18	21
$Ru_5(CO)_7$	4n-16	18	$Rh_6(CO)_7$	4n-16	20
$Ru_5(CO)_8$	4n-14	17	$Rh_6(CO)_8$	4n-14	19
$Ru_5(CO)_9$	4n-12	16	$Rh_6(CO)_9$	4n-12	18
$Ru_5(CO)_{10}$	4n-10	15	$Rh_6(CO)_{10}$	4n-10	17
$Ru_5(CO)_{11}$	4n-8	14	$Rh_6(CO)_{11}$	4n-8	16
$Ru_5(CO)_{12}$	4n-6	13	$Rh_6(CO)_{12}$	4n-6	15
$Ru_5(CO)_{13}$	4n-4	12	$Rh_6(CO)_{13}$	4n-4	14
$Ru_5(CO)_{14}$	4n-2	11	$Rh_6(CO)_{14}$	4n-2	13
$Ru_5(CO)_{15}$	4n+0	10	$Rh_6(CO)_{15}$	4n+0	12
$Ru_5(CO)_{16}$	4n+2	9	$Rh_6(CO)_{16}$	4n+2	11
$Ru_5(CO)_{17}$	4n+4	8	$Rh_6(CO)_{17}$	4n+4	10
$Ru_5(CO)_{18}$	4n+6	7	$Rh_6(CO)_{18}$	4n+6	9

Table 6 continued

CLUSTER	SERIES	k VALUE	CLUSTER	SERIES	k VALUE
$Ru_5(CO)_{19}$	4n+8	6	$Rh_6(CO)_{19}$	4n+8	8
$Ru_5(CO)_{20}$	4n+10	5	$Rh_6(CO)_{20}$	4n+10	7
$Ru_5(CO)_{21}$	4n+12	4	$Rh_6(CO)_{21}$	4n+12	6
$Ru_5(CO)_{22}$	4n+14	3	$Rh_6(CO)_{22}$	4n+14	5
$Ru_5(CO)_{23}$	4n+16	2	$Rh_6(CO)_{23}$	4n+16	4
$Ru_5(CO)_{24}$	4n+18	1	$Rh_6(CO)_{24}$	4n+18	3
$Ru_5(CO)_{25}$	4n+20	0	$Rh_6(CO)_{25}$	4n+20	2
			$Rh_6(CO)_{26}$	4n+22	1
			$Rh_6(CO)_{27}$	4n+24	0

Although the series and the skeletal numbers predict that each of the Re skeletal atoms except one should have a hydrogen atom, the structural determination indicates all the hydrogen atoms are bridging (Miessler, et al, 2014) as observed in borane clusters.

Table 7. The k Values Generated by Adding CO ligands to a Large Naked Metallic Fragment

CLUSTER	SERIES	k VALUE	CLUSTER	SERIES	k VALUE	CLUSTER	SERIES	k VALUE
Pd_{23}	4n-92	92						
$Pd_{23}(CO)$	4n-90	91	$Pd_{23}(CO)_{16}$	4n-60	76	$Pd_{23}(CO)_{31}$	4n-30	61
$Pd_{23}(CO)_2$	4n-88	90	$Pd_{23}(CO)_{17}$	4n-58	75	$Pd_{23}(CO)_{32}$	4n-28	60
$Pd_{23}(CO)_3$	4n-86	89	$Pd_{23}(CO)_{18}$	4n-56	74	$Pd_{23}(CO)_{33}$	4n-26	59
$Pd_{23}(CO)_4$	4n-84	88	$Pd_{23}(CO)_{19}$	4n-54	73	$Pd_{23}(CO)_{34}$	4n-24	58
$Pd_{23}(CO)_5$	4n-82	87	$Pd_{23}(CO)_{20}$	4n-52	72	$Pd_{23}(CO)_{35}$	4n-22	57
$Pd_{23}(CO)_6$	4n-80	86	$Pd_{23}(CO)_{21}$	4n-50	71	$Pd_{23}(CO)_{36}$	4n-20	56
$Pd_{23}(CO)_7$	4n-78	85	$Pd_{23}(CO)_{22}$	4n-48	70	$Pd_{23}(CO)_{37}$	4n-18	55
$Pd_{23}(CO)_8$	4n-76	84	$Pd_{23}(CO)_{23}$	4n-46	69	$Pd_{23}(CO)_{38}$	4n-16	54
$Pd_{23}(CO)_9$	4n-74	83	$Pd_{23}(CO)_{24}$	4n-44	68	$Pd_{23}(CO)_{39}$	4n-14	53
$Pd_{23}(CO)_{10}$	4n-72	82	$Pd_{23}(CO)_{25}$	4n-42	67	$Pd_{23}(CO)_{40}$	4n-12	52
$Pd_{23}(CO)_{11}$	4n-70	81	$Pd_{23}(CO)_{26}$	4n-40	66	$Pd_{23}(CO)_{41}$	4n-10	51
$Pd_{23}(CO)_{12}$	4n-68	80	$Pd_{23}(CO)_{27}$	4n-38	65	$Pd_{23}(CO)_{42}$	4n-8	50
$Pd_{23}(CO)_{13}$	4n-66	79	$Pd_{23}(CO)_{28}$	4n-36	64	$Pd_{23}(CO)_{43}$	4n-6	49
$Pd_{23}(CO)_{14}$	4n-64	78	$Pd_{23}(CO)_{29}$	4n-34	63	$Pd_{23}(CO)_{44}$	4n-4	48
$Pd_{23}(CO)_{15}$	4n-62	77	$Pd_{23}(CO)_{30}$	4n-32	62	$Pd_{23}(CO)_{45}$	4n-2	47

Table 7 continued

CLUSTER	SERIES	k VALUE	CLUSTER	SERIES	k VALUE	CLUSTER	SERIES	k VALUE
$Pd_{23}(CO)_{46}$	4n+0	46	$Pd_{23}(CO)_{62}$	4n+32	30	$Pd_{23}(CO)_{78}$	4n+64	14
$Pd_{23}(CO)_{47}$	4n+2	45	$Pd_{23}(CO)_{63}$	4n+34	29	$Pd_{23}(CO)_{79}$	4n+66	13
$Pd_{23}(CO)_{48}$	4n+4	44	$Pd_{23}(CO)_{64}$	4n+36	28	$Pd_{23}(CO)_{80}$	4n+68	12
$Pd_{23}(CO)_{49}$	4n+6	43	$Pd_{23}(CO)_{65}$	4n+38	27	$Pd_{23}(CO)_{81}$	4n+70	11
$Pd_{23}(CO)_{50}$	4n+8	42	$Pd_{23}(CO)_{66}$	4n+40	26	$Pd_{23}(CO)_{82}$	4n+72	10
$Pd_{23}(CO)_{51}$	4n+10	41	$Pd_{23}(CO)_{67}$	4n+42	25	$Pd_{23}(CO)_{83}$	4n+74	9
$Pd_{23}(CO)_{52}$	4n+12	40	$Pd_{23}(CO)_{68}$	4n+44	24	$Pd_{23}(CO)_{84}$	4n+76	8
$Pd_{23}(CO)_{53}$	4n+14	39	$Pd_{23}(CO)_{69}$	4n+46	23	$Pd_{23}(CO)_{85}$	4n+78	7
$Pd_{23}(CO)_{54}$	4n+16	38	$Pd_{23}(CO)_{70}$	4n+48	22	$Pd_{23}(CO)_{86}$	4n+80	6
$Pd_{23}(CO)_{55}$	4n+18	37	$Pd_{23}(CO)_{71}$	4n+50	21	$Pd_{23}(CO)_{87}$	4n+82	5
$Pd_{23}(CO)_{56}$	4n+20	36	$Pd_{23}(CO)_{72}$	4n+52	20	$Pd_{23}(CO)_{88}$	4n+84	4
$Pd_{23}(CO)_{57}$	4n+22	35	$Pd_{23}(CO)_{73}$	4n+54	19	$Pd_{23}(CO)_{89}$	4n+86	3
$Pd_{23}(CO)_{58}$	4n+24	34	$Pd_{23}(CO)_{74}$	4n+56	18	$Pd_{23}(CO)_{90}$	4n+88	2
$Pd_{23}(CO)_{59}$	4n+26	33	$Pd_{23}(CO)_{75}$	4n+58	17	$Pd_{23}(CO)_{91}$	4n+90	1
$Pd_{23}(CO)_{60}$	4n+28	32	$Pd_{23}(CO)_{76}$	4n+60	16	$Pd_{23}(CO)_{92}$	4n+92	0
$Pd_{23}(CO)_{61}$	4n+30	31	$Pd_{23}(CO)_{77}$	4n+62	15			

Table 8. The Capping Series obtained from Stripping $Pd_{23}(CO)_{46}$ Cluster

Cluster	Series	Capping Symbol	Cluster	Series	Capping Symbol
$Pd_{23}(CO)_{46}$	4n+0	$C^1C[M-22]$	$Pd_{23}(CO)_{28}$	4n-36	$C^{19}C[M-4]$
$Pd_{23}(CO)_{45}$	4n-2	$C^2C[M-21]$	$Pd_{23}(CO)_{27}$	4n-38	$C^{20}C[M-3]$
$Pd_{23}(CO)_{44}$	4n-4	$C^3C[M-20]$	$Pd_{23}(CO)_{26}$	4n-40	$C^{21}C[M-2]$
$Pd_{23}(CO)_{43}$	4n-6	$C^4C[M-19]$	$Pd_{23}(CO)_{25}$	4n-42	$C^{22}C[M-1]$
$Pd_{23}(CO)_{42}$	4n-8	$C^5C[M-18]$	$Pd_{23}(CO)_{24}$	4n-44	$C_{23}C[M-0]$
$Pd_{23}(CO)_{41}$	4n-10	$C^6C[M-17]$	$Pd_{23}(CO)_{23}$	4n-46	$C^{24}C[M^{-1}]$
$Pd_{23}(CO)_{40}$	4n-12	$C^7C[M-16]$	$Pd_{23}(CO)_{22}$	4n-48	$C^{25}C[M^{-2}]$
$Pd_{23}(CO)_{39}$	4n-14	$C^8C[M-15]$	$Pd_{23}(CO)_{21}$	4n-50	$C^{26}C[M^{-3}]$
$Pd_{23}(CO)_{38}$	4n-16	$C^9C[M-14]$	$Pd_{23}(CO)_{20}$	4n-52	$C^{27}C[M^{-4}]$
$Pd_{23}(CO)_{37}$	4n-18	$C^{10}C[M-13]$	$Pd_{23}(CO)_{19}$	4n-54	$C^{28}C[M^{-5}]$
$Pd_{23}(CO)_{36}$	4n-20	$C^{11}C[M-12]$	$Pd_{23}(CO)_{18}$	4n-56	$C^{29}C[M^{-6}]$
$Pd_{23}(CO)_{35}$	4n-22	$C^{12}C[M-11]$	$Pd_{23}(CO)_{17}$	4n-58	$C^{30}C[M^{-7}]$
$Pd_{23}(CO)_{34}$	4n-24	$C^{13}C[M-10]$	$Pd_{23}(CO)_{16}$	4n-60	$C^{31}C[M^{-8}]$
$Pd_{23}(CO)_{33}$	4n-26	$C^{14}C[M-9]$	$Pd_{23}(CO)_{15}$	4n-62	$C^{32}C[M^{-9}]$
$Pd_{23}(CO)_{32}$	4n-28	$C^{15}C[M-8]$	$Pd_{23}(CO)_{14}$	4n-64	$C^{33}C[M^{-10}]$
$Pd_{23}(CO)_{31}$	4n-30	$C^{16}C[M-7]$	$Pd_{23}(CO)_{13}$	4n-66	$C^{34}C[M^{-11}]$
$Pd_{23}(CO)_{30}$	4n-32	$C^{17}C[M-6]$	$Pd_{23}(CO)_{12}$	4n-68	$C^{35}C[M^{-12}]$
$Pd_{23}(CO)_{29}$	4n-34	$C^{18}C[M-5]$	$Pd_{23}(CO)_{11}$	4n-70	$C^{36}C[M^{-13}]$

Table 8 continued

Cluster	Series	Capping Symbol
$Pd_{23}(CO)_{10}$	4n-72	$C^{37}C[M^{-14}]$
$Pd_{23}(CO)_9$	4n-74	$C^{38}C[M^{-15}]$
$Pd_{23}(CO)_8$	4n-76	$C^{39}C[M^{-16}]$
$Pd_{23}(CO)_7$	4n-78	$C^{40}C[M^{-17}]$
$Pd_{23}(CO)_6$	4n-80	$C^{41}C[M^{-18}]$
$Pd_{23}(CO)_5$	4n-82	$C^{42}C[M^{-19}]$
$Pd_{23}(CO)_4$	4n-84	$C^{43}C[M^{-20}]$
$Pd_{23}(CO)_3$	4n-86	$C^{44}C[M^{-21}]$
$Pd_{23}(CO)_2$	4n-88	$C^{45}C[M^{-22}]$
$Pd_{23}(CO)_1$	4n-90	$C^{46}C[M^{-23}]$
Pd_{23}	4n-92	$C^{47}C[M^{-24}]$

2.5.3 Capping and De-capping Series

The addition of carbonyl ligands(CO) to M_x, that is, $M_x + yCO \rightarrow M_x(CO)_y$ corresponds to the decrease in de-capping process whereas the opposite $M_x(CO)_y \rightarrow M_x + yCO$ corresponds to the increase in the capping process. These opposing processes are reflected in Tables 7 and 8. The removal of CO ligands from transition metal carbonyl clusters has been an intense field of research (Critchley, et al, 1999, Douglas, et al, 2001, Butcher, et al, 2002, 2003, Crawford, et al, 2006, Henderson, et al, 2009). The silicon clusters $Si_n(n=4-10)$ were theoretically studied(Slee, et al, 1989) using extended Hückel calculations. According to series, Si and C skeletal atoms belong to the same series, $S=4n+0$, $Cp=C^1C$(mono-capped series). Thus, Si_4, $Cp=C^1C[M-3]$, Si_5, $Cp=C^1C[M-4]$, Si_6, $Cp=C^1C[M-5]$, Si_7, $Cp=C^1C[M-6]$, Si_8, $Cp=C^1C[M-7]$, $Si_9,Cp=C^1C[M-8]$and $Si_{10},Cp=C^1C[M-9]$. The cluster, Si_7, $Cp=Cp=C^1C[M-6]$,is expected to have a mono-capped octahedral skeletal shape similar to that of $Os_7(CO)_{21}$ or $Rh_7(CO)_{16}^{3-}$(Hughes, et al, 2000). The complex, $Ru_6Pd_6(CO)_{24}^{2-}$(Ru, k=5,Pd,k=4);k(cluster)=6(5)+6(4)-24-1=29;q/2=2n-k, n=12, q/2=2(12)-29=-5, q=-10, S = 4n-10, $Cp = C^1+C^5=C^6C[M-6]$. This means 6 of the skeletal metal atoms are capped around an inner octahedral nucleus as was reported (Butcher, et al, 2003). Another good example regarding capping is $Ni_{38}Pt_6(CO)_{48}(H)^{5-}$. The k value of the cluster is given by k =38(4)+6(4)-48-0.5-2.5=125. From the knowledge of series we know that k =2n-q/2 and hence q/2=2n-k, n=38+6=44, q/2=2(44)-125=-37, q=-74. Hence $S=4n+q=4n-74$, $Cp=C^1+C^{37}=C^{38}C[M-6]$. This means the giant cluster of 44 skeletal metal atoms, 6 will constitute an octahedral nucleus and the remaining 38 will form a capping shell which is also found to form a tetrahedral structure (Rossi and Zanello, 2011). What is also interesting is that all the 6 nuclear atoms are platinum and the shell comprises of the nickel atoms. The skeletal numbers can also be applied to a cluster with a mixture of main group and transition metal skeletal elements. A good example is $Te_2Ru_4(CO)_{10}Cu_2Cl_2^{2-}$. The skeletal numbers of the atoms the charge in the cluster, Te(k=1), Ru(k=5), Cu(k=3.5), CO(k=-1), and Cl(k=-0.5), and 2-(k=-1). Hence k=2(1)+4(5)+2(3.5)-10-1-1=17, q/2=2n-k, n=2+4+2=8, q/2= 2(8)-17=-1, q=-2. The cluster series will be $S=4n-2(n=8)$, $Cp=C^1+C^1=C^2C[M-6]$. The ideal skeletal shape is a bi-capped octahedron and geometry has been observed (Shieh, et al, 2012). The valence electron content, $Ve=4n-2=4(8)-2=30$. V_M $=30+10(6)=90$. Thus, $V_M=Ve+10n$ where n=number of metal atoms present in a cluster.

2.5.4 Application of Skeletal Numbers to Isolobal Relationship

The isolobal principle (Hoffmann, 1982) simply refers to chemical fragments or molecules which are chemically similar in terms of structures. The isolobal relationship can readily be verified using skeletal numbers. Consider the following fragments, $Mn(CO)_5$, k=5.5-5=0.5, CH_3, k=2-1.5=0.5. Hence, $Mn(CO)_5 \leftrightarrow CH_3$; $Fe(CO)_4$, k=5-4=1, CH_2, k=2-1=1→ $Fe(CO)_4 \leftrightarrow CH_2$; $Ir(CO)_3$, k=4.5-3=1.5, CH, k=2-0.5=1.5→ $Ir(CO)_3 \leftrightarrow CH$, $Fe(CO)_3$, k=5-3=2,Co(Cp), k=4.5-2.5=2, C, k=2→ $Fe(CO)_3 \leftrightarrow Co(Cp) \leftrightarrow C$. Some of the fragments can be handled in the same manner.

Table 9. Examples to Illustrate the Principle of Conservation of Skeletal Value Content of the Naked Parent Cluster Fragment

$M_1(K=5)$			$M_2(K = 10)$			$M_3(15)$		
K_S	$K_L(CO)$	K_T	K_S	$K_L(CO)$	K_T	K_S	$K_L(CO)$	K_T
5	0	5	10	0	10	15	0	15
4	1	5	9	1	10	14	1	15
3	2	5	8	2	10	13	2	15
2	3	5	7	3	10	12	3	15
1	4	5	6	4	10	11	4	15
0	5	5	5	5	10	10	5	15
			4	6	10	9	6	15
			3	7	10	8	7	15
			2	8	10	7	8	15
			1	9	10	6	9	15
			0	10	10	5	10	15
						4	11	15
						3	12	15
						2	13	15
						1	14	15
						0	15	15

2.6 The Conservation of Cluster Skeletal Linkage Content Principle

A naked transition metal element possesses an inherent number of skeletal linkages by virtue of its valence electrons. When the skeletal element reacts with suitable ligands such as CO and H some or all the linkages are used up, the fragment develops a tendency towards the attainment of the ultimate 18-electron rule. The 18-electron rule implies the maximization of all the atomic orbitals of the orbital set [s(1),p(3)and d(5)](Pauling,1977). Since $1[:CO]$ neutralizes 1 skeletal k unit, it is proposed that we assign it a value of $k=-1$. We also know that it donates 2 electrons, and so every ligand donor of 2 electrons may similarly be assigned a k value of -1 and for 1 electron donor such as H and Cl or a unit negative charge ligands are correspondingly assigned a numerical value of $k=-0.5$. As can be seen from Table 9, $|k_L|+k_S =k_T$ where k_L represents the used up k-values of the initial skeletal fragment, k_S = the skeletal linkages still available and k_T= the original skeletal linkages of the parent naked fragment. This result is very important because if we know the value of k_T and $|k_L|$, then we can deduce the value of k_S and hence use it as a guiding tool in designing and predicting the shape of the carbonyl cluster. This important principle is hereby expressed in an equation form. This relationship is well illustrated in Table 9.

$$k_T = |k_L|+ k_S$$

$\qquad$ ....(i)

Taking the example of $Os_3(CO)_{12}$, $k_{Os}=3\times5=15=k_T$, $k_{CO}=12(-1)=-12=k_L$. Therefore $k_S=15-12=3$. This means that $Os_3(CO)_{12}$ will have a skeletal triangular shape comprising of the three osmium skeletal elements. We also note that $|k_L|+k_S=12+3=15=k_T$. The negative sign of k_L indicates the skeletal linkages used up by the ligands and k_S refers to the remaining linkages of the skeletal elements. Let us look at some selected examples of osmium carbonyls to test the above principle. Examples are $Os_3H_2(CO)_{10}$; $k_T=3\times5=15$, $k_L=-2(0.5)-10(1)=-11$. Hence $k_S=15-11=4$. The sum of $|k_L|$and $k_S=15=k_T$. Other examples include, $Os_3(CO)_{12}(k_T=3\times5=15$, $k_L=-12\rightarrow K_S=3$,triangle), $Os_4(CO)_{14}(k_T=4\times5=20$, $k_L=-14\rightarrow k_S=6$, tetrahedral); $Os_4(CO)_{16}$, $(k_T=20$, $k_L=-16$, $k_S=4$, square);$Os_5(CO)_{16}$, $(k_T=25$, $k_L=-16$, $k_S=9$, trigonal bipyramid) and $Os_5(CO)_{19}$, $(k_T=25$, $k_L=-19$, $k_S=6$, two triangles joined at one of the edges. The she shapes of these carbonyls are well known (Hughes and Wade, 2000; Greenwood and Earnshaw (1998).

2.7 Applications of Skeletal Numbers to Predict Shapes of Carbonyl Cluster Structures Including the Tentative Distribution of the Ligands on Skeletal Elements

2.7.1 Possible Shapes of the Skeletal Linkages

The main concept is that a skeletal atom possesses naturally inherent linkages as deduced by 4n series. When a ligand is attached to the skeletal atom, it utilizes or neutralizes some of those linkages depending upon the number of electrons that ligand donates to the skeletal element. For every electron donated by the ligand or a negative charge, 0.5 of the linkage is utilized or removed from the skeletal element. Since a:CO ligand donates 2 electrons, it utilizes or removes 1 skeletal value from the element. On this basis, a CO ligand is assigned a value of $k=-1$. Similarly we can assign H• atom a value of $k=-0.5$ and one negative charge, $k=-0.5$. The implications of this is that for 1 Fe(k=5) when combined with 5CO[$k=5x(-1)=-5$] ligands, the net k value of the cluster becomes zero $(k = 0)$. Hence $Fe(CO)_5$, $k =0$. The k values for transition elements may be regarded as the number of electron pairs needed for the element to obey the 18 electron rule. Hence Sc, k=7.5 pairs=7.5 x2=15 electrons required for it to obey the 18 electron rule. Accordingly, the other transition metal elements in the same period will require the following electrons, Ti(k=7), 14 electrons; V(k=6.5), 13; Cr(k=6), 12; Mn(k=5.5),11; Fe(k=5), 10; Co(k=4.5), 9; Ni(k=4), 8; Cu(k=3.5), 7 and Zn(k=3),6. We can also tentatively assign possible shapes of the skeletal linkages to individual skeletal atoms. This is shown in Figure 2.

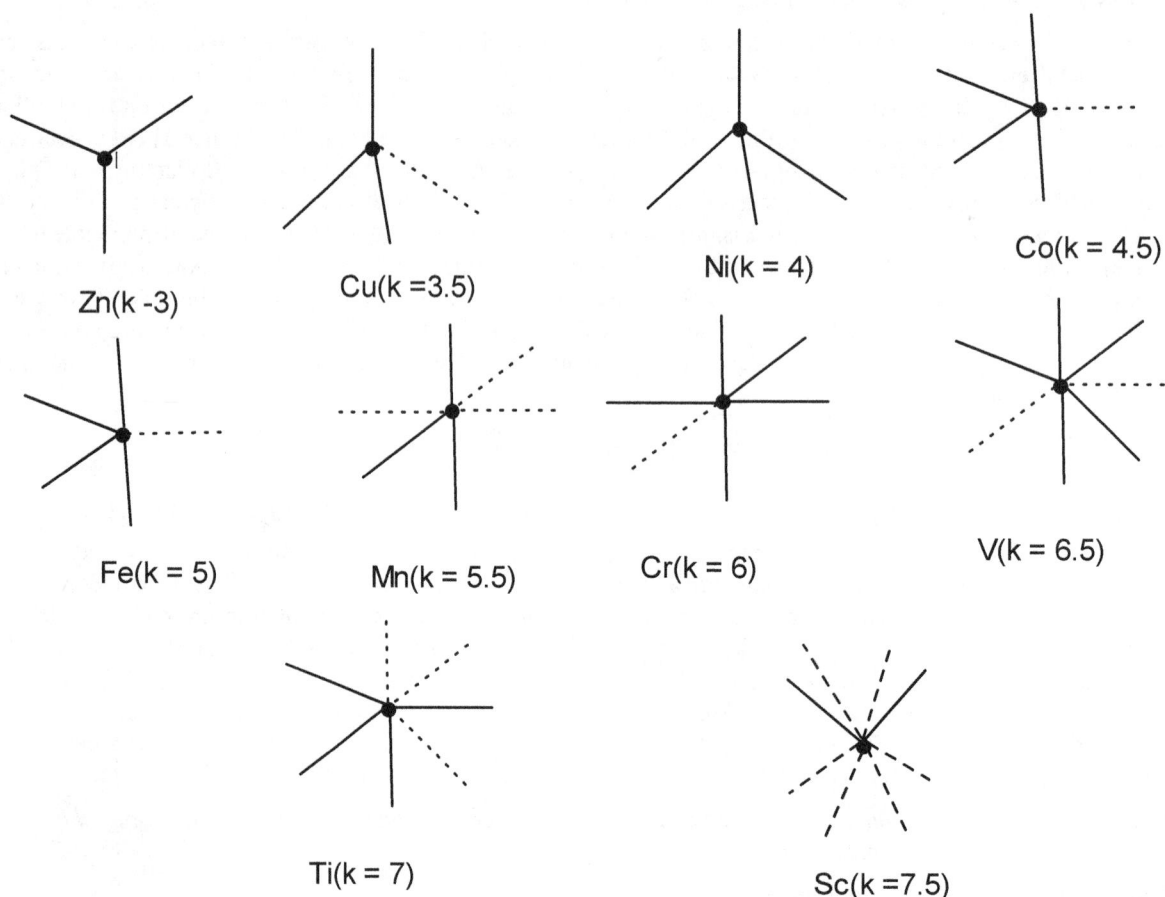

Zn(k -3)   Cu(k =3.5)   Ni(k = 4)   Co(k = 4.5)

Fe(k = 5)   Mn(k = 5.5)   Cr(k = 6)   V(k = 6.5)

Ti(k = 7)   Sc(k =7.5)

Figure 2. Proposed tentative shapes of the skeletal linkages of first row transition metals

2.7.2 Shapes of Clusters

For complexes with two or more skeletal elements, the skeletal numbers can be used to determine the k value of the cluster and hence the possible skeletal shape. Take the example of $Re_4(CO)_{16}^{2-}$; Re(k=5.5), CO(k=-1), 1charge(k=-0.2). Therefore, the total skeletal linkages of the Re atoms=4(5.5)=22. These have to be 'neutralized' by the ligands and the charge present. The remaining ones will constitute the skeletal bonds or linkages which are remaining to bind the skeletal elements. Hence k value of the complex will be given by k=22+16(-1)+2(-0.5)=22-17=5. The possible skeletal ideal shape of one of its isomers is shown in the labeled example 1(Ex-1) below. A possible ideal isomer shape is a square or rectangle with a diagonal. Using the labeled diagram as a basis, we can also use the skeletal numbers to deduce the possible number of CO ligands needed to complete the remaining skeletal linkages on the skeletal atom so as to enable it fulfill the 18-electron rule. The skeletal linkages available are given the labels k1 to k4. From the sketch, the atom labeled 1, has two bonds connected to it. This means it is receiving one electron donation from each of the bonds linked to it. These two electrons will neutralize a k value by 2(-0.5)=-1. Hence k1 =5.5-1=4.5. This means that atom 1 will have 4.5 CO ligands. In essence, there will be 4 CO ligands and the fractional component will represent one of the negative charges. Other k values, k2=4, k3=k1=4.5 and k4=4 were similarly calculated. In this way a possible isomer of the cluster can be sketched. This is also shown in Ex-1 below. More examples (Ex-2 to Ex-4) are well explained and provided. As the use skeletal numbers as a concept to predict possible shapes of clusters is being introduced for the first time, more well explained examples have been worked out. These are given in Schemes 4-16.

2.7.3 K-Isomerism of Clusters

Let us take the cluster $Rh_6(CO)_{16}$ as an illustration. The k value for the octahedral cluster is 11. Using skeletal numbers tentative distribution of carbonyl ligands on skeletal rhodium atoms can be sketched. Some of the selected isomers are given in Figure 3. The calculated k values on each rhodium atoms indicate the number of carbonyl ligands that can be accommodated according to 4n series approach using the skeletal numbers. Isomer-1 has already been given in F-9b. It is given here for comparison purposes.

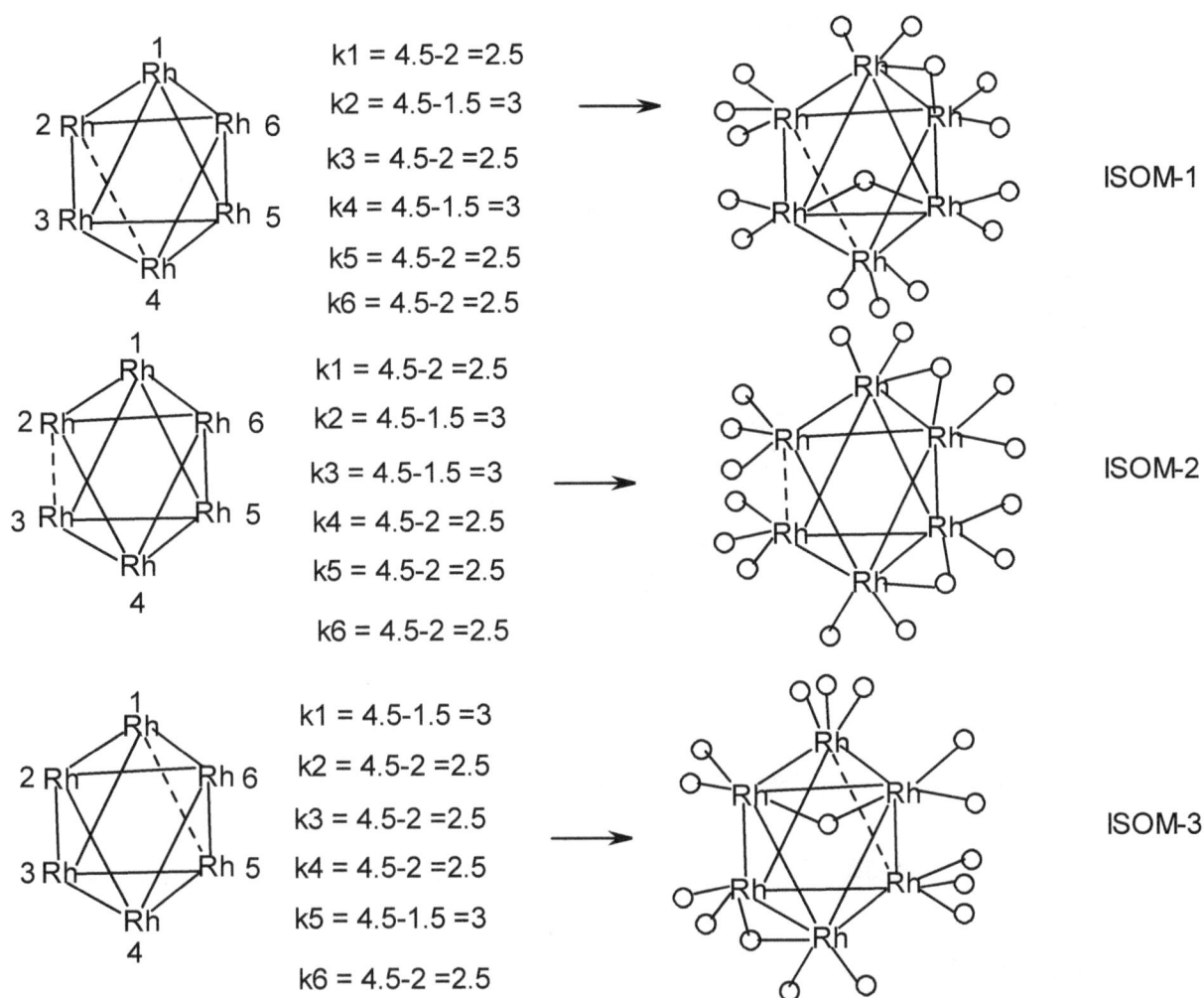

$k1 = 4.5-2 = 2.5$
$k2 = 4.5-1.5 = 3$
$k3 = 4.5-2 = 2.5$
$k4 = 4.5-1.5 = 3$
$k5 = 4.5-2 = 2.5$
$k6 = 4.5-2 = 2.5$

ISOM-1

$k1 = 4.5-2 = 2.5$
$k2 = 4.5-1.5 = 3$
$k3 = 4.5-1.5 = 3$
$k4 = 4.5-2 = 2.5$
$k5 = 4.5-2 = 2.5$
$k6 = 4.5-2 = 2.5$

ISOM-2

$k1 = 4.5-1.5 = 3$
$k2 = 4.5-2 = 2.5$
$k3 = 4.5-2 = 2.5$
$k4 = 4.5-2 = 2.5$
$k5 = 4.5-1.5 = 3$
$k6 = 4.5-2 = 2.5$

ISOM-3

Figure 3. Sketches of selected possible isomers of $Rh_6(CO)_{16}$.

Ex-1. $Re_4(CO)_{16}^{2-}$ Re(v=7), S = 14n-7(n = 1) $\longrightarrow$ 4n-7(n = 1), k = 2n+3.5 = 2(1)+3.5 = 5.5

k = 4(5.5)   -16-1 = -17 $\longrightarrow$ 22-17 = 5

k1 = 5.5-2(0.5) = 5.5-1 = 4.5 = k3   k2 = 5.5-3(0.5) = 5.5-1.5 = 4 = k4

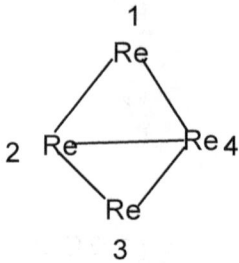

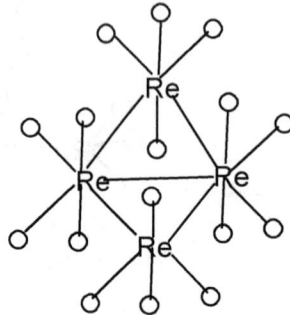

Ex-1

Ex-2. $Os_5(CO)_{18}$ Os(k =5) , k(CO) = -1

k = 5(5)-18 = 25-18 = 7

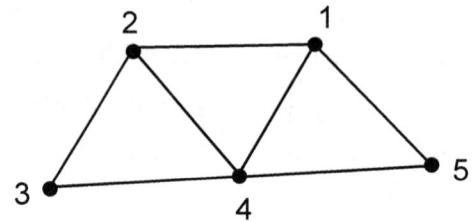

k1 = 5-3(0.5) = 5-1.5 = 3.5 = k2

k3 = 5-2(0.5) = 5-1 = 4 = k5   k4 = 5-4(0.5) = 5-2 = 3

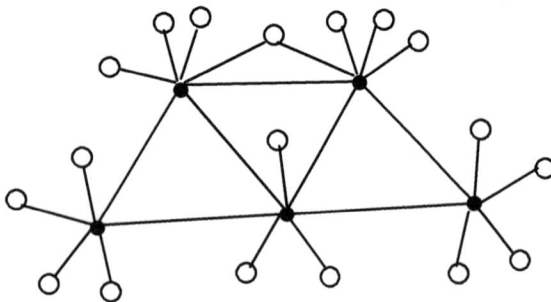

Ex-2

Ex-3. $H_2Os_5(CO)_{16}$ $Os(k = 5)$, $CO(k = -1)$, $H(k = 0.5)$

$k = 5(5)-2(0.5)-16(1) = 25-17 = 8$ $k1 = 5-1.5 = 3.5$, $k2 = 5-2 = 3$, $k3 = 5-1 = 4$, $k4 = 5-2 = 3$,
$k5 = 5-1.5 = 3.5$

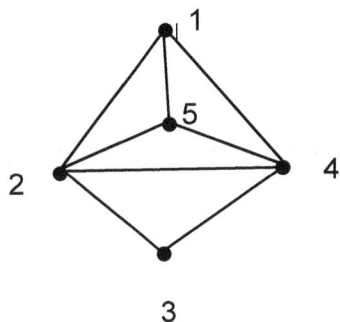

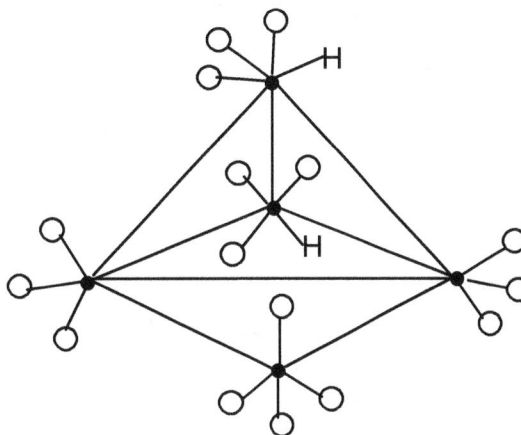

Ex-3

Ex-4. $H_2Os_6(CO)_{18}$ $Os(k = 5)$, $CO(k = -1)$, $H(k = -0.5)$

$k = 6×5-18(1)-2(0.5) = 30-19 = 11$

$k1 = 5-1.5 = 3.5$ $k2 = 5-2 = 3$ $k3 = 5-1.5 = 3.5$
$k4 = 3.5$ $k5 = 3$ $k6 = 5-2.5 = 2.5$

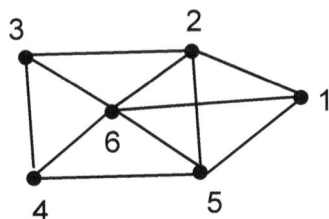

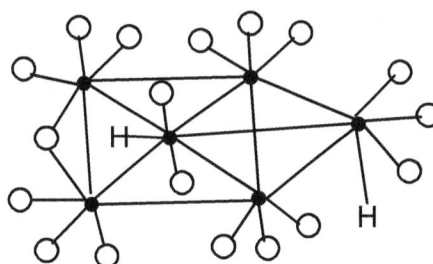

Ex-4

1. $Mn_2(CO)_{10}$ $k_{mn} = 2 \times 5.5 = 11$ $\longrightarrow$ $k_s = 11-10 = 1$

                         $k_{co} = 10(-1) = -10$

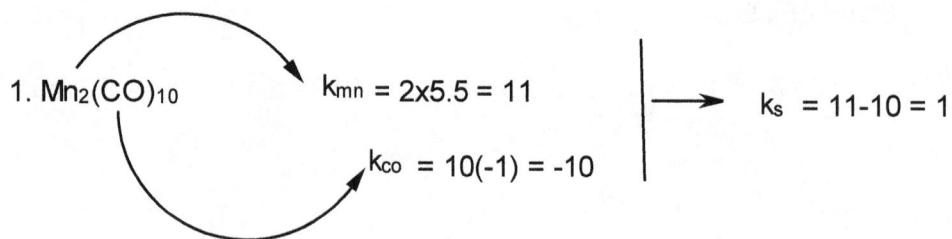

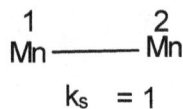

$\underset{Mn}{\overset{1}{}} \text{——} \underset{Mn}{\overset{2}{}}$ $k_1 = k_2 = 5.5 + 1(-0.5) = 5 \longrightarrow$ Five skeletal linkages available
                                                on each Mn atom.
$k_s = 1$

Possible structure   shown in F-1.

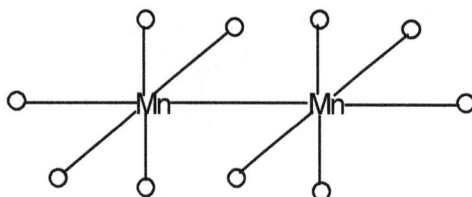

F-1

Scheme 4. Derivation of $Mn_2(CO)_{10}$ structure using skeletal numbers

2. $Co_2(CO)_8$ $k_{Co} = 2 \times 4.5 = 9$ $k_s = 9 - 8 = 1$

$k_{co} = 8(-1) = -8$

$$\overset{1}{Co} \rule{3em}{0.4pt} \overset{2}{Co}$$
$$k_s = 1$$

$k_1 = k_2 = 4.5 + 1(-0.5) = 4 \longrightarrow$ Four skeletal linkages available on each Co atom.

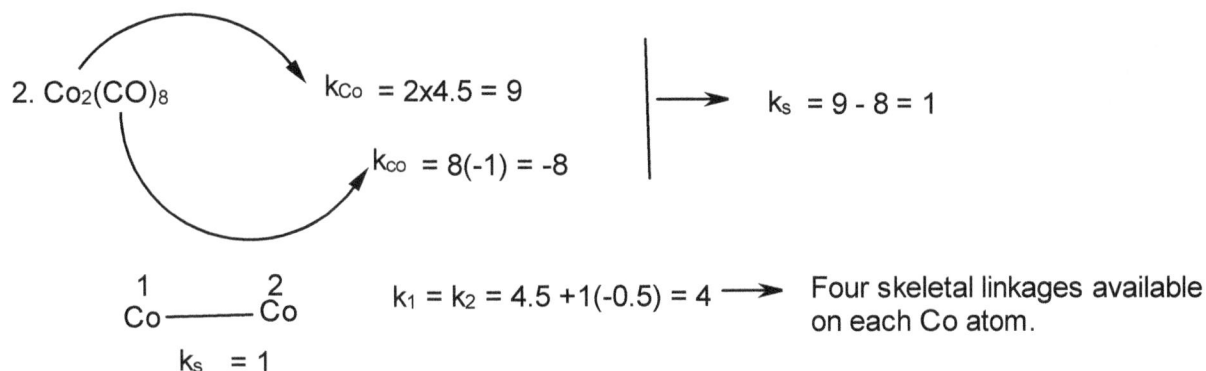

Possible structure shown in F-2.

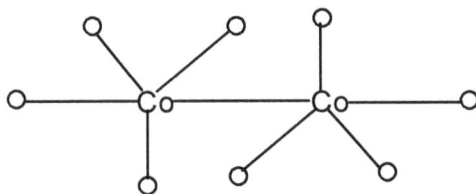

F-2

Scheme 5. Derivation of $Co_2(CO)_8$ structure using skeletal numbers

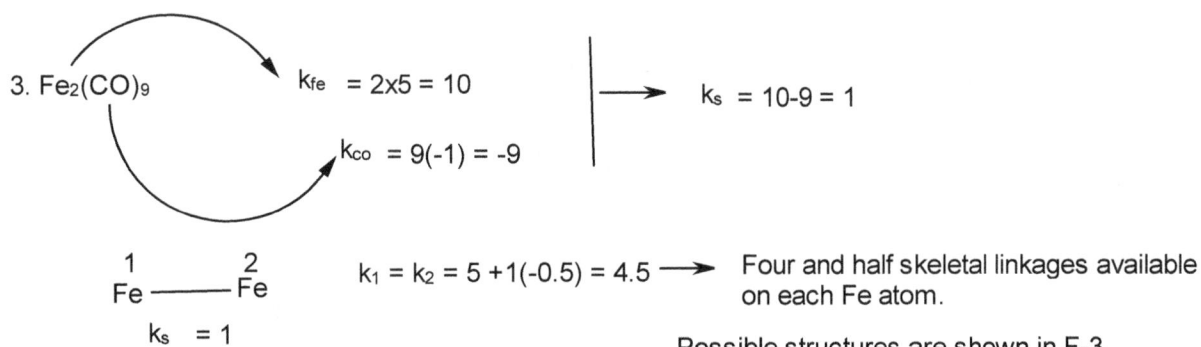

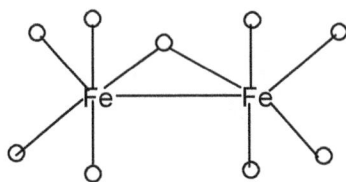

3. $Fe_2(CO)_9$ $k_{fe} = 2 \times 5 = 10$ $k_s = 10 - 9 = 1$

$k_{co} = 9(-1) = -9$

$$\overset{1}{Fe} \rule{3em}{0.4pt} \overset{2}{Fe}$$
$$k_s = 1$$

$k_1 = k_2 = 5 + 1(-0.5) = 4.5 \longrightarrow$ Four and half skeletal linkages available on each Fe atom.

Possible structures are shown in F-3.

F-3a

Predicted possible isomer of $Fe_2(CO)_9$

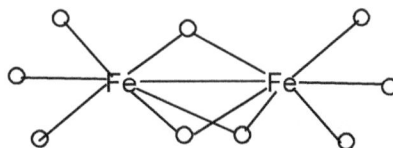

F-3b

Another possible isomer of $Fe_2(CO)_9$

Scheme 6. Derivation of $Fe_2(CO)_9$ structure using skeletal numbers

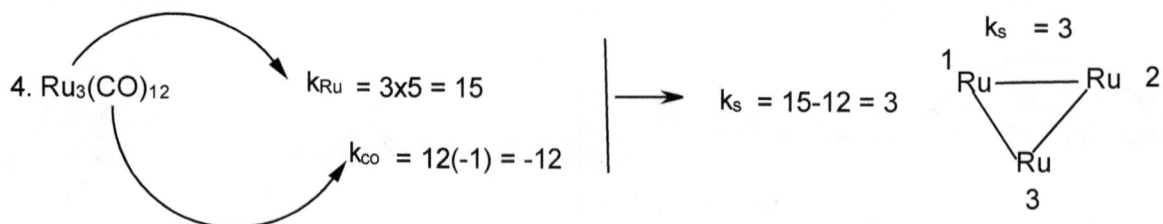

$4.\ Ru_3(CO)_{12}$ $k_{Ru} = 3 \times 5 = 15$ $\longrightarrow$ $k_s = 15-12 = 3$

$k_{co} = 12(-1) = -12$

$k_s = 3$

$k_1 = k_2 = k_3 = 5 + 2(-0.5) = 5-1 = 4$

This implies that each Ru atom will have 4 CO ligands attached to it.

Possible isomer is shown in F-4.

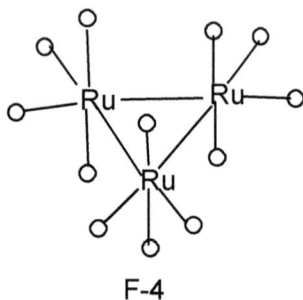

F-4

Scheme 7. Derivation of $Ru_3(CO)_{12}$ structure using skeletal numbers

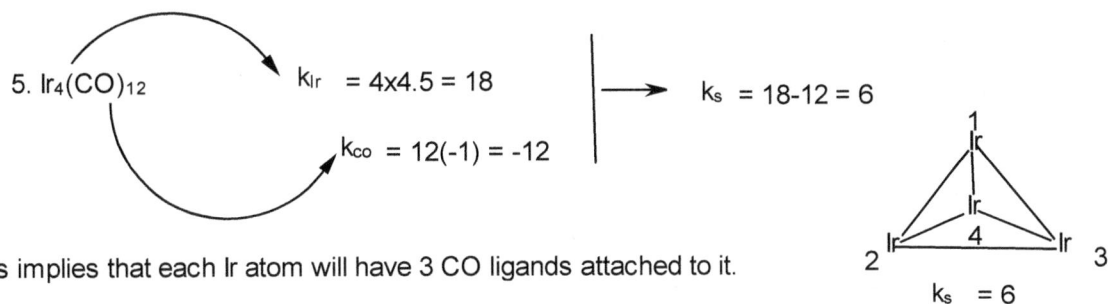

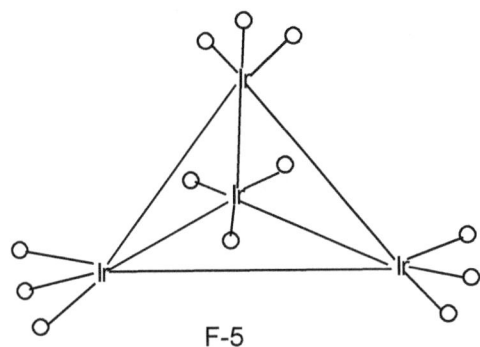

5. $Ir_4(CO)_{12}$ k_{Ir} = 4x4.5 = 18 k_s = 18-12 = 6

k_{co} = 12(-1) = -12

This implies that each Ir atom will have 3 CO ligands attached to it.

k_s = 6

$k_1 = k_2 = k_3 = k_4$ = 4.5 +3(-0.5) = 4.5-1.5 = 3

Possible isomer is shown in F-5.

F-5

Scheme 8. Derivation of $Ir_4(CO)_{12}$ structure using skeletal numbers

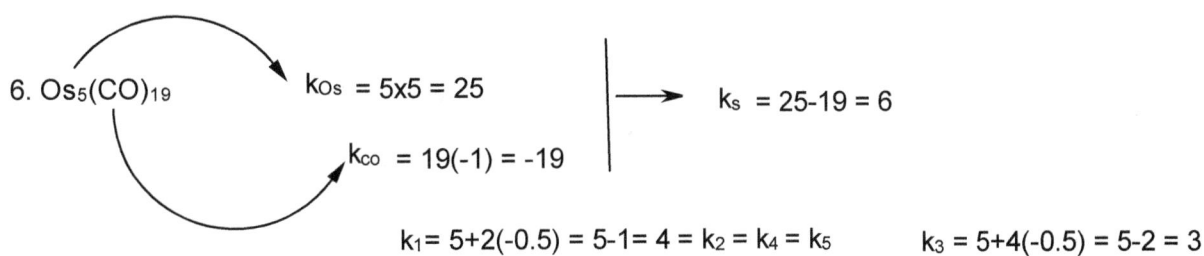

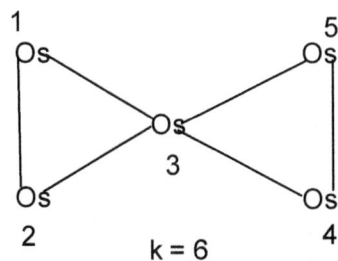

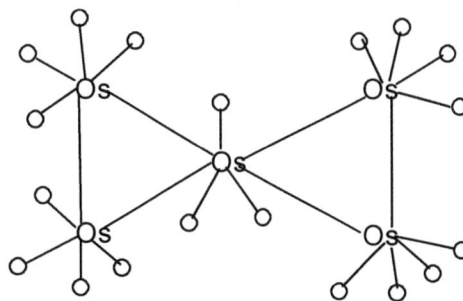

6. $Os_5(CO)_{19}$ k_{Os} = 5x5 = 25 k_s = 25-19 = 6

k_{co} = 19(-1) = -19

k_1= 5+2(-0.5) = 5-1= 4 = k_2 = k_4 = k_5 k_3 = 5+4(-0.5) = 5-2 = 3

k = 6

Possible isomer is shown in F-6.

Scheme 9. Derivation of $Os_5(CO)_{19}$ structure using skeletal numbers

7. $Fe_4(C)(CO)_{12}{}^{2-}$ $\longrightarrow$ $k_S = 4(5)-1(2)-12(1)-2(0.5) = 5$. The skeletal shape is sketched in F-7a.

If the carbon atom is taken as a skeletal atom, then $kS = 4(5)+1(2)-12-1 = 22-13 = 9$. This means that the 5 skeletal atoms are linked by 9 lines. The sketch appears as in F-7b. The tentative distribution of carbonyl ligands is shown in F-7c.

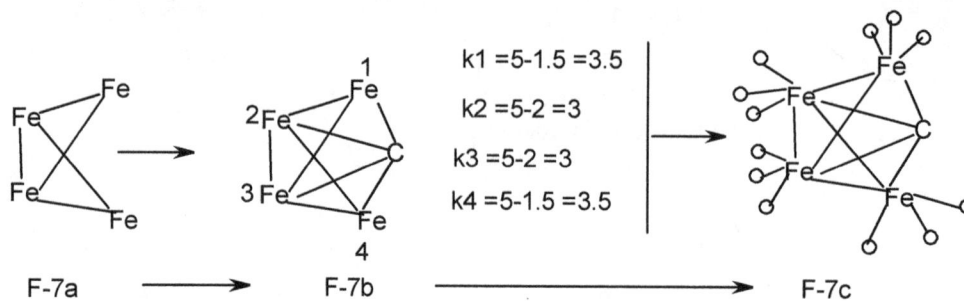

$k1 = 5-1.5 = 3.5$

$k2 = 5-2 = 3$

$k3 = 5-2 = 3$

$k4 = 5-1.5 = 3.5$

F-7a $\longrightarrow$ F-7b $\longrightarrow$ F-7c

The $k1 = k4 = 3.5$ means that there 3 carbonyl ligands and the fractions represent the negative charges.

Scheme 10. Derivation of the structure of $Fe_4(C)(CO)_{12}{}^{2-}$ using skeletal numbers

Converting the k value of $Fe_4(C)(CO)_{12}^{2-}$ into series

We have learnt that the series S = 4n+q has a corresponding k value given by

k=2n-q/2. Since there are 5 skeletal elements if we include the carbon atom, then n=5 and k=9. Hence, k=9=2(5)-q/2; q/2=10-9=1. Therefore, q=2.

Hence the cluster series S=4n+2. This is CLOSO cluster.

Its borane equivalent F_B=[BH](5)+2 =$B_5H_5^{2-}$. The hydrocarbon equivalent cluster will be given by F_{CH}=[C](5)+2H=C_5H_2.

Scheme 11. Transforming the k value of a given cluster into series

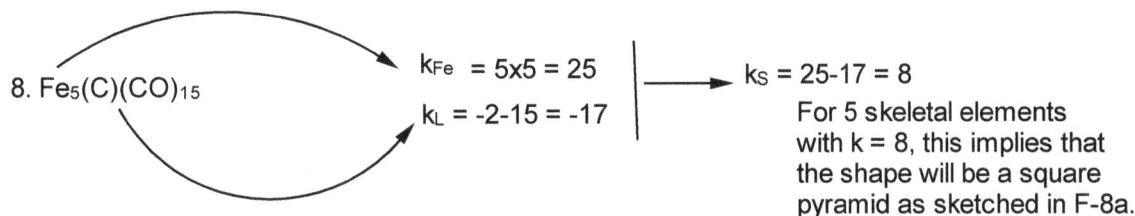

8. $Fe_5(C)(CO)_{15}$

k_{Fe} = 5x5 = 25

k_L = -2-15 = -17

k_S = 25-17 = 8

For 5 skeletal elements with k = 8, this implies that the shape will be a square pyramid as sketched in F-8a.

If we include the carbon atom as a skeletal atom, then its contribution to cluster k value is 2. hence, the net k = 25+2 = 27. The k_L = -15. Hence, k_S = 27-15 = 12.

The difference between the k-values = 12-8 = 4 will be due to the C linkages.

The modified skeletal shape will be as in shown in F-8b.

Using F-8b as a basis, k_1 = 5-2 = 3, k_2 = 5-2 = 3, k_3 = 5-2 = 3, k_4 =5-2 =3, k_5 =5-2 =3.

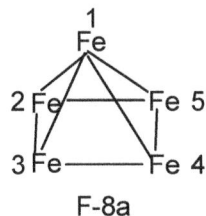

F-8a

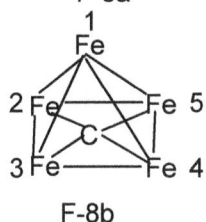

F-8b

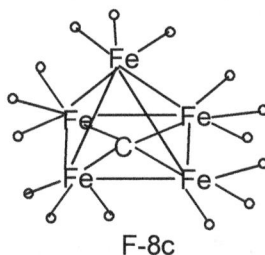

F-8c

Scheme 12. Derivation of structure of $Fe_5(C)(CO)_{15}$ cluster using skeletal numbers

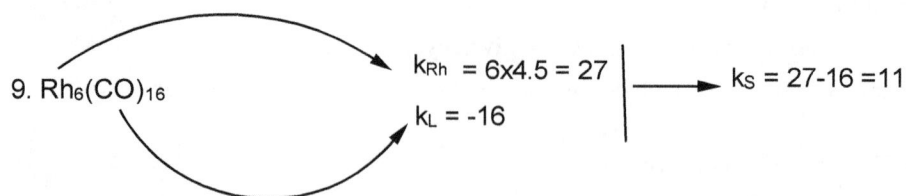

$$k_{Rh} = 6 \times 4.5 = 27$$

$$k_L = -16$$

$$k_S = 27 - 16 = 11$$

The k-value of 11 for 6 skeletal elements is characteristic of an octahedral geometry. Baed on this, we can conch struct an octahedral sketch as shown in F-9a.

$k_1 = 4.5 + 4(-0.5) = 4.5 - 2 = 2.5$, $k_2 = 4.5 + 3(-0.5) = 4.5 - 1.5 = 3$, $k_3 = 4.5 - 4(0.5) = 4.5 - 2 = 2.5$,

$k_4 = 4.5 + 3(-0.5) = 4.5 - 1.5 = 3$, $k_5 = 4.5 + 4(-0.5) = 4.5 - 2 = 2.5$,

$k_6 = 4.5 + 4(-0.5) = 4.5 - 2 = 2.5$

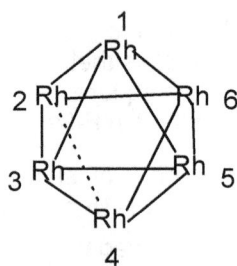

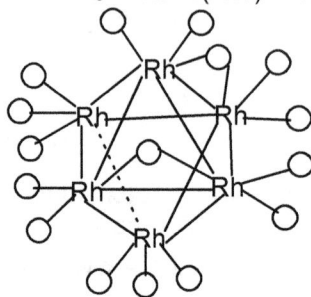

The dotted line is not used in the calculation since we are using k =11 for an O_h symmetry.

Final sketch of one of the possible isomer is given in F-9b.

F-9a

F-9b

Scheme 13. Derivation of structure of $Rh_6(CO)_{16}$ cluster using skeletal numbers

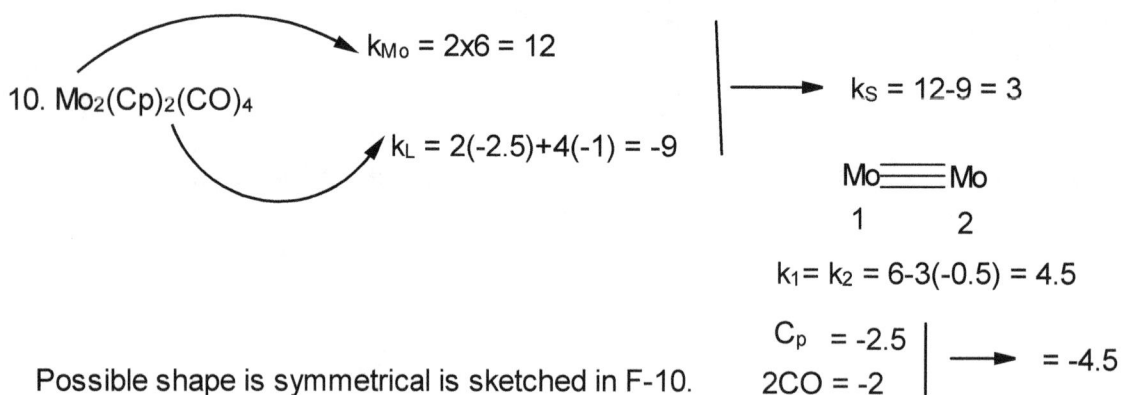

$$k_{Mo} = 2 \times 6 = 12$$

10. $Mo_2(Cp)_2(CO)_4$

$$k_L = 2(-2.5) + 4(-1) = -9$$

$$k_S = 12 - 9 = 3$$

$$Mo \equiv\equiv Mo$$
$$1 \qquad 2$$

$$k_1 = k_2 = 6 - 3(-0.5) = 4.5$$

$$C_p = -2.5$$
$$2CO = -2$$

$$= -4.5$$

Possible shape is symmetrical is sketched in F-10.

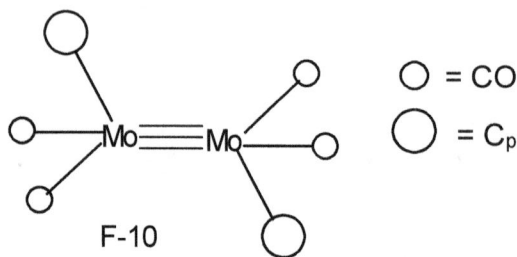

$\bigcirc = CO$

$\bigcirc = C_p$

F-10

Scheme 14. Derivation of structure of $Mo_2(Cp)_2(CO)_4$ using skeletal numbers

11. $SnFe_4(CO)_{16}$

$k_{SnFe} = 2+4(5) = 22$

$k_{CO} = 16(-1) = -16$

$k_S = 22-16 = 6$

$k_1 = 5-1 = 4 = k_2 = k_3 = k_4$

This means eound to each of the iron skeletal atoms will be linked to 4 CO ligands.

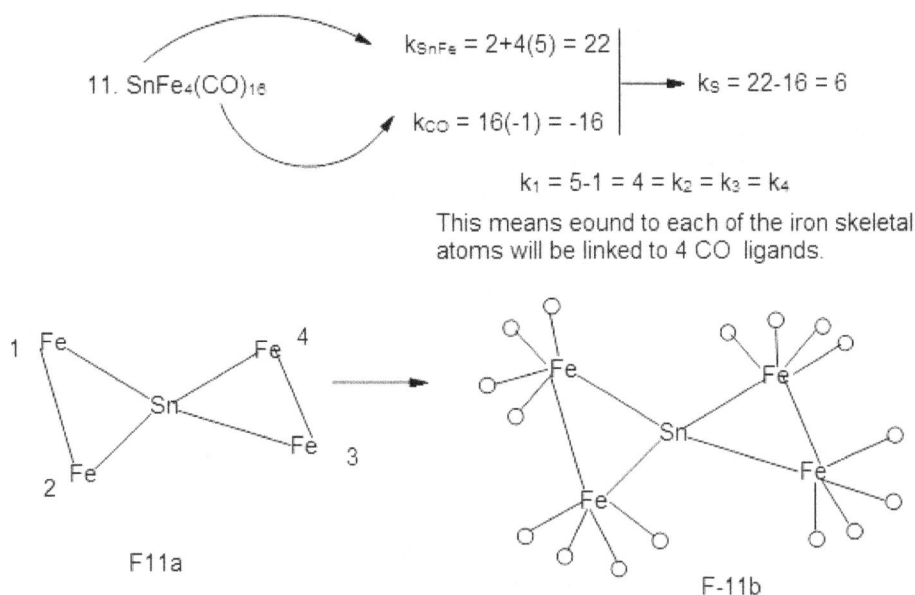

F11a   F-11b

Scheme 1   Derivation of possible structure of $Mo_2(Cp)_2(CO)_4$ using skeletal numbers

12. Another way of Redistributing Ligands in $Fe_5(C)(CO)_{15}$

$Fe(k = 5), C(k = -2), CO(k = -1)$

$k = 5(5)-2-17 = 8$ $k1 = 5-1.5 = 3.5$ $k2 = k1 = k3 = k4$ $k5 = 5-2 = 3$

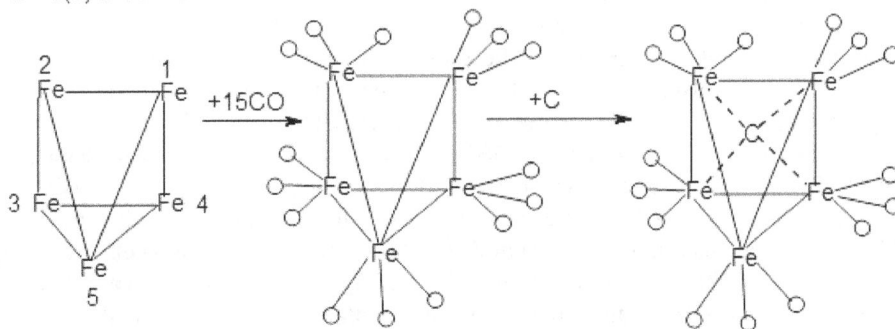

Scheme 1 . Stepwise derivation of the structure of $Fe_5(C)(CO)_{15}$ using skeletal numbers

2.8 Consolidation of the Skeletal Linkages Principle

Arising from the analysis of the series of generated from adding carbonyl ligands to transition metal fragments, that we can regard the parent skeletal metal fragment as the provider of the total skeletal numbers (see Table 2)of the cluster in question. When the carbonyl and other ligands are added to the parent skeletal metal fragment, they use up some of the skeletal numbers. The remaining skeletal numbers or linkages hold the skeletal fragment of the cluster together. Take a mono-skeletal fragment such as rhodium Rh with skeletal value of k = 4.5. When the ligands, CO, 2PPh₃, and Cl are added, to form the Vaska's complex(Cotton, et al, 1980), $Rh(Cl)(PPh_3)_2(CO)$, the CO utilizes one(1) skeletal value, the 2 PPh₃ use up two(2) and the Cl takes (0.5) skeletal numbers. Hence, the k value of the cluster will be given by k=4.5-1(1)-2(1)-1(0.5) = 1. This means the cluster is missing a pair of valence electrons to complete the 18-electron rule. Thus, the cluster is a 16-valence electron system. Hence, it is no surprise to see the Vaska's complex undergo the oxidative addition reactions such as $Rh(Cl)(PPh_3)_2(CO)+H_2 \rightarrow Rh(Cl)(PPh_3)_2(CO)(H)_2(k=0)$, and
$Rh(Cl)(PPh_3)_2(CO)+MeI \rightarrow Rh(Cl)(PPh_3)_2(CO)(Me)(I)(k =0)$. The addition of 8CO ligands to $Co_2(k=2x4.5=9)$ skeletal fragment produces $Co_2(CO)_8$ complex with k=9-8=1. This means that the 8 carbonyl ligands use up 8 of the 9 skeletal values of Co_2 leaving behind k =1. Therefore the cobalt carbonyl complex $Co_2(CO)_8$ is held by one Co-Co bond. This

approach is indirectly applying equation (i). That is, in order to determine the number of skeletal linkages (k_S) of a given cluster, the equation (i) becomes extremely useful. The k_S value is thus given by $k_S = k_T - k_L$. This equation simply means that if the k values utilized by the ligands are subtracted from the original k values of the naked skeletal elements, the remaining k values represent the skeletal linkages of the luster. Let us illustrate this using the following few examples. Consider the complex, $Re_2(Cp)_2(CO)_5$. Since Re has the skeletal number, k=5.5, the skeletal number $k_S = 2(5.5) - 2(2.5) - 5(1) = 11 - 10 = 1$ and S=4n+q, k=2n-q/2. Hence, q/2 = 2n-k = 2(2)-1=3, q=6 and S=4n+6. The number of valence electrons, Ve = 4n+6 = 4(2)+6 = 14. Since S = 4n+6 $\leftrightarrow$14n+6, the corresponding valence electrons for the transition metal cluster V_M will be given by $V_M = 14 + 10(2) = 34$. This means that the rhenium complex will have a single Re-Re bond around which the 5CO and the 2 cyclopentadienyl ligands will be bound. For the complex, $Ir_4(CO)_{12}$, Ir , k=4.5(see Table 2), hence k_S = 4(4.5)-12(1) = 6. Using this value of k =6, we can derive the series and then use the series to deduce the number of valence electrons. Thus, k=6, q/2=2(4)-6 = 2, q=4, S=4n+4, Ve=4(4)+4 =20, and $V_M = 20 + 10(4) = 60$. The k value of 6 for 4 iridium skeletal atoms implies that they will take up an ideal tetrahedral skeletal shape. Let us consider the rhenium complex $Re_5(C)(CO)_{16}(H)^{2-}$. Its cluster number is given by k=5(5.5)-1(2)-16(1)-1(0.5)-2(0.5)=8, q/2 =2(5)-8=2, q=4 and S=4n+4(nido series), Ve=4n+4=4(5)+4=24. Hence, the valence electrons of the corresponding transition metal carbonyl cluster will be given by $V_M = 24 + 10(5) = 54$. The k value of 8 for 5 five skeletal elements is characteristic of a square pyramid geometry. Another example is $Re_6(C)(CO)_{19}^{2-}$; k=6(5.5)-2-19-1=11,q/2=2(6)-11=1, q=2, S=4n+2(closo series), Ve=4n+2=4(6)+2=26, and $V_M = 26 + 10(6) = 86$. The k value of 11 for 6 skeletal elements and closo series is characteristic of an octahedral geometry. Applying the same concept of assuming that all the skeletal values are supplied by the skeletal elements and that the carbonyl and other ligands consume some of the skeletal values and that the remaining k values hold the skeletal atoms together, then we can calculate the k values of clusters, derive their corresponding series and valence electrons and predict the shapes of some of the cluster complexes. The following examples in Schemes 17 and 18 illustrate this concept. Due to the fact that the use of skeletal numbers is so easy and flexible, it has been extended to many more clusters carbonyl complexes taken mainly from various reviews (King, 1986; Lewis and Johnson, 1982; Zanello, 2002; Teo and Zhang, 1990, 1991; Belyakova and Slovokhov, 2003; Hughes and Wade,2000; Rossi and Zanello,2011). The results are given in Table 10. The table summarizes the analysis of nearly 50 complexes. In addition, 15 more examples are illustrated in Schemes 17 and 18 in order to enable the readers be exposed to a wide range of clusters analyzed using the skeletal numbers. Furthermore, these examples clearly demonstrate the fact that the valence electron counts can readily be derived from the corresponding 4n series of the clusters. The results agree with many of the known clusters but also provide more insights on clusters and new interpretations based on the type of 4n series and k values. Let us look at some few examples of categorization of a cluster using its k value. These include(Greenwood, et al,1998) $Co_2(CO)_8$; k=2(4.5)-8=9-8=1(single metal-metal bond), q/2=2n-k, n=2, q/2=2(2)-1=3, q=6, S=4n+6(arachno series),Ve=4n+6 =4(2)+6 =14,V_M=14+10(2)=34, $Ir_4(CO)_{12}$; k= 4(4.5)-12=6(six skeletal linkages, tetrahedral skeletal shape),q/2 =2n-k, n =4, q/2 =2(4)-6=2, q=4, S=4n+4(nido series), Ve=4n+4=4(4)+4=20, V_M=20+10(4)=60, $M_6(CO)_{16}$,M=Co, Rh, Ir; k=6(4.5)-16=11, q/2=2n-k, n=6, q/2 =2(6)-11=1, q=2, S=4n+2(closo), for n=6 and cluster belongs to closo series, this corresponds to $B_6H_6^{2-}$ cluster which has an ideal octahedral skeletal shape. This is what is observed (Greenwood, et al, 1998). The valence electron count for the octahedral system, Ve = 4n+2, n=6, Ve=4(6)+2=26 and V_M=26+10(6)=86. The Ve=26 is the valence electron count of the main group octahedral geometry and VM=86 is the corresponding valence electron count for a transition metal complex.. Hence, $B_6H_6^{2-}$will have Ve=26. However its counterpart transition metal complex will have its valence electron count of V_M=26+10(6)=86. This is a consequence of a type of isolobal relationship S=4n+q (Main Group Element cluster)$\leftrightarrow$S =14n+q(Transition Metal Cluster). The difference between the two systems is simply ±10n. The examples in Table 10 and Schemes 17 and 18 have included the valence electron counts of clusters to demonstrate that valence electron counts are a direct consequence of 4n series. The valence electron counts which are associated with skeletal shapes of clusters are also a direct consequence of the 4n series (Fehlner, et al, 2007).

$$S = 4n+q, \; k = 2n-q/2, \; q/2 = 2n-k$$

1. Co $\xrightarrow[+4CO]{+H}$ **HCo(CO)$_4$**

$k = 4.5$

$k = 4.5-0.5-4(1) = 0$

$n = 1, \; q/2 = 2(1)-0 =2; \; q=4, \; S =4n+4; \; Ve = 4(1)+4 =8.$
Hence $V_M = 8+10(1) = 18$

2. Co$_2$ $\xrightarrow{+8CO}$ **Co$_2$(CO)$_8$**

$k = 2(4.5) = 9$

$k = 9-8(1) = 1$

$n = 2, \; q/2 = 2(2)-1 =3; \; q =6, \; S =4n+6; \; Ve = 4(2)+6 =14.$
Hence $V_M = 14+10(2) = 34$

3. Rh$_2$ $\xrightarrow[+2C_p]{+2CO}$ **Rh$_2$(C$_p$)$_2$(CO)$_2$**

$k = 2(4.5) = 9$

$k = 9-2(2.5)-2(1) = 2$

$n = 2, \; q/2 = 2(2)-2 =2; \; q =4, \; S =4n+4; \; Ve = 4(2)+4 =12.$
Hence $V_M = 12+10(2) = 32$

4. Mo$_2$ $\xrightarrow[+2C_p]{+4CO}$ **Mo$_2$(C$_p$)$_2$(CO)$_4$**

$k = 2(6) = 12$

$k = 12-2(2.5)-4(1) = 3$

$n = 2, \; q/2 = 2(2)-3 =1; \; q =2, \; S =4n+2; \; Ve = 4(2)+2 =10.$
Hence $V_M = 10+10(2) = 30$

5. Os$_3$ $\xrightarrow{+ 12CO}$ **Os$_3$(CO)$_{12}$**

$k = 3(5) = 15$

$k = 15-12(1) = 3$

$n = 3, \; q/2 = 2(3)-3 =3; \; q =6, \; S =4n+6; \; Ve = 4(3)+6 =18.$
Hence $V_M = 18+10(3) = 48$

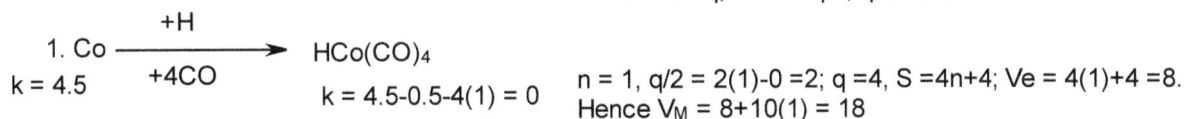

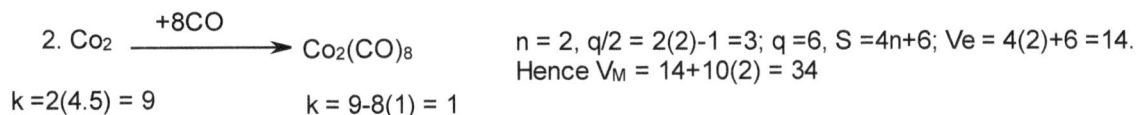

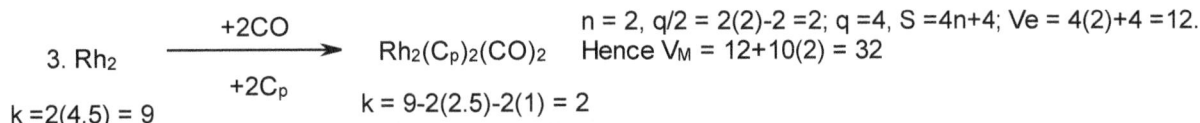

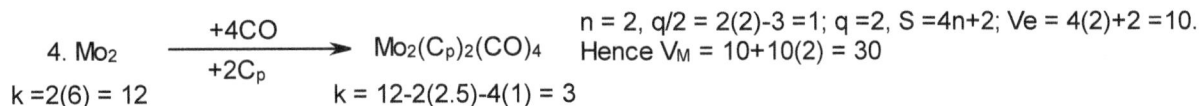

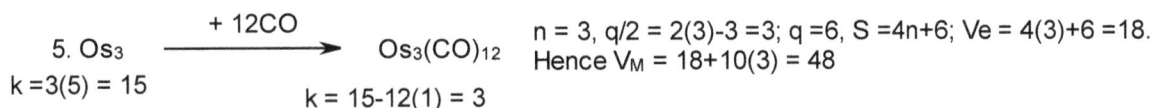

Scheme 17. Examples showing Derivation of series and Valence Electron Counts using the series

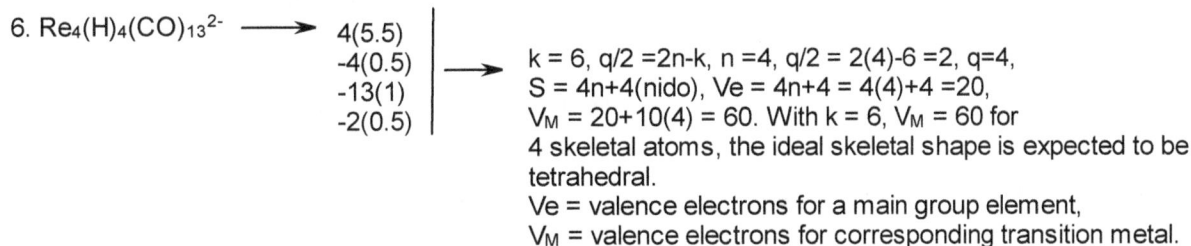

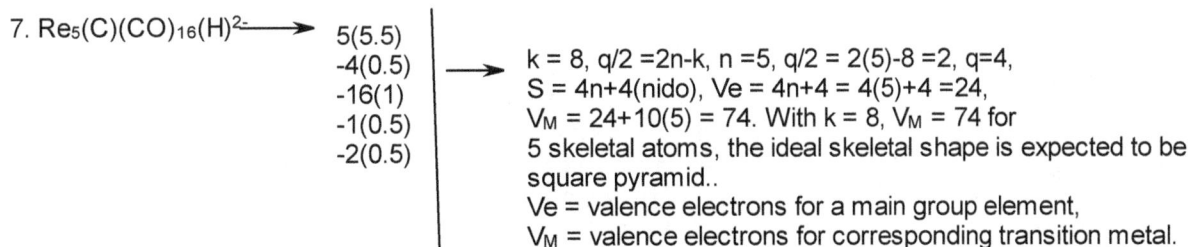

6. Re$_4$(H)$_4$(CO)$_{13}{}^{2-}$ $\longrightarrow$

$4(5.5)$
$-4(0.5)$
$-13(1)$
$-2(0.5)$

$\longrightarrow$

$k = 6, \; q/2 =2n-k, \; n =4, \; q/2 = 2(4)-6 =2, \; q=4,$
$S = 4n+4 (nido), \; Ve = 4n+4 = 4(4)+4 =20,$
$V_M = 20+10(4) = 60.$ With $k = 6, \; V_M = 60$ for
4 skeletal atoms, the ideal skeletal shape is expected to be
tetrahedral.
Ve = valence electrons for a main group element,
V_M = valence electrons for corresponding transition metal.

7. Re$_5$(C)(CO)$_{16}$(H)$^{2-}$ $\longrightarrow$

$5(5.5)$
$-4(0.5)$
$-16(1)$
$-1(0.5)$
$-2(0.5)$

$\longrightarrow$

$k = 8, \; q/2 =2n-k, \; n =5, \; q/2 = 2(5)-8 =2, \; q=4,$
$S = 4n+4 (nido), \; Ve = 4n+4 = 4(5)+4 =24,$
$V_M = 24+10(5) = 74.$ With $k = 8, \; V_M = 74$ for
5 skeletal atoms, the ideal skeletal shape is expected to be
square pyramid..
Ve = valence electrons for a main group element,
V_M = valence electrons for corresponding transition metal.

8. $Re_6(C)(CO)_{19}{}^{2-} \longrightarrow$

6(5.5)
-4(0.5)
-19(1)
-2(0.5)

$\longrightarrow$ k = 11, q/2 =2n-k, n =6, q/2 = 2(6)-11 =1, q=2,
S = 4n+2(closo), Ve = 4n+2 = 4(6)+2 =26,
V_M = 26+10(6) = 86. With k = 11, V_M = 86 for
6 skeletal atoms, the ideal skeletal shape is expected to be octahedral..
Ve = valence electrons for a main group element,
V_M = valence electrons for corresponding transition metal.

9. $Re_7(C)(CO)_{21}{}^{3-} \longrightarrow$

7(5.5)
-2
-21
-1.5

$\longrightarrow$ k = 14, q/2 =2n-k, n =7, q/2 = 2(7)-14 =0, q=0,
S = 4n+0(mono-capped closo, $C^1C[M-6]$),
Ve = 4n+0 = 4(7)+0 =28,
V_M = 28+10(7) = 98. With k = 14, V_M = 98 for
7 skeletal atoms, the ideal skeletal shape is expected to be a mono-capped octahedral..
Ve = valence electrons for a main group element,
V_M = valence electrons for corresponding transition metal.

10. $Re_8(C)(CO)_{24}{}^{2-} \longrightarrow$

8(5.5)
-2
-24
-1

$\longrightarrow$ k = 17, q/2 =2n-k, n =8, q/2 = 2(8)-17 =-1, q=-2,
S = 4n-2(bi-capped closo, $C^2C[M-6]$),
Ve = 4n-2 = 4(8)-2 =30,
V_M = 30+10(8) = 110. With k = 17, V_M =110 for
8 skeletal atoms, the ideal skeletal shape is expected to be a bi-capped octahedral..
Ve = valence electrons for a main group element,
V_M = valence electrons for corresponding transition metal.

11. $B_5H_9 \longrightarrow$

5(2.5)

-9(0.5)

$\longrightarrow$ k = 8, q/2 = 2n-k, n=5, q/2 =2(5)-8 =2, q = 4,
S = 4n+q = 4n+4(nido). With k =8 for 5 skeletal atoms of nido series, the expected ideal shape is suare pyramid.

12. $B_6H_6{}^{2-} \longrightarrow$

6(2.5)

-6(0.5)
-2(0.5)

$\longrightarrow$ k = 11, q/2 = 2n-k, n=6, q/2 =2(6)-11 =1, q = 2,
S = 4n+q = 4n+2(closo). With k =11 for 6 skeletal atoms of closo series, the expected ideal shape is octahedral.

13. $C_2B_{10}H_{12} \longrightarrow$ k = 2(2)+10(2.5)-12(0.5) =23
q/2 = 2n-k, n =2+10 = 12, q/2 = 2(12)-23 = 1, q = 2
S =4n+q = 4n+2 (closo series).
Since n = 12 and belongs to closo series, the ideal shape of the cluster will be similar to $B_{12}H_{12}{}^{2-}$.
Ve = 4n+2 = 4(12)+2 = 50

14. $(Cp)_3Co_3B_3H_5 \longrightarrow$ k =3(4.5)+3(2.5)-3(2.5)-5(0.5) = 11,q/2 =2n-k = 2(6)-11=1,
S = 4n+2 (closo). For k =11 and cluster belongs to closo series with = 6. the predicted shape of the cluster will be similar to that of that of $B_6H_6{}^{2-}$ which is octahedral.

15. $Ru_6Pd_6(24)^{2-}$

$k = 6(5)+6(4)$
-24
-1

$\longrightarrow$ k=29, q/2 =2n-k, n = 6+6 = 12, q/2 =2(12)-29=-5, q = -10
S = 4n-10, Cp = C^1+C^5 = $C^6C[M-6]$, heca-capped octahedron.
Ve = 4n-10 =4(12)-10=38, V_M = 38+10(12) =158.

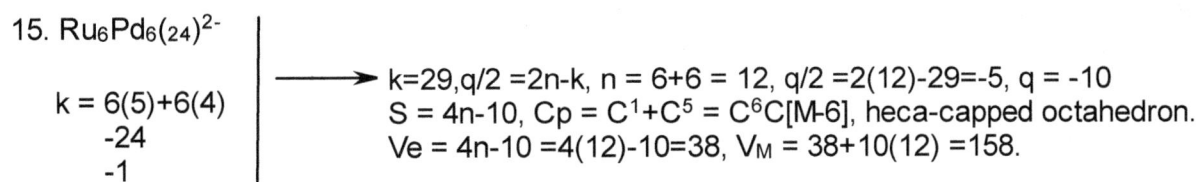

Scheme 18. Using skeletal numbers to derive series and valence electrons

Table 10. Derivation of 4n Series and Valence Electron Counts Using the k-Values of Clusters

Cluster	k Value	q value	Series (S)	Category	n value	Borane equivalent	Valence Electrons
$Fe_4(C)(CO)_{12}^{2-}$	5	6	4n+6	arachno	4	B_4H_{10}	22+40 =62
$(H)Os_3(CO)_{10}(AuL)$	7	2	4n+2	closo	4	$B_4H_4^{2-}$	18+40 =58
$(H)FeCo_3(CO)_9L_3$	6	4	4n+4	nido	4	B_4H_8	20+40 =60
$Re_4(H)_4(CO)_{12}$	8	0	4n+0	$C^1C[M-3]$	4	B_4H_4	16+40 =56
$Fe_4(CO)_4(\eta^5-C_5H_5)_4$	6	4	4n+4	nido	4	B_4H_8	20+40 =60
$Ir_4(CO)_{12}$	6	4	4n+4	nido	4	B_4H_8	20+40 =60
$Fe_5(N)(CO)_{14}^-$	8	4	4n+4	nido	5	B_5H_9	24+50 =74
$Ni_5(CO)_{12}^{2-}$	7	6	4n+6	arachno	5	B_5H_{11}	26+50 = 76
$Rh_5(CO)_{15}^-$	7	6	4n+6	arachno	5	B_5H_{11}	26+50 = 76
$Ni_3Cr_2(CO)_{16}^{2-}$	7	6	4n+6	arachno	5	B_5H_{11}	26+50 = 76
$Os_5(CO)_{16}$	9	2	4n+2	closo	5	$B_5H_5^{2-}$	22+50 =72
$Se_2Cr_3(CO)_{10}^{2-}$	9	2	4n+2	closo	5	$B_5H_5^{2-}$	22+30 =52
$PtRh_4(CO)_{12}^{2-}$	9	2	4n+2	closo	5	$B_5H_5^{2-}$	22+50 =72
$Se_2Fe_3(CO)_9$	8	4	4n+4	nido	5	B_5H_9	24+30 =54
$Ru_6(CO)_{18}^{2-}$	11	2	4n+2	closo	6	$B_6H_6^{2-}$	26+60 =86
$Fe_6(C)(CO)_{16}^{2-}$	11	2	4n+2	closo	6	$B_6H_6^{2-}$	26+60 =86
$Os_6(P)(CO)_{18}^-$	9	6	4n+6	arachno	6	B_6H_{12}	30+60 = 90
$Os_6(CO)_{18}$	12	0	4n+0	$C^1C[M-5]$	6	B_6H_6	24+60 = 84
$Os_4(CO)_{12}(H)_2(AuL)_2$	12	0	4n+0	$C^1C[M-5]$	6	B_6H_6	24+60 = 84
$Se_2Mn_4(CO)_{12}^{2-}$	11	2	4n+2	closo	6	$B_6H_6^{2-}$	26+40 =66
$Ru_6(C)(CO)_{17}$	11	2	4n+2	closo	6	$B_6H_6^{2-}$	26+60 =86
$Os_6(CO)_{18}^{2-}$	11	2	4n+2	closo	6	$B_6H_6^{2-}$	26+60 =86
$Rh_6(C)(CO)_{15}^{2-}$	9	6	4n+6	arachno	6	B_6H_{12}	30+60 = 90
$Pt_6(CO)_{12}^{2-}$	11	2	4n+2	closo	6	$B_6H_6^{2-}$	26+60 =86
$Co_6(C)(CO)_{15}^{2-}$	9	6	4n+6	arachno	6	B_6H_{12}	30+60 = 90
$Ru_6(C)(CO)_{16}^{2-}$	11	2	4n+2	closo	6	$B_6H_6^{2-}$	26+60 =86
$Os_7(CO)_{21}$	14	0	4n+0	$C^1C[M-6]$	7	B_7H_7	28+70 = 98
$Co_7(N)(CO)_{15}^{2-}$	13	2	4n+2	closo	7	$B_7H_7^{2-}$	30+70 =100
$Ru_4(CO)_{12}(AuL)_3(H)$	15	-2	4n-2	$C^2C[M-5]$	7	B_7H_5	26+70 = 96
$Rh_7(CO)_{16}^{3-}$	14	0	4n+0	$C^1C[M-6]$	7	B_7H_7	28+70 = 98
$Os_8(CO)_{22}^{2-}$	17	-2	4n-2	$C^2C[M-6]$	8	B_8H_6	30+80 = 110
$Ru_8(P)(CO)_{22}^-$	15	2	4n+2	closo	8	$B_8H_8^{2-}$	34+80 =114
$Re_8(C)(CO)_{24}^{2-}$	17	-2	4n-2	$C^2C[M-6]$	8	B_8H_6	30+80 = 110
$Cu_2Ru_6(C)(CO)_{16}$	19	-6	4n-6	$C^4C[M-4]$	8	B_8H_2	26+80 = 106
$Co_8(C)(CO)_{18}^{2-}$	15	2	4n+2	closo	8	$B_8H_8^{2-}$	34+80 =114
$Ni_8(C)(CO)_{16}^{2-}$	13	6	4n+6	arachno	8	B_8H_{14}	38+80 =118
Bi_8^{2-}	13	6	4n+6	arachno	8	B_8H_{14}	38+80 =118
$Co_6Ni_2(C)_2(CO)_{16}^{2-}$	14	-4	4n-4	$C^3C[M-5]$	8	B_8H_4	28+80 =108
$Ni_8(L)_6(CO)_8$	12	8	4n+8	hypho	8	B_8H_{16}	40+80 =120
$Ni_8(CO)_{18}^{2-}$	13	-2	4n-2	$C^2C[M-6]$	8	B_8H_6	30+80 = 110
Ge_9^{2-}	17	2	4n+2	closo	9	$B_9H_9^{2-}$	38
$Ni_9(CO)_{18}^{2-}$	17	2	4n+2	closo	9	$B_9H_9^{2-}$	38+90=128
$Rh_9(CO)_{19}^{3-}$	20	-8	4n-8	$C^5C[M-4]$	9	B_9H_1	28+90 =118
$Ni_9(C)(CO)_{17}^{2-}$	16	4	4n+4	nido	9	B_9H_{13}	40+90 =130
$Ni_9(CO)_{18}^{2-}$	17	2	4n+2	closo	9	$B_9H_9^{2-}$	38+90=128
L = PPh3							
$Ru_{10}(P)(CO)_{24}^{2-}$	23	-6	4n-6	$C^4C[M-6]$	10	$B_{10}H_4$	34+100 =134
$Os_{10}(C)(CO)_{24}^{2-}$	23	-6	4n-6	$C^4C[M-6]$	10	$B_{10}H_4$	34+100 = 134
$Rh_{10}(P)(CO)_{22}^{3-}$	19	2	4n+2	closo	10	$B_{10}H_{10}^{2-}$	42+100=148

3. Conclusion

A skeletal transition metal atom possesses inherent skeletal linkages. The linkages are derived from the valence electrons of the element. They correspond to the number of pairs of electrons needed to enable the metal atom obey the eighteen electron rule. The k-values derived are as follows: Group 3, Sc family, k=7.5; Group 4, Ti family, k=7.0; Group 5, V family, k=6.5; Group 6, Cr family, k=6; Group 7, Mn family, k=5.5; Group 8, Fe family, k=5, Group 9, Co family, k=4.5, Group 10, Ni family, k=4, Group 11, Cu family, k=3.5 and Group 12, Zn family, k=3. Ligands have been assigned negative k values as deduced from the 4n series. It is proposed that a single electron donor be assigned a k value of -0.5 and a two electron donor k=-1. The use of skeletal numbers greatly facilitates the categorization of simple to medium large clusters in a simple manner. Furthermore, it is possible to predict the shapes of some clusters. The skeletal numbers can also be utilized as a guide to assigning the ligands and charges to specific skeletal elements of clusters. The method makes the testing of the 18-electron rule, the understanding of some catalytic processes and the isolobal principle much easier. The skeletal values which have now been introduced for the first time in chemistry and the atoms of the main group and transition metal elements can be arranged into groups based on k values. Nearly 80 clusters of different types have been analyzed using skeletal numbers to demonstrate the ease and flexibility of applying skeletal numbers. This paper introduces a fundamental principle of viewing a naked skeletal cluster of elements as being a reservoir of inherent skeletal linkages which are subject to change when it is gets bound to electron donor ligands. The observed linkages or bonds are just remnants of those skeletal linkages which were not utilized by the ligands. This could be viewed as a form of conservation of skeletal cluster linkages.

Acknowledgements

The author wishes to express gratitude to the University of Namibia for the provision of facilities and NAMSOV, Namibia for the financial support, Mrs Merab Kambamu Kiremire for proof-reading the paper.

Dedication

This work is dedicated to Prof. Charles Alfred Coulson,(UK) whose brief teaching at Makerere University, Uganda in the late 1960s has had a life-long inspiration. His style of teaching valence in chemistry in a simple and inspiring manner left inerasable fond memories. I also wish to dedicate this work to Prof. Frank Bottomley who was my enthusiastic PhD mentor in inorganic chemistry at the University of New Brunswick, Fredericton, Canada in the late 1970s.

References

Butcher, C. P. G., Dyson, P. J., Johnson, B. F. G., Khimyak, T., & McIndoe, J. S. (2003). Fragmentation of Transition Metal Carbonyl Cluster Anions: Structural Insights from Mass Spectrometry. *Chem. Eur. J., 9*(4), 944-950. http://dx.doi.org/10.1002/chem.200390116

Butcher, C. P. G., Johnson, B. F. G., & McIndoe, J. S. (2002). Collision-Induced dissociation and photo-detachment of singly and doubly charged anionic polynuclear transition metal carbonyl clusters: $Ru_3Co(CO)_{13}^-$, $Ru_6(C)(CO)_{16}^{2-}$ and $Ru_6(CO)_{18}^{2-}$. *J. Chem. Phys., 116*(15), 6560-6566. http://dx.doi.org/10.1063/1.1462579

Cotton, F. A., & Wilkinson, G. (1980). *Advanced Inorganic Chemistry, 4th Ed.*, John Wiley and Sons, New York, 1980.

Crawford, E., Dyson, P. J., Forest, O., Kwok, S., & McIndoe, J. S. (2006). Energy-dependent Electrospray Ionization Mass Spectrometry of Carbonyl Clusters. *J. Cluster Science, 17(1)*, 47-63. http://dx.doi.org/10.1007/s10876-005-0043-8

Critchley, G., Dyson, P. J., Johnson, B. F. G., McIndoe, J. S., O'Reilly, R. K., & Langridge-Smith, P. R. R. (1999). Reactivity and Characterization of Transition Metal Carbonyl Clusters Using UV Laser Desorption Mass Spectrometry. *Organometallics, 18*, 4090-4097. http://dx.doi.org/10.1021/om990075f

Douglas, B., McDaniel, D., & Alexander, J. (1994). *Advanced Concepts and Models of Inorganic Chemistry, 3rd Ed.*, John-Wiley& Sons, New York.

Dyson, P. J., Hearly, A. K., Johnson, B. F. G., McIndoe, J. S., & Whyte, P. R. R. L. S. C. (2001). Combining energy-dependent Electrospray Ionization with tandem Mass Spectrometry for the analysis of inorganic compounds. *Rapid Commun. Mass Spectrom, 15*, 895-897. http://dx.doi.org/10.1002/rcm.314

Felner, T. P., Halet, J. F., & Saillard, J. Y. (2007). *Molecular Clusters, a bridge to Solid-State Chemistry,* Cambridge University Press, UK.

Freund, H. J., & Hohlneicher, G. (1979). Calculation of Transition Metal Compounds Using an Extension of the CNDO Formalism Method of calculation and Application to Mono and Di- and Tetracarbonyl Compounds. *Theoret. Chim. Acta(Berl.), 51*, 145-162. http://dx.doi.org/10.1007/BF00554098

Gardner, T. G., & Girolami, G. S. (1987). Seven Coordinate Titanium and Vanadium Carbonyl. Synthesis and X-ray Crystal Structures of [t-BuSi(CH$_2$PMe$_2$)$_3$]Ti(CO)$_4$ and [t-BuSi(CH$_2$PMe$_2$)$_3$V(CO)$_3$(H). *Organometallics, 6,* 2551-2556. http://dx.doi.org/10.1021/om00155a016

Greenwood, N. N., & Earnshaw, A. (1998). *Chemistry of the Elements, 2nd Ed.* Butterworth, Oxford.

Henderson, M. A., Kwok, S., & McIndoe, J. S. (2009). Gas-Phase Reactivity of Ruthenium Carbonyl Cluster Anions. *J. Am. Soc. Mass Spectrom., 20,* 658-666. http://dx.doi.org/10.1016/j.jasms.2008.12.006

Henderson, W., McIndoe, J. S., Nicholson, B. K., & Dyson, P. J. (1998). Electrospray Mass Spectrometry of Metal Carbonyl Complexes. *J. Chem. Soc., Dalton Trans.,* 519-525. http://dx.doi.org/10.1039/a707868d

Hoffmann, R. (1982). Building Bridges Between Inorganic and Organic Chemistry. *Angew. Chem. Int. Ed. Engl., 21,* 711-724. http://dx.doi.org/10.1002/anie.198207113

Housecroft, C. E., & Sharpe, A. G. (2005). Inorganic Chemistry, 2nd Edition, Pearson, Prentice-hall, Essex, England.

Hughes, H. K., & Wade, K. (2000). Metal-metal and metal-ligand bond strengths in metal carbonyl clusters. *Coord. Chem. Rev., 197,* 191-229. http://dx.doi.org/10.1016/S0010-8545(99)00208-8

Ignatyev, I. S., Schaefer, III, H. F., King, R. B., & Brown, S. T. (2000). Binuclear Homoleptic Nickel Carbonyls: Incorporating Ni-Ni Single, Double and Triple Bonds Ni$_2$(CO)x, x=5, 6, 7. *J. Am. Chem. Soc., 122,* 1989-1994. http://dx.doi.org/10.1021/ja9914083

Jemmis, E. D., & Balakrishnarajan, M. M. (2000). Electronic Requirements of Polyhedral Boranes. *J. Am. Chem. Soc., 122,* 4516-4517. http://dx.doi.org/10.1021/ja994199v

Jemmis, E. D., & Balakrishnarajan, M. M. (2001a). Polyhedral Boranes and Elemental Boron. Direct Structural Relations and Diverse Electronic Requirements. *J. Am. Chem. Soc., 123,* 4324-4330.

Jemmis, E. D., & Balakrishnarajan, M. M. (2002). Electronic Requirements of Macropolyhedral Boranes. *Chem. Rev., 102*(1), 93-144. http://dx.doi.org/10.1021/cr990356x

Jemmis, E. D., Balakrishnarajan, M. M., & Pancharatna, P. D. (2001b). A Unifying Electron-Counting Rule for Macropolyhedral Boranes, Metallaboranes and Metallocenes. *J. Am. Chem. Soc., 123*(18), 4313-4323.

Jensen, W. B. (1978). Electron-Orbital Counting Rules and Bonding Topology. *De Educacion Quimica, 3,* 210-222.

King, R.B. (1986). Metal Cluster Topology. Osmium Carbonyls. *Inorg. Chim. Acta, 116,* 99-107. http://dx.doi.org/10.1016/S0020-1693(00)82162-3

Kiremire, E. M. (2015a). Isolobal Series of Chemical Fragments. Orient. *J. Chem., 31*(Spl. Edn), 59-70.

Kiremire, E. M. R. (2015b). Capping and Decapping Series of Boranes. *Int. J. Chem., 7*(2), 186-197.

Kiremire, E. M. R. (2016a). Carbonyl Chalcogenide Clusters Existing as Disguised Forms of Hydrocarbon Isomers. *Int. J. Chem., 8*(3), 35-49.

Kiremire, E. M. R. (2016b). The Categorization and Structural Prediction of Transition metal Carbonyl Clusters Using 14n Series Numerical Method. *Int. J. Chem., 8*(1), 109-125.

Kiremire, E. M. R. (2016c). The Application of the 4n Series Method to Categorize Metalloboranes. *Int. J. Chem., 8*(3), 62-73.

Kiremire, E. M. R. (2016d). Unusual Underground Capping Carbonyl Clusters of Palladium. *Int. J. Chem., 8*(1), 145-157.

Kiremire, E. M. R. (2016e). Classification of Zintl Ion Clusters Using 4n Series Approach. *Orient. J. Chem., 32*(4), 1731-1738.

Lewis, J., & Johnson, B. F. G. (1982). The Chemistry of Some Carbonyl Cluster Compounds of Osmium. *Pure and Applied Chem., 54*(1), 97-112. http://dx.doi.org/10.1351/pac198254010097

Lipscomb, W. N. (1976). *Chemistry,* 224-245.

Miessler, G., Fischer, P., & Tarr, D. (2014). *Inorganic Chemistry, 5th Edition,* Pearson Education, Inc., Upper Saddle River.

Mingos, D. M. P. (1972). A General Theory for Cluster and Ring Compounds of the Main Group and Transition Elements. *Nature(London), Phys. Sci., 236,* 99-102. http://dx.doi.org/10.1038/physci236099a0

Mingos, D. M. P. (1984). *Polyhedral Skeletal Electron Pair Approach. Acc. Chem. Res., 17*(9), 311-319. http://dx.doi.org/10.1021/ar00105a003

Mingos, D. M. P. (1991). Theoretical aspects of metal cluster chemistry. *Pure and Appl. Chem., 83*(6), 807-812. http://dx.doi.org/10.1351/pac199163060807

Pauling, L. (1977). Structure of transition-metal cluster compounds: use of an additional orbital resulting from f, g character spd bond orbitals. *Proc. Natl. Acad. Sci. USA, 74*, 5235-5238. http://dx.doi.org/10.1073/pnas.74.12.5235

Rossi, F., & Zanello, P. (2011). Electron Reservoir Activity of High Nuclearity Transition Metal Carbonyl Clusters. *Portugaliae Electrochimica Acta, 29*(5), 309-327. http://dx.doi.org/10.4152/pea.201105309

Schaefer, III, H. F., & King, R. B. (2001). Unsaturated Binuclear Homoleptic Metal Carbonyls $M_2(CO)x$, M = Fe, Co, Ni, x = 5,6,7,8: Are multiple bonds between transition metals possible for these molecules?. *Pure Appl. Chem. 73*(7), 1059-1073. http://dx.doi.org/10.1351/pac200173071059

Shaik, S., Danovich, D., Wu, W., Su, P., Rzepa, H. S., & Hiberty, P. C. (2012). Quadruple bonding in C_2 and analogous eight valence electron species. *Nature Chemistry, 4*, 195-200. http://dx.doi.org/10.1038/nchem.1263

Shieh, M., Miu, C. Y., Chu, Y. Y., & Lin, C. N. (2012). Recent progress in the chemistry of anionic groups 6-8 carbonyl chacogenide clusters. *Coord. Chem. Rev., 256*, 637-694. http://dx.doi.org/10.1016/j.ccr.2011.11.010

Slee, T., Zhenyang, L., & Mingos, D. M. P. (1989). Polyhedral Skeletal Electron Pair Theory of Bare Clusters. Small Silicon Clusters. *Inorg. Chem.*, 2256-2261.

Teo, B. K., & Zhang, H. (1990). High Nuclearity Metal Clusters: Miniature Bulk of Unusual Structures and Properties. *J. Cluster Science, 1*(2), 155-186. http://dx.doi.org/10.1007/BF00702718

Teo, B. K., Longoni, G., & Chung, F. R. K. (1984). Applications of topological electron-counting theory to polyhedral metal clusters. *Inorg. Chem., 23*(9), 1257-1266. http://dx.doi.org/10.1021/ic00177a018

Wade, K. (1971). The structural significance of the number of skeletal bonding electron-pairs in carboranes, the higher boranes and borane ions and various transition metal carbonyl cluster compounds. *Chem. Commun.*, 792-793. http://dx.doi.org/10.1039/c29710000792

Wales, D. J. (2005). Electronic Structure of Clusters in *Encyclopedia of Inorganic Chemistry, 2nd Edition, Vol III. Edited*, R. B. King, John Wiley and Sons, Ltd., Chichester, UK, 1506-1525.

Wheeler, R. A., & Hoffmann, R. (1986). A New Magic Cluster Electron-Count and Metal-Metal Multiple Bonding. *J. Am. Chem. Soc., 108*, 6605-6610. http://dx.doi.org/10.1021/ja00281a025

Zanello, P. (2002). Structure and Electrochemistry of Transition Metal Carbonyl Clusters with Interstitial or Semi-Interstitial Atoms: Contrast between Nitrides or Phosphides and Carbides. Unusual Structures and Physical Properties in Organometallic Chemistry, Edited, Gielen, M., Willem, R., Wrackmeyer, John Wiley and Sons, Ltd,1-49.

8

Creation of Photoactive Inorganic/Organic Interfaces Using Occlusion Electrodeposition Process of Inorganic Nanoparticles during Electropolymerization of 2,2′:5′,2″-Terthiophene

Kasem K. Kasem[1], Christopher Santuzzi[1], Nick Daanen[1], Kortany Baker[1]

[1]School of Sciences, Indiana University Kokomo, Kokomo, IN, 46904, USA

Correspondence: Kasem K. Kasem, School of Sciences, Indiana University Kokomo, Kokomo, IN, 46904, USA.

E-mail: kkasem@iuk.edu

Abstract

Photoactive (IOI) inorganic/organic interface assemblies were prepared using an occlusion electrodeposition method. Poly-2,2′:5′,2″-Terthiophene (PTTh) were the organic thin films that occluded each of CdS, TiO_2, and Zn-doped WO_3 nanoparticles. The energy band gap structures were investigated using spectroscopic and electrochemical techniques. The obtained assemblies were investigated in aqueous solutions under both dark and illuminated conditions. The results were compared with the behavior of PTTh thin film. Oxygen played an important role in minimizing electron/hole recombination as was evident by observed very low photocurrent when oxygen was removed by nitrogen purge. Results show that PTTh/CdS gave the greatest photocurrent, followed by PTTh/Zn-WO$_3$ and PTTh/TiO$_2$.

Keywords: Photoelectrochemistry, Interface, Occlusion, Inorganic, Organic semiconductors

1. Introduction

Surface modification can create or eliminate defects and alter the energy band structure of the modified surface, consequently altering the donor /acceptor character of the modified surfaces. Some modifications require co-immobilization of several substances on the electrode surface. However, when the semiconductor's particle preparation and immobilization/deposition steps are separated, the intrinsic nature of the semiconductor particles (e.g., size, morphology, and crystal structure) are greatly altered or compromised by the deposition process itself. A search for simple and direct methods for immobilization is very important (Withers J. C et al 1961). One of the simple methods with fewer steps that can enhance intrinsic properties is occlusion. Occlusion involves immobilization of semiconductor particles in the matrix of an electrochemically synthesized substance.

Composite films containing occluded TiO_2 (Tomaszewski et al 1963, Tomaszewski et al 1969, and Hovestad A. et al 1995) or CdS (Tacconi NR de et al 1997) particles in a Ni matrix were prepared. Other metal matrices such as Ni, Cu, Ag, In (Zhou M. et al 1996) or in poly-pyrrole (Beck P. et al 1992) have been utilized for immobilizing the TiO_2 particles.

Occlusion electrodeposition of ZnO and carbon nanotubes has been reported (Haining Chen, et al 2011). Further studies (Santos, M. J. L et al 2009) show that CdS and CdS/ZnS enhance the photoelectrochemical behavior of poly-terthiophene. Poly-terthiophene was used to modify the surface of CdS nanoparticles by adsorption of the monomer on CdS particles, followed by photopolymerization of the adsorbed monomer. The modified CdS particles were studied in nanopowder form or in solid thin film composite over ITO (Kasem K. et al 2012). However the modification of CdS surface involved several separate steps (especially the photopolymerization step) that may negatively affected the intrinsic properties of the CdS/PTTh assembly.

In this paper we explored the use of the occlusion electrodeposition (OE) process for creation of an Inorganic /organic interface (IOI) between poly terthiophene (PTTh) and CdS, TiO_2, and Zn-WO$_3$ (Zn-doped WO$_3$) to judge the effects on the photoelectron-chemical behavior of the created IOI assemblies.

2. Experimental

2.1 Reagents

All the reagents were of analytical grade. All of the solutions were prepared using deionized water, unless otherwise stated.

2.2 Preparations

A- Electropolymerization of PTTh:

Polymer thin films were generated electrochemically using cyclic voltammetry (CV) technique by repetitive cycling of the FTO electrode potential at a scan rate 0.10V/s between -1.0 and 2.0 V vs Ag/AgCl in acetonitrile 1 mM of the monomer and 0.5M LiClO$_4$.

B- (Occlusion Method):

Thin films of IOI assemblies were generated electrochemically using cyclic voltammetry (CV) technique by repetitive cycling of the FTO electrode between -1.0 and 2.0V vs Ag/AgCl in acetonitrile suspension of each CdS, TiO$_2$ and Zn-WO$_3$ and 1.0 mM of the monomer and 0.5M LiClO$_4$.

C- Electrodeposition of CdS thin solid film:

CdS was prepared in two steps: first, Cd was deposited by applying -0.850V vs Ag/AgCl on a FTO working electrode in three-electrode cell containing 0.2 M CdSO$_4$ for 10 minutes, then thin solid film of Cd was deposited on a FTO and transferred to another electrochemical cell containing 2.0 M Na$_2$S, where it was subjected to a linear sweep volumetric between -1.0 V and 0.700 V vs Ag/AgCl at 0.01V/s scan rate. A Canarias yellow film appeared after the first scan. The generated CdS film was annealed at 150 °C for 60 minutes.

2.3 Instrumentation

All electrochemical experiments were carried out using a conventional three-electrode cell consisting of a Pt wire as a counter electrode, Ag/AgCl as a reference electrode, and FTO as the working electrode. Photoelectrochemical studies on thin solid films were performed on an experimental set up as illustrated in Figure 1 inset C. A BAS 100W electrochemical analyzer (Bioanalytical Co.) was used to perform the electrochemical studies. Steady state reflectance spectra were performed using Shimadzu UV-2101 PC. Irradiation was performed with a solar simulator 300 watt xenon lamp (Newport) with an IR filter.

3. Results and Discussion

3.1 Occlusion Electrodeposition of CdS, Zn-WO$_3$, and TiO$_2$ Nanoparticles in PTTh

Repetitive cycling of the FTO electrode between -1.0 and 2.0 V vs Ag/AgCl in an acetonitrile suspension of CdS nanoparticles and 1.0 mM of the monomer TTh and 0.5M LiClO$_4$ generated a homogenous thin film where CdS particles were entrapped into the film's matrix. Figure 1A shows a positive shift of the anodic peak corresponding to the polymerization of TTh monomer. Furthermore, the growth of both anodic and cathodic peaks current is indication of film build up. The coexistence of insoluble CdS nanoparticles with the monomer did not affect the electrochemical behavior of the monomer as CdS nanoparticles are electrochemically inactive under the experimental conditions. The occlusion electrodeposition of Zn-doped WO$_3$ was similar to that displayed in Figure 1A. On the other hand, the CV for occlusion of TiO$_2$ nanoparticles is displayed in Figure 1B, which again show similar behavior to that shown in Figure 1A. This is consistent with the facts that both Zn-Doped WO$_3$ and TiO$_2$ nanoparticles are electrochemically inactive under the experimental conditions.

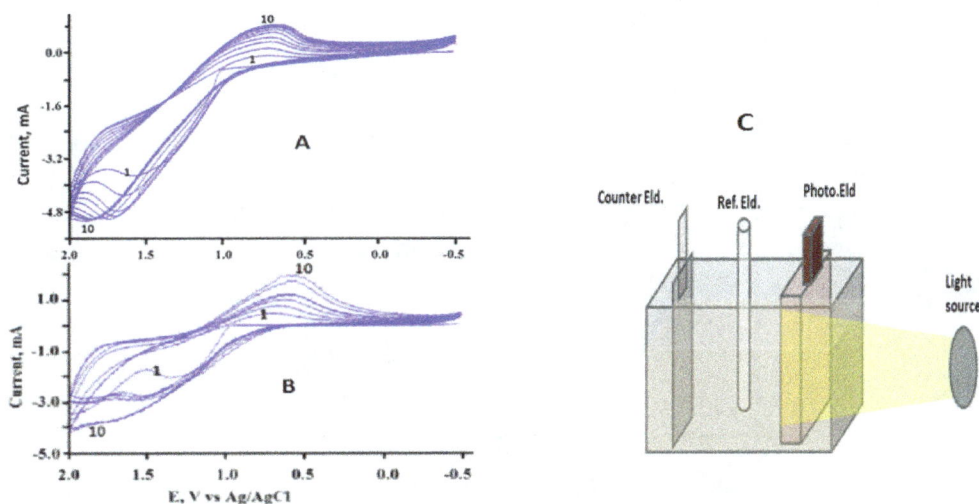

Figure 1. Electrochemical occlusion deposition of A) CdS nanoparticles, and B) TiO$_2$ during the polymerization of terthiophene, and C) a diagram for the Photoelectrochemical cell.

3.2 Absorption Spectra of PTTh/ Inorganic Assemblies Interfaces

Absorption spectra of the polymer PTTh and its assemblies with, CdS and Zn-Doped WO_3 were studied and the results are displayed in Figure 2. Figure 2b and 2c show the overlapping absorption peaks around the approximate band gaps of both CdS, Zn-WO_3 and PTTh. On the other hand Figure 2a shows a gradual increase of absorption that reaches maximum around TiO_2 band gap (≈ 3.1 eV). Furthermore, Figure 2 shows more absorption of a visible region. This can be attributed to fact that each of the occluded materials acted as absorbent beside the host polymer, resulting in the broad absorption spectra shown in Figure 2. The absorption spectra of these assemblies were subject to further analysis using Tauc equations (Robert, V. L. et al 1995, and Bhatt R. 2012). The results are listed in Table 1. The data listed in Table 1 suggest the existence of direct and indirect band gaps structures in these assemblies. The existence of a direct and indirect band gaps suggests the possibility of creating a hybrid band at the organic/inorganic interface. This would allow for band alignment (Chiatzun G. et al 2007, and Blumstengel S. et al 2008) that facilitates better charge separation and transfer processes, further leading to more efficient photochemical process.

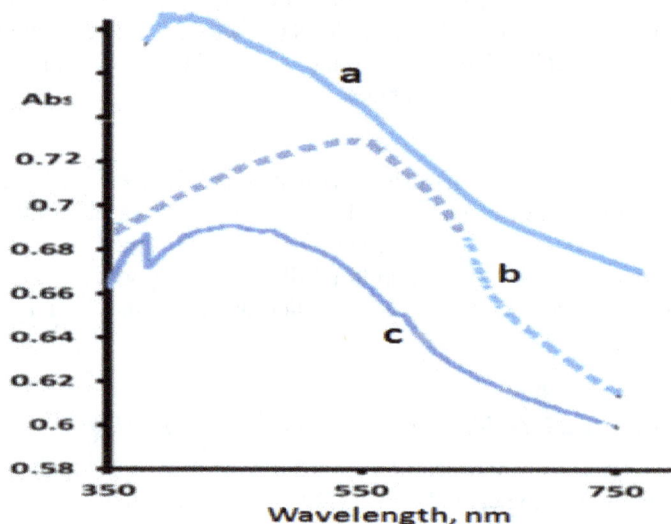

Figure 2. Absorption spectra for PTTh Occluded with a) TiO_2, b) Zn-WO_3, and c) CdS

3.3 Electrochemical Behavior of the FTO/PTTh /Inorganic Assemblies in Aqueous Phosphate Buffer

The electrochemical behaviors of FTO/PTTh/ with each of TiO_2, CdS and Zn-WO_3 CdS were investigated by cycling the potential of FTO modified with each OII assembly in the dark and light at a scan rate 0.050V/s, between -1.2 to 1.8 V vs Ag/AgCl in nitrate buffer (pH 7). Figure 3A shows that the photocurrent recorded for PTTh/TiO_2

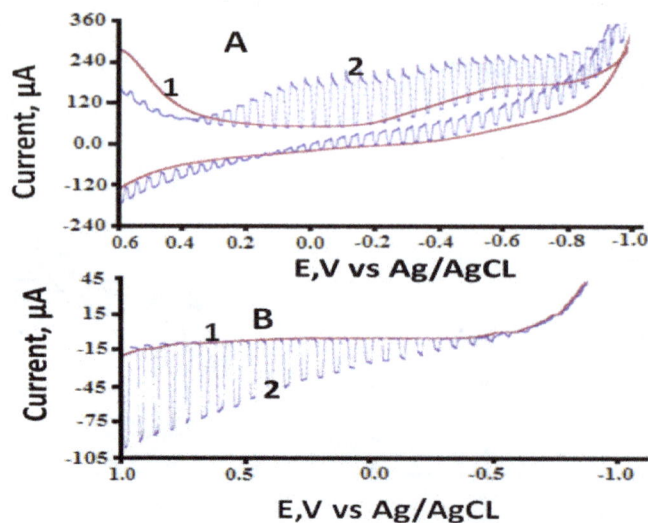

Figure 3. CV in 0.1 M KNO_3 scan rate 50 mV/S of A) FTO/PTTh/TiO_2, and B) FTO/TiO_2 1: CV Under Dark, 2):CV during illumination.

assembly is greater than that recorded under the dark conditions in both the cathodic scan between at ≈0.30 vs Ag/AgCl and the anodic scan at ≈ 0.2V vs Ag/AgCl. These potential values are more positive than that recorded for TiO_2 film only (Figure 3 B). These observations indicate that the approximate E_{fb} (flat band potential) of the assembly is at ≈ 0.2 V vs Ag/AgCl or 0.4 V vs SHE. We infer that is the value of the hybrid sub-band created upon occlusion of TiO_2 nanoparticles in PTTh. Likewise, Figures 4 and 5 show the photocurrent recorded for PTTh/ CdS (Figure 4) and PTTh/Zn-WO_3 (Figure 5) indicating that the E_{fb} in the assembly of PTTh and occluded nanoparticles are shifted to more positive potentials than in the cases of the thin films of CdS or Zn-WO_3 respectively. This again indicates the formation of hybrid sub band in these assemblies.

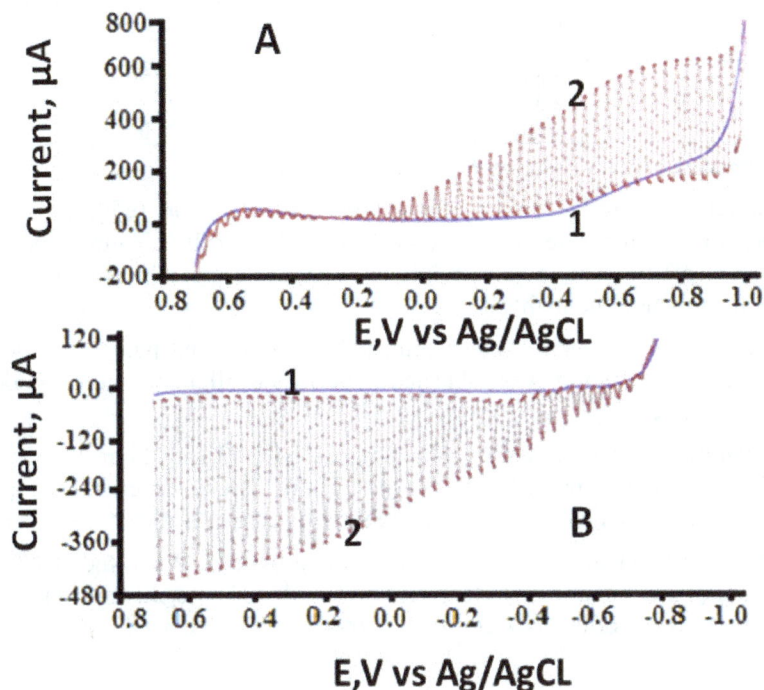

Figure 4. CV in 0.1 M KNO_3 scan rate 50 mV/S of A) FTO/PTTh/CdS, and B) FTO/CdS 1: CV in Dark, and 2: CV during illumination.

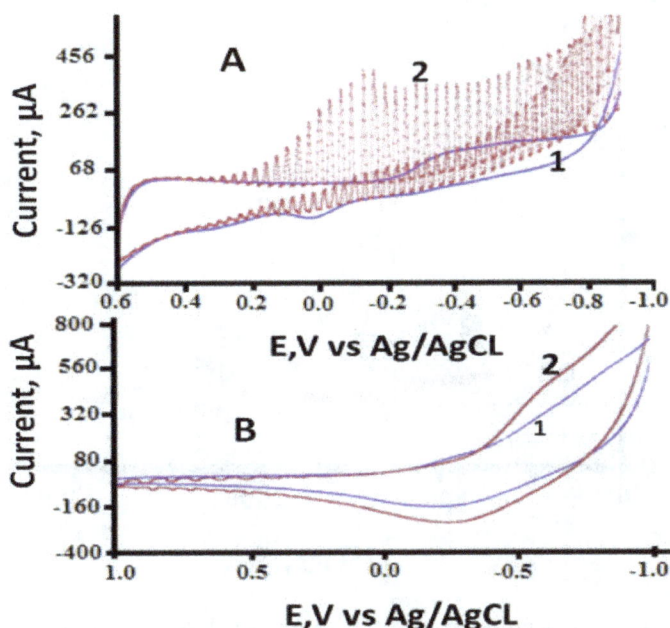

Figure 5. CV in 0.1 M KNO_3 scan rate 50 mV/S of A) FTO/PTTh/Zn-WO3, and B) FTO/Zn-WO_3 1: CV Under Dark, and 2: CV during illumination.

Because phosphate buffer is widely used in photoelectrochemical studies and the role of HPO_4^{2-} as a hole scavenger was previously studied (Abdullah M. et al 1990) , cyclic voltammetry studies was performed for the assemblies FTO/PTTh (TiO_2, CdS or Zn-WO$_3$) in phosphate buffer. These CV were similar but not identical to those displayed in figures 3,4, and 5. (CVs are not shown) as large photocurrent was observed in during the cathodic scan between 0.2 and -1.0 V. Table 1 lists the approximate flat band potential for each of these assemblies in KNO_3 electrolyte. It can be noticed that the occlusion process causes a positive shift in the flat band potential of the inorganic semiconductor when it occluded by PTTh film. This can be attributed to the role of the PTTh in creation of a hybrid sub-band at these assemblies/electrolyte interface.

Table 1. Band gap data and flat band potential (in KNO_3) for the studies at O/I/I

Assembly	Band Gap, eV	Direct band Gap, eV	Indirect band gap, eV	E_{fb} vs Ag/AgCl	E_{fb} pot. For inorganic only
PTTh/CdS	2.45	2.32	2.770	0.180 V	-0.70 V
PTTh/TiO$_2$	3.1	2.5, 2.8	3.1	0.30 V	-0.60 V
PTTh/Zn-doped WO$_3$	2.7	2.68	2.7	0.30 V	-0.40 V

3.4 Role of Oxygen in Suppression of Hole/Electron Recombination

As the generated photocurrent is a reflection of efficient charge separation (e/h), it is important to study the photocurrent –time relationship by monitoring the photocurrent under longer illumination time at constant potential. We have chosen phosphate buffer (pH 6) to be the electrolyte to study these assemblies photocurrent-time behavior. We performed these studies in the presence and in the absence of O_2.

In the photocurrent –time study, each assembly was illuminated under constant potential which was identified by the CV under illumination studies. The assembly was subjected to periods of illumination and darkness and the generated photocurrent was recorded over time.

Figure 6 shows photocurrent-time curve generated by subjecting the FTO/PTTh/CdS assemblies to a constant potentials under illumination for a long duration. Upon illumination of FTO/PTTh/CdS assemblies (Figure 6A) in oxygenated (aired) phosphate buffer (pH 6), a sudden increase in the current followed by slow increase in the

Photocurrent, as indicated by the formation of a positively slopping plateau. When the electrolyte was deoxygenated (after N_2 purge), the illumination generated less photocurrent and exhibited a negatively sloping plateau (Figure 6B red trace) with a steady value closer to that of the dark current was recorded. Such behavior was

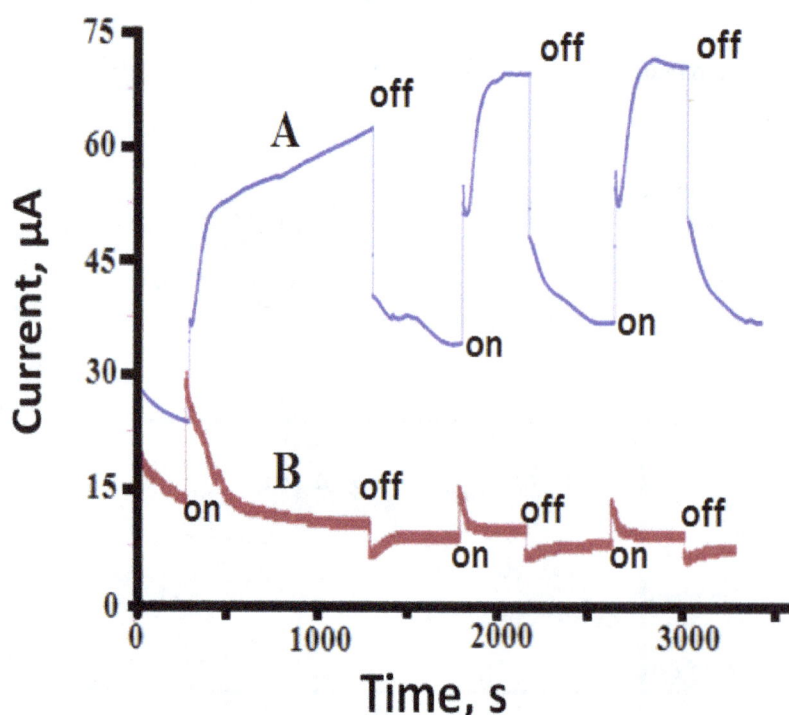

Figure 6. FTO/PTTh//CdS in Phosphate buffer at -0.600 V vs Ag/AgCl, A) In presence of O_2, and B) Deoxygenated solution in the presence of N_2.

reproducible as indicated during the illumination/dark processes. These photocurrent curves generated in presence and in absence of O_2, indicate again that O_2 plays an important role in enhancing charge separation, as demonstrated by the steady increase in photocurrent during the illumination period. In the case of FTO/PTTh/TiO$_2$ assembly (Figure 7A), in oxygenated phosphate buffer, a sudden increase in the current was followed by a slow increase in the photocurrent. When the electrolyte was deoxygenated (Figure 7B), the illumination generated a very low photocurrent with a slow increase in its value.

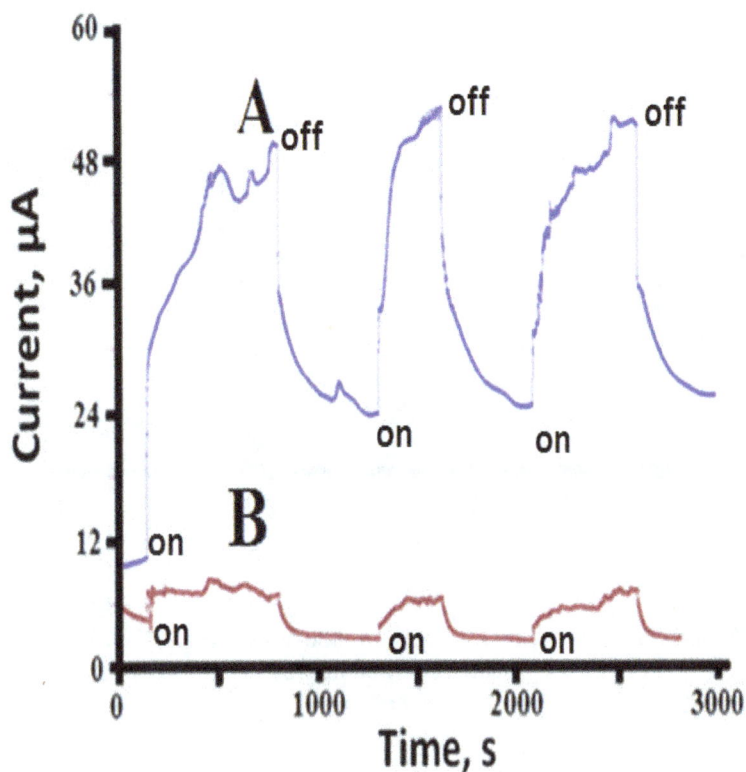

Figure 7. FTO/PTTh//TiO2 in Phosphate buffer at-0.580 V vs Ag/AgCl. A) In presence of O_2 and B) in deoxygenated solution in presence of N_2.

The behavior of FTO/PTTh/Zn-WO$_3$ in phosphate buffer in the presence (Figure 8A) and absence of oxygen (Figure 8B) was more or less similar to that of FTO/PTTh /CdS. It is apparent from these data that the presence of oxygen plays important role in the mechanism of charge separation and transfer at the IOI/ electrolyte interface.

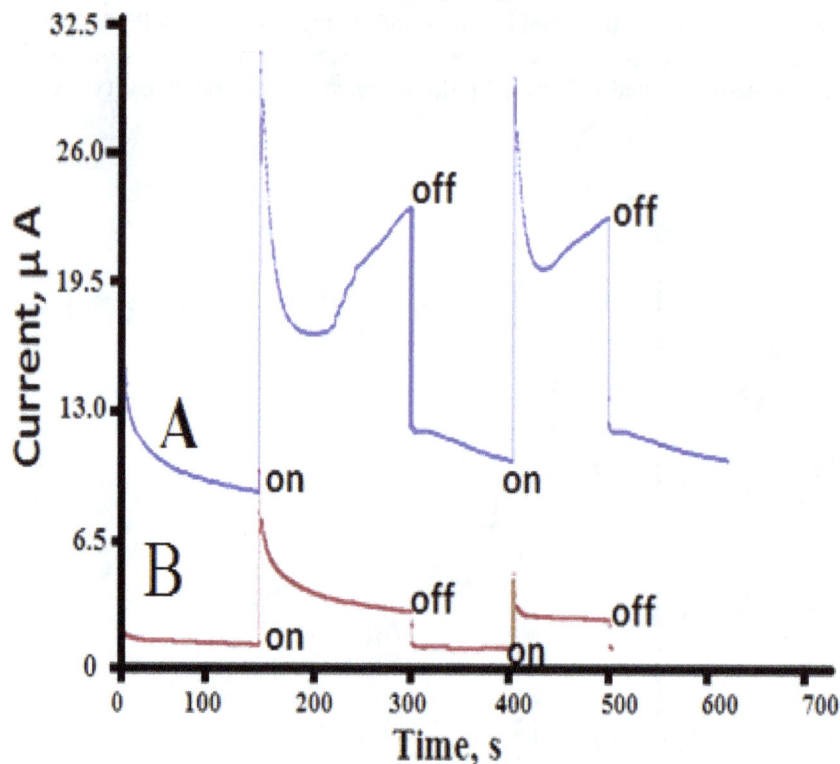

Figure 8. FTO/PTTh/Zn-WO3 in Phosphate buffer at -0.500 V vs Ag/AgCl. A) In presence of O_2 and B) in Deoxygenated solution in presence of N_2.

4. Conclusion

A photoactive inorganic /organic interface was created using occlusion electrodeposition. The reproducible photoactivities of the studied assemblies suggests that occlusion electrodeposition method is reliable and effective method for the creation of inorganic /organic interfaces. The magnitude of the photocurrent produced by each assembly is controlled by the band alignments between each assembly components. Creation of hybrid sub band between PTTh and inorganic compound was suggested as supported by previous studies (Abdullah M et al, 1990, and Chiatzun G. et al 2007).

Acknowledgment

The authors acknowledge the great contribution of Indiana University Kokomo that supported this research work.

References

Abdullah, M., Low, G. K. C., & Matthews, R. W. (1990). Effects of common inorganic anions on rates of photocatalytic oxidation of organic carbon over illuminated titanium dioxide. *J. Phys. Chem., 94*(17), 6820. http://dx.doi.org/10.1021/j100380a051

Beck, P., Dahhaus, M., & Zahedi, N. (1992). Anodic codeposition of polypyrrole and dispersed TiO_2. *Electrochem. Acta, 37*, 1265. http://dx.doi.org/10.1016/0013-4686(92)85066-T

Bhatt, R., Bhaumik, J., Ganesamoorthy, S., Karnal, A. K., Swami, M. K., Patel, H. S., & Gupta P. K. (2012). Urbach tail and bandgap analysis in near stoichiometric $LiNbO_3$ crystals. *Physics Status Solidi A, 209*(1), 176-180. http://dx.doi.org/10.1002/pssa.201127361

Blumstengel, S., Sadofev, S., Xu, C., Puls, J., Johnson, R. L., Glowatzki, H., Koch, N., & Henneberger, F. (2008). Electronic coupling in organic-inorganic semiconductor hybrid structures with type-II energy level alignment. *Physical Review* B, *77*, 085323. http://dx.doi.org/10.1103/PhysRevB.77.085323

Chiatzun, G., Scully, S. R., & McGehee, M. D. (2007). Effects of molecular interface modification in hybrid organic-inorganic photovoltaic cells. *Journal of Applied Physics, 101*, 114503. http://dx.doi.org/10.1063/1.2737977

de Tacconi, N. R., Wenren, H., & Rajeshwar, K. (1997). Photoelectrochemical behavior of nanocomposite films of cadmium sulfide, or titanium dioxide, and nickel. *J. Electrochem. Soc.*, *144*(9), 3159-3163. http://dx.doi.org/10.1149/1.1837975

Hovestad, A., & Janssen, L. J. J. (1995). Electrochemical codeposition of inert particles in a metallic matrix. *J. of Applied Electrochem.*, *25*, 519-527. http://dx.doi.org/10.1007/BF00573209

Chen, H. N., Li, W. P., Hou, Q., Liu, H. C., & Zhu, L. Q. (2011), A general deposition method for ZnO porous films: Occlusion electrosynthesis., *Electrochimica Acta, 56*, 9459–9466. http://dx.doi.org/10.1016/j.electacta.2011.08.037

Kasem, K., & Zia, N. (2012), Photoelectrochemical studies at CdS/PTTH nanoparticles interface. *Mat. Sciences and application, 3*, 719-727.

Robert, V. L., Hung, C. J., Kammler, D. R., & Switzer, J. A. (1995). Optical and Electronic Transport Properties of Electrodeposited Thallium (III) Oxide Films. *J. Phys. Chem., 99, p* 15247.

Santos, M. J. L., Ferreira, J., Radovanovic, E., Romano, R., Alves, O. L., & Girotto, E. M. (2009). Enhancement of the photoelectrochemical response of poly (terthiophenes) by CdS(ZnS) core-shell nanoparticles. *Thin Solid Films, 517*(18), 5523-5529. http://dx.doi.org/10.1016/j.tsf.2009.03.170

Tomaszewski, T. W., Clauss, R. J., & Brown, H. (1963). Satin nickel by codeposition of finely dispersed solids. *Tech. Proc. Am. Electroplater's Soc., 50*, 169.

Tomaszewski, T. W., Tomaszewski, L. C., & Brown, H. (1969). Codeposition of finely dispersed particles with metals. *Plating, 56*, 1234 .

Withers, J. C. (1961). Protective Coatings for Refractory Metals. *Tech Rep, 60.*

Zhou, M., Lin, W. Y., Tacconi, N. R., & Rajeshwar, K. (1996). Metal /semiconductor electrocomposite photoelectrodes. *J. Electroanal Chem., 402*, 221-224. http://dx.doi.org/10.1016/0022-0728(95)04368-3

9

Density-Functional Study of Structural and Electronic Properties of the $Zr_nAl^{\pm m}$ Clusters

Jun-Zai Yu[1], Feng-Qi Zhao[2], Si-Yu Xu[2], Lu-Jing Sun[3], Xue-Hai Ju[1]

[1]Key Laboratory of Soft Chemistry and Functional Materials of MOE, School of Chemical Engineering, Nanjing University of Science and Technology, Nanjing 210094, P. R. China

[2]Laboratory of Science and Technology on Combustion and Explosion, Xi'an Modern Chemistry Research Institute, Xi'an 710065, P. R. China

[3]School of Chemistry and Materials Science, Nanjing Normal University, Nanjing 210094, P. R. China

Correspondence: Xue-Hai Ju, School of Chemical Engineering, Nanjing University of Science and Technology, Nanjing, P. R. China. E-mail: xhju@njust.edu.cn

Abstract

The geometries, stabilities, and electronic properties of $Zr_nAl^{\pm m}$ ($n = 1 - 7$ and $m = 0, 1$) clusters were investigated at the UB3LYP/LANL2DZ level. The variations of structural and electronic properties with the changes of n and m were probed. Several possible multiplicities of each cluster were tested. The multiplicity of the most stable neutral clusters is 4 instead of 2. For all the three differently charged of Zr_nAl clusters, the lowest-energy geometry is in favor of three-dimensional structure when $n \geq 3$. The Zr_3Al^+, Zr_4Al^-, Zr_5Al^+, Zr_6Al^- and Zr_7Al clusters are more stable than their corresponding differently charged species of the same size. Moreover, the odd-even oscillations are found in the fragmentation energy and the second-order difference of total energies for Zr_nAl^- clusters. The Zr_2Al^+ cluster is more inert to chemical reaction than others judged by the HOMO-LUMO gaps. NBO analysis was done to analysis the electronic properties.

Keywords: $Zr_nAl^{\pm m}$ clusters, UB3LYP/LANL2DZ, odd-even oscillations, NBO, three-dimensional structures

1. Introduction

Recently, a large number of experimental and theoretical studies of clusters were performed (Alexandrova et al, 2004),(Zhai et al, 2003), (Lei, 2011), (Hua et al, 2013), (Addicoat et al, 2007), because they have many unique physical and chemical properties in the terms of the geometry and electronic properties (Schmidt et al, 1998), (Herry, 2012), but also have potential applications in catalysis (Yamazoe et al, 2014), (Tang et al, 2014) , hydrogen storage (Ramos-Castillo et al, 2015), (Wu et al, 2015) *etc*. Moreover, the study of clusters plays a key role in understanding the growth behavior of microscopic particles of their bulk.

As we all know, zirconium atom is a rare metal, a $4d$ transition-metal (TM), and has an electronic configuration of $4d^25s^2$. The Zr material used in the nuclear industry for cladding fuel elements, and own to a lower absorption cross section for neutrons. It is very resistant to corrosion by many common acids, alkalis, and sea water. Therefore, the metal, which is utilized as an alloying agent in steel and for making surgical appliances, is developed extensively by the chemical industry where corrosive agents are employed (Zhao et al, 2009), (Wang, 2006). Due to these special properties of TM Zr clusters, a number of research groups have been striving to investigate the geometrical structures and electronic properties for X-doped (X = metals) zirconium and the pure zirconium clusters recently. (Zhao et al, 2009), (Wang, 2006), (Sengupta et al, 2016), (Yang et al, 2008). (Lekka, 2010) investigated the bonding characteristics and mechanical properties of Cu–Zr and Cu–Zr–Al clusters by density functional theory (DFT), they found the most abundant microstructural units on the $Cu_{60}Zr_{40}$ cluster. Zhao and co-workers studied the structural, electronic, and magnetic properties of the Zr_nCr ($n = 2–14$) clusters, showed that the Zr_6Cr, Zr_8Cr and $Zr_{12}Cr$ clusters are more stable than their neighbors. (Zhao et al, 2009). The chemisorption of molecular hydrogen on small Zr_n clusters ($n = 2–15$) was performed by (Sheng et al, 2008). The preferred adsorption sites for H_2 reacting with the Zr_n clusters are the bridge sites. As for Al-doped Zr clusters, (Du et al, 2010) were studied the geometrical and electronic properties of neutral $Zr_{n-1}Al$ clusters and the pure Zr_n clusters ($n = 2–8$) with hybrid HF/DFT functional. From the above reports, although there are a lot of researches for X-doped Zr clusters, but few researches are systematically performed for the neutral and positively/negatively charged Al-doped zirconium clusters. What is the difference between the neutral Zr_nAl and $Zr_nAl^{\pm}$ clusters? Can we find the "odd-even alteration" phenomenon in $Zr_nAl^{\pm m}$ clusters as in MgB_n clusters (Wu et al, 2014) for some properties? To explore these, we investigated the geometric structures, stabilities and electronic properties of $Zr_nAl^{\pm m}$ ($n = 1-7$ and $m = 0, 1$) clusters.

2. Computational Methods

All the clusters were optimized by the B3LYP method (Lee et al, 1988), (Becke, 1993) in combination with the LANL2DZ basis set (Hay & Wadt, 1985). This basis set is modified by the relativistic effective core potential, therefore, is suitable for the transition metals (TM) Zr clusters (Yao et al, 2008), (Ge et al, 2012). In order to check the correctness of this method used for the investigation of the Zr_nAl clusters, we first accomplished the calculation on Al_2 and Zr_2 dimers. The bond length of the Al_2 (2.64 Å) and Zr_2 (2.33 Å) dimers are in nice accordance with experimental value of 2.70 Å (Fu et al, 1990) and 2.24 Å (Doverstål et al, 1998), respectively.

We have also considered the spin multiplicities for the initial configurations of $Zr_nAl^{\pm m}$ clusters ($n = 1 - 7$ and $m = 0, 1$). To investigate the relative stability of differently charged Zr_nAl clusters, we calculated the average binding energy, fragmentation energy and the second-order difference in total energies. The average binding energy (E_b) for $Al_{13}B_n$ clusters can be defined by the following formula:

$$E_b(n)^{\pm m} = [\, nE(Zr) + E(Al)^{\pm m} - E(Zr_nAl)^{\pm m}]/(n+1) \tag{1}$$

The fragmentation energy (E_f) can be defined by the following formula:

$$E_f(n)^{\pm m} = E(Zr) + E(Zr_{n-1}Al)^{\pm m} - E(Zr_nAl)^{\pm m} \tag{2}$$

The second-order difference of total energies (Δ_2E) can be defined by the following formula:

$$\Delta_2E(n)^{\pm m} = E(Zr_{n+1}Al)^{\pm m} + E(Zr_{n-1}Al)^{\pm m} - 2E(Zr_nAl)^{\pm m} \tag{3}$$

where $E(Zr_{n+1}Al)^{\pm m}$, $E(Zr_{n-1}Al)^{\pm m}$ and $E(Zr_nAl)^{\pm m}$ represent the energies of the most stable of $Zr_{n+1}Al^{\pm m}$, $Zr_{n-1}Al^{\pm m}$ and $Zr_nAl^{\pm m}$ clusters, respectively. $E(Zr)$ and $E(Al)^{\pm m}$ represent the total energies of the Zr and $Al^{\pm m}$ atoms, respectively. For the electronic properties, we calculated HOMO-LUMO (highest occupied molecular orbital-lowest unoccupied molecular orbital) gaps energies and the chemical hardness for the most stable $Zr_nAl^{\pm m}$ clusters. Chemical hardness is defined as the resistance of chemical potential to a change in the number of electrons: (Pearson, 2005), (Suh et al, 2015)

$$\eta = (I - A)/2 \tag{4}$$

where I and A are the ionization potential and electron affinity, respectively.

The natural bond orbital (NBO) (Carpenter & Weinhold, 1988), (Reed et al, 1988) analysis was carried out for the most stable structures in order to study the chemical bonding characteristics. All the computations were performed through Gaussian 09 package (Frisch et al, 2010).

3. Results and Discussion

3.1 Stable Geometric Structure

The fully optimized structures with lowest-energy and low-lying metastable isomers of $Zr_nAl^{\pm m}$ clusters were shown in Figure 1. Through this figure we can fully understand the characteristics of different charged Zr_nAl clusters enough. For $n = 2$, the lowest-energy structures of $AlZr_2$ and $AlZr_2^-$ are isosceles triangle (C_{2v}), whose apex angle (Zr-Al-Zr) are 50.26° and 53.75°, respectively, and the Al–Zr bond length are slightly different (See Table 1). While the most stable structure of Zr_2Al^+ cluster is a chain, the energy of Zr_2Al^+ cluster is 0.09 eV lower than its isosceles triangle structure. This phenomenon of structures for $Zr_2Al^{\pm m}$ clusters is similar to $B_2Mg^{\pm m}$ clusters (Wu et al, 2014). As for $Zr_3Al^{\pm m}$ clusters, we got the distorted tetrahedron geometry (3a) with C_1 symmetry for Zr_3Al cluster, but the lowest-energy structures of $Zr_3Al^{\pm}$ clusters are trigonal pyramid with C_{2v} symmetry. The second most stable structure for the cation Zr_3Al^+ cluster is a quadrilateral structure with C_{2v} symmetry whose total energy is 1.04 eV higher than the lowest-energy structure. We also tried to construct some planar and quasi-planar structures as initial configuration, but these initial configurations change to the three-dimensional (3D) structures for Zr_3Al and Zr_3Al^- clusters during the geometrical optimization. The V-like structures appear in Zr_3Al and Zr_3Al^- clusters, whose energy are 0.11 eV and 0.36 eV larger than the most stable ones, respectively.

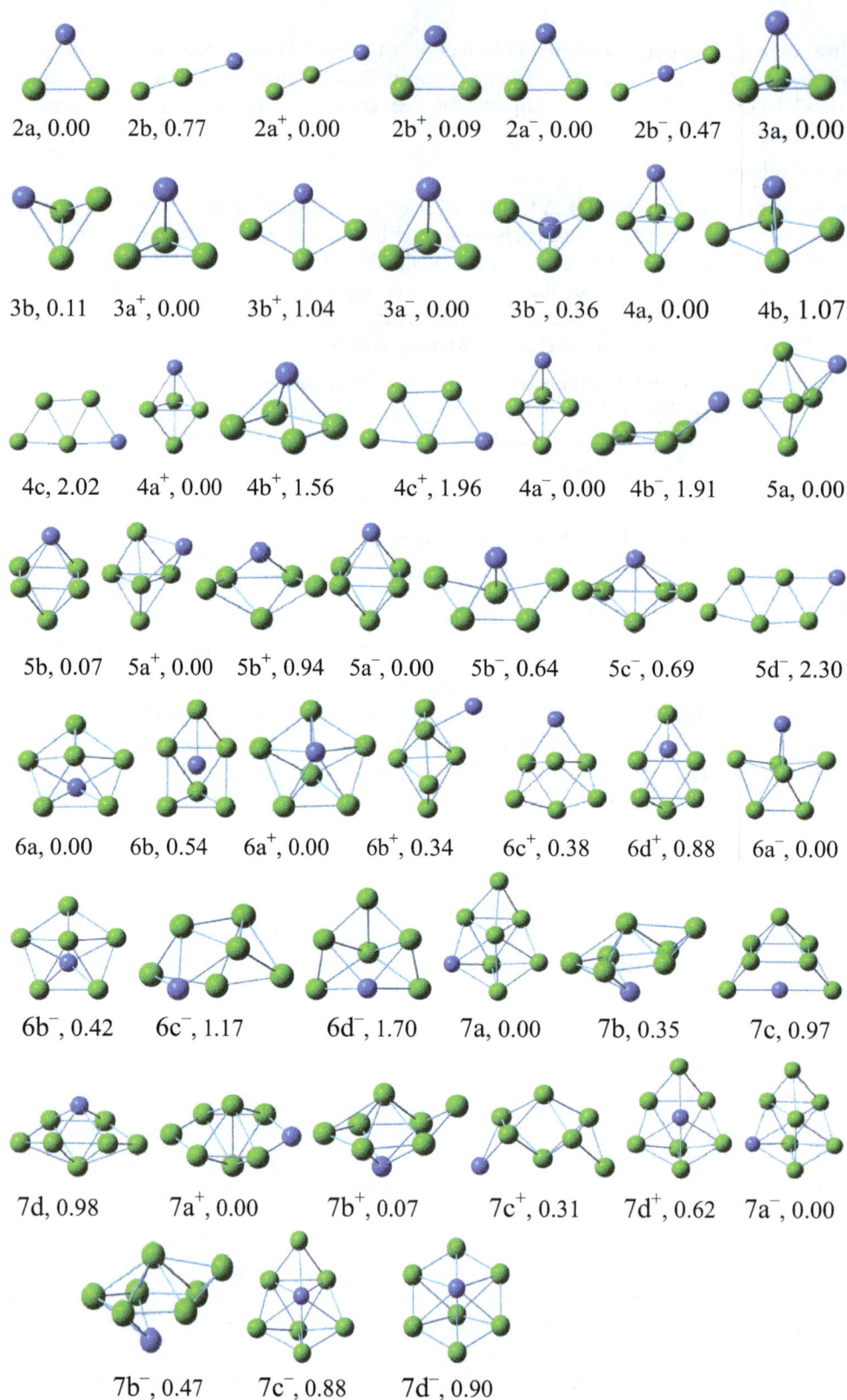

Figure 1. Lowest-energy and low-lying structures of $Zr_nAl^{\pm m}$ ($n = 1 - 7$ and $m = 0, 1$) clusters. (The first digit denotes the number of zirconium, the letter (a, b, c, d) is the structure label (from lower to higher energy), superscript + or - denotes a positive/negative charge, and the 0.00 represents the relative energy in eV. Others are similar)

Table 1. Shortest Zr-Al bond lengths ($R_{\text{Zr-Al, min}}$), shortest Zr-Zr bond lengths ($R_{\text{Zr-Zr, min}}$), average Zr-Zr bond lengths ($R_{\text{Zr-Zr, ave}}$), multiplicity, symmetry and electronic state of the $Zr_nAl^{\pm m}$ ($n = 1 - 7$, $m = 0, 1$) clusters for the lowest-energy structures [a]

Cluster	Multi.	Symmetry	$R_{\text{Al-Zr,min}}$	$R_{\text{Zr-Zr,min}}$	$R_{\text{Zr-Zr,ave}}$
AlZr	4	$D_{\infty h}$	2.58	—	—
AlZr$_2$	4	C_{2v}	2.88	2.45	2.45
AlZr$_3$	4	C_1	2.73	2.63	2.76
AlZr$_4$	4	C_{3v}	2.83	2.79	2.82
AlZr$_5$	4	C_s	2.74	2.76	2.87
AlZr$_6$	4	C_s	2.76	2.72	2.84
AlZr$_7$	4	C_1	2.85	2.74	2.87
AlZr$^+$	1	$D_{\infty h}$	2.42	—	—
AlZr$_2^+$	1	$D_{\infty h}$	3.07	2.30	2.30
AlZr$_3^+$	3	C_{3v}	2.78	2.77	2.78
AlZr$_4^+$	3	C_{3v}	2.83	2.79	2.85
AlZr$_5^+$	3	C_1	2.79	2.74	2.87
AlZr$_6^+$	3	C_s	2.82	2.86	2.88
AlZr$_7^+$	1	C_{2v}	2.71	2.61	2.71
AlZr$^-$	3	$D_{\infty h}$	2.67	—	—
AlZr$_2^-$	3	C_{2v}	2.79	2.52	2.52
AlZr$_3^-$	1	C_{3v}	2.85	2.75	2.75
AlZr$_4^-$	3	C_{3v}	2.84	2.82	2.82
AlZr$_5^-$	1	C_{4v}	2.93	2.79	2.79
AlZr$_6^-$	1	C_{2v}	2.83	2.60	2.60
AlZr$_7^-$	1	C_1	2.85	2.66	2.86

[a] Bond lengths in Å.

The $Zr_4Al^{\pm m}$ clusters favor trigonal bipyramid (TBP) structures with C_{3v} symmetry (4a, 4a$^+$, 4a$^-$) as the lowest-energy structures. It is noteworthy that there is a tetragonal pyramid isomer (4b$^+$) with high C_{4v} symmetry, but the relative energy (ΔE) is 1.56 eV. The most stable structures for Zr_5Al and Zr_5Al^+ clusters are in 3D configurations (similar to Al-caped add the top right of TBP structure), and with high spin multiplicity and C_1 symmetry. Meanwhile, the minimum Al–Zr bond lengths are 2.74 Å and 2.79 Å for Zr_5Al and Zr_5Al^+ clusters, respectively. For Zr_5Al^- cluster, the most stable structures is a rectangular bipyramid with the Al atom located at an apex of the bipyramid (5a$^-$, C_{4v}). As for $n = 6$ and 7, we have optimized the isomers for $Zr_6Al^{\pm m}$ and $Zr_7Al^{\pm m}$ clusters, all the stable structures are 3D geometry. This phenomenon has also been found for AlB_n clusters.(Feng & Luo, 2007)

3.2 Relative Stability

We plotted the binding energies (E_b) for the most stable structures of $Zr_nAl^{\pm m}$ clusters in Figure 2, The E_b generally increases with increasing cluster size. No odd-even oscillations are exhibited. As a whole, the average binding energies of the Zr_nAl clusters are smaller than those of the Zr_nAl^+ and Zr_nAl^- clusters from $n \geq 2$, indicating that the Zr_nAl clusters gain stability after the gain or loss of one electron, this phenomenon is similar to $MgB_n^{\pm m}$ clusters (Wu et al, 2014). It can be observed that the E_b essentially increases sharply when n goes from 1 to 4, but increasing smoothly when $n \geq 4$.

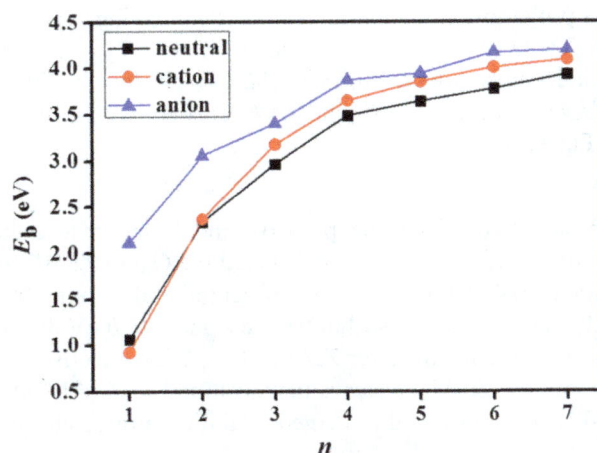

Figure 2. Clusters size (n) dependence of the average binding energy (E_b) of $Zr_nAl^{\pm m}$ clusters

As show in Figure 3, the fragmentation energies (E_f) for the anionic Zr_nAl clusters have an obvious odd-even oscillation with increasing size of clusters. The local maximum fragmentation energies appear at $n = 4$ and 6 for the anionic Zr_nAl clusters, indicating that $AlZr_4^-$ and $AlZr_6^-$ clusters are more stable than their neighbors of Zr_nAl^- clusters. The local minimum E_f appears at Zr_5Al, Zr_5Al^- and Zr_7Al^+ clusters, it means that these clusters are less stable than their neighbors.

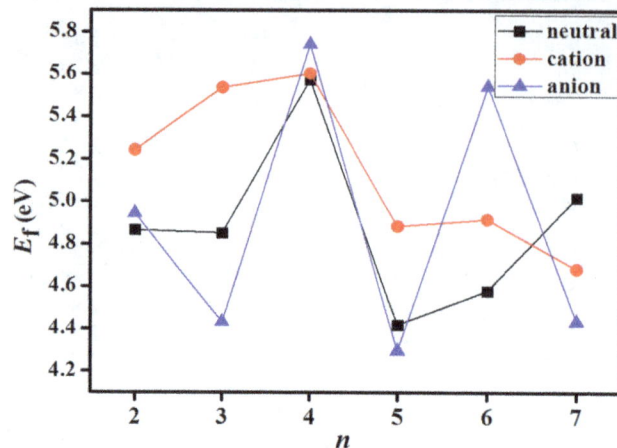

Figure 3. Cluster size (n) dependence of the fragmentation energy (E_f) for $Zr_nAl^{\pm m}$ clusters

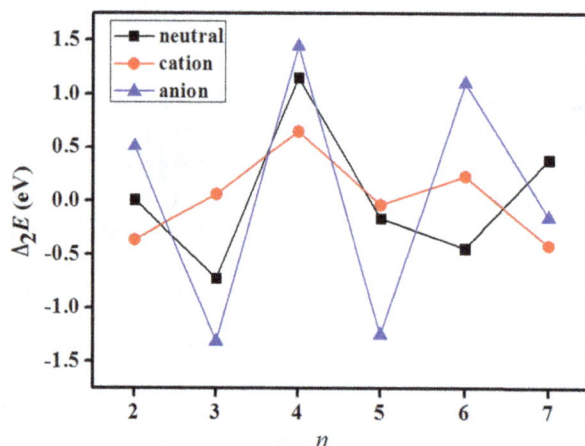

Figure 4. Cluster size (n) dependence of the second-order difference energy (Δ_2E) for $Zr_nAl^{\pm m}$ clusters

There is a sensitive quantity that can reflect the relative stability of clusters (Wang et al, 2010), this parameter is the second-order difference of total energies (Δ_2E) in Figure 4 and it was calculated with Eq. (3). The values of Δ_2E are the largest at $n = 4$ for all three kinds of clusters, indicating that these clusters present the highest stabilities with respect to the clusters with the same charge. This is probably due to the $Zr_4Al^{\pm m}$ clusters having high symmetry (C_{3v}) and more charge transfer from Zr atoms to Al atom (Table 2). This viewpoint could be partially supported by Ref (Feng & Luo, 2007). Additionally, we analyzed the relationship among different charged clusters in Figure 4. The second-order difference of total energies of Zr_3Al^+, Zr_4Al^-, Zr_5Al^+, Zr_6Al^- and Zr_7Al clusters are larger than their corresponding species, this phenomenon also appears in Figure 3.

3.3 Electronic Property

The HOMO-LUMO gap (G_{HL}) is the prototypical electronic property and is an invaluable parameter in clusters electronic property analysis. (Du et al, 2010). Therefore, we have calculated the G_{HL} of $Zr_nAl^{\pm m}$ clusters in Figure 5. The G_{HL} values of the cationic Zr_nAl clusters are usually larger than those of neutral and anionic Zr_nAl clusters except at n = 4 and 7, and the gaps are close to each other at $n = 5$ and 7 within 0.072eV and 0.030 eV. It should be pointed out that the gaps of all clusters have minimum and maximum values for ZrAl and Zr_2Al^+, respectively, indicating that the ZrAl cluster has the highest chemical activities, but the Zr_2Al^+ cluster is the most inert in $Zr_nAl^{\pm m}$ clusters. In this figure, we found no correlation between the HOMO-LUMO gaps and the energetic stability of these clusters. (Feng & Luo, 2007)

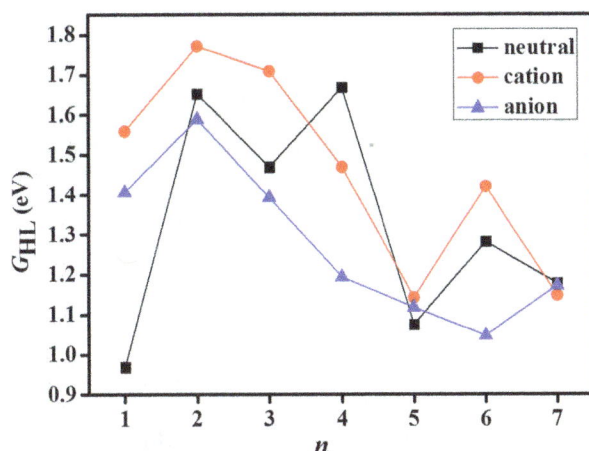

Figure 5. Cluster size (n) dependence of HOMO-LUMO gaps (G_{HL}) energy for $Zr_n Al^{\pm m}$ clusters

We calculated the chemical hardness (η) for neutral $Zr_n Al$ clusters and the results are displayed in Figure 6. The maximum hardness principle can be understood from eq 4. Bigger η means larger ionization potential and smaller electron affinity, which implies that the system has a smaller tendency to accept electrons and/or a smaller tendency to give away electrons. (Parr & Zhou, 1993). As can be seen in Figure 6, the trend of η is decreasing with the increase of size from $n = 2$ to 6, it means that the neutral $Zr_n Al$ clusters have a larger tendency to accept or give away electrons as the clusters size increases.

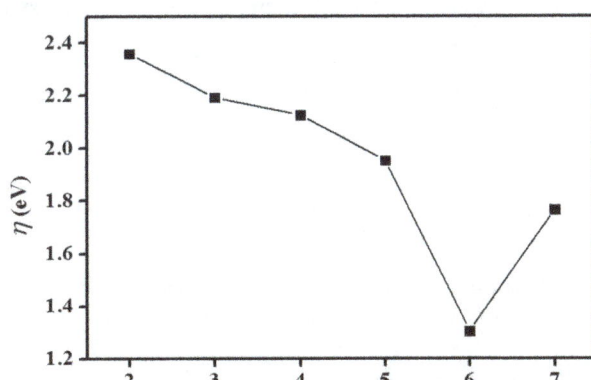

Figure 6. Cluster size (n) dependence of chemical hardness (η) for neutral $Zr_n Al$ clusters

The NBO method provides an effective electronic configuration for each atom in a cluster owing to the localized orbital constructed from the occupancy-weighted symmetric orthogonalized natural atomic orbital (Wu et al, 2009). Therefore, we studied the atomic charges (Q_{Al}) of Al atom versus the size n of clusters in Figure 7 and the natural electron configuration of Al atom was listed in Table 2. Zr_4Al, Zr_4Al^+ and Zr_4Al^- clusters are chosen as examples. The values of Q_{Al} for $Zr_4Al^{\pm m}$ clusters are negative, indicating that the charges transfer from Zr atoms to Al atom. On the contrary, charges in other clusters transfer from Al atom to Zr atoms. This could be further demonstrated by the natural electron configuration of Al atom being $3S^{0.76}3P^{0.74}$, $3S^{0.79}3P^{0.74}$ and $3S^{0.74}3P^{0.81}$ for Zr_4Al, Zr_4Al^+ and Zr_4Al^- clusters, respectively. Therefore, the electrostatic interaction between the Al atom and Zr atoms of the $Zr_4Al^{\pm m}$ clusters are stronger than that of other clusters, indicating that the $Zr_4Al^{\pm m}$ clusters are more stable than other clusters. (Feng & Luo, 2007)

Table 2. The natural electron configuration of Al atom in $Zr_n Al^{\pm m}$ ($n = 1 - 7$, $m = 0, 1$) clusters

n	neutral	cation	anion
1	$3s^{0.84}3p^{0.42}$	$3s^{0.85}3p^{0.37}$	$3s^{0.90}3p^{0.67}$
2	$3s^{0.86}3p^{0.26}$	$3s^{0.96}3p^{0.23}$	$3s^{0.81}3p^{0.68}$
3	$3s^{0.76}3p^{0.69}$	$3s^{0.75}3p^{0.63}$	$3s^{0.74}3p^{0.78}$
4	$3s^{0.76}3p^{0.74}$	$3s^{0.79}3p^{0.74}$	$3s^{0.74}3p^{0.81}$
5	$3s^{0.70}3p^{0.61}$	$3s^{0.72}3p^{0.52}$	$3s^{0.68}3p^{0.73}$
6	$3s^{0.52}3p^{0.86}$	$3s^{0.53}3p^{0.80}$	$3s^{0.65}3p^{0.72}$
7	$3s^{0.62}3p^{0.69}$	$3s^{0.61}3p^{0.69}$	$3s^{0.63}3p^{0.70}$

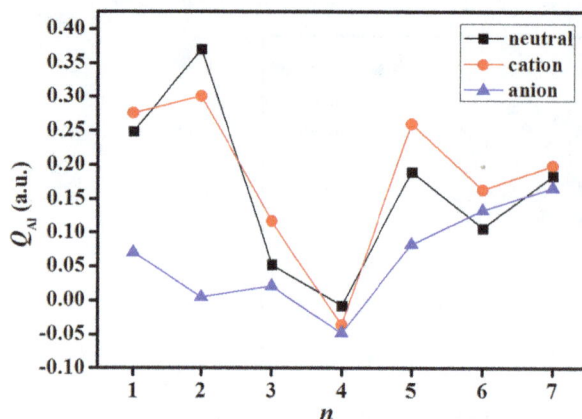

Figure 7. Cluster size (n) dependence of average atomic charges (Q_{Al}) of Al in $Zr_nAl^{\pm m}$ clusters

4. Conclusions

The geometries, relative stability and electronic property of the $Zr_nAl^{\pm m}$ clusters were investigated at the B3LYP/LANL2DZ level. For $Zr_6Al^{\pm m}$ clusters and $Zr_7Al^{\pm m}$ clusters, all stable structures are three-dimensional. The average binding energies of $Zr_nAl^{\pm m}$ clusters increase as the size n increases. It is noticeable that the Zr_3Al^+, Zr_4Al^-, Zr_5Al^+, Zr_6Al^- and Zr_7Al clusters are more stable than their corresponding differently charged species judged by the fragmentation energies and the second-order difference energies. The HOMO-LUMO energy gaps indicate that the AlZr cluster has the highest chemical activity, while the $AlZr_2^+$ cluster is the most inert in $Zr_nAl^{\pm m}$ clusters. Natural electron populations show that the electrons transfer from the Zr atoms to Al atom only in $Zr_4Al^{\pm m}$ clusters.

Acknowledgments

We gratefully acknowledge the project funded by the Priority Academic Program Development of Jiangsu Higher Education Institutions for partial financial support.

References

Addicoat, M. A., Buntine, M. A., & Metha, G. F. (2007). BFW: A Density Functional for transition metal clusters. *J. Phys. Chem. A 111*, 2625-2628. http://dx.doi.org/10.1021/jp0677521

Alexandrova, A. N., Zhai, H. J., Wang, L. S., & Boldyrev, A. I. (2004). Molecular wheel B_8^{2-} as a new inorganic ligand. photoelectron spectroscopy and ab initio characterization of LiB_8^-. *Inorg. Chem., 43*, 3552-3554. http://dx.doi.org/10.1021/ic049706a

Becke, A. D. (1993). Density-functional thermochemistry. III. The role of exchange, *J. Chem. Phys., 98*, 5648-5652.

Carpenter, J. E., & Weinhold, F. (1988). Analysis of the geometry of the hydroxymethyl radical by the "different hybrids for different spins" natural bond orbital procedure. *J. Mol. Struct. (THEOCHEM), 169*, 41-62. http://dx.doi.org/10.1016/0166-1280(88)80248-3

Doverstål, M., Karlsson, L., Lindgren, B., & Sassenberg, U. (1998). Resonant two-photon ionization spectroscopy studies of jet-cooled Zr_2. *J. Phys. B: At. Mol. Opt. Phys., 31*, 795–803. http://dx.doi.org/10.1088/0953-4075/31/4/025

Du, J. G., Sun, X. Y., & Jiang, G. (2010). Structures, chemical bonding, magnetisms of small Al-doped zirconium clusters. *Phys. Lett. A 374*, 854–860. http://dx.doi.org/10.1016/j.physleta.2009.12.009

Feng, X. J., & Luo, Y. H. (2007). Structure and stability of Al-doped boron clusters by the density-functional theory. *J. Phys. Chem. A, 111*(12), 2420-2425. http://dx.doi.org/10.1021/jp0656429

Frisch, M. J., Trucks, G. W., Schlegel, H. B., Scuseria, G. E., Robb, M. A., Cheeseman, J. R., ... Foresman, J. B. (2010). Gaussian 09. Gaussian, Inc., Wallingford.

Fu, Z. W., Lemire, G. W., Bishea, G. A., & Morse, M. D. (1990). Spectroscopy and electronic structure of jet-cooled Al_2. *J. Chem. Phys., 93*(12), 8420–8441. http://dx.doi.org/10.1063/1.459280

Ge, G. X., Jing, Q., Cao, H. B., & Yan, H. X. (2012). Structural, electronic, and magnetic properties of MB_n (M = Y, Zr, Nb, Mo, Tc, Ru, $n \le 8$) clusters. *J. Clust. Sci., 23*(2), 189–202. http://dx.doi.org/10.1007/s10876-011-0419-x

Hay, P. J., & Wadt, W. R. (1985). Ab initio effective core potentials for molecular calculations. Potentials for K to Au including the outermost core orbitals. *J. Chem. Phys., 82*, 299-310. http://dx.doi.org/10.1063/1.448975

Herry, D. J. (2012). Structures and stability of doped gallium nanoclusters. *J. Phys. Chem. C 116* (46), 24814-24823. http://dx.doi.org/10.1021/jp307555r

Hua, Y. W., Liu, Y. L., Jiang, G., Du, J. G., & Chen, J. (2013). Geometric transition and electronic properties of titanium-doped aluminum clusters: Al_nTi ($n = 2 - 24$). *J. Phys. Chem. A, 117*, 2590–2597. http://dx.doi.org/10.1021/jp309629y

Lee, C., Yang, W. T., & Parr, R. G. (1988). Development of the Colle-Salvetti correlation-energy formula into a functional of the electron density. *Phys. Rev., B 37*, 785. http://dx.doi.org/10.1103/physrevb.37.785

Lei, X. L. (2011). Geometrical and electronic properties of neutral and anionic Al_nB_m ($n + m =13$) clusters. *J. Clust. Sci., 22*, 159–172. http://dx.doi.org/10.1007/s10876-011-0370-x

Lekka, C. E. (2010). Cu–Zr, and Cu–Zr–Al clusters: Bonding characteristics and mechanical properties. *J. Alloy. Compd, 504*, 190–193. http://dx.doi.org/10.1016/j.jallcom.2010.02.067

Parr, R. G., & Zhou, Z. X. (1993). Absolute Hardness: Unifying concept for identifying shells and subshells in nuclei, atoms, molecules, and metallic clusters. *Acc. Chem. Res, 26*, 256-258. http://dx.doi.org/10.1021/ar00029a005

Pearson, R. G. (2005). Chemical hardness and density functional theory. *J. Chem. Sci., 117*(5), 369–377. http://dx.doi.org/10.1007/BF02708340

Ramos-Castillo, C. M, Reveles, J. U., Zope, R. R., & de Coss, R. (2015). Palladium clusters supported on graphene monovacancies for hydrogen storage. *J. Phys. Chem. C, 119*(15), 8402–8409. http://dx.doi.org/10.1021/acs.jpcc.5b02358

Reed, A. E., Curtiss, L. A., & Weinhold, F. (1988). Intermolecular interactions from a natural bond orbital, Donor-Acceptor Viewpoint. *Chem. Rev., 88*(6), 899–926. http://dx.doi.org/10.1021/cr00088a005

Schmidt, M., Kusche, R., Issendorff, B. V., & Haberland, H. (1998). Irregular variations in the melting point of size-selected atomic clusters. *Nature, 393*, 238-240. http://dx.doi.org/10.1038/30415

Sengupta, T., Das, S., & Pal, S. (2016). Transition Metal Doped Aluminum Clusters: An Account of Spin. *J. Phys. Chem. C 120*, 10027−10040. http://dx.doi.org/10.1021/acs.jpcc.6b00510

Sheng, X. F., Zhao, G. F., & Zhi, L. L. (2008). Evolution of small Zr clusters and dissociative chemisorption of H_2 on Zr clusters. *J. Phys. Chem. C 112*, 17828–17834. http://dx.doi.org/10.1021/jp8072602

Suh, Y. J., Chae, J. W., Jang, H. D., & Cho, K. (2015). Role of chemical hardness in the adsorption of hexavalent chromium species onto metal oxide nanoparticles. *Chem. Eng. J., 273*, 401-405. http://dx.doi.org/10.1016/j.cej.2015.03.095

Tang, X., Bumueller, D., Lim, A., Schneider, J., Heiz, U., Ganteför, G., Fairbrother, D. H., & Bowen, K. H. (2014). Catalytic dehydration of 2-Propanol by Size-Selected $(WO_3)_n$ and $(MoO_3)_n$ metal oxide clusters. *J. Phys. Chem. C 118*(50), 29278-29286. http://dx.doi.org/10.1021/jp505440g

Wang, C. C. (2006). Geometries and magnetisms of the Zr_n ($n = 2 – 8$) clusters: The density functional investigations. *J. Chem. Phys., 124*, 194301. http://dx.doi.org/10.1063/1.2200346

Wang, J., Liu, Y., & Lia, Y. C. (2010). Au@Sin: Growth behavior, stability and electronic structure. *Phys. Lett. A 374*(27), 2736-2742. http://dx.doi.org/10.1016/j.physleta.2010.04.068

Wu, Y. Y., Xu, S. Y., Zhao, F. Q., & Ju, X. H. (2015). Adsorption and dissociation of H_2 on B_n and MgB_n ($n = 2 − 7$) clusters: A DFT investigation. *J. Clust. Sci., 26*(3), 983–999. http://dx.doi.org/10.1007/s10876-014-0791-4

Wu, Y. Y., Zhao, F. Q., & Ju, X. H. (2014). DFT study on structure and stability of $MgB_n^{\pm m}$ clusters. Comput. *Theor. Chem., 1027*, 151–159. http://dx.doi.org/10.1016/j.comptc.2013.11.001

Wu, Z. K., Gayathri, C., Gri, R. R., & Jin, R. (2009). Probing the structure and charge state of Glutathione-Capped $Au_{25}(SG)_{18}$ clusters by NMR and Mass Spectrometry. *J. Am. Chem. Soc., 131*(18), 6535-6542. http://dx.doi.org/10.1021/ja900386s

Yamazoe, S., Koyasu, K., & Tsukuda, T. (2014). Nonscalable Oxidation Catalysis of Gold Clusters. *Acc. Chem. Res, 47*(3), 816-824. http://dx.doi.org/10.1021/ar400209a

Yang, C. L., Wang, M. S., Sun, M. Y., Wang, D. H., Ma, X. G., & Gong, Y. B. (2008). Dominant role of the interstitial $4d$ transition-metal in $TM@Zr^Z_{12}$ (TM = Y–Cd, Z = 0, ±1) icosahedral cages. *Chem. Phys. Lett, 457*, 49–53. http://dx.doi.org/10.1016/j.cplett.2008.04.023

Yao, J. G., Wang, X. W., Wang, Y. X. (2008). A theoretical study on structural and electronic properties of Zr-doped B clusters: ZrB_n ($n = 1$–12). *Chem. Phys., 351*, 1–6. http://dx.doi.org/10.1016/j.chemphys.2008.03.020

Zhai, H. J., Kiran, B., Li, J., & Wang, L. S. (2003). Hydrocarbon analogues of boron clusters-planarity, aromaticity and antiaromaticity. *Nature Mater, 2*, 827-833. http://dx.doi.org/10.1038/nmat1012

Zhao, G. F., Sheng, X. F., Zhi, L. L., Sun, J. M., & Gu, Y. Z. (2009). Density-functional study of structural, electronic, and magnetic properties of the Zr_nCr ($n = 2$–14) clusters. *J. Mol. Struct. (THEOCHEM), 908*, 40–46. http://dx.doi.org/10.1016/j.theochem.2009.04.041

10

Synthesis and Structure of Organic-Inorganic Hybrid Semi-interpenetrating Polymer Network Gels

Naofumi Naga[1], Yukie Uchiyama[1], Yuri Takahashi[1], Hidemitsu Furukawa[2]

[1]Department of Applied Chemistry, Materials Science Course, College of Engineering, Shibaura Institute of Technology, 3-7-5 Toyosu, Koto-ku, Tokyo 135-8548, Japan

[2]Department of Mechanical Systems Engineering, Graduate School of Science and Engineering, Yamagata University, 4-3-16 Jonan, Yonezawa, Yamagata 992-8510, Japan

Correspondence: Naofumi Naga, Department of Applied Chemistry, Materials Science Course, College of Engineering, Shibaura Institute of Technology, 3-7-5 Toyosu, Koto-ku, Tokyo 135-8548, Japan

Abstract

Semi-interpenetrating polymer network (semi-IPN) gels have been synthesized using a hydrosilylation reaction of 1,3,5,7-tetramethylcyclotetrasiloxane (TMCTS) as a joint molecule, and α,ω-nonconjugated dienes, 1,5-hexadiene (HD) or 1,9-decadiene (DD) as linker molecules in the presence of polystyrene (PS) as a liner polymers in toluene or cyclohexane. Network structure, mesh size and mesh size distribution, of the resulting semi-IPN gels was quantitatively characterized by means of a scanning microscopic light scattering (SMILS). The relaxation peaks derived from three kinds of structures were detected in the semi-IPN gels prepared in toluene by the SMILS analysis. One was derived from the mesh formed by TMCTS/α,ω-nonconjugated dienes about 1-2 nm. Others were derived from transition networks about 20-150 nm and large clustered liner polymer chains about 700-2300 nm. Effect of concentration and molecular weight of the liner polymer on the network structure of the semi-IPN gels in toluene was investigated. The relaxation peaks derived from transition networks or random coils formed by aggregated PS chains were detected in the semi-IPN gels containing high concentration or high molecular weight PS. The semi-IPN gels containing PS were also prepared in cyclohexane as a poor solvent for PS at 40°C, which was a higher temperature than the upper critical solution temperature (UCST = 34°C) of PS in cyclohexane. The network structure of the semi-IPN gels was traced by SMILS on the cooling process. In the semi-IPN gel with the short linker molecule of HD, the relaxation peak derived from clustered PS chains was detected over the UCST. By contrast, the relaxation peak derived from transition network was observed in the semi-IPN gel with the long linker molecule of DD.

Keywords: semi-interpenetrating gel, organic-inorganic hybrid gel, polystyrene

1. Introduction

Organic-inorganic hybrid polymers having network structures have been developed due to their characteristic properties: high transparency, high thermal stability, good mechanical strength, excellent solvent resistance, low dielectric constant, and so on. The organic-inorganic hybrid polymers have been prepared by some effective methods (Mark et al., 1995; Mascia, 1995; Novak, 1993; Provatas et al., 1997; Laine et al., 1998; Vaia et al., 1996; Sogah et al., 1999; Saegusa et al., 1991; Landry st al., 1993; Yamada et al., 1997; Chujo et al., 2003; Laine et al., 2003; Galiastsatos et al., 2001). Hydrosilylation reaction of multi-functionalized crosslinking reagents containing Si-H or vinyl group is one of the effective methods to yield the organic-inorganic network polymers containing Si (Yoshida et al., 2004, 2001; Tsumura et al., 1998, 1999, 2000; Michalczyk et al., 1993; Lercher et al., 2002; Laine et al., 1998; Schaefer et al., 2003). The authors recently attained effective synthesis of organic-inorganic hybrid gels by means of a hydrosilylation reaction of cyclic siloxane or cubic silsesquioxane with α,ω-nonconjugated dienes. Characterization of the network structures of the gels was quantitatively investigated by a scanning microscopic light scattering (SMILS) system (Naga et al., 2006). The SMILS analysis of the gels from cyclic siloxane or cubic silsesquioxane with α,ω-nonconjugated dienes showed the extremely narrow distribution of mesh size in the gels. Further more, the mesh size of the gels could be controlled by the length of the α,ω-nonconjugated dienes used (Naga et al., 2006). We also reported co-gelation of siloxane or silsesquioxane and α,ω-nonconjugated dienes to control the mesh size of these organic-inorganic hybrid gels precisely (Naga et al., 2007). One of the next steps of this study should be modification and application development of the organic-inorganic hybrid gel.

We focused on concept of interpenetrating polymer network (IPN) for the modification of the hybrid gel. The IPN is composed of two kinds of network structures which are entangled each other. The IPNs have been developed as one of the useful methods to compound the polymers which are immiscible. The IPN is also effective to add some functions and/or to improve its thermal and mechanical properties. Some kinds of gels have been developed based on the concept of the IPN. For example, a hydrogel prepared by in-site polymerization of acrylaminde within a crosslinked poly(2-acrylamide-2-methyl-1-propane sulfonic acid), so called "double network gel", which shows extraordinary strong mechanical properties (Gong, et al., 2003, 2004, 2005, 2007). We came up with an idea to synthesize the organic-inorganic hybrid gel in the presence of liner polymers, namely semi-IPN gel. Well-defined network structure of the present organic-inorganic gel should be useful to develop the semi-IPN gels with functionalities derived from the linear polymer. For a preliminary examination, we selected atactic-polystyrene (PS) as a linear polymer. PS is soluble in kinds of organic solvents which do not poise the Pt catalyst used to synthesize the present organic-inorganic hybrid gel. In this basic study, we conducted a hydrosilylation reaction of TMCTS with HD or DD in the presence of PS to form the semi-IPN gels in toluene, which is a good solvent for PS, as shown in Scheme 1. The semi-IPN gels with PS were also prepared in cyclohexane, which is a poor solvent for PS. The network structure of the resulting semi-IPN gels was investigated by SMILS to study the effects of the solvent, molecular length of dienes, concentration & molecular weight of PS, or temperature on the network structure.

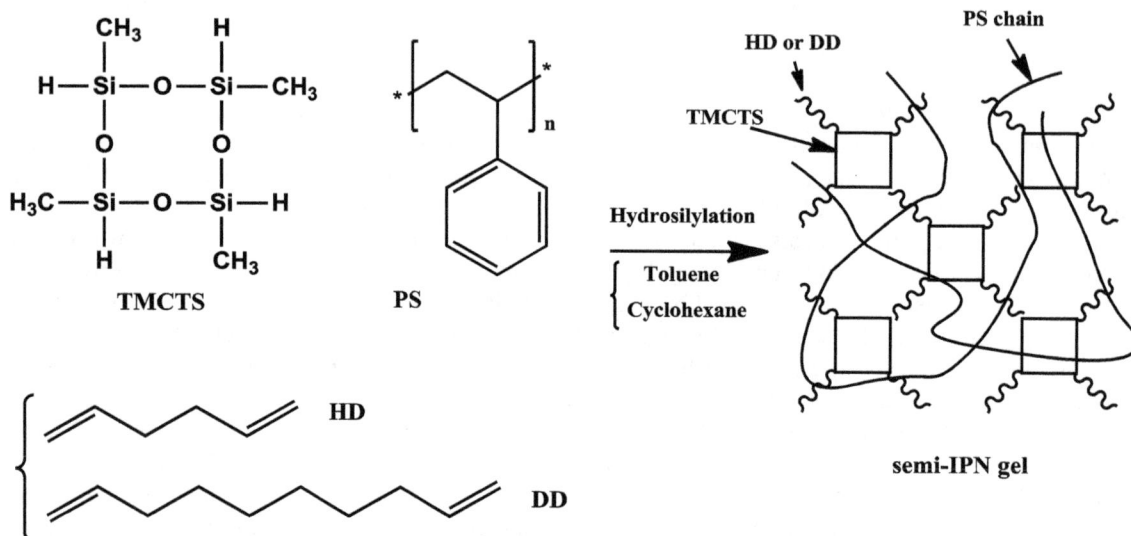

Scheme 1. Synthesis of organic-inorganic hybrid semi-IPN gels by means of a hydrosilylation reaction of TMCTS with HD or DD in the presence of PS.

2. Method

2.1 Materials

1,5-Hexadien (HD) and 1,9-decadiene (DD) (Tokyo Kasei Kogyo Co.) were dried over calcium hydride and distilled under nitrogen atmosphere before use. 1,3,5,7-Tetramethylcyclotetrasiloxane (TMCTS) (Chisso Co. Ltd.), illustrated in Scheme 1, were used without further purification. Platinum-divinyltetramethyldisiloxane complex (1) was purchased from Chisso Co. Ltd., and used without purification. Toluene and cyclohexane were dried over calcium hydride under refluxing for 6 h and distilled under nitrogen atmosphere before use. The platinum complex 1 was dissolved in distilled toluene or cyclohexane (0.6 mM), and stored under nitrogen atmosphere. PS samples were synthesized by anionic polymerization of styrene with n-butyllithium in cyclohexane at room temperature. Molecular weight and molecular weight distribution, determined with gel-permeation chromatography, of the obtained PS samples are as follows, PS1: $M_n = 15,000$, $M_w/M_n = 1.4$, PS2: $M_n = 69,000$, $M_w/M_n = 1.2$, PS3: $M_n = 347,000$, $M_w/M_n = 1.6$.

2.2 Synthesis of the semi-IPN gels

The molar ratio of vinyl group in α,ω-nonconjugated diene to Si-H group in TMCTS was adjusted to 1.0. Molar ratio of diene to catalyst 1 was 5000 mol/mol in the reaction system.

2.2.1 Synthesis of TMCTS-HD semi-IPN gel in toluene (12 wt% gel containing 5.0 wt% of PS1)

In a sample tube of 4 mm diameter, PS1 10.1 mg, TMCTS 14.4 mg (0.06 mmol) and   HD 9.9 mg (0.12 mmol) were dissolved in a 155.0 μL of toluene. Then a 40.0 μL of toluene solution of the catalyst 1 (0.6 mM) was added and mixed at room temperature.   It was placed without stirring, and the whitely nebulous gel was generated presently. The semi-IPN gels with other PS concentrations and higher molecular weight PS (PS2) were also synthesized by the same procedure.

2.2.2 Synthesis of TMCTS-HD semi-IPN gel in cyclohexane (12 wt% gel containing 0.1 wt% of PS3)

In a sample tube of 4 mm diameter, TMCTS 14.4 mg (0.06 mmol) and HD 9.9 mg (0.12 mmol) were dissolved in a 129.6 μL of cyclohexane, and a 42.4 μL of 0.5 wt% cyclohexane solution of PS3 (0.17 mg) was added at a fixed temperature (28 or 40°C). Then a 42.4 μL of cyclohexane solution (0.6 mM) of the catalyst **1** was added and mixed (TMCTS + HD = 12.0 wt%, PS = 0.1 wt% in the reaction system).   It was placed without stirring, and the whitely nebulous gel was generated presently. Synthesis of the semi-IPN gel with DD (TMCTS + DD = 10.0 wt%) was also synthesized by the same procedure.

2.3 Measurements

Quantitative determination of minute mesh size of the gels was performed with the scanning microscopic light scattering (SMILS) system (Furukawa et al., 2003, 2006), which was developed for the detailed characterization of network structure in polymer gels. The SMILS system enables us to scan and measure the light scattering at many different positions in a gel, in order to rigorously determine a time- and space-averaged, i.e. ensemble-averaged, (auto-)correlation function of fluctuating mesh size in the gel. Analysis of the ensemble-averaged function makes it possible to quantitatively characterize the mesh-size distribution of network structure in the gel. Scanning measurement was performed at more than 25 points on a sample to determine ensemble-averaged dynamic structure factor. The determined correlation function was transformed to the distribution function of relaxation time by using numerical inversed Laplace transform calculation. A few peaks of relaxation modes were observed in the distribution function in the present types of organic-inorganic hybrid gels (Naga et al., 2006). Based on the observation of scattering-angle, q, dependence of the relaxation modes, all the observed modes usually have q^2-dependence, which correspond to translational diffusion. In the following, all the results were determined at the scattering angle fixed at 90°. The observed modes, as assigned to the cooperative diffusion of gel network, were used for the determination of radius of mesh (mesh size) (ξ; m) with Einstein-Stokes formula.

$$\xi = \frac{16\pi n^2 \tau_R K_B \sin^2 \frac{\theta}{2}}{3\eta\lambda^2}$$

where n, τ_R, K_B, θ, η, and λ are refractive index of toluene or cyclohexane, Ensemble-averaged relaxation time (s), Boltzmann constant (1.38×10^{-23} JK^{-1}), scattering angle (90°), viscosity coefficient of solvent at the measured temperature, wave length of incident ray (6.328×10^{-7} m), respectively.

Mechanical properties of the gels were investigated by compression test with Tensilon RTE-1210 (ORIENTEC Co. LTD.). The test samples were cut to 1 cm cube, and pressed at a rate of 0.5 mm/min at room temperature.

3. Results and Discussion

3.1 TMCTS/α,ω-nonconjugated diene-PS semi-IPN gels In Toluene

3.1.1 Effect of PS Concentration on Network Structure of the Semi-IPN Gels

TMCTS/HD or DD-PS semi-IPN gels were synthesized in the presence of PS1 (M_n = 15,000) in toluene. Effect of PS1 concentration in the semi-IPN gels on the network structure was studied by SMILS. Ensemble-averaged relaxation-time distribution as a function of the relaxation time of the TMCTS/HD-PS semi-IPN gels is shown in Figure 1, and the results are summarized in Table 1. A structure model of TMCTS/HD-PS semi IPN gel is shown in Scheme 2. The peak at the fast relaxation time at around $\tau_R = 10^{-6}$ s is derived from the mesh of TMCTS/HD network about 1.6-2.1 nm as detected in the gel without PS (Run 1), as shown in Scheme 2 (a). The relaxation peak was detected at around $\tau_R = 10^{-2}$ s corresponding to the size of 1100-1300 nm in the semi-IPN gels containing 2.5 and 5.0 wt% of PS-1. These relaxation peaks should be derived from the clustered PS chains, as shown in Scheme 2 (c). In the semi-IPN gel containing 2.5 wt% of PS1, another relaxation peak was observed at $\tau_R = 8.8 \times 10^{-5}$ s corresponding to the size of 90 nm. The relaxation peak was not detected in the semi-IPN gels containing 5.0 wt% of PS1. This phenomenon should be derived from lightly clustered PS chains, termed "transition network" concerned with overlapping concentration (C*), as shown in Scheme 2 (b) (Simha et al., 1960; de Gennes, 1976, 1979), and we shall return to this subject later.

Figure 2 shows the ensemble-averaged relaxation time distribution as a function of the relaxation time of the TMCTS/DD-PS semi-IPN gels, and the results are summarized in Table 2. The relaxation peaks observed at around $\tau_R = 6$-10×10^{-6} s were derived from the mesh of TMCTS/DD network about 1.1-1.6 nm. The relaxation peaks at around $\tau_R = 1.5$-2.1×10^{-2} s, which corresponded to the size about 3000 nm, were observed in the semi-IPN gels containing 0.5 and 2.5 wt% of PS1. The peak was also observed in TMCTS/DD gel without PS, as shown in Figure 2 (a). The relaxation peak should be derived from large defects in the network structure, which was formed by intramolecular cyclization or entanglement of long flexible DD, as previously reported. A relaxation peak detected at $\tau_R = 2.1 \times 10^{-4}$ s, which corresponds to the size of 37 nm, in the semi-IPN gels containing 5.0 wt% of PS-1 should be derived from the transitional network of PS chains as observed in the TMCTS/HD-PS1 semi-IPN gel. The semi-IPN gel containing 5.0 wt% of PS1 also showed a relaxation peak at $\tau_R = 4.3 \times 10^{-3}$ s. The relaxation peak corresponds to the size of 770 nm, and would be derived from the clustered PS chains.

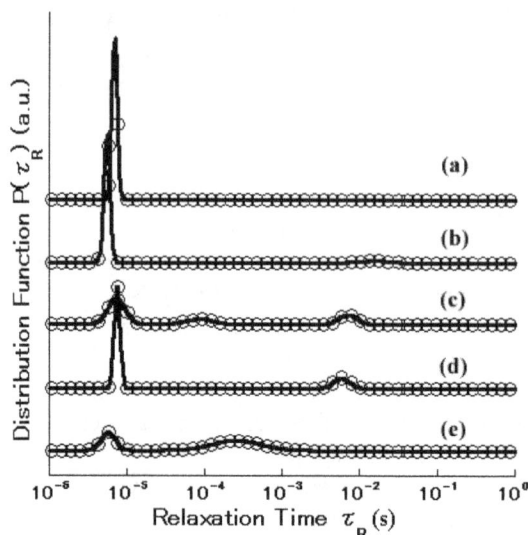

Figure 1. Ensemble-averaged relaxation time distributions as a function of the relaxation time of TMCTS/HD-PS semi-IPN gels in toluene, monomer concentration = 12.0 wt%; (a) without PS, (b) 0.5 wt% of PS1, (c) 2.5 wt% of PS1, (d) 5.0 wt% of PS1, and (e) 0.5 wt% of PS2.

Table 1. Network structure of TMCTS/HD-PS semi-IPN gel in toluene determined by SMILS[a]

Run	PS	wt%	$\tau_R^b \times 10^{-6}$ s	σ^c	Mesh size nm
1	---	0	6.9	0.03	1.6
2	PS1	0.5	8.7	0.04	1.0
3	PS1	2.5	7.3	0.04	1.3
			87.9	0.19	15.7
			7440	0.12	1290
4	PS1	5.0	7.5	0.08	2.1
			5990	0.11	1070
5	PS2	0.5	5.8	0.08	1.0
			267	0.32	47.8

[a]Monomer concentration; TMCTS+HD = 12.0 wt%. [b] Relaxation time. [c] Standard deviation of a peak of the ensemble-averaged relaxation time distribution.

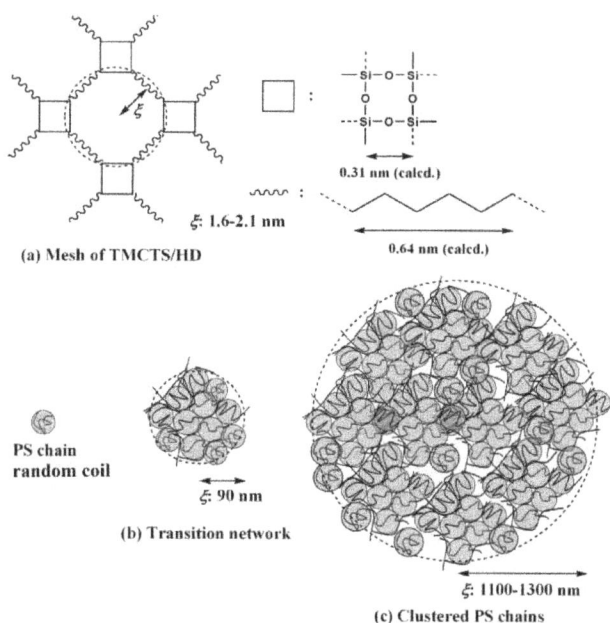

Scheme 2. Structure models of TMCTS/HD mesh (a) and PS chains in the semi-IPN gel; transition network (b), clustered structure (c).

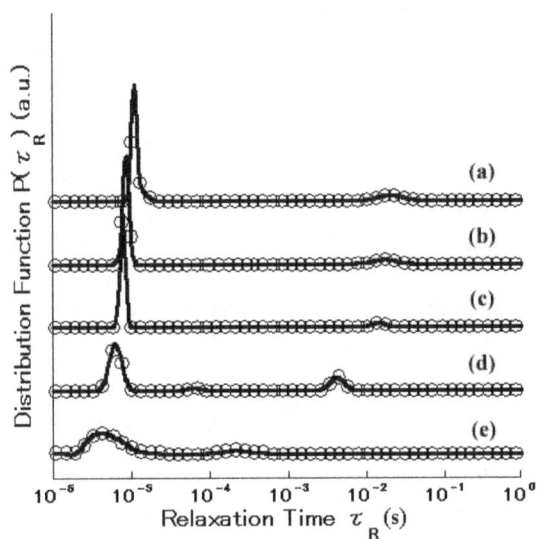

Figure 2. Ensemble-averaged relaxation time distributions as a function of the relaxation time of TMCTS/DD-PS semi-IPN gels in toluene, monomer concentration = 10.0 wt%; (a) without PS, (b) 0.5 wt% of PS1, (c) 2.5 wt% of PS1, (d) 5.0 wt% of PS1, and (e) 0.5 wt% of PS2.

Table 2. Network structure of TMCTS/DD-PS semi-IPN gel in toluene determined by SMILS [a]

Run	PS	wt%	$\tau_R{}^{b} \times 10^{-6}$ s	$\sigma{}^{c}$	Mesh size nm
6	---	0	11.1	0.04	2.0
			21300	0.04	3810
7	PS1	0.5	8.7	0.04	1.6
			17900	0.14	3190
8	PS1	2.5	7.9	0.03	1.4
			15000	0.09	2670
9	PS1	5.0	6.3	0.08	1.1
			34.3	0.09	11.5
			4310	0.09	769
10	PS2	0.5	4.6	0.20	0.8
			225	0.14	40.1

[a] Monomer concentration; TMCTS+DD = 10.0 wt%. [b] Relaxation time. [c] Standard deviation of a peak of the ensemble-averaged relaxation time distribution.

3.1.2 Effect of Molecular Weight of PS on the Network Structure

Effect of molecular weight of PS on the network structure of the semi-IPN gel was studied by SMILS. We prepared the semi-IPN gels containing 0.5 wt% of high molecular weight PS, PS2 M_n = 69,000 (Runs 19 and 24). The ensemble-averaged relaxation time distributions as a function of the relaxation time of TMCTS/HD-PS2 and TMCTS/DD-PS2 semi-IPN gels are illustrated in Figure 1 (e) and Figure 2 (e), respectively. In the case of the TMCTS/HD-PS2 semi-IPN gel, a broad relaxation peak derived from the transitional network of clustered PS chains was observed at around $\tau_R = 2.7 \times 10^{-4}$ s corresponding to the size of 48 nm. Furthermore, the relaxation peak derived from the mesh of TMCTS/HD network at around $\tau_R = 5.8 \times 10^{-6}$ s, corresponding to the size of 1.0 nm, became broad in comparison with that of the semi-IPN gel containing low molecular weight PS1, as shown in Figure 1 (b). These phenomena were also observed in the relaxation time distributions as a function of the relaxation time of the TMCTS/DD-PS2 semi-IPN gels, as shown in Figure 2 (e). In addition, the mesh size of the TMCTS/DD network in the semi-IPN gel containing PS2 (0.83 nm) was smaller than that of the corresponding gel containing PS1 (1.55 nm).

We can then go on to consider the network model of the semi-IPN gels in toluene base on the observations of the relaxation peaks of both the TMCTS/α,ω-nonconjugated diene and PS chains. The relaxation peaks derived from the transitional network or the clustered PS chains were detected in both the TMCTS/HD-PS and TMCTS/DD-PS semi-IPN gels by SMILS analysis. The transition network would be formed by lightly clustered PS chains (Scheme 2 (b)) when the concentration of PS is around the overlapping concentration (C*). This state should correspond to the TMCTS/HD-PS semi-IPN gels containing 2.5 wt% of PS1 and 0.5 wt% of PS2, Figures 1 (c) and (e), and the TMCTS/DD-PS semi-IPN gels containing 5.0 wt% of PS1 and 0.5 wt% of PS2, Figures 2 (d) and (e). Further aggregation of the transition networks would form larger cluster (Scheme 2 (c)), as detected in Figure 1 (c) and Figure 2 (d). In the case of the TMCTS/HD-PS semi-IPN gel, the relaxation peaks derived from the transition network and clustered PS chains were detected in the semi-IPN gels containing low concentration of PS1 (2.5 wt%). Whereas, these relaxation peaks were not detected in the corresponding TMCTS/DD-PS semi-IPN gel containing 2.5 wt% of PS1. The difference would indicate that the PS chains may cluster easier in the semi-IPN gel with HD than in that with DD.

Clear correlation was not observed between the mesh size derived from TMCTS/HD network and the concentration or molecular weight of PS in the semi-IPN gels. On the other hand, the mesh size derived from TMCTS/DD network slightly decreased with increasing of the PS concentration. Increase of the molecular weight of PS decreased the mesh size of the TMCTS/DD network in the same way. These results can be explained by a following model, as shown in Scheme 3. Some part of PS chains would penetrate into the TMCTS/DD network or entangle with the TMCTS/DD mesh during formation process of the semi-IPN gel. After the gelation, the aggregation of PS chains may decrease the mesh size of the TMCTS/DD network. The aggregation would be accelerated by increasing of the concentration and molecular weight of PS.

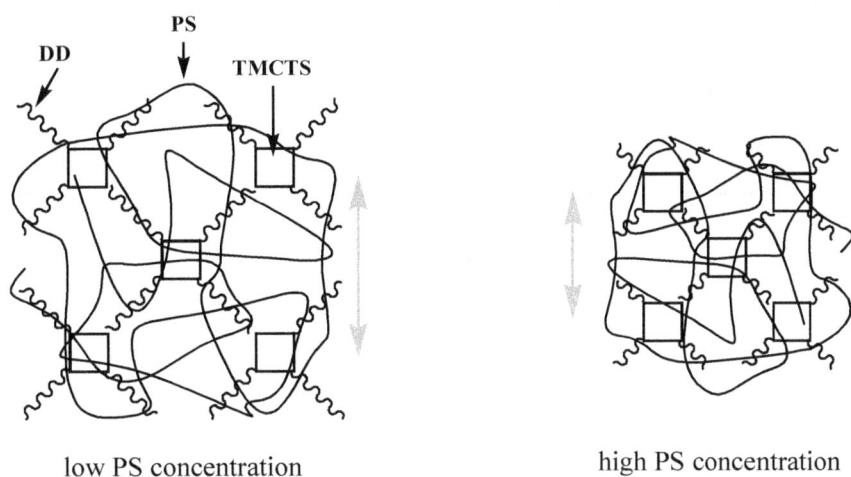

low PS concentration   high PS concentration

Scheme 3. Plausible model of TMCTS/DD-PS semi-IPN gel in toluene

3.2 TMCTS/α,ω-nonconjugated diene-PS semi-IPN gels in Cyclohexane

3.2.1 Effect of Temperature on Network Structure of Semi-IPN Gels

The TMCTS/HD or DD-PS semi-IPN gels containing 0.1 wt% of PS3 were also synthesized in cyclohexane, which is a poor solvent for PS. PS3 was used in this experiment, because the sample showed UCST at 34°C in cyclohexane. The UCST is a little higher than room temperature and convenient to investigate the effect of the temperature on the network structure of the semi-IPN gels by SMILS. The TMCTS/HD-PS3 semi-IPN gel was prepared at 40°C, and the network structure of the resulting semi-IPN gel was traced with SMILS on the cooling process. The ensemble-averaged relaxation-time distributions as a function of the relaxation time are shown in Figure 3, and the results are summarized in Table 3. The relationship between the temperature and mesh size of the TMCTS/HD-PS3 semi-IPN gel is shown in Figure 4. In the SMILS analysis of the TMCTS/HD-PS3 semi-IPN gel at 40°C, a relaxation peak was detected at $\tau_R = 5.2 \times 10^{-6}$ s corresponding to 1.2 nm of size. This size was close to that of the mesh size of the TMCTS/HD gel (1.1 nm) prepared without PS. The size derived from the TMCTS/HD mesh increased with decreasing of the temperature over the UCST (> 34°C). There was a rapid decrease in the mesh size below the UCST. Another relaxation peak was detected at around $\tau_R = 10^{-4}$-10^{-3} s corresponding to the size of 900-2300 nm derived from the large clustered PS chains. The size was almost fixed about 2000-2300 nm independent of the temperature over the UCST (40-34 °C). The relaxation peak also rapidly sifted to short time at around $\tau_R = 10^{-4}$-10^{-3} s corresponding to the size of 930-980 nm, and its intensity drastically decreased below the UCST (< 32 °C).

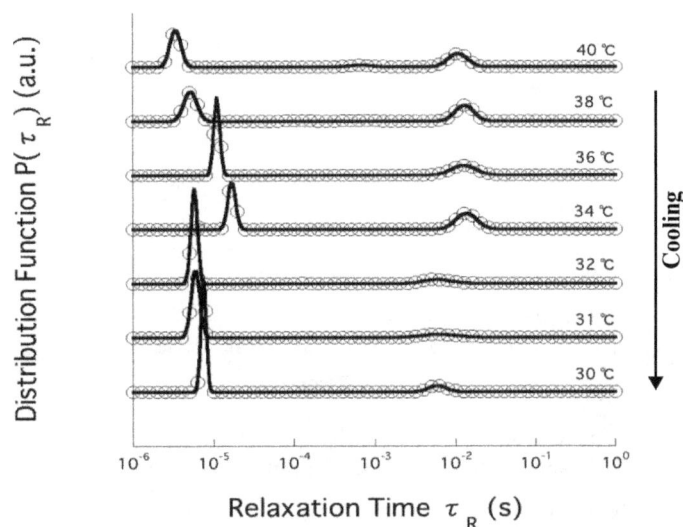

Figure 3 Ensemble-averaged relaxation-time distributions as a function of the relaxation time on the cooling process of TMCTS/HD-PS3 semi-IPN gel prepared at 40°C on the cooling process.

Table 3. Network structure of TMCTS/HD-PS3 semi-IPN gel in cyclohexane determined by SMILS on the cooling process[a]

Run	Temperature °C	$\tau_R^{\text{b}} \times 10^{-6}$ s	σ^{c}	Mesh size nm
11	40	3.4	0.07	0.6
		10600	0.12	2000
12	38	5.2	0.08	1.0
		12800	0.11	2340
13	36	11.1	0.04	2.0
		12400	0.15	2180
14	34	17.1	0.05	2.9
		13400	0.13	2270
15	32	5.8	0.04	0.9
		5870	0.17	960
16	31	6.0	0.05	1.0
		6130	0.23	982
17	30	7.5	0.03	1.2
		5920	0.11	928

[a] PS3 = 0.1 wt%, Monomer concentration; TMCTS+HD = 12.0 wt%. [b] Relaxation time. [c] Standard deviation of a peak of the ensemble-averaged relaxation time distribution.

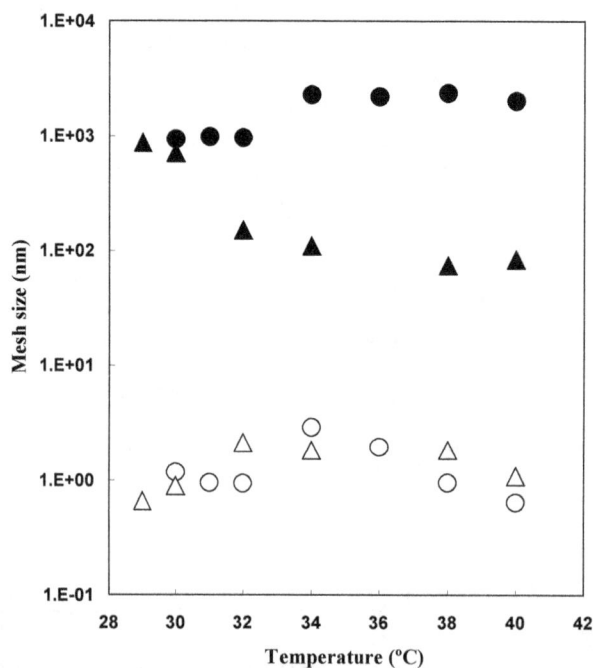

Figure 4 Temperature dependence of mesh size of TMCTS/HD-PS3 semi-IPN gel (circle) and TMCTS/DD-PS3 semi-IPN gel (triangle) on the cooling process.

The TMCTS/DD-PS3 semi-IPN gel was also prepared at 40°C, and the network structure of the semi-IPN gel was investigated with SMILS on the cooling process. The ensemble-averaged relaxation-time distributions as a function of the relaxation time are shown in Figure 5, and the results are summarized in Table 4. The relationship between the temperature and mesh size of the TMCTS/DD-PS3 semi-IPN gel is also plotted in Figure 4. In the ensemble-averaged relaxation-time distribution acquired at 40°C, a relaxation peak was detected at $\tau_R = 5.8 \times 10^{-6}$ s corresponding to 1.1 nm derived from the TMCTS/DD mesh. The mesh size slightly increased with decreasing of the temperature over the UCST (34°C), whereas it returned to the original size below the UCST. Another relaxation peak was detected at slow relaxation time at around $\tau_R = 10^{-4}$ s corresponding to the size about 80 nm derived from the transition network over the UCST, and the size increased to 710-870 nm below the UCST (30, 29°C). The result indicates that aggregation of the transition network would form the clustered PS chains under the UCST.

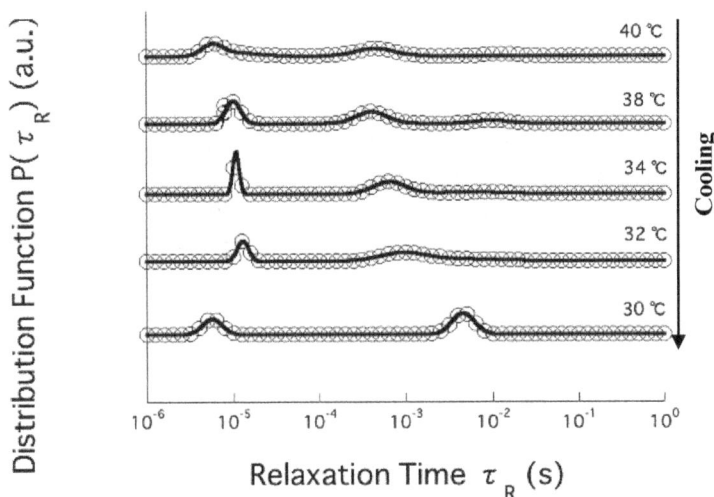

Figure 5. Ensemble-averaged relaxation-time distributions as a function of the relaxation time of TMCTS/DD-PS3 semi-IPN gel prepared at 40°C on the cooling process.

Table 4. Network structure of TMCTS/DD-PS3 semi-IPN gel in cyclohexane on the cooling process [a]

Run	Temperature °C	τ_R [b] $\times 10^{-6}$ s	σ [c]	Mesh size nm
18	40	5.8	0.07	1.1
		446	0.23	84.3
19	38	10.1	0.09	1.8
		409	0.18	74.6
20	34	10.8	0.04	1.8
		650	0.19	110
21	32	13.0	0.07	2.1
		920	0.27	150
22	30	5.7	0.11	0.9
		4530	0.13	711
23	29	4.3	0.07	0.7
		5654	0.07	871

[a] PS3 = 0.1 wt%, Monomer concentration; TMCTS+DD = 10.0 wt%. [b] Relaxation time. [c] Standard deviation of a peak of the ensemble-averaged relaxation time distribution.

We are now ready to consider the effect of the temperature on the network structure of the semi-IPN gels on the cooling process. In the case of the TMCTS/HD-PS3 semi-IPN gel, a relaxation peak corresponding to the size about 2000 nm derived from the large clustered PS chains was detected over the UCST. When the temperature dropped below the UCST, the semi-IPN gel became muddy due to the insolubility of the PS chains, and which would gather with the TMCTS/HD network. A part of PS chains would form clusters about 1000 nm.

In the case of TMCTS/DD-PS3 semi-IPN gel, the clustered PS chains formed the transition network about 100 nm over the UCST. The transition networks would gradually aggregate with decreasing the temperature over the UCST. When the temperature dropped below the UCST, the size drastically increased due to the acceleration of the aggregation of the transition networks to form the clustered PS chains.

The size of small mesh derived from both the TMCTS/HD and TMCTS/DD networks slightly increased with decreasing of the temperature over the UCST. When the temperature dropped below the UCST, the mesh size returned to almost the original size at prepared temperature (40°C). The temperature dependence of the mesh size was observed only in the semi-IPN gels. Although the mechanism has not cleared yet, the interaction between the network and PS chains should induce the phenomena.

3.3 Mechanical Properties of the Semi-IPN gels

The mechanical properties of the semi-IPN gels prepared in toluene were investigated by the compression test. Stress-strain curves of TVMCTS/HD, DD-PS1 gels, and TVMCTS/HD, DD gels for reference, are shown in Figure 6. Young's modulus, breaking stress, and breaking point, of the prepared gels are summarized in Table 5. Although all the gels were fragile, the semi-IPN gels showed higher Young's modulus and breaking stress than the gels without PS1. The PS in the network structure of the gels should enhance the mechanical properties of the gels.

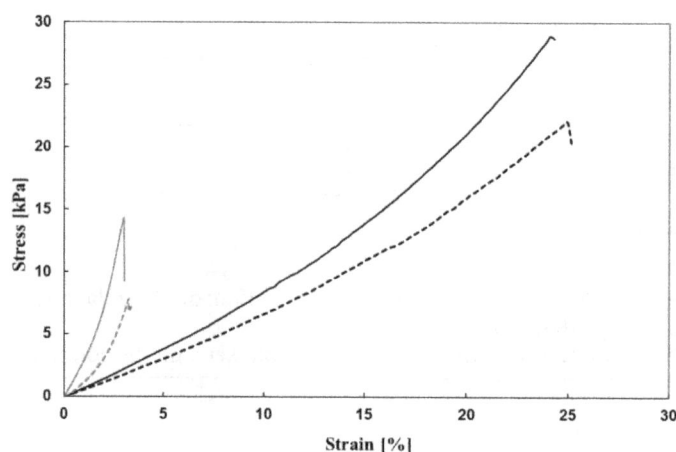

Figure 3 Stress-strain curves of TMCTS/HD gel (gray dot line), TMCTD/HD-PS1 semi-IPN (gray line), TMCTS/DD gel (black dot line), and TMCTS/DD-PS1 semi-IPN (black line) in toluene, monomer concentration = 10 wt% (TMCTS/HD), 12 wt% (TMCTS/DD).

Table 5. Compression test of TMCTS-HD, DD gels, monomer concentration 12wt%.

Run	Network	PS1 [wt%]	Young's modulus [kPa]	Breaking stress [kPa]	Breaking point [%]
1	TMCTS/HD	0	104.8	7.8	3.3
4	TMCTS/HD	5	250.1	14.3	3.0
6	TMCTS/DD	0	45.1	22.1	25.2
9	TMCTS/DD	5	91.2	28.9	24.3

4. Conclusion

The TMCTS/HD or DD-PS semi-IPN gels were successfully synthesized by means of a hydrosilylation reaction using a Pt catalyst, and network structure of the semi-IPN gels was quantitatively investigated by SMILS. Features of the solvent and molecular length of the α,ω-nonconjugated dienes strongly affected the network structures of the resulting semi-IPN gels. The following phenomena were cleared by the structure analysis of the semi-IPN gels by SMILS. In the case of the semi-IPN gels in toluene, the transition network or cluster of PS chains was formed in the semi-IPN gels containing high concentration or high molecular weight PS. These structures derived from PS chains were easily formable in the semi-IPN gels with DD. In the case of the semi-IPN gels in cyclohexane, the transition networks or clustered PS chains were detected in the semi-IPN gels containing low concentration, 0.1 wt%, of PS chains. In the semi-IPN gels with HD, a relaxation peak derived large clustered PS chains about 2000-3000 nm size was detected in the SMILS analysis over the UCST. On the other hand, the SMILS analysis of the semi-IPN gels with DD indicated that the existence of the transition networks about 75-150 nm size. The PS chains in the semi-IPN gels in cyclohexane were educed below the UCST, and formed relatively small clustered structure about 710-870 nm. These results indicate that PS chains can be aggregated easier in the semi-IPN gels with HD than those with DD. The PS chains would penetrate into the TMCTS/DD network or entangle with network in the TMCTS/DD-PS semi-IPN gel due to the large mesh size of the TMCTS/DD gel, as previously reported. The PS chains in the semi-IPN gels were effective to improve the mechanical properties of the hybrid gels.

The fundamental studies of semi-IPN gel composed of multi-functional siloxane and α,ω-nonconjugated dienes by means of a hydrosilylation reaction in the presence of a linear polymer should be useful for the application of the functionalized organic-inorganic hybrid gel. Developments of the organic-inorganic hybrid semi-IPN gels with photo-function, electrical-function, or excellent mechanical property are now proceeding, and the results will be reported elsewhere.

References

Choi, J., Yee, A. F., & Laine, R. M. (2003). Organic/Inorganic Hybrid Composites from Cubic Silsesquioxanes. Epoxy Resins of Octa(dimethylsiloxyethylcyclohexylepoxide) Silsesquioxane. Macromolecules, 36, 5666-5682. http://pubs.acs.org/doi/abs/10.1021/ma030172r

de Gennes, P. G. (1976). Dynamics of Entangled Polymer Solutions. I. The Rouse Model. *Macromolecules, 9*, 587-593. http://pubs.acs.org/doi/abs/10.1021/ma60052a011

de Gennes, P. G. (1976). Dynamics of Entangled Polymer Solutions. II. Inclusion of Hydrodynamic Interactions. *Macromolecules, 9*, 594-598. http://pubs.acs.org/doi/abs/10.1021/ma60052a012

de Gennes, P. G. (1979). *Scaling Concepts in Polymer Physics*, Ithaca, NY, Cornell University Press. http://www.cornellpress.cornell.edu/book/?GCOI=80140100785990

Furukawa, H., Horie K., Nozaki, R., & Okada, M. (2003). Swelling-induced modulation of static and dynamic fluctuations in polyacrylamide gels observed by scanning microscopic light scattering. *Phys. Rev. E, Statistical, Nonlinear, and Soft Matter Phys. 68*, 031406-1. http://journals.aps.org/pre/abstract/10.1103/PhysRevE.68.031406

Furukawa, H., Kobayashi, M., Miyashita, Y., & Horie, K. (2006). End-Crosslinking Gelation of Poly(amide acid) Gels studied with Scanning Microscopic Light Scattering. High Performance Polymers, 18, 837-847. http://hip.sagepub.com/content/18/5/837.abstract

Galiastsatos, V., Subramanian, P. R., & Klein-Castner, L. (2001). Designing heterogeneity into bimodal elastomeric PDMS networks. *Macromol. Symp., 171*, 97-104. http://onlinelibrary.wiley.com/doi/10.1002/1521-3900(200106)171:1%3C97::AID-MASY97%3E3.0.CO;2-1/abstract

Gong, J. P., Katsuyama, Y., Kurokawa, T., & Osada, Y. (2003). Double-Network Hydrogels with Extremely High Mechanical Strength. *Adv. Mater., 15*, 1155-1158. http://onlinelibrary.wiley.com/doi/10.1002/adma.200304907/full

Kim, K. M., & Chujo, Y. (2003). Organic–inorganic hybrid gels having functionalized silsesquioxanes. *J. Mater. Chem., 13*, 1384-1391. http://pubs.rsc.org/en/content/articlelanding/2003/jm/b211030j#!divAbstract

Krishnamoorti, R., Vaia, R. A., & Giannelis, E. P. (1996). Structure and Dynamics of Polymer-Layered Silicate Nanocomposites. *Chemistry of Materials, 8*, 1728-1734. http://pubs.acs.org/doi/abs/10.1021/cm960127g

Laine, R. M., Zhang, C., Sellinger, A., & Viculis, L. (1998). Polyfunctional cubic silsesquioxanes as building blocks for organic/inorganic hybrids. *Appl. Organomet. Chem., 12*, 715-723. http://onlinelibrary.wiley.com/doi/10.1002/(SICI)1099-0739(199810/11)12:10/11%3C715::AID-AOC778%3E3.0.CO;2-0/abstract

Landry, C. J. T., Coltrain, B. K., Landry, M. R., & Long V. K. (1993). Poly(vinyl acetate)/silica-filled materials: material properties of in situ vs fumed silica particles. Macromolecules, 26, 3702-3712. http://pubs.acs.org/doi/abs/10.1021/ma00066a032?journalCode=mamobx

Mark, J. E., Lee, C., & Bianconi P. A. (1995). Hybrid Organic-Inorganic Composites. *ACS Symposium Series 585*. http://pubs.acs.org/isbn/9780841231481

Mascia, L. (1995). Developments in organic-inorganic polymeric hybrids: Ceramers. *Trends in Polymer Science, 3*, 61–66.

Michalczyk, M. J., Farneth, W. E., & Vega, A. J. (1993). High temperature stabilization of crosslinked siloxanes glasses. Chem. Mater., 5, 1687-1689. http://pubs.acs.org/doi/abs/10.1021/cm00036a001

Na, Y. H., Kurokawa, T., Katsuyama, Y., Tsukeshiba, H., Gong, J. P., Osada, Y., Okabe, S., Karino, T., & Shibayama, M. (2004). Structural Characteristics of Double Network Gels with Extremely High Mechanical Strength. Macromolecules, 37, 5370-5374. http://pubs.acs.org/doi/abs/10.1021/ma049506i

Naga, N., Oda, E., Toyota, A., & Furukawa, H. (2007). Mesh Size Control of Organic-Inorganic Hybrid Gels by Means of a Hydrosilylation Co-Gelation of Siloxane or Silsesquioxane and α, ω-Non-Conjugated Dienes. *Macromol. Chem. Phys., 208*, 2331-2338. http://onlinelibrary.wiley.com/doi/10.1002/macp.200700184/abstract

Naga, N., Oda, E., Toyota, A., Horie, K., & Furukawa, H. (2006). Tailored Synthesis and Fundamental Characterization of Organic-Inorganic Hybrid Gels by Means of a Hydrosilylation Reaction. *Macromol. Chem. Phys., 207*, 627-635. http://onlinelibrary.wiley.com/doi/10.1002/macp.200500501/abstract

Nakayama, A., Kakugo, A,, Gong, J. P., Osada, Y., Takai, M., & Erata, T. (2004). High Mechanical Strength Double-Network Hydrogel with Bacterial Cellulose. *Adv. Func. Mater., 14*, 1124-1128. http://onlinelibrary.wiley.com/doi/10.1002/adfm.200305197/abstract

Novak, B. M. (1993). Hybrid Nanocomposite Materials-between inorganic glasses and organic polymers. *Adv. Mater., 5*, 422-433. http://onlinelibrary.wiley.com/doi/10.1002/adma.19930050603/pdf

Pan, G., Mark, J. E., & Schaefer D. W. (2003). Synthesis and characterization of fillers of controlled structure based on polyhedral oligomeric silsesquioxane cages and their use in reinforcing siloxane elastomers. *J. Polym. Sci. Part B: Polym. Phys., 41*, 3314-3323. http://onlinelibrary.wiley.com/doi/10.1002/polb.10695/full

Pinho, R. O., Radovanovic, E., Torriani, I. L., & Yoshida, I. V. P. (2004). Hybrid materials derived from divinylbenzene and cyclic siloxane. Eur. Polym. J., 40, 615-622. http://www.sciencedirect.com/science/article/pii/S0014305703002489

Provatas, A., & Matisons, J. G. (1997). Silsesquioxanes: synthesis and applications. *Trends in Polymer Science, 5*, 327-332. http://cat.inist.fr/?aModele=afficheN&cpsidt=2828673

Redondo, S. U. A., Radovanovic, E., Torriani, I. L., & Yoshida, I. V. P. (2001). Polycyclic silicone membranes. Synthesis, characterization and permeability evaluation. Polymer, 42, 1319-1327. http://www.sciencedirect.com/science/article/pii/S0032386100005206

Saegusa. T., & Chujo. Y. (1991). Macromolecular engineering on the basis of the polymerization of 2-oxazolines. *Makromol. Chem. Macromol. Symp., 51*, 1-10. http://onlinelibrary.wiley.com/doi/10.1002/masy.19910510103/abstract;jsessionid=4332BBF608B5946C6B7DE190327D4E1D.f02t03

Simha, R., & Zakin, J. L. (1960). Compression of Flexible Chain Molecules in Solution. *J. Chem. Phys., 33*, 1791. http://scitation.aip.org/content/aip/journal/jcp/33/6/10.1063/1.1731504

Su, R. Q., Müller, T. E., Procházka, J., & Lercher J. (2002). A New Type of Low-ϰ Dielectric Films Based on Polysilsesquioxanes. Adv. Mater., 14, 1369-1373. http://onlinelibrary.wiley.com/doi/10.1002/1521-4095(20021002)14:19%3C1369::AID-ADMA1369%3E3.0.CO;2-I/epdf

Tominaga, T., Tirumala, V. R., Lin E. K., Gong, J. P., Furukawa, H., Osada, Y., & Wu, W. L.(2007). The molecular origin of enhanced toughness in double-network hydrogels: A neutron scattering study. Polymer, 48, 7449-7454. http://www.sciencedirect.com/science/article/pii/S0032386107009858

Tsumura, M., & Iwahara, T. (1999). Synthesis and Properties of Crosslinked Polycarbosilanes by Hydrosilylation Polymerization. Polymer J., 31, 452-457. http://www.nature.com/pj/journal/v31/n5/abs/pj199985a.html

Tsumura, M., & Iwahara, T. (2000). Crosslinked Polycarbosilanes. Synthesis and Properties. Polymer J., 32, 567-573. http://www.nature.com/pj/journal/v32/n7/abs/pj2000104a.html

Tsumura, M., & Iwahara, T. (2000). Silicon-based materials prepared by IPN formation and their properties. *J. Appl. Polym. Sci., 78*, 724-731. http://onlinelibrary.wiley.com/doi/10.1002/1097-4628(20001024)78:4%3C724::AID-APP50%3E3.0.CO;2-9/abstract

Tsumura, M., Ando, K., Kotani, J., Hiraishi, M., & Iwahara, T. (1998). Silicon-Based Interpenetrating Polymer Networks (IPNs): Synthesis and Properties, Macromolecules, 31, 2716-2723. http://pubs.acs.org/doi/abs/10.1021/ma971308m

Weimer, M. W., Chen, H., & Giannelis, E. P., Sogah, D. Y. (1999). Direct Synthesis of Dispersed Nanocomposites by in Situ Living Free Radical Polymerization Using a Silicate-Anchored Initiator. J. Am. Chem. Soc., 121, 1615–1616. http://pubs.acs.org/doi/abs/10.1021/ja983751y

Yamada, N., Yoshinaga, I., & Katayama, S.(1997). Synthesis and dynamic mechanical behaviour ofinorganic–organic hybrids containing various inorganiccomponents. *J. Mater. Chem., 7*, 1491-1495. http://pubs.rsc.org/en/Content/ArticleLanding/1997/JM/a700793k#!divAbstract

Yasuda, K., Gong, J. P., Katsuyama, Y., Nakayama, A., Tanabe, Y., Kondo, E., Ueno, M., & Osada, Y. (2005). Biomechanical properties of high-toughness double network hydrogels. Biomaterials, 26, 4468-4475. http://www.sciencedirect.com/science/article/pii/S0142961204010038

Zhang, C., Babonneau, F., Bonhomme, C., Laine, R. M., Soles, C. L., Hristov, H. A., & Yee, A. F. (1998). Highly Porous Polyhedral Silsesquioxane Polymers. Synthesis and Characterization. J. Am. Chem. Soc., 120, 8380-8391. http://pubs.acs.org/doi/abs/10.1021/ja9808853

11

Structural Modeling of Glutathiones Containing Selenium and Tellurium

Valter A. Nascimento[1], Petr Melnikov[1], André V. D. Lanoa[1], Anderson F. Silva[1], Lourdes Z. Z. Consolo[1]

[1]School of Medicine of the Federal University of Mato Grosso do Sul/UFMS, Brazil

Correspondence: Valter A. Nascimento, School of Medicine, Federal University of Mato Grosso do Sul/UFMS, Caixa Postal 549, Campo Grande/MS, Brazil. E-mail: aragao60@hotmail.com

Abstract

The comparative structural modeling of reduced and oxidized glutathiones, as well as their derivatives containing selenium and tellurium in chalcogen sites (Ch = Se, Te) has provided detailed information about the bond lengths and bond angles, filling the gap in the structural characteristics of these tri-peptides. The investigation using the molecular mechanics technique with good approximation confirmed the available information on X-ray refinements for the related compounds. It was shown that Ch-H and Ch-C bond lengths grow in parallel with the increasing chalcogen ionic radii. Although the distances C-C, C-O, and C-N are very similar, the geometry of GChChG glutathiones is rich in conformers owing to the possibility of rotation about the bridge Ch-Ch. It is confirmed that the distances Ch-Ch are essentially independent of substituents in most of chalcogen compounds from elemental chalcogens to oxydized glutathiones. The standard program Hyperchem 7.5 has proved to be an appropriate tool for the structural description of less-common bioactive compositions when direct X-ray data are missing.

Keywords: glutathione, selenium, tellurium, structural modeling

1. Introduction

Glutatione is a tri-peptide with a molecular mass 307.3 g mol^{-1} composed from the amino acids L-cysteine, L-glutamic acid and glycine (Figure 1). This is a major antioxidant acting as a free radical scavenger that protects the cell from reactive oxygen species (ROS). In addition, GSH is involved in nutrient metabolism and regulation of cellular metabolic functions, ranging from DNA and protein synthesis to signal transduction, cell proliferation, and apoptosis (Masella & Mazza, 2009).

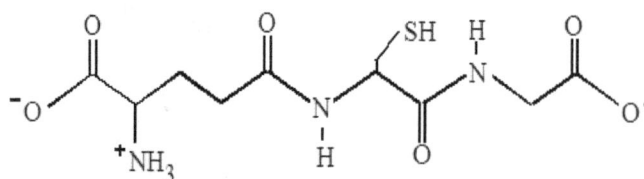

Figure 1. Schematic representation of GSH

This complex network of roles, functions, and effects makes GSH and sulfur amino acids a fascinating subject for protein chemists, biochemists, nutritionists, and pathologists. The sulfur atom of glutathione, being a good donor of electrons is responsible for its biochemical activity. Inside the cells, most of glutathione is in the cytosol, where it permanently exists in a reduced form, GSH. The latter can be reversibly transformed to form an oxidized derivative, GSSG. Their interrelationship is given by equation (1):

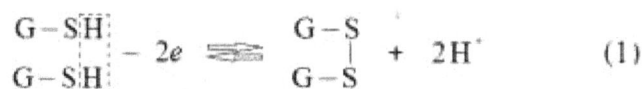

$$\begin{matrix} G-SH \\ G-SH \end{matrix} - 2e \rightleftharpoons \begin{matrix} G-S \\ | \\ G-S \end{matrix} + 2H^{\cdot} \qquad (1)$$

So it should be pointed out that GSSG is not really a "form" of GSH, but a product of condensation of the two GSH mol with the formation of S-S bridge (Figure 2). The interconversion between thiols and disulfide groups is a redox reaction, the thiol corresponding to the reduced state, and the disulfide to the oxidized. That is why the high GSH/GSSG ratio provides the essential reducing environment inside the cell (Reed et al., 2008). The S-S bond between the two divalent sulfur atoms plays an important role as the main stabilizer of the tertiary structure of many proteins.

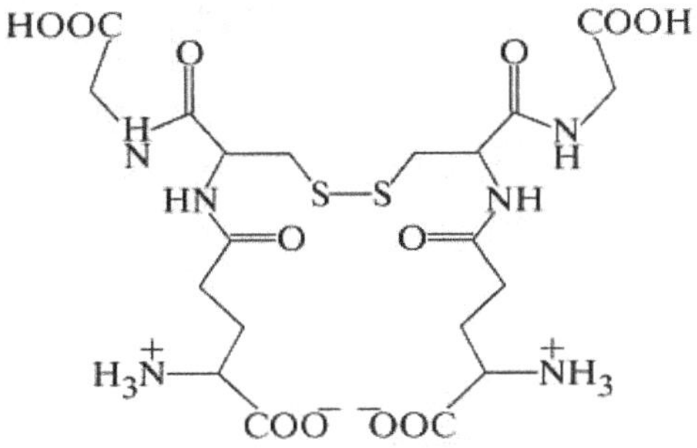

Figure 2. Schematic representation of GSSG

The crystal arrangement of both glutathione varieties has been the subject of several structural and spectroscopic studies at both room temperature (Nicolet, 1930; Qian & Krimm, 1994), and at 120 K (Gorbitz, 1987), in order to obtain further knowledge on hydrogen bonds involving the SH group. It was concluded that GSH and GSSG crystallize in the orthorhombic system, the space group $P2_12_12_1$. The latter contains a crystallographic twofold symmetry axis through the middle of the disulfide bridge. At the same time virtually nothing is known about glutathione analogs containing selenium and tellurium, although there are grounds to suggest that the substitution for sulfur may lead to uncommon biochemical and physico-chemical properties (Soriano-Garcia, 2004). Further, selenium-containing molecules might be expected to reduce H_2O_2 owing to their established anti-oxidant credentials (Ramoutar & Brumaghim, 2010). The simple di-selenide PhSeSePh proved more effective than sulfur derivatives against a panel of 116 pathogenic fungi with greatest inhibitory activity evident against *Candida albicans, Candida dubliniensis, Aspergillus spp.* and *Fusarium spp.* (Tiekink, 2012). Selenium can act as an effective radiosensitizer to enhance the anticancer efficacy through induction of cancer cell apoptosis (Xie, He, Lai, Zheng, & Chen, 2014). Motivated by the potential utility of selenium and tellurium against a number of diseases and pathological conditions, we have undertaken this study in order to fill the gap in structural characteristics of the two substituted tri-peptides. The purpose of this publication is to perform their structural simulations using the modern molecular modeling software to elucidate the similarities and differences between the substituted derivatives and natural glutathione. All methods use empirical data to determine individual force constants, in particular, bond lengths and bond angles. Herein we consider both glutathiones as independent unities, and not as a part of numerous enzymes they are attached to, particularly glutathione peroxidases, which contain selenium in the form of selenocysteine (SeCys), and their chemistry is still obscure.

2. Methods

The geometry optimization was carried out in Cartesian coordinates using the Berny optimization algorithm, adjusting the parameters until a stationary point on the potential surface was found. That means that for a small displacement the energy does not change within a certain amount, and the placements are successfully converged. It should be pointed out that we did not perform any systematic energy sampling for searching conformational energy space. Angles and interatomic distances were evaluated by using special features of the program. The experimental parameters used for comparisons were taken from databases and publications on structural X-ray refinements of related compounds containing sulfhydryl and selenohydril groups, as well as disulfide bridges.

3. Results and Discussion

A number of techniques exist for computerized modeling and calculating the potential energy of molecular systems as a function of coordinates of their atomic nuclei, neglecting explicit treatment of electrons. In this work, both varieties of glutathione were simulated using the standard HYPERCHEM 7.5 software package employing MM+ force field and PM3 semi-empirical (Hyperchem for windows, 2003). In vacuo calculations would bring out most of the underlying conformations without being side-tracked by the solvent used in the study or the limitations imposed by the densest packing. Strictly speaking, no conformational search routine guarantees that all conformers have been found, so the strategy chosen in this work was to study a reasonably representative set of the optimized geometries.

3.1 Reduced Form

To the best of our knowledge, glutathiones containing either selenium (GSeH) or tellurium (GTeH) have never been described. As a result, the reference compounds with the structural data available for comparisons are limited to GSH (Wright, 1958; Gorbitz, 1987), selenocysteine (Nascimento, Melnikov, & Zanoni, 2011) and, to some extent, to the simplest chalcogen hydrides H_2S, H_2Se and H_2Te (National Institute of Standards and Techonology [NIST], 2014). Three structurally similar conformers, one for each chalcogen, were built and oriented in a comparable way. The corresponding models are represented in Figure 3.

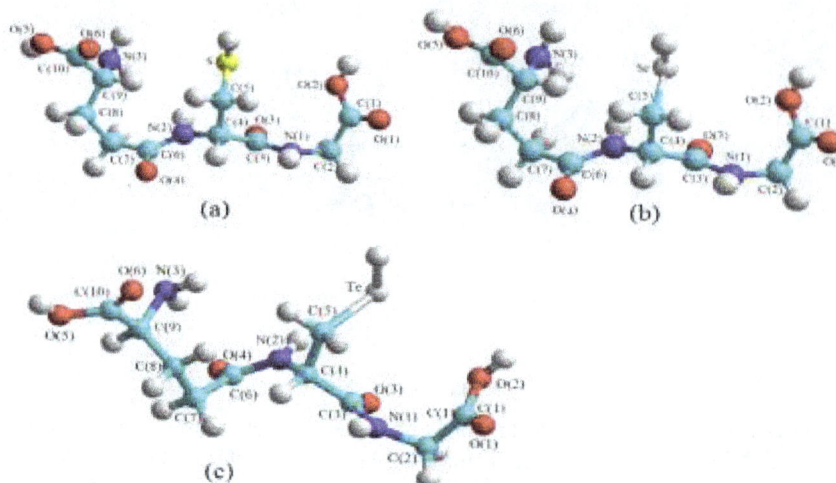

Figure 3. Structural models proposed for reduced glutathiones. (a) GSH; (b) GSeH; (c) GTeH

The geometries can be analyzed using the set of interatomic distances and angles listed in Table 1. Although the carbon atoms make an irregular chain, the interatomic distances C(1)-C(2), C(3)-C(4), C(4)-C(5), C(6)-C(7), C(7)-C(8), C(8)-C(9), C(9)-C(10), are the same (1.51–1.55 Å) as the typical C-C bond lengths in isolated molecules. They neither increase along the chain from C(1) to C(9) nor depart in a significant manner from the normal single-bond value of 1.542 Å as in the solid aminoacids and polypeptides (Protein Interatomic Distance Distribution Database [PIDD] (2014)). That is why they are not presented in Table 1. The same applies to the C-N and C-O bonds.

As for the key bonds Ch-H and Ch-C(5), the former are in good agreement with Se-H distance in SeCys, but larger than in crystalline GSH (Table 2), while the latter are practically the same as in SeCys and GSH. When plotted *vs* the chalcogen ionic radii the bond lengths Ch-H (Figure 4) showed a net linear dependence on this parameter and practically coincide with the analogous dependence of H_2Ch on the same radii (Figure 4). This finding unambiguously suggests that selenol GSeH and tellurol GTeH are polar compounds comparable to the simple hydrides, with all the biochemical consequences in regard to the reduction of cytosolic pH and protein glutathionylation. An important role of GSH is the detoxification of xenobiotics, electrophilic compounds which are eliminated by conjugation to GSH at sulfur site. Hence, selenium and tellurium analogs might enhance this function.

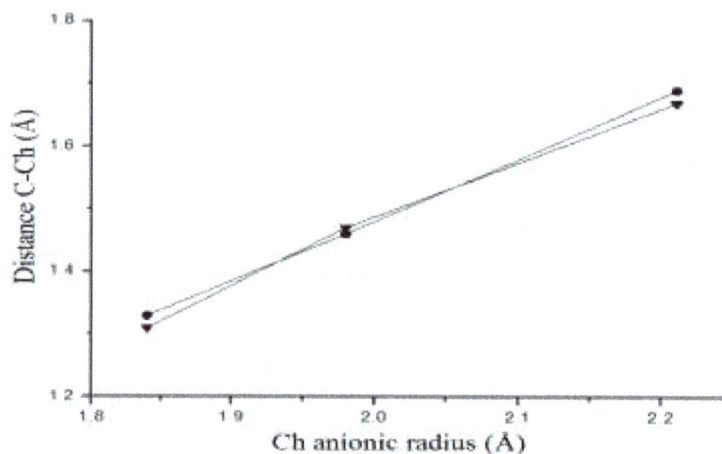

Figure 4. Dependence of the distances Ch-H in GChH (●) and in H_2Ch (▼) on the chalcogen ionic radii

It is worth noting that the calculated distance S-C(5) is practically the same as in the crystalline GSH, a fact which to a certain degree supports predictions made for Se-C and Te-C distances. If these distances are arranged in a row S-Se-Te, there is a similar parallelism with the ionic radii as in the case of the aforementioned Ch-H bond lengths. The comparison between the data obtained using MM+ force field and PM3 semi-empirical quantum mechanical methods shows that the former provides better approximations.

The angles of the main chain in γ-L-glutamyle residue do not appear to be sensitive to the chalcogen nature, being within a range of 111.1–114.7° for C(6)-C(7)-C(8) and C(7)-C(8)-C(9). The angles C(8)-C(9)-C(10) are always smaller, that is 108.3–109.7°. The remaining calculated angles of GSH are similar, but not identical with those determined by X-ray diffraction. A more pronounced dispersion in angles can be easily explained by the existence of a number of conformationl isomers with slightly different values of potential energy (Table 1) due to the simultaneous rotation about the bonds C(1)-C(2), C(3)-C(4), C(4)-C(5), C(6)-C(7) and C(8)-C(9). It is to be born in mind that only the energies calculated using identical techniques can be compared. Naturally, all these considerations are true for isolated molecules as in the solid state such rotation is hindered.

Table 1. Interatomic distances (Å), angles (°) and minimal potential energies (kcal/mol) calculated for GSH, GSeH and GTeH

Parameters	MM+			PM3		
	S	Se	Te	S	Se	Te
Distances	1. 35	1.354	1.35	1.33	1.35	1.35
O(2)-C(1)	1.79	1.934	2.13	1.82	1.95	2.22
S/Se/Te-C(5)	1.46	1.461	1.45	1.51	1.48	1.48
C(4)-N(2)	1.41	1.41	1.42	1.50	1.42	1.42
N(2)-C(6)	1.33	1.480	1.68	1.34	1.46	1.67
Bond angles	120.22	120.22	118.87	123.50	116.60	116.48
O(1)-C(1)-O(2)	119.58	119.58	121.22	121.50	127.90	116.56
O(2)-C(1)-C(2)	120.18	120.18	119.89	114.86	115.38	126.94
O(1)-C(1)-C(2)	109.51	109.50	109.54	109.43	112.43	115.50
C(1)-C(2)-N(1)	112.99	112.61	111.43	108.29	103.22	112.96
C(4)-C(5)-S(Se)	111.11	110.88	112.01	111.70	109.40	107.14
C(3)-C(4)-C(5)	120.91	120.90	117.47	121.42	116.01	115.79
N(2)-C(6)-O(4)	120.22	120.24	124.41	118.82	119.98	120.39
N(2)-C(6)-C(7)	122.62	122.59	124.83	110.84	119.60	120.56
	-15.74	-15.75	-15.82	-3619.9	-3624.7	-3611.3

Energy

Table 2. Main literature data on interatomic distances (Å) and angles (°) in SeCys and GSH

Parameters	Reference compounds			
	Glutathione [4]	Glutathione [7]	Cysteine and selenocysteine [13]	
	S	S	S	Se
O(2)-C(1)	1.30	1.30	1.35	1.35
S/Se/-C(5)	1.78	1.82	1.79	1.93
C(4)-N(2)	1.46	1.45	1.48	1.48
N(2)-C(6)	1.31	1.34	-	-
S/Se/-H	-	1.21	1.34	1.47
O(1)-C(1)-O(2)	122.8	123.1	119.71	119.74
O(2)-C(1)-C(2)	121.5	121.2	120.44	120.40
O(1)-C(1)-C(2)	115.6	115.6	119.68	119.71
C(1)-C(2)-N(1)	109.4	116.8	110.71	110.61
C(4)-C(5)-S(Se)	116.7	114.7	113.65	112.39
C(3)-C(4)-C(5)	109.1	109.6	111.18	110.54
N(2)-C(6)-O(4)	122.6	121.1	-	-
N(2)-C(6)-C(7)	117.9	117.1	-	-
C(4)-N(2)-C(6)	121.9	120.9	-	-

3.2 Oxidized Form

Hitherto, oxidized glutathiones containing either selenium (GSeSeG) or tellurium (GTeTeG) have never been described. As a result, the reference compounds with the structural data available for comparisons are limited to GSSG [5], cysteine (Chaney & Steinrauf, 1974; Hameka, Jensen, Ong, Samuels & Vlahacos, 1998) and, to some extent, to the simplest chalcogen hydrides H_2S_2, H_2Se_2 and H_2Te_2 (Boyd, Perkyns & Ramani, 1983; Elemental., 2003). Three structurally similar conformers, one for each chalcogen, were built and oriented in a comparable way. Calculations were carried out by analogy with the reduced form. The corresponding models are represented in Figure 5 showing the upper and lower moieties of glutathiones under consideration.

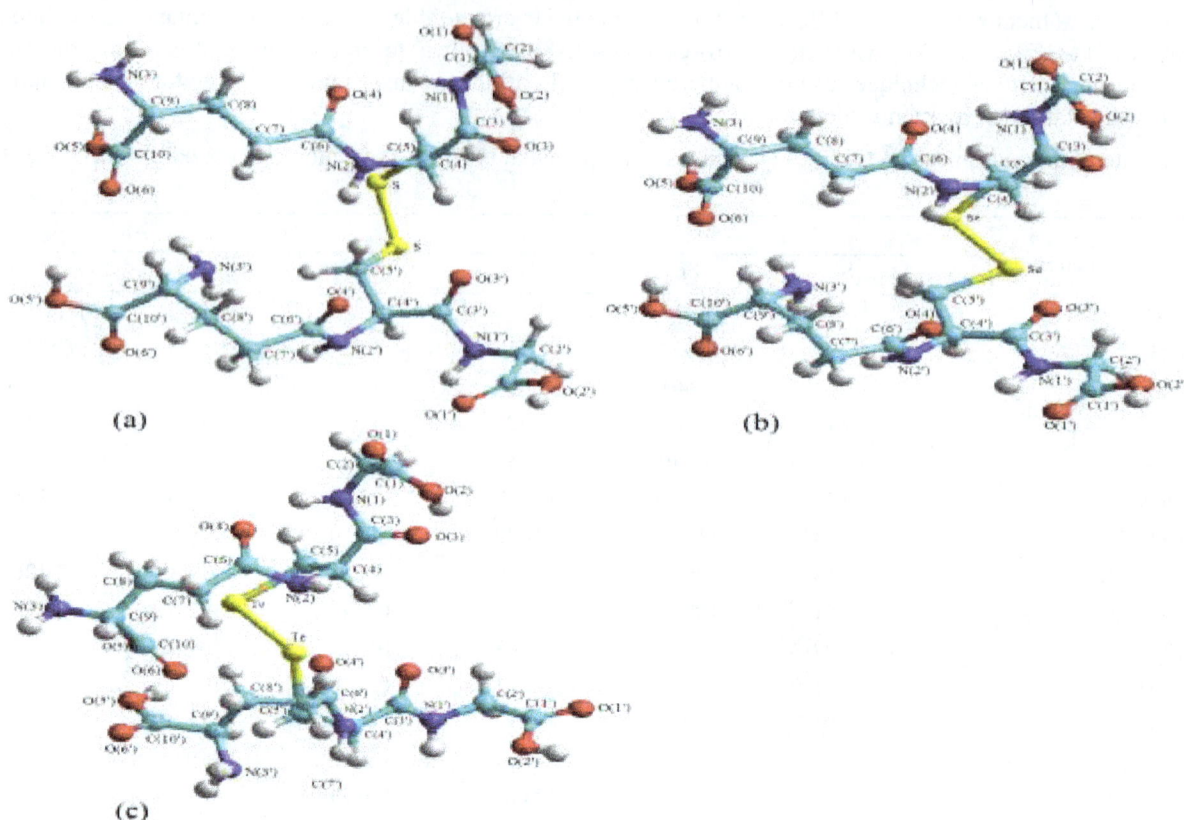

Figure 5. Structural models proposed for oxydized glutathiones. (a) GSSG; (b) GSeSeG; (c) GTeTeG

As can be seen from Table 3, the interatomic distances in both parts of the molecules practically coincide with each other, and simultaneously with those in the reduced portion of glutathione, including the distances Ch-C(5). As for the angles, the coincidences are not so precise, but that fact can be easily explained by the existence of a number of conformational isomers. First of all we are interested in the newly formed bridge Ch-Ch which connects the starting molecules. These distances (means calculated by MM+ and PM3), as obtained in this work are 2.02, 2.35 and 2.72 Å for GSSG, GSeSeG and GTeTeG, respectively. At the same time, the bond length within the S-S bridge as determined from X-ray diffraction for the oxidized form of glutathione is 2.043 Å (Jelsh & Didierjean, 1999). Moreover, the same distance in the H_2S_2 calculated using Gaussian programs is 2.065Å, so our value for GSSG must be correct. As concerns the values calculated for GSeSeG and GTeTeG in the present work, they match the corresponding distances determined experimentally for eight-membered homocycles S_8 and Se_8 which are a common structural motif of the chalcogens (Handbook., 2007), as well as for a number of related organic compounds (Allen et al., 1987): 2.34 for Se-Se and 2.704 Å for Te-Te, respectively. These comparisons suggest that the distances Ch-Ch are essentially independent of substituents in most of chalcogen compounds from elemental chalcogens oxydized glutathiones.

Table 3. Selected interatomic distances (Å) and minimal potential energies (kcal/mol) calculated for the "upper" moiety of GSSG, GSeSeG and GTeTeG

Parameters	MM+			PM3		
	S	Se	Te	S	Se	Te
O(1)-C(1)	1.20	1.20	1.20	1.21	1.21	1.21
O(2)-H	0.96	0.96	0.96	0.96	0.96	0.96
O(2)-C(1)	1.33	1.33	1.33	1.34	1.34	1.34
C(1)-C(2)	1.52	1.52	1.52	1.52	1.52	1.52
C(2)-N(1)	1.44	1.44	1.44	1.47	1.47	1.48
C(3)-N(1)	1.37	1.37	1.37	1.41	1.40	1.37
C(3)-C(4)	1.53	1.53	1.53	1.54	1.53	1.52
C(3)-O(3)	1.20	1.20	1.20	1.22	1.23	1.26
C(4)-C(5)	1.54	1.54	1.54	1.52	1.51	1.51
S/Se/Te-C(5)	1.82	1.93	2.13	1.82	1.94	2.19
C(4)-N(2)	1.45	1.45	1.45	1.48	1.48	1.48
N(2)-C(6)	1.37	1.37	1.37	1.41	1.42	1.42
N(3)-H	1.01	1.01	1.01	0.99	0.99	0.99
Energy	-34.63	-34.76	-34.60	-7118.1	-7186.6	-7150.6

Table 4. Selected interatomic distances (Å) and minimal potential energies (kcal/mol) calculated for the "lower" moiety of GSSG, GSeSeG and GTeTeG

Parameters	MM+			PM3		
	S	Se	Te	S	Se	Te
O(1')-C(1')	1.20	1.20	1.20	1.21	1.21	1.21
O(2')-H	0.97	0.97	0.97	0.95	0.95	0.95
O(2')-C(1')	1.33	1.33	1.33	1.35	1.34	1.35
C(1')-C(2')	1.52	1.52	1.52	1.51	1.52	1.51
C(2')-N(1')	1.44	1.44	1.44	1.47	1.48	1.47
C(3')-N(1')	1.37	1.37	1.37	1.42	1.37	1.39
C(3')-C(4')	1.53	1.53	1.53	1.54	1.53	1.54
C(3')-O(3')	1.20	1.20	1.20	1.21	1.26	1.25
C(4')-C(5')	1.53	1.53	1.53	1.53	1.51	1.50
S/Se/Te-C(5')	1.82	1.96	2.14	1.83	1.97	2.203
C(4')-N(2')	1.45	1.45	1.45	1.49	1.48	1.49
N(2')-C(6')	1.38	1.38	1.38	1.44	1.44	1.43
N(3')-H	1.01	1.01	1.01	0.99	0.99	1.007
Energy	-34.63	-34.76	-34.60	-7118.1	-7186.6	-7150.6

From a comparison of Tables 3 and 4 follows that the main interatomic distances in both parts of the models are identical. It should also be remembered that after condensation this system is provided with a new degree of freedom due to free rotation about the bond Ch-Ch. This enables the oxydized form with the possibility of *cis-trans* isomerism in respect to the glutathione moieties. Naturally, this ability makes available a large population of intermediate conformers. However, it is not the aim of this article to give an overview of their geometry and stabilities since these aspects have been extensively considered in the work dedicated to conformations of simple disulfides and L-cystine (Boyd et al., 1983). Thus our data appear to be appropriate to fill the gap in structural characteristics of seleno–and telluroglutathiones.

4. Conclusion

The comparative structural modeling of reduced and oxidized glutathiones and their derivative containing selenium and tellurium in chalcogen sites (Ch = Se, Te) has provided detailed information about the bond lengths and bond angles, filling the gap in the structural characteristics of these tri-peptides. The investigation using the molecular mechanics technique with good approximation confirmed the available information on X-ray refinements for the related compounds. It was shown that Ch-H and Ch-C bond lengths grow in parallel with the increasing chalcogen ionic radii. Although the distances C-C, C-O, and C-N are very similar, the geometry of GChChG conformers is richer in conformers owing to the possibility of rotation about the bridge Ch-Ch. It is confirmed that the distances Ch-Ch are essentially independent on substituents in most of chalcogen compounds from elemental chalcogens to oxidized glutathiones. The standard program Hyperchem has proved to be an appropriate tool for the structural description of less-common bioactive compositions when direct X-ray data are missing.

Acknowledgments

The authors are indebted to CNPq and FUNDECT (Brazilian agencies) for financial support.

References

Allen, F. H., Kennard, O., Watson D. G., Brammer, L., Orpen, A. G., & Taylor, R. (1987). Tables of bond lengths determined by X-ray and neutron diffraction. Part 1. Bond lengths in organic compounds. *Journal of the Chemical Society, 2*, S1-S19. http://dx.doi.org/10.1039/P298700000S1

Boyd, R. J., Perkyns, J. S., & Ramani, R. (1983). Conformations of simple disulfides and L-cystine. *Canadian Journal of Chemistry, 61*, 1082-1085. http://dx.doi.org/10.1139/v83-191

Chaney, M. O., & Steinrauf, L. K. (1974). The crystal and molecular structure of tetragonal L-cystine. *Acta Crystallographica Section B, 30*. http://dx.doi.org/10.1107/S0567740874003566

Devillanova, F. A. (Ed.) (2007). *Handbook of Chalcogen Chemistry: New Perspectives in Sulfur, Selenium and Tellurium.* Cambridge: The Royal Society of Chemistry.

Gorbitz, C. H. (1987). A redetermination of the crystal and molecular structure of glutathione (γ-L-glutamyl-L-cysteineglycine) at 120K. *Acta Chemica Scandinavica, 41*, 362-368.

Hameka, H. F., Jensen, J. O., Ong, K. K., Samuels, A. C., & Vlahacos, C. P. (1998). Fluorescence of cysteine and cystine. *Journal of Physical Chemistry, 102*, 361-367. http://dx.doi.org/10.1021/jp971631r

Hyperchem for windows (2003) Hyperchem for windows tools for molecular modeling, Release 7.5, Hypercube Inc., 1115 4th street, Gainville, Florida, 32601;UD\USA.

Jelsh, C., & Didierjean, C. (1999). The oxidized form of glutathione. *Acta Crystallographica Section C, 55,* 1538-1540. http://dx.doi.org/10.1107/S0108270199007039

Masella, R., & Mazza, G. (Eds.) (2009). *Glutathione and Sulfur Amino Acids in Human Health and Disease.* John Wiley & Sons, Inc., Hoboken, New Jersey.

Nascimento, V. A., Melnikov, P., & Zanoni, L. Z. (2011). Comparative structural modeling of cysteine and selenocysteine. *Journal of Solids and Structures, 5,* 153-161.

National Institute of Standards and Techonology [NIST] (2014). In *NIST Chemistry WebBook: Standard* Reference Data. Retrived November 10, 2014, from http://webbook.nist/chemistry

Nicolet, B. H. (1930). The structure of glutathione. *Science, 71,* 589-590. http://dx.doi.org/10.1126/science.71.1849.589

Protein Interatomic Distance Distribution Database [PIDD] (2014). In *Protein Interatomic Distance Distribution Database.* Retrived November 12, 2014, from http://math.iastate.edu/pidd

Qian, W., & Krimm, S. (1994). Vibrational analysis of glutathione. *Biopolymers, 34,* 1377-1394. http://dx.doi.org/10.1002/bip.360341009

Ramoutar, R. R., & Brumaghim, J. L. (2010). Antioxidant and anticâncer properties and mechanisms of inorganic selenium, oxo-sulfur, and oxo-sulfur-selenium compounds. *Cell Biochemistry and Biophisics, 58,* 1-12. http://dx.doi.org/10.1007/s12013-010-9088-x

Rauf, S. (Ed.) (2003). *Elemental sulfur and sulfur-rich compounds II.* Springer-Verlag Berlin, Heindelberg.

Reed, M. C., Thomas, R. L., Pavisic, J., James, S. J., Ulrich, C. M., & Nijhout, H. F. (2008). A mathematical model of glutathione metabolism. *Theorical Biological and Medical Modelling, 5,* 1-16. http://dx.doi.org/10.1186/1742-4682-5-8

Soriano-Garcia, M. (2004). Organoselenium compounds as potential therapeutic and chemopreventive agents: a review. *Current Medical Chemistry, 11,* 1657-1669. http://dx.doi.org/10.2174/0929867043365053

Tiekink, E. R. (2012). Therapeutic potential of selenium and tellurium compounds: opportunities yet unrealized. *Dalton Transactions, 41,* 6390-6395. http://dx.doi.org/10.1039/c2dt12225a

Wright, B. H. (1958). The crystal structure of glutathione. *Acta Crystallographica, 11,* 632-642. http://dx.doi.org/10.1107/S0365110X58001699

Xie, Q., He, L., Lai, H., Zheng, W., & Chen, T. (2014). Selenium substitution endows cysteine with radiosensitization activity against cervical cancer cells. *RSC Advances, 4,* 34210-34216. http://dx.doi.org/10.1039/C4RA07031C

12

Preparation of Nano-Sized Mesoporous Silica and Its Application in Immobilization of β-galactosidase from Aspergillus Oryzae

Jiao Yang[1], Zhenzhen Liu[1], Huan Ge[1], Sufang Sun[1]

[1]College of Chemistry and Environmental Science, Hebei University, Baoding, China

Correspondence: Sufang Sun, College of Chemistry and Environmental Science, Hebei University, Baoding, China.
E-mail: hebeiborate@163.com

Abstract

In alkaline conditions, monodisperse nano-sized mesoporous silica was synthesized using cetyl trimethyl ammonium bromide (CTAB) as template and tetraethoxysilane (TEOS) silica as source in ethanol / water cosolvent conditions. Using method of nitrogen adsorption, specific surface area of the dried monodisperse nano-sized mesoporous silica was about 1591 m^2/g and the pore size was about 3.8 nm. The field-emission scanning electron microscope (SEM) micrographs showed that the silica particles obtained were spherical with an approximate diameter of 160 nm and of good dispersion. Transmission electron microscopy (TEM) revealed that the carrier had an excellent cellular structure with disordered multi-channels and smooth surface. The nano-sized mesoporous silica above was employed to immobilize β-galactosidase from aspergillus oryzae for the first time. At the experimental conditions in section 2.4, the enzyme activity and the activity yield were 535.11 U/g dry carrier and 79.63%, respectively. Kinetic data of the immobilized enzyme such as optimum temperature, pH, and thermal and pH stability among other valuable results were also determined.

Keywords: nano-sized mesoporous silica, immobilization, β-galactosidase, CTAB, TEOS

1. Introduction

In 1990s, many researches showed the possibility of synthesizing pore-size distribution of uniform nano-sized mesoporous silica particles using appropriate surfactants such as template agent (He *et al.*, 2009; Zhu *et al.*, 2013). Various effective methods for synthesizing nanomaterials such as hydrothermal reaction, room temperature synthesis, microwave radiation synthesis, precipitation synthesis and sol-gel synthesis have been documented (Zhang *et al.*, 2001; Zhao *et al.*, 1999). The quantities of the mesoporous nanomaterials are diffrenet when the experimental environments are varied (Kresge *et al.*, 1992; Beck *et al.*, 1992; Huang & Kruk, 2015). Among them, more attention has been paid to the synthesis of well-defined mesoporous silica spheres, because their applications are very promising in chromatography, catalysis, cosmetics, and photonic crystals (Huo *et al.*, 1997; Grün *et al.*, 1997; Qi *et al.*, 1998; Boissiere *et al.*, 2001; Kosuge & Singh, 2001; Kosuge *et al.*, 2003; Kosuge *et al.*, 2004). However, there were few reports on the solution synthesis of monodisperse mesoporous silica spheres with nanometer-sized diameters (Yang *et al.*, 2006).

In this study, monodisperse nano-sized mesoporous silica was synthesized in aqueous methanol using cetyl trimethyl ammonium bromide (CTAB) as template and tetraethoxysilane (TEOS) as silica source. Its surface area and pore size were determined by using nitrogen adsorption method. The diameter and structure of the spheres were characterized by the field-emission scanning electron microscope (SEM) and transmission electron microscopy (TEM). The silica spheres obtained were initially used to immobilize β-galactosidase (β-D-galactosidase galactohydrolase, EC 3.2.1.23) from aspergillus oryzae (Bayramoğlu *et al.*, 2005). At the experimental conditions, the enzyme activity and the activity yields were obtained. Similarly, the kinetic data such as temperature and pH optima, thermal and pH stabilities were tested and compared with those of the free enzyme. Finally, the operational stability of the immobilized enzyme was investigated.

2. Experiment

2.1 Reagents and Apparatus

Reagents: All materials were of analytical grade and were used without any further purification. Cetyl trimethyl ammonium bromide (CTAB, Sigma-Aldrich Co. Ltd.), β-galactosidase (From aspergillus oryzae, Sigma-Aldrich Co.

Ltd.), O-nitrophenyl-β-D-galactopyranoside (ONPG, > 99.0%, Bio Basic Inc.), Sodium Cyanoborohydride (95.0%, Shanghai Macklin Biochemical Co. Ltd.), Glutaraldehyde (≥ 50.0%, Tianjin Kemiou Chemical Reagent Co. Ltd.), Diethanolamine (≥ 99.0%, Tianjin Kemiou Chemical Reagent Co. Ltd.) and HCl (37%, Tianjin Kemiou Chemical Reagent Co. Ltd.) were used as received.

Apparatus: Ultraviolet Visible Spectrophotometer (T6 New Century), Vacuum Pump with Circulated Water System (SHZ-D (III)), Digital pH Meter (PHS-3C), Vacuum Desiccator (DZF-6020), Water Constant Temperature Oscillator (SHA-B), Magnetic Hotplate-Stirrer with Timer (MS-H-Pro^T), Scanning Electron Microscope (JSM-7500) and Transmission electron microscope (Tecnai G2 F20 S-TWIN) were used.

2.2 Preparation of Nano-sized Mesoporous Silica Spheres

Monodisperse nano-sized mesoporous silica was synthesized by aqueous methanol (Meng *et al.*, 2009; Trofimova *et al.*, 2012). 3.9400 g of Cetyl trimethyl ammonium bromide (CTAB) was weighed and dissolved in 800 g of aqueous methanol (40%), then 2.3 mL of sodium hydroxide solution (1.0 mol/L) was added into it. After the solution was stirred for 30 minutes, 1.3 mL of TEOS was slowly dropped and the mixture was stirred overnight at room temperature. The precipitate was collected and washed sufficiently with distilled water and ethanol, and then it was put into round-bottom flask which had 20 mL of ethanol and 4 mL of concentrated hydrochloric acid and heated to reflux at 80 °C for 24 h in order to remove the template agent (Anderson *et al.*, 1998). The solid particles, the nano-sized mesoporous silica spheres, were obtained by centrifugation, fully washed with deionized water and dried at 60 °C under vacuum.

2.3 Preparation of the Activated Nano-sized Mesoporous Silica Spheres

3.0000 g of nano-sized mesoporous silica spheres were weighed and put into the solution of 60.0 mL toluene and 4.5 mL aminopropyl triethoxysilane. The mixture was refluxed with constant stirring at 110 °C for 12 h in a nitrogen atmosphere (Zhang *et al.*, 2007). Then, the solid particles, i.e. the amino mesoporous silica, were obtained by centrifugation, washed with toluene and acetone thoroughly, and dried for 12 h at 80 °C vacuum, then put in a desiccator to be used in the next step.

Then, with 2.0 mL glutaraldehyde (25 %) and 8.0 mL diethanolamine mixed together and stirred for 30 minutes at 20 °C, 1.0000 g amino silica obtained above was added (Tu *et al.*, 1999), the precipitate was obtained with constant stirring for 4 h at 30 °C, after being washed completely with distilled water, it was put into HCl solution (0.1 mol/L). The activated mesoporous silica was obtained after 30 minutes with stirring. After being washed with distilled water and dried at 60 °C, it was stored to be used later.

2.4 Immobilization of β-galactosidase

0.0500 g of the activated nano-sized mesoporous silica spheres as carriers was put into 0.5 mL β-galactosidase solution (4.0 mg/mL, prepared with 0.1 mol/L, pH 5.0 phosphate buffer), and process was carried out in the Water Constant Temperature Oscillator at 35 °C for 4 h. The immobilized enzyme was then filtered and washed with distilled water to wash the excess of β-galactosidase.

2.5 Activity Assay of β-galactosidase

The activities of the free and the immobilized enzyme were determined using ONPG (1.5 mg/mL) as the substrate (Jun *et al.*, 2000). For the activity of free enzyme, the free enzyme (0.1 mL) was added to the phosphate buffer (pH 6.0, 0.1 mol/L, 0.9 mL). Then ONPG (1.5 mg/mL, 0.2mL) was added to the mixture. After incubating at 55 °C for 15 minutes, the reaction was stopped by adding Na$_2$CO$_3$ solution (1.0 mol/L, 2.0 mL), and the amount of ONPG was measured directly at 405 nm. For the activity of immobilized enzyme, the immobilized enzyme (0.0500 g) was added to the phosphate buffer (0.1 mol/L, pH 6.0, 1.0 mL). The reaction was carried out and analyzed as above. All activity assays were repeated three times.

One unit (1U) of β-galactosidase activity was defined as the amount of enzyme that released 1 μmol ONPG in 1 minute at 55 °C. The activity yield was calculated as the ratio of immobilized enzyme activity to the total enzyme activity subjected to immobilization, which can be expressed by the equation (1):

$$\text{The enzyme activity yield (\%)} = \frac{\text{immobilize enzyme activity}}{\text{the total enzyme activity subjected to immobilization}} \quad (1)$$

2.6 Operational Stability

The operational stability of the immobilized enzyme was determined according to the following procedures. 0.0500 g of the immobilized enzyme was taken and soaked in 1.8 mL phosphate buffer overnight. After the mixture was incubated at 55 °C for 15 minutes, the reaction was initiated by adding 0.2 mL 1.5 mg/mL ONPG, and then the reactive mixture was analyzed as above. Afterward, the solid was filtered and washed thoroughly with distilled water and the above experiment was repeated under the same conditions.

3. Results and Discussion

3.1 Morphology of Nano-sized Mesoporous Silica Particles

Scanning electron microscopy (SEM) was used to characterize the surface structures and the diameters of the silica spheres and the results were illustrated in Figure 1. The measured diameter of the mesoporous silica particle from the figure was about 160 nm. As is also revealed in Figure 1, the mesoporous silica particles have an excellent dispersibility. Transmission electron microscopy (TEM) had also been done in order to determine the structures of the silica particles obtained. Results in Figure 2 showed that a cellular structure with disordered multi-channels was found in the mesoporous silica, which would be very beneficial to immobilize the enzyme due to the increase in the surface area (Bayramoğlu et al., 2005).

3.2 Surface Area and Pore Size of Mesoporous Silica Particle

As shown in Figure 3, nitrogen adsorption-desorption linear isotherms of mesoporous materials obtained from Quantachrome Instruments version 3.01 were typical Langmuir isotherm of type IV having H3 hysteresis loop (Liang et al., 2014), which indicated the presence of disordered mesopores.

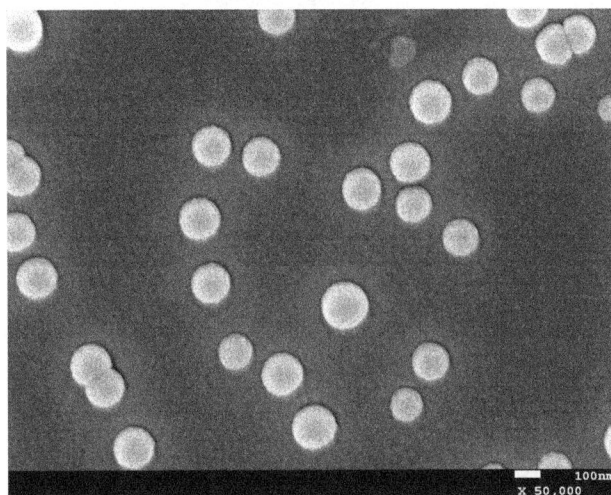

Figure 1. Scanning electron microscope image of nano-sized mesoporous silica

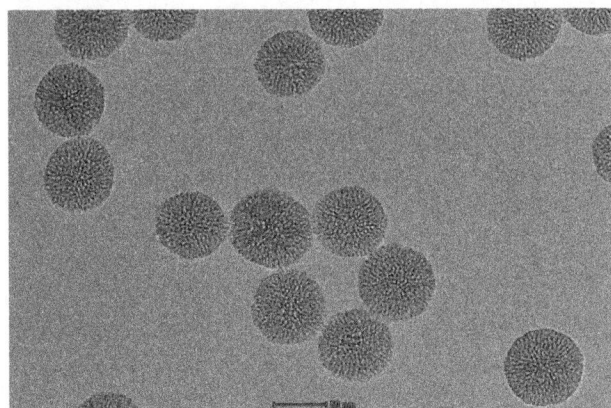

Figure 2. Transmission electron microscopy photograph of nano-sized mesoporous silica

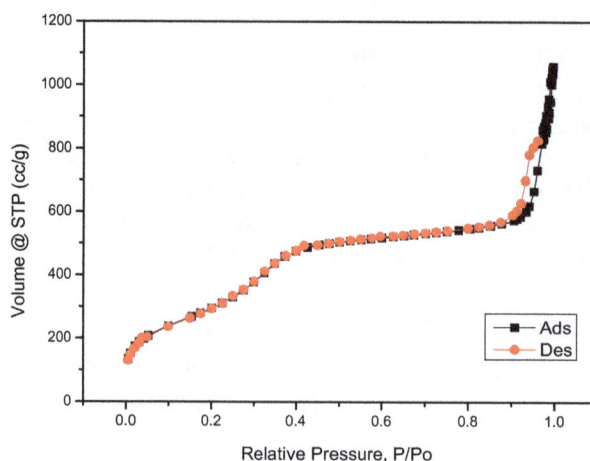

Figure 3. Nitrogen adsorption desorption linear isotherms of mesoporous materials at 77 K

The pore size distribution was determined by DFT method. As shown in Figure 4, a peak appeared at the pore width of 3.8 nm, which means the mainly existing pore size of the synthetic mesopores silica was about 3.8 nm in diameter.

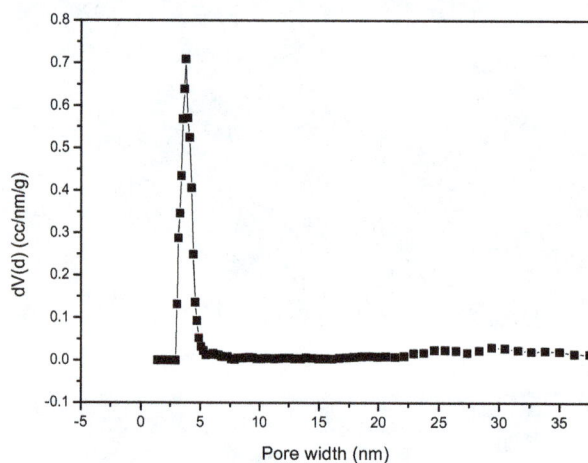

Figure 4. Pore size distribution of synthetic mesoporous materials by DFT method at 77 K

According to Multi-Point BET Plot (Figure 5), the BET surface area was calculated by Quantachrome Instruments Version 3.01 using equation (2) (Fagerlund, 1973):

$$\frac{1}{X(p_0/p-1)} = \frac{1}{X_m C} + \frac{C-1}{V_m C}(\frac{p}{p_0}) \tag{2}$$

Therein, X is the total mass of the adsorbed gas per unit gram carrier at adsorption equilibrium when the gas pressure is p, X_m is the total mass of the adsorbed gas per gram carrier when its surface is fully covered by a single molecular layer, p_0 is the saturated vapor pressure of nitrogen at 77 K, and C is the adsorption coefficient.

According to Figure 5 and Equation (2), X_m could be determined and its value can be substituted into equation (3) (Fagerlund, 1973):

$$S_{BET} = \frac{X_m}{M} N A_m \tag{3}$$

Therein, M is the molecular weight of adsorbate, N is Avogadro's number and A_m is the cross-sectional area of nitrogen and its value is 1.62×10^{-19} m^2 at 77 K.

The specific surface area calculated for nano-sized mesoporous silica particles according to equation (3) was 1591 m^2/g.

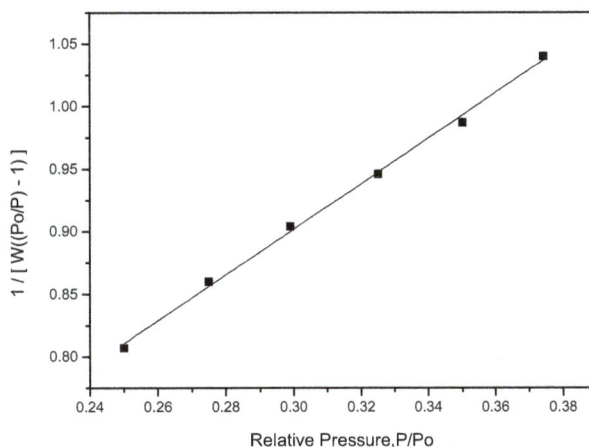

Figure 5. Multi-Point BET Plot of mesoporous materials at 77 K

3.3 Enzyme Activity and the Activity Yield

In the experimental conditions listed in section 2.3, β-galactosidase from aspergillus oryzae was immobilized on the nano-sized mesoporous silica obtained. The enzyme activity and the activity yield determined were 535.11 U/g dry carrier and 79.63%, respectively.

3.4 Properties of the Immobilized β-galactosidase

3.4.1 Optimum Temperature and Thermostability

In Figure 6, the enzyme activities were determined by ONPG as a substrate at temperature range of 40-65 °C at pH of 5.0 for 15 minutes. The results showed that the optimum temperature for free enzyme and immobilized enzyme is 55 °C.

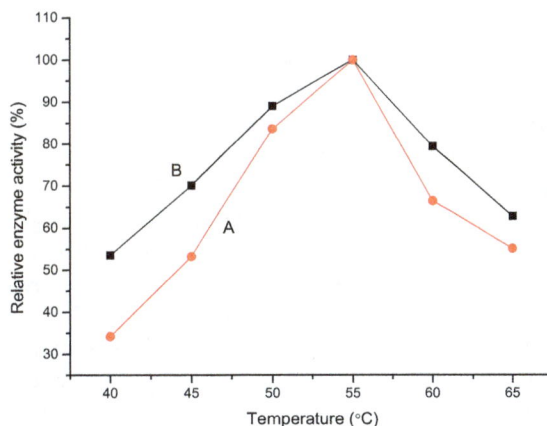

Figure 6. Effect of temperature on the activity of free and immobilized enzyme (A: free enzyme; B: immobilized enzyme)

The thermal stabilities of the free and immobilized enzyme were measured at 50 °C and 60 °C, respectively. After 8 h at 50 °C, the remaining activity of the free enzyme was 49.81% and the immobilized enzyme was 52.63%. After 80 minutes at 60 °C, the residual activity of the former was 19.88% and the latter was 25.15%. So the residual activity of the immobilized enzyme was slightly higher than the free enzyme in 50 °C or 60 °C, and the conclusion was in agreement with the reference (Tanriseven & Doğan, 2002).

3.4.2 Optimum pH and pH Stability

The enzyme activities were tested at 55 °C for 5 minutes in various pH ranging from 3.0 to 8.0. The experimental data indicated that both free and immobilized enzyme attained their maximum activities (100%) in pH 6.0.

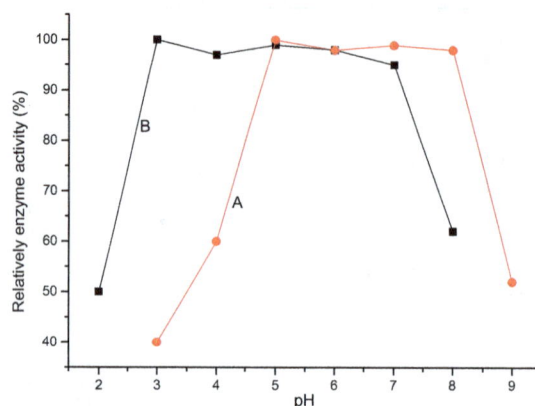

Figure 7. The PH stability of free and immobilized enzyme (A: free enzyme; B: immobilized enzyme)

The pH stabilities of the enzymes were also determined by changing the pH of the phosphate buffer. The results in Figure 7 described that the free enzyme was very stable in the pH range of 5.0-8.0, while that of the immobilized enzyme were 3.0-7.0, the data showed that the acid resistance of immobilized enzyme was greatly improved, but its activity decreased greatly when pH > 7.0, this could be caused by the instability of the silica gel in the base.

3.5 Operational Stability

The experiment was repeated five times by using the procedures mentioned in section 2.6 with the same immobilized enzyme at the same initial concentration of ONPG. In Table 1, all of the measurements of the enzyme activity were almost identical and the maximum deviation was 0.5%. The results showed that there is no insignificant decrease for activity of immobilized enzyme after being used five times, which means that almost no enzyme fell off from the nano-sized mesoporous silica (Sun *et al.*, 1999). Therefore, the immobilized enzyme had a good operational stability.

Table 1. Enzyme activity and recovery rate of immobilized β-galactosidase for five times

Times	1	2	3	4	5
Activity yield (%)	100	99.9	99.7	99.6	99.5

4. Conclusion

In this paper, the monodisperse, nano-sized mesoporous silica was synthesized in methanol aqueous solution with CTAB as template and TEOS as silica source. Scanning electron microscopy (SEM) micrographs showed that the diameter of the silica material obtained was about 160 nm. Nooney et al (Nooney, *et al.*, 2002) discovered that the diameter of the silica material obtained was 740 nm, the discrepancy may be brought due to the dosage of ammonium hydroxide or the different experimental conditions, which need to be further verified. In addition, a cellular structure with disordered multi-channels was also presented by transmission electron microscope (TEM). The specific surface area of the mesoporous silica was about 1591 m^2/g and the pore size was about 3.8 nm by nitrogen adsorption isotherm. At the experimental conditions, β-galactosidase from aspergillus oryzae was immobilized on carrier above and the enzyme activity and the activity yield determined were 535.11 U/g dry carrier and 79.63%, respectively. Kinetic data determined from the immobilized enzyme were as follows: the optimal temperature and pH was 55 °C and 5.0, separately, and the activity was very stable when pH ranged from 3.0 to 7.0, the thermal stability of the immobilized was slightly better than that of the free enzyme. Operational stability revealed that almost no enzyme fell off after being used five times. Considering all the facts described above, the nano-sized mesoporous silica, which has a great specific surface area, is potential to immobilize enzyme and needs to be further studied.

References

Anderson, M. T., Martin, J. E., Odinek, J. G., & Newcomer, P. P. (1998). Surfactant-Templated Silica Mesophases Formed in Water: Cosolvent Mixtures. *Chemistry of Materials*, *10*(1), 311-321. http://dx.doi.org/10.1021/cm9704600

Bayramoğlu, G., Kaya, B., & Arica, M. Y. (2005). Immobilization of Candida rugosa lipase onto spacer-arm attached poly (GMA-HEMA-EGDMA) microspheres. *Food Chemistry*, *92*(2), 261-268. http://dx.doi.org/10.1016/j.foodchem.2004.07.022

Beck, J. S., Vartuli, J. C., Roth, W. J., Leonowicz, M. E., Kresge, C. T., Schmitt, K. D., ... Schlenker, J. L. (1992). A new family of mesoporous molecular sieves prepared with liquid crystal templates. *Journal of the American Chemical Society*, *114*(27), 10834-10843. http://dx.doi.org/10.1021/ja00053a020

Boissiere, C., Kümmel, M., Persin, M., Larbot, A., & Prouzet, E. (2001). Spherical MSU-1 Mesoporous Silica Particles Tuned for HPLC. *Advanced Functional Materials*, *11*(2), 129-135. http://dx.doi.org/10.1002/1616-3028(200104)11:2<129::AID-ADFM129>3.0.CO;2-W

Fagerlund, G. (1973). Determination of specific surface by the BET method. *Matériaux et Construction*, *6*(3), 239-245. http://dx.doi.org/10.1007/BF02479039

Grün, M., Lauer, I., & Unger, K. K. (1997). The synthesis of micrometer- and submicrometer-size spheres of ordered mesoporous oxide MCM-41. *Advanced Materials*, *9*, 254-257. http://dx.doi.org/10.1002/adma.19970090317

He, Q. J., Shi, J. L., Zhao, J. J., Chen, Y., & Chen, F. (2009). Bottom-up tailoring of nonionic surfactant-templated mesoporous silica nanomaterials by a novel composite liquid crystal templating mechanism. *Journal of Materials Chemistry*, *19*(36), 6498-6503. http://dx.doi.org/10.1039/B907266G

Huang, L., & Kruk, M. (2015). Versatile Surfactant/Swelling-Agent Template for Synthesis of Large-Pore Ordered Mesoporous Silicas and Related Hollow Nanoparticles. *Chemistry of Materials*, *27*(3), 679-689. http://dx.doi.org/10.1021/cm5028749

Huo, Q. S., Feng, J. L., Schüth, F., & Stucky, G. D. (1997). Preparation of Hard Mesoporous Silica Spheres. *Chemistry of Materials*, *9*, 14-17. http://dx.doi.org/10.1021/cm960464p

Jun, S., Joo, S. H., Ryoo, R., Kruk, M., Jaroniec, M., Liu, Z., ... Terasaki, O. (2000). Synthesis of New, Nanoporous Carbon with Hexagonally Ordered Mesostructure. *Journal of the American Chemical Society*, *122*(43), 10712-10713. http://dx.doi.org/10.1021/ja002261e

Kosuge, K., & Singh, P. S. (2001). Rapid Synthesis of Al-Containing Mesoporous Silica Hard Spheres of 30-50 μm Diameter. *Chemistry of Materials*, *13*(8), 2476-2482. http://dx.doi.org/10.1021/cm000623b

Kosuge, K., Kikukawa, N., & Takemori, M. (2004). One-Step Preparation of Porous Silica Spheres from Sodium Silicate Using Triblock Copolymer Templating. *Chemistry of Materials*, *16*(21), 4181-4186. http://dx.doi.org/10.1021/cm0400177

Kosuge, K., Murakami, T., Kikukawa, N., & Takemori, M. (2003). Direct Synthesis of Porous Pure and Thiol-Functional Silica Spheres through the S⁺X⁻I⁺ Assembly Pathway. *Chemistry of Materials*, *15*(16), 3184-3189. http://dx.doi.org/10.1021/cm030225j

Kresge, C. T., Leonowicz, M. E., Roth, W. J., Vartuli, J. C., & Beck, J. S. (1992). Ordered mesoporous molecular sieves synthesized by a liquid-crystal template mechanism. *Nature*, *359*, 710-712.

Liang, J. X., Zhang, X. K., Li, X., Huo, H. F., & Wang, C. (2014). Preparation and Characterization of Ordered Mesoporous SiO₂ Under Certain Acidity. *Journal of capital normal university*, *35*(3), 25-29.

Meng, H., Xia T., George, S. J., & Nel, A. E. (2009). A predictive toxicological paradigm for the safety assessment of nanomaterials. *American chemistry society nano*, *3*(7), 1620-1627. http://dx.doi.org/10.1021/nn9005973

Nooney, R. I., Thirunavukkarasu, D., Chen, Y. M., Josephs, R., & Ostafin, A. E. (2002). Synthesis of Nanoscale Mesoporous Silica Spheres with Controlled Particle Size. *Chemistry of Materials*, *14*(11), 4721-4728. http://dx.doi.org/10.1021/cm0204371

Qi, L. M., Ma, J. M., Cheng, H. M., & Zhao, Z. G. (1998). Micrometer-Sized Mesoporous Silica Spheres Grown under Static Conditions. *Chemistry of Materials*, *10*(6), 1623-1626. http://dx.doi.org/10.1021/cm970811a

Sun, S. F., Li, X. Y., Nu, S. L., & You, X. (1999). Immobilization and Characterization of â-Galactosidase from the Plant Gram Chicken Bean (Cicer arietinum). Evolution of Its Enzymatic Actions in the Hydrolysis of Lactose. *Journal of Agricultural and Food Chemistry*, *47*(3), 819- 823. http://dx.doi.org/10.1021/jf980442i

Tanriseven, A., & Doğan Ş. (2002). A novel method for the immobilization of β-galactosidase. *Process Biochemistry*, *38*(1), 27-30. http://dx.doi.org/10.1016/S0032-9592(02)00049-3

Trofimova, E. Y., Grudinkin, S. A., Kukushkina, Y. A., Kurdyukov, D. A., Medvedev, A. V., Yagovkina, M. A., & Golubev, V. G. (2012). Fluorescent monodisperse spherical particles based on mesoporous silica containing rhodamine 6G. *Physics of the Solid State*, *54*(6), 1298-1305. http://dx.doi.org/10.1134/S1063783412060339

Tu, W. X., Sun, S. F., Nu, S. L., & Li, X. Y. (1999). Immobilization of β-galactosidase from *Cicer arietinum* (gram chicken bean) and its catalytic actions. *Food Chemistry*, *64*(4), 495-500. http://dx.doi.org/10.1016/S0308-8146(98)00141-1

Yang, L. M., Wang, Y. J., Sun, Y. W., Luo, G. S., & Dai, Y. Y. (2006). Synthesis of micrometer-sized hard silica spheres with uniform mesopore size and textural pores. *Journal of Colloid and Interface Science*, *299*(2), 823-830.

Zhang, H. J., Wu, J., Zhou, L. P., Zhang, D. Y., & Qi, L. M. (2007). Facile Synthesis of Monodisperse Microspheres and Gigantic Hollow Shells of Mesoporous Silica in Mixed Water-Ethanol Solvents. *Langmuir*, *23*(3), 1107-1113. http://dx.doi.org/10.1021/la0625421

Zhang, X. Z., Le, Y. H., & Gao, Z. (2001). Studies on 12-Tungstophosphoric Heteropolyacid Supported SBA-15 Catalysts. *Journal of advanced chemistry*, *22*(7), 1169-1172.

Zhao, S. L., Zhang, Y. J., Sun, G. D., & Zhai, Y. C. (1999). The Characterization and Synthesis of Mesoporous Molecular Sieve MCM-41 by Microwave Radiation. *petrochemical industry*, *28*(3), 139-141.

Zhu, S. J., Meng, Q. G., Wang, L., Zhang, J. H., Song, Y. B., Jin, H., … Yang, B. (2013). Highly Photoluminescent Carbon Dots for Multicolor Patterning, Sensors, and Bioimaging. *Angewandte Chemie International Edition*, *52*(14), 3953-3957. http://dx.doi.org/10.1002/anie.201300519

13

Structural Refinements and Thermal Properties of L(+)-Tartaric, D(−)-Tartaric, and Monohydrate Racemic Tartaric Acid

Takanori Fukami[1], Shuta Tahara[1], Chitoshi Yasuda[1], Keiko Nakasone[1]

[1]Department of Physics and Earth Sciences, Faculty of Science, University of the Ryukyus, Japan

Correspondence: Takanori Fukami, Department of Physics and Earth Sciences, Faculty of Science, University of the Ryukyus, Okinawa 903-0213, Japan. E-mail: fukami@sci.u-ryukyu.ac.jp

Abstract

Differential scanning calorimetry, thermogravimetric-differential thermal analysis, and X-ray diffraction measurements were performed on single crystals of L(+)-tartaric, D(−)-tartaric, and monohydrate racemic (MDL-) tartaric acid. The exact crystal structures of the three acids, including the positions of all hydrogen atoms, were determined at room temperature. It was pointed out that one of O–H–O hydrogen bonds in MDL-tartaric acid has an asymmetric double-minimum potential well along the coordinate of proton motion. The weight losses due to thermal decomposition of L- and D-tartaric acid were observed to occur at 443.0 and 443.2 K, respectively, and at 306.1 and 480.6 K for MDL-tartaric acid. The weight losses for L- and D-tartaric acid during decomposition were probably caused by the evolution of $3H_2O$ and $3CO$ gases. By considering proton transfer between two possible sites in the hydrogen bond, we concluded that the weight losses at 306.1 and 480.6 K for MDL-tartaric acid were caused by the evaporation of half the bound water molecules in the sample, and by the evaporation of the remaining water molecules and the evolution of $3H_2O$ and $3CO$ gases, respectively.

Keywords: tartaric acid $C_4H_6O_6$, monohydrate racemic tartaric acid $C_4H_6O_6 \cdot H_2O$, double-minimum potential, thermal decomposition, DSC, TG-DTA, X-ray diffraction

1. Introduction

Tartaric acid (chemical formula: $C_4H_6O_6$; systematic name: 2,3-dihydroxybutanedioic acid) is found in grapes, currants, gooseberries, oranges, apples, and in most acidulous fruits, and is widely used in food, medicine, chemistry, light industry, etc. Many tartrate compounds are formed by the reaction of tartaric acid with various positive ions and are used in numerous industrial applications for transducers and in linear and non-linear mechanical devices due to their excellent dielectric, ferroelectric, piezoelectric, and nonlinear optical properties (Abdel-Kader et al., 1991; Desai & Patel, 1988; Firdous et al., 2010; Torres et al., 2002). Several types of tartaric acid crystals, such as potassium hydrogen tartrate, $KC_4H_5O_6$, and calcium tartrate, $CaC_4H_4O_6$, develop naturally in bottled wine and are the major cause of wine's natural and harmless sediment (Boese & Heinemann, 1993; Buschmann & Luger, 1985; Derewenda, 2008; Hawthorne et al., 1982). The sediment of tartrate crystals is a by-product of the wine industry and has to be removed from the wine after yeast fermentation of the grape juice.

Tartaric acid has two asymmetric carbon atoms in a molecule, which provides for four possible different forms: L(+)-tartaric, D(−)-tartaric, racemic (DL-) tartaric, and meso-tartaric acid (Bootsma & Schoone, 1967; Derewenda, 2008; Nie et al., 2001; Okaya et al., 1966; Parry, 1951; Song et al., 2006; Stern & Beevers, 1950). The most common form in nature is L-tartaric acid; meso-tartaric acid is human-made and does not occur in nature. Solutions of L- and D-tartaric acid rotate the plane of polarized light to the left and to the right, respectively, whereas of DL- and meso-tartaric acid show no rotation of plane-polarized light. These properties of optically active molecules derived from tartaric acid were discovered by Biot (Lowry, 1923). Louis Pasteur first separated the two enantiomers of sodium ammonium tartrate by utilizing the asymmetric habit of their crystals (Gal, 2008, 2013; Pasteur, 1848; Tobe, 2003). He also discovered the change in optical rotation induced by the different structures of the enantiomers in water solution. The discovery of enantiomers has played an important role in advancing the scientific understanding of molecular chirality.

The only effective method of establishing the absolute configuration of molecules, by assessing the anomalous scattering in an X-ray diffraction experiment, was proposed by Bijvoet et al. (1951). Thereafter, the absolute crystal structures of L- and D-tartaric acid were determined to be monoclinic with space group $P2_1$ by Stern and Beevers

(1950), and by Okaya et al. (1966), respectively. However, the exact crystal structure of L-tartaric acid, including the positions of all hydrogen atoms, has not yet been determined. As reported in recently published papers, the crystal structures of monohydrate D-tartaric and monohydrate racemic (MDL-) tartaric acid were determined by means of single-crystal X-ray diffraction (Nie et al., 2001; Song et al., 2006). However, the positions of all hydrogen atoms in MDL-tartaric acid have not been refined (Nie et al., 2001). These structures are very different from those of anhydrate L- and D-tartaric acid described in previous papers (Okaya et al., 1966; Stern & Beevers, 1950).

The purpose of this study is to determine the exact crystal structures of L(+)-tartaric, D(–)-tartaric, and MDL-tartaric acid, including the positions of all hydrogen atoms, at room temperature using X-ray diffraction measurements and to report the thermal properties of these acid crystals by differential scanning calorimetry (DSC) and thermogravimetric-differential thermal analysis (TG-DTA) measurements.

2. Experimental

2.1 Crystal Growth

Single crystals of L(+)-tartaric and D(–)-tartaric acid were grown at room temperature by slow evaporation from aqueous solutions in a desiccator over P_2O_5. DL-tartaric acid crystals were also grown by the same method in air under ambient conditions at room temperature. The solution used for DL-tartaric acid crystals was prepared from a solution of L-tartaric acid sealed in an autoclave and maintained at 448 K for 48 hours. Some of the acid in the sealed solution was converted to D-tartaric acid. The DL-tartaric acid, which consists of 50% L-tartaric and 50% D-tartaric acid, has a lower solubility in water than L- and D-tartaric acid. Thus, the growth of DL-tartaric acid crystals from the solution was more rapid than that of L- or D-tartaric acid crystals.

2.2 X-ray Crystal Structure Determination

The X-ray diffraction measurements were carried out using a Rigaku Saturn CCD X-ray diffractometer with graphite-monochromated Mo K_α radiation ($\lambda = 0.71073$ Å). The diffraction data were collected at 299 K using an ω scan mode with a crystal-to-detector distance of 40 mm, and processed using the CrystalClear software package. The samples used were spherical with diameters of 0.32–0.38 mm. The intensity data were corrected for Lorentz polarization and absorption effects. The crystal structures were solved with direct methods using the SIR2011 program and refined on F^2 by full-matrix least-squares methods using the SHELXL-2013 program in the WinGX package (Burla et al., 2012; Farrugia, 2012; Sheldrick, 2015).

2.3 Thermal Measurements

DSC and TG-DTA measurements were respectively carried out in the temperature ranges of 105–380 K and 300–760 K, using DSC7020 and TG-DTA7300 systems from Seiko Instruments Inc. Aluminum open pans with no pan cover were used as the measuring vessels and reference pans for the DSC and TG-DTA measurements. Fine powder samples prepared from crushed single crystals were used for the measurements. The sample amount varied between 3.22 and 7.04 mg, and the heating rates were 5 or 10 K min^{-1} under a dry nitrogen gas flow.

3. Results and Discussion

3.1 Crystal Structures

The crystal structures of L(+)-tartaric, D(–)-tartaric, and DL-tartaric acid were determined at room temperature by X-ray diffraction. The lattice parameters calculated from all observed reflections for L- and D-tartaric acid indicated that both crystals belong to a monoclinic system. The systematic extinctions in the observed reflections revealed that the possible space group of both acids is $P2_1$ or $P2_1/m$. Furthermore, the intensity statistics of the reflections strongly indicated that the crystals belong to an acentric point group. Thus, the space groups of L- and D-tartaric acid were determined to be monoclinic $P2_1$. The lattice parameters calculated from all observed reflections for DL-tartaric acid indicated that the crystal belongs to a triclinic system. The intensity statistics strongly revealed that the crystal belongs to a centric point group. Thus, the space group of DL-tartaric acid was determined to be triclinic $P\bar{1}$.

The atomic coordinates and thermal parameters for these acid crystals, including the positions of all hydrogen atoms, were determined at room temperature. The observed crystal of DL-tartaric acid in this work was confirmed to be monohydrate racemic (MDL-) tartaric acid, $C_4H_6O_6 \cdot H_2O$. Final R-factors of 2.80%, 2.82%, and 4.73% for L-tartaric, D-tartaric, and MDL-tartaric acid, respectively, were calculated for 2619, 2703, and 3162 unique observed reflections. The relevant crystal data, as well as a summary of the intensity data collection, and structure refinement parameters are given in Table 1. The positional parameters in fractions of a unit cell, and the thermal parameters are listed in Table 2. Selected bond lengths (in Å) and angles (in degrees) are given in Table 3. The hydrogen-bond geometry (in Å and degrees) is presented in Table 4.

Figure 1 shows the projections of the crystal structures of L- and D-tartaric acid along the b-axis at room temperature.

The observed structures of L- and D-tartaric acid, including the positions of all hydrogen atoms, are almost the same as that of D-tartaric acid described by Okaya et al. (1966). The observed structure of D-tartaric acid is very different from that of monohydrate D-tartaric acid crystal (orthorhombic $P2_12_12_1$) reported by Song et al. (2006). Although analyses of their structure indicate that an additional water molecule is present in D-tartaric acid, the water molecule is not present in the structure of D-tartaric acid used in this study and in the previous paper (Okaya et al., 1966). The single crystals of D-tartaric acid reported by Song et al. (2006) were grown by the slow cooling method by reducing the temperature from 343 K. The sample crystals used in this study were grown by the slow evaporation method at room temperature. Therefore, it is considered that the difference in water molecules between the D-tartaric acid crystals is caused by the difference in crystal growth conditions.

Table 1. Crystal data, intensity collection and structure refinement for (a) L-tartaric acid ($C_4H_6O_6$), (b) D-tartaric acid ($C_4H_6O_6$), and (c) MDL-tartaric acid (monohydrate racemic tartaric acid, $C_4H_6O_6 \cdot H_2O$) crystals at room temperature.

	(a) L-tartaric acid	(b) D-tartaric acid	(c) MDL-tartaric acid
Compound	$C_4H_6O_6$	$C_4H_6O_6$	$C_4H_6O_6 \cdot H_2O$
M_r	150.09	150.09	168.10
Crystal shape, color	Plate, colorless	Plate, colorless	Prism, colorless
Crystal system, space group	Monoclinic, $P2_1$	Monoclinic, $P2_1$	Triclinic, $P\bar{1}$
Lattice constants	$a = 7.7230(6)$ Å	$a = 7.7281(5)$ Å	$a = 4.8701(4)$ Å
	$b = 6.0056(3)$ Å	$b = 6.0025(3)$ Å	$b = 8.0586(8)$ Å
	$c = 6.2134(4)$ Å	$c = 6.2113(4)$ Å	$c = 9.1550(10)$ Å
			$\alpha = 109.289(3)$ °
	$\beta = 100.176(3)$ °	$\beta = 100.143(2)$ °	$\beta = 99.846(3)$ °
			$\gamma = 96.104(2)$ °
V, Z	283.65(3) Å^3, 2	283.63(3) Å^3, 2	328.97(6) Å^3, 2
D(cal.)	1.757 Mg m^{-3}	1.757 Mg m^{-3}	1.697 Mg m^{-3}
μ(Mo K_α)	0.173 mm^{-1}	0.173 mm^{-1}	0.169 mm^{-1}
$F(000)$	156	156	176
Sample shape	Sphere	Sphere	Sphere
Size in diameter	$2r = 0.36$ mm	$2r = 0.32$ mm	$2r = 0.38$ mm
θ range for data collection	$3.33 - 37.74$ °	$3.33 - 37.86$ °	$2.42 - 37.92$ °
Index ranges	$-13 \leq h \leq 13$	$-13 \leq k \leq 13$	$-8 \leq l \leq 8$
	$-10 \leq h \leq 10$	$-10 \leq k \leq 10$	$-13 \leq l \leq 13$
	$-10 \leq h \leq 10$	$-10 \leq k \leq 10$	$-15 \leq l \leq 15$
Reflections collected	8176	8151	9411
Unique	2911 [R(int)=0.0236]	2955 [R(int)=0.0213]	3393 [R(int)=0.0273]
Completeness to θ_{max}	97.2 %	97.6 %	95.3 %
Absorption correction type	Spherical	Spherical	Spherical
Transmission factor $T_{min}-T_{max}$	$0.8614 - 0.8625$	$0.8614 - 0.8625$	$0.8614 - 0.8625$
Date [$I > 2\sigma(I)$]	2619	2703	3162
Parameter	116	116	133
Final R indices	$R_1 = 0.0280$	$R_1 = 0.0282$	$R_1 = 0.0473$
	$wR_2 = 0.0673$	$wR_2 = 0.0743$	$wR_2 = 0.1189$
R indices (all data)	$R_1 = 0.0317$	$R_1 = 0.0316$	$R_1 = 0.0508$
	$wR_2 = 0.0689$	$wR_2 = 0.0765$	$wR_2 = 0.1217$
Factors a and b in weighting[*]	$a = 0.0369, b = 0$	$a = 0.0445, b = 0$	$a=0.0489, b=0.0647$
Goodness-of-fit on F^2	1.012	1.038	1.109
Extinction coefficient	0.158(17)	0.134(19)	0.177(18)
Largest diff. peak and hole	0.263, -0.199 eÅ^{-3}	0.245, -0.196 eÅ^{-3}	0.454, -0.347 eÅ^{-3}
Flack parameter	-0.1(3)	0.0(3)	

[*]Weighting scheme $w = 1/[\sigma^2(F_o^2) + (aP)^2 + bP]$, $P = (F_o^2 + 2F_c^2)/3$

Table 2. Atomic coordinates and thermal parameters ($\times 10^4$ Å^2) at room temperature for (a) L-tartaric acid, (b) D-tartaric acid, and (c) MDL-tartaric acid crystals with standard deviations in brackets. The anisotropic thermal parameters are defined as $\exp[-2\pi^2 \, (U_{11}a^{*2}h^2 + U_{22}b^{*2}k^2 + U_{33}c^{*2}l^2 + 2U_{23}b^*c^*kl + 2U_{13}a^*c^*hl + 2U_{12}a^*b^*hk)]$. The isotropic thermal parameters (Å^2) for H atoms are listed under U_{11}.

Atom	x	Y	z	U_{11}	U_{22}	U_{33}	U_{23}	U_{13}	U_{12}
(a) L-tartaric acid									
C(1)	0.0217(1)	0.4564(2)	0.2500(2)	142(4)	302(5)	217(4)	6(3)	47(3)	-21(3)
C(2)	0.2003(1)	0.4808(2)	0.1811(2)	133(3)	239(4)	179(4)	1(3)	41(3)	0(3)
C(3)	0.2941(1)	0.6928(2)	0.2824(2)	145(3)	230(4)	221(4)	-22(3)	61(3)	-7(3)
C(4)	0.4635(1)	0.7145(2)	0.1906(2)	169(3)	208(3)	246(4)	-21(3)	75(3)	-29(3)
O(1)	-0.0774(1)	0.6283(2)	0.1801(2)	165(3)	454(5)	450(5)	154(4)	116(3)	86(3)
O(2)	-0.0215(1)	0.3009(2)	0.3499(2)	279(4)	433(5)	522(6)	171(5)	188(4)	-13(4)
O(3)	0.30657(9)	0.2902(1)	0.2341(1)	196(3)	251(3)	225(3)	-15(3)	55(2)	30(3)
O(4)	0.3378(1)	0.6748(2)	0.5104(1)	231(3)	412(4)	205(3)	-73(3)	72(3)	-17(3)
O(5)	0.4293(1)	0.7414(2)	-0.0221(1)	232(3)	463(5)	251(3)	50(3)	100(3)	-50(3)
O(6)	0.60889(9)	0.7012(2)	0.3009(1)	152(3)	414(4)	314(4)	-56(4)	60(3)	-23(3)
H(1)	-0.167(3)	0.620(4)	0.232(3)	0.059(6)					
H(2)	0.339(2)	0.282(3)	0.375(3)	0.030(4)					
H(3)	0.247(3)	0.703(4)	0.560(3)	0.049(5)					
H(4)	0.530(2)	0.748(4)	-0.082(3)	0.045(5)					
H(5)	0.179(2)	0.496(3)	0.028(2)	0.020(3)					
H(6)	0.221(2)	0.825(3)	0.234(2)	0.017(3)					
(b) D-tartaric acid									
C(1)	0.4785(1)	0.4347(2)	0.2502(2)	142(3)	307(5)	213(4)	8(3)	46(3)	-18(3)
C(2)	0.2996(1)	0.4103(2)	0.3188(1)	132(3)	240(4)	177(3)	-1(3)	38(2)	-3(3)
C(3)	0.2058(1)	0.1982(2)	0.2175(2)	147(3)	231(4)	219(3)	-19(3)	62(2)	-8(3)
C(4)	0.0365(1)	0.1766(2)	0.3094(2)	173(3)	206(3)	242(4)	-20(3)	76(3)	-30(3)
O(1)	0.5775(1)	0.2626(2)	0.3199(2)	167(3)	460(5)	446(5)	157(4)	114(3)	87(3)
O(2)	0.5215(1)	0.5905(2)	0.1502(2)	280(4)	434(5)	519(6)	171(5)	189(4)	-15(4)
O(3)	0.19333(9)	0.6010(1)	0.2658(1)	195(3)	254(3)	222(3)	-13(2)	54(2)	33(3)
O(4)	0.1623(1)	0.2163(2)	-0.0105(1)	230(3)	415(5)	197(3)	-70(3)	69(2)	-11(3)
O(5)	0.0708(1)	0.1497(2)	0.5224(1)	233(3)	467(5)	245(3)	49(3)	98(3)	-54(3)
O(6)	-0.10885(9)	0.1899(2)	0.1991(1)	154(3)	416(4)	310(4)	-54(3)	58(2)	-24(3)
H(1)	0.668(3)	0.268(4)	0.267(3)	0.049(6)					
H(2)	0.160(2)	0.609(3)	0.126(3)	0.030(4)					
H(3)	0.251(2)	0.185(4)	-0.063(3)	0.042(5)					
H(4)	-0.030(3)	0.140(4)	0.579(3)	0.048(5)					
H(5)	0.320(2)	0.396(3)	0.471(3)	0.024(3)					
H(6)	0.280(2)	0.065(3)	0.267(2)	0.018(3)					
(c) MDL-tartaric acid									
C(1)	0.3471(2)	-0.2953(1)	0.10734(9)	278(3)	228(3)	233(3)	41(2)	82(2)	96(2)
C(2)	0.2171(2)	-0.1335(1)	0.18337(9)	260(3)	227(3)	236(3)	62(2)	74(2)	108(2)
C(3)	0.4410(2)	0.0373(1)	0.24760(9)	276(3)	230(3)	250(3)	64(2)	100(2)	107(2)
C(4)	0.3078(2)	0.1980(1)	0.3221(1)	306(3)	207(3)	310(4)	79(3)	123(3)	94(2)
O(1)	0.5115(2)	-0.2696(1)	0.0164(1)	488(4)	281(3)	404(4)	61(3)	272(3)	124(3)
O(2)	0.2909(2)	-0.43576(9)	0.1294(1)	518(4)	260(3)	471(4)	132(3)	271(3)	182(3)
O(3)	0.0959(1)	-0.16165(9)	0.30456(8)	300(3)	290(3)	327(3)	113(2)	158(2)	149(2)
O(4)	0.6667(1)	0.0229(1)	0.35914(8)	283(3)	349(3)	285(3)	29(2)	53(2)	149(2)
O(5)	0.1053(2)	0.2210(1)	0.22026(9)	432(4)	312(3)	357(3)	105(3)	100(3)	208(3)
O(6)	0.3878(2)	0.2889(1)	0.46123(9)	503(4)	333(3)	338(3)	-3(3)	81(3)	194(3)
O(7)	0.8819(2)	0.4944(1)	0.3164(1)	543(5)	310(3)	546(5)	159(3)	290(4)	210(3)
H(1)	0.563(6)	-0.382(4)	-0.039(3)	0.107(9)					
H(2)	-0.050(4)	-0.103(2)	0.319(2)	0.052(4)					
H(3)	0.668(4)	0.090(3)	0.448(2)	0.065(6)					
H(4)	0.030(5)	0.324(3)	0.268(3)	0.075(6)					
H(5)	0.077(3)	-0.125(2)	0.096(2)	0.031(3)					
H(6)	0.508(3)	0.055(2)	0.158(2)	0.029(3)					
H(7)	0.801(5)	0.539(3)	0.396(3)	0.076(6)					
H(8)	0.977(5)	0.586(3)	0.308(3)	0.080(7)					

Table 3. Selected interatomic distances (in Å) and angles (in degrees) for (a) L-tartaric acid, (b) D-tartaric acid, and (c) MDL-tartaric acid crystals at room temperature.

(a) L-tartaric acid			
O(1)–C(1)	1.313(1)	O(2)–C(1)	1.201(1)
O(3)–C(2)	1.413(1)	O(4)–C(3)	1.401(1)
O(5)–C(4)	1.311(1)	O(6)–C(4)	1.210(1)
C(1)–C(2)	1.522(1)	C(2)–C(3)	1.543(1)
C(3)–C(4)	1.523(1)		
O(1)–C(1)–O(2)	125.73(9)	O(1)–C(1)–C(2)	109.61(8)
O(2)–C(1)–C(2)	124.65(9)	C(1)–C(2)–O(3)	111.98(8)
C(1)–C(2)–C(3)	110.37(8)	O(3)–C(2)–C(3)	111.22(7)
C(2)–C(3)–O(4)	111.15(8)	C(2)–C(3)–C(4)	106.87(7)
O(4)–C(3)–C(4)	108.43(7)	C(3)–C(4)–O(5)	110.77(8)
C(3)–C(4)–O(6)	123.69(9)	O(5)–C(4)–O(6)	125.50(8)
(b) D-tartaric acid			
O(1)–C(1)	1.313(1)	O(2)–C(1)	1.201(1)
O(3)–C(2)	1.414(1)	O(4)–C(3)	1.401(1)
O(5)–C(4)	1.313(1)	O(6)–C(4)	1.211(1)
C(1)–C(2)	1.524(1)	C(2)–C(3)	1.543(1)
C(3)–C(4)	1.522(1)		
O(1)–C(1)–O(2)	125.79(9)	O(1)–C(1)–C(2)	109.67(8)
O(2)–C(1)–C(2)	124.53(9)	C(1)–C(2)–O(3)	112.05(8)
C(1)–C(2)–C(3)	110.47(8)	O(3)–C(2)–C(3)	111.16(7)
C(2)–C(3)–O(4)	111.10(8)	C(2)–C(3)–C(4)	106.89(7)
O(4)–C(3)–C(4)	108.46(7)	C(3)–C(4)–O(5)	110.76(8)
C(3)–C(4)–O(6)	123.73(9)	O(5)–C(4)–O(6)	125.47(9)
(c) MDL-tartaric acid			
O(1)–C(1)	1.298(1)	O(2)–C(1)	1.225(1)
O(3)–C(2)	1.412(1)	O(4)–C(3)	1.409(1)
O(5)–C(4)	1.310(1)	O(6)–C(4)	1.210(1)
C(1)–C(2)	1.520(1)	C(2)–C(3)	1.535(1)
C(3)–C(4)	1.520(1)		
O(1)–C(1)–O(2)	124.94(7)	O(1)–C(1)–C(2)	113.54(7)
O(2)–C(1)–C(2)	121.50(7)	C(1)–C(2)–O(3)	108.53(6)
C(1)–C(2)–C(3)	110.55(6)	O(3)–C(2)–C(3)	111.63(6)
C(2)–C(3)–O(4)	110.25(7)	C(2)–C(3)–C(4)	109.97(6)
O(4)–C(3)–C(4)	110.55(7)	C(3)–C(4)–O(5)	112.48(7)
C(3)–C(4)–O(6)	121.03(8)	O(5)–C(4)–O(6)	126.48(8)

Table 4. Hydrogen bond distances (in Å) and angles (in degrees) for (a) L-tartaric acid, (b) D-tartaric acid, and (c) MDL-tartaric acid crystals at room temperature.

D–H···A	D–H	H···A	D···A	<D–H···A
(a) L-tartaric acid				
O(1)–H(1)···O(6)[a]	0.82(2)	1.92(2)	2.696(1)	159(2)
O(3)–H(2)···O(6)[b]	0.87(2)	2.04(2)	2.897(1)	168(2)
O(4)–H(3)···O(2)[c]	0.83(2)	2.01(2)	2.836(1)	173(2)
O(5)–H(4)···O(3)[d]	0.92(2)	1.73(2)	2.633(1)	169(2)
C(2)–H(5)	0.94(2)			
C(3)–H(6)	0.99(1)			
(b) D-tartaric acid				
O(1)–H(1)···O(6)[e]	0.83(2)	1.90(2)	2.696(1)	161(2)
O(3)–H(2)···O(6)[f]	0.86(2)	2.05(2)	2.896(1)	168(2)
O(4)–H(3)···O(2)[g]	0.83(2)	2.01(2)	2.836(1)	172(2)
O(5)–H(4)···O(3)[h]	0.92(2)	1.73(2)	2.632(1)	168(2)
C(2)–H(5)	0.94(2)			
C(3)–H(6)	1.00(2)			
(c) MDL-tartaric acid				
O(1)–H(1) ···O(2)[i]	0.96(3)	1.72(3)	2.6789(9)	171(2)
O(3)–H(2)···O(4)[j]	0.90(2)	1.82(2)	2.7123(9)	177(2)
O(4)–H(3)···O(6)	0.82(2)	2.20(2)	2.6539(9)	115(2)
O(4)–H(3)···O(3)[k]	0.82(2)	2.21(2)	2.881(1)	140(2)
O(5)–H(4)···O(7)[j]	0.95(2)	1.59(2)	2.524(1)	170(2)
O(7)–H(7)···O(6)[l]	0.88(2)	1.99(2)	2.823(1)	158(2)
O(7)–H(8)···O(3)[m]	0.86(2)	2.07(2)	2.899(1)	162(2)
C(2)–H(5)	0.99(1)			
C(3)–H(6)	0.98(1)			

Symmetry codes: (a) $x-1, y, z$, (b) $-x+1, y-1/2, -z+1$, (c) $-x, y+1/2, -z+1$, (d) $-x+1, y+1/2, -z$, (e) $x+1, y, z$, (f) $-x, y+1/2, -z$, (g) $-x+1, y-1/2, -z$, (h) $-x, y-1/2, -z+1$, (i) $-x+1, -y-1, -z$, (j) $x-1, y, z$, (k) $-x+1, -y, -z+1$, (l) $-x+1, -y+1, -z+1$, (m) $x+1, y+1, z$.

The crystal structures of L- and D-tartaric acid are related by mirror symmetry, as shown in Fig. 1. The intramolecular bond lengths and angles in the $C_4H_6O_6$ molecule, and the hydrogen bond lengths and angles between adjacent $C_4H_6O_6$ molecules, for L-tartaric acid are almost the same as those for D-tartaric acid, respectively, as shown in Tables 3 and 4. The angle between the two least-squares planes of atoms C(1)C(2)O(1)O(2)O(3) and C(3)C(4)O(4)O(5)O(6) for L-tartaric acid was calculated to be 56.33(5)°, whereas that for D-tartaric acid was calculated to be 56.34(5)°. The structures of both acids consist of hydrogen-bonded networks, which are formed by four different types of O–H–O hydrogen bonds between adjacent $C_4H_6O_6$ molecules, forming layers parallel to the ac-plane. Thus, it is concluded that the crystal structure of L-tartaric acid is almost exactly the same as that of D-tartaric acid.

Figure 2 shows the projection of the MDL-tartaric acid crystal structure along the a-axis at room temperature. The positional parameters of all atoms in the MDL-tartaric acid crystal, including the positions of all hydrogen atoms, were determined in this study. The observed structure is almost the same as that reported by Nie et al. (2001). There is one bound water molecule in the structure of MDL-tartaric acid, which is different from the structures of L- and D-tartaric acid, as mentioned above. There are five different types of O–H–O hydrogen bonds in the structure, forming layers parallel to the bc-plane and chains along the a-axis. The layers of hydrogen-bonded networks consist of an O(1)–H(1)···O(2) hydrogen bond between adjacent $C_4H_6O_6$ molecules, and O(5)–H(4)···O(7)–H(7)···O(6) and O(5)–H(4)···O(7)–H(8)···O(3) bonds between two $C_4H_6O_6$ and H_2O molecules, as shown in Fig. 2. Moreover, the chains along the a-axis consist of O(3)–H(2)···O(4) and O(4)–H(3)···O(3) hydrogen bonds between $C_4H_6O_6$ molecules. These hydrogen-bonded networks are also very different from those in the structures of L- and D-tartaric acid. The angle between the two least-squares planes of atoms C(1)C(2)O(1)O(2)O(3) and C(3)C(4)O(4)O(5)O(6) for MDL-tartaric acid was calculated to be 72.79(4)°. The intramolecular bond lengths and angles in the $C_4H_6O_6$ molecule of MDL-tartaric acid are very similar to those of L- and D-tartaric acid, as shown in Table 4. However, the value of the angle is widely different from those of L- and D-tartaric acid [56.33(5)° and 56.34(5)°, respectively], and the torsion in the molecule of MDL-tartaric acid is larger than those of L- and D-tartaric acid. When the single crystals of MDL-tartaric acid were kept under low humidity, the crystal surface exhibited chalky white opacities. It is considered that the white opacities on the crystal surface are caused by the breaking of hydrogen bonds, followed by water evaporation from the crystal surface.

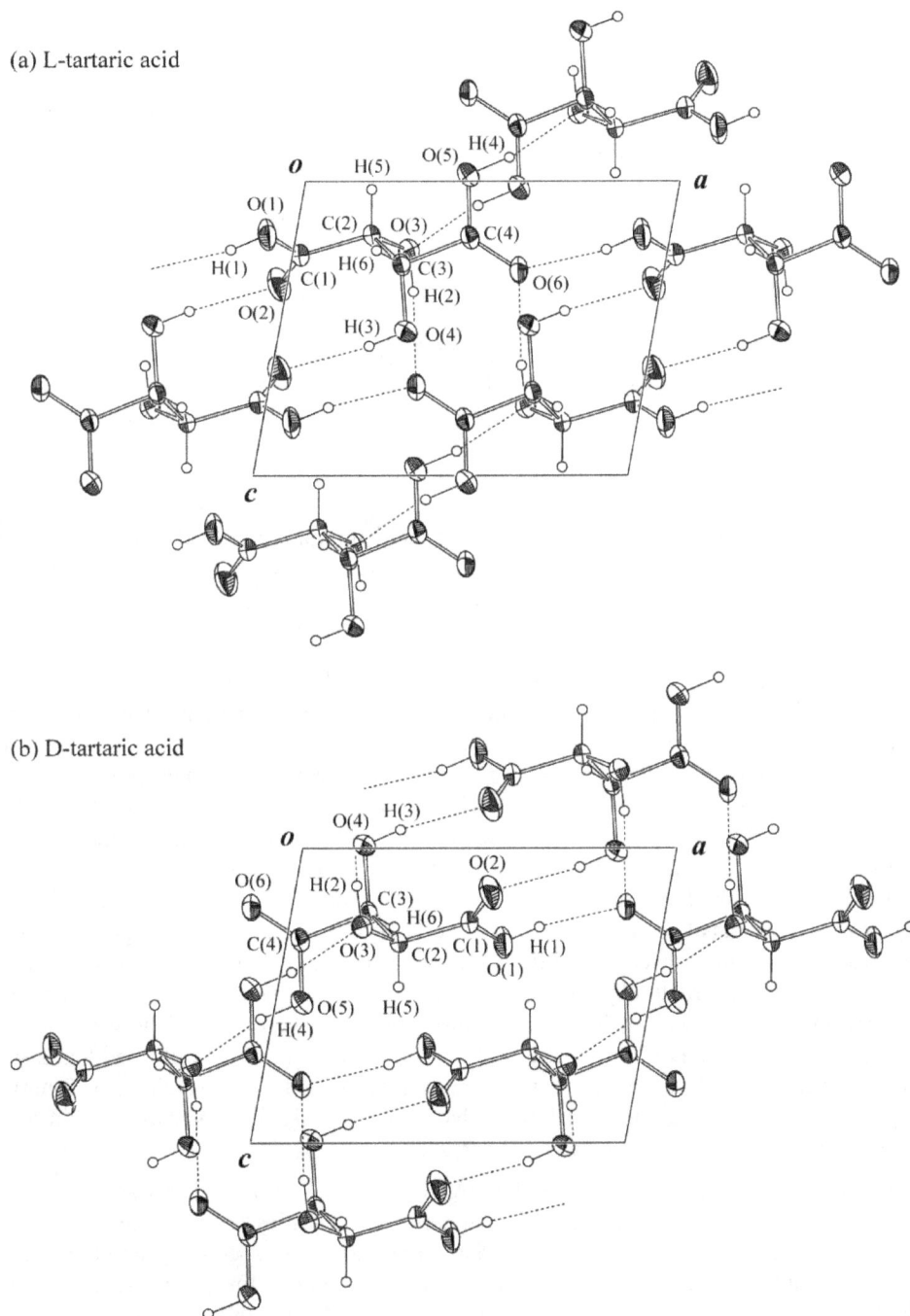

Figure 1. Projections of the crystal structures of (a) L-tartaric and (b) D-tartaric acid along the *b*-axis at room temperature with 50% probability-displacement thermal ellipsoids. The dashed short lines show O–H···O hydrogen bonds, as shown in Table 4.

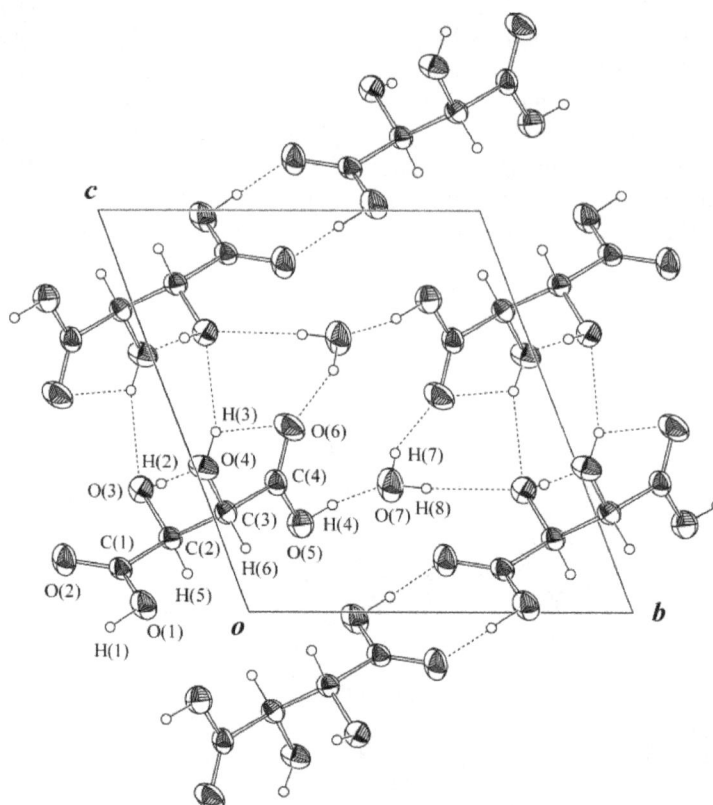

Figure 2. Projection of the crystal structure of MDL-tartaric acid along the *a*-axis at room temperature with 50% probability-displacement thermal ellipsoids. The dashed short lines show O–H···O hydrogen bonds, as shown in Table 4.

3.2 Thermal Analyses

Figure 3 shows the DSC curve of MDL-tartaric acid upon heating in the temperature range from 105 to 380 K. The weight of the sample (powder) used for the measurement was 4.36 mg, and the heating rate was 5 K min^{-1} under a dry nitrogen gas flow of 40 ml min^{-1}. A large endothermic peak is clearly seen in the DSC curve at 321.2 K, with an onset temperature of 306.1 K. The transition enthalpy, ΔH, and transition entropy, ΔS, associated with the large peak were determined to be 47.6 kJ mol^{-1} and 18.7R, respectively, where R is the gas constant (8.314 JK^{-1}mol^{-1}). Moreover, no significant endothermic or exothermic peaks were observed in the DSC curves of L- and D-tartaric acid in the temperature range of 105–380 K. Generally, it is believed that a clear peak in the DSC curve is attributed to a change of exchange energy at phase transition. Thus, these results indicate that the transition of MDL-tartaric acid takes place at 306.1 K, and there is no phase transition of L- and D-tartaric acid in the temperature range between 105 and 380 K. Table 5 shows the peak temperature, onset temperature (transition temperature), ΔH, and ΔS obtained from the DSC curve. Owing to the presence of structurally bound water molecules in MDL-tartaric acid mentioned above, it is expected that the large endothermic peak at 321.2 K is caused by the evaporation of the water. Furthermore, a very small endothermic peak is seen in the DSC curve at 272 K, as shown in Fig. 3. We confirmed that a small endothermic peak (~0.5 mW) is observed at 260 K in the DSC curve, representing the solution (weight of 0.73 mg) of MDL-tartaric acid used for the growth medium. Thus, the small peak observed at 272 K in MDL-tartaric acid is considered to be derived from aqueous-solution infiltration into the sample crystal.

Figure 4 shows the TG, differential TG (DTG), and DTA curves for L-tartaric, D-tartaric, and MDL-tartaric acid crystals in the temperature range of 300–760 K. The weights of the samples (powder) of L-tartaric, D-tartaric, and MDL-tartaric acid used for the measurements were 6.60, 5.82, and 6.51 mg, respectively, and the heating rates were 10 K min^{-1} under a dry nitrogen gas flow of 300 ml min^{-1}. The DTA curve for L-tartaric acid exhibited one large endothermic peak at 444.2 K and two rather small endothermic peaks at 496.9 and 532.1 K. The D-tartaric acid crystal also showed one large and two small endothermic peaks at 444.6, 495.5, and 529.4 K. Moreover, peaks in the DTG curve of L-tartaric acid were seen at 444.3, 493.2, and 531.0 K, and those of D-tartaric acid were seen at 444.5, 496.5, and 528.7 K. The onset temperatures of the large endothermic peaks in the DTA curves of L- and D-tartaric acid were determined to be 443.0 and 443.2 K, respectively. The endothermic peaks in the DTA curve of both acids correspond to the peaks in the DTG curve, respectively. The DTG curve, which is the first derivative of the TG curve, reveals the temperature dependence of the rate of weight loss. Therefore, the endothermic peaks on the DTA curve are associated with the rate of weight loss on the TG curve due to thermal decomposition of the sample.

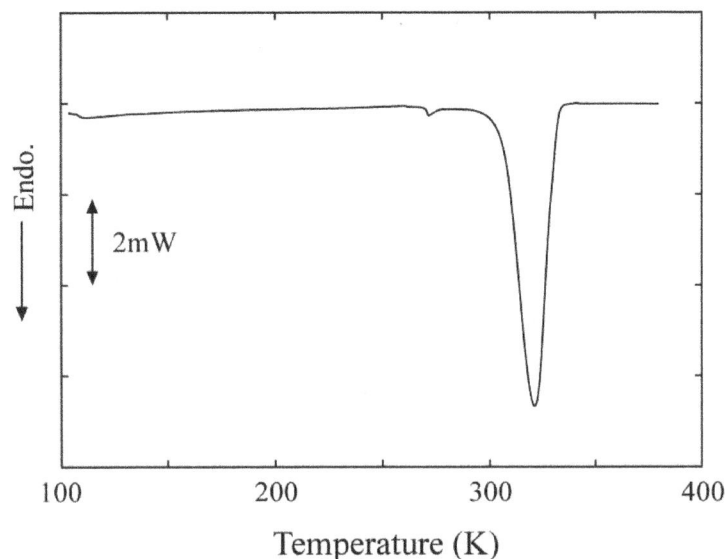

Figure 3. DSC curve for MDL-tartaric acid crystal on heating. The sample weight (powder) was 4.36 mg, and the heating rate was 5 K min^{-1} under a dry N$_2$ gas flow of 40 ml min^{-1}.

Table 5. Peak temperatures, onset temperatures (transition temperatures), transition enthalpy ΔH, and transition entropy ΔS for (a) L-tartaric acid, (b) D-tartaric acid, and (c) MDL-tartaric acid crystals obtained from DSC, DTA, and DTG curves.

(a) L-tartaric acid	DTA	Peak temp. (K)		444.2	496.9	532.1
		Onset temp. (K)		443.0		
	DTG	Peak temp. (K)		444.3	493.2	531.0
(b) D-tartaric acid	DTA	Peak temp. (K)		444.6	495.5	529.4
		Onset temp. (K)		443.2		
	DTG	Peak temp. (K)		444.5	496.5	528.7
(c) MDL-tartaric acid	DSC	Peak temp. (K)	321.2			
		Onset temp. (K)	306.1			
		ΔH (kJ mol^{-1})	47.6			
		$\Delta S/R$	18.7			
	DTA	Peak temp. (K)	324.2	483.8	511.7	526.1
		Onset temp. (K)	313.1	480.6		
	DTG	Peak temp. (K)	322.8	483.7	513.3	525.5

Gas constant R = 8.314 JK^{-1}mol^{-1}

The DTA curve of MDL-tartaric acid exhibited two large endothermic peaks at 324.2 and 483.8 K and two small endothermic peaks at 511.7 and 526.1 K. The DTG curve of MDL-tartaric acid has peaks at 322.8, 483.7, 513.3, and 525.5 K, and these DTG peaks correspond to the respective DTA peaks. The observed DTA peak at 324.2 K corresponds to the above-mentioned DSC peak at 321.2 K. The slight difference of 3 K between the peak temperatures is probably caused by differences in heating rate (5 or 10 K min^{-1}) and start temperature (105 or 300 K) for the measurements. The onset temperatures of the two large endothermic peaks at 324.2 and 483.8 K were determined to be 313.1 and 480.6 K, respectively. The peak at 324.2 K was not observed in the DSC and DTA curves of L- and D-tartaric acid. The start temperature of the weight loss (onset temperature of 480.6 K) in the TG curve of MDL-tartaric acid is higher than those of L- and D-tartaric acid (443.0 or 443.2 K, respectively), as shown in Fig. 4 and Table 5. This temperature of MDL-tartaric acid is approximately 40 K higher than those of L- and D-tartaric acid. The end temperatures (approximately 550 K) of the weight loss of the three acids are approximately same, as shown in the TG curves of Fig. 4. The temperature range of the weight loss in L- and D-tartaric acid is 88 and 85 K, respectively, and in MDL-tartaric acid is 42 K. This range of MDL-tartaric acid is thus approximately half those of L- and D-tartaric acid. The results indicate that the MDL-tartaric acid crystal at high temperature is thermally more stable than the L- and D-tartaric acid crystals.

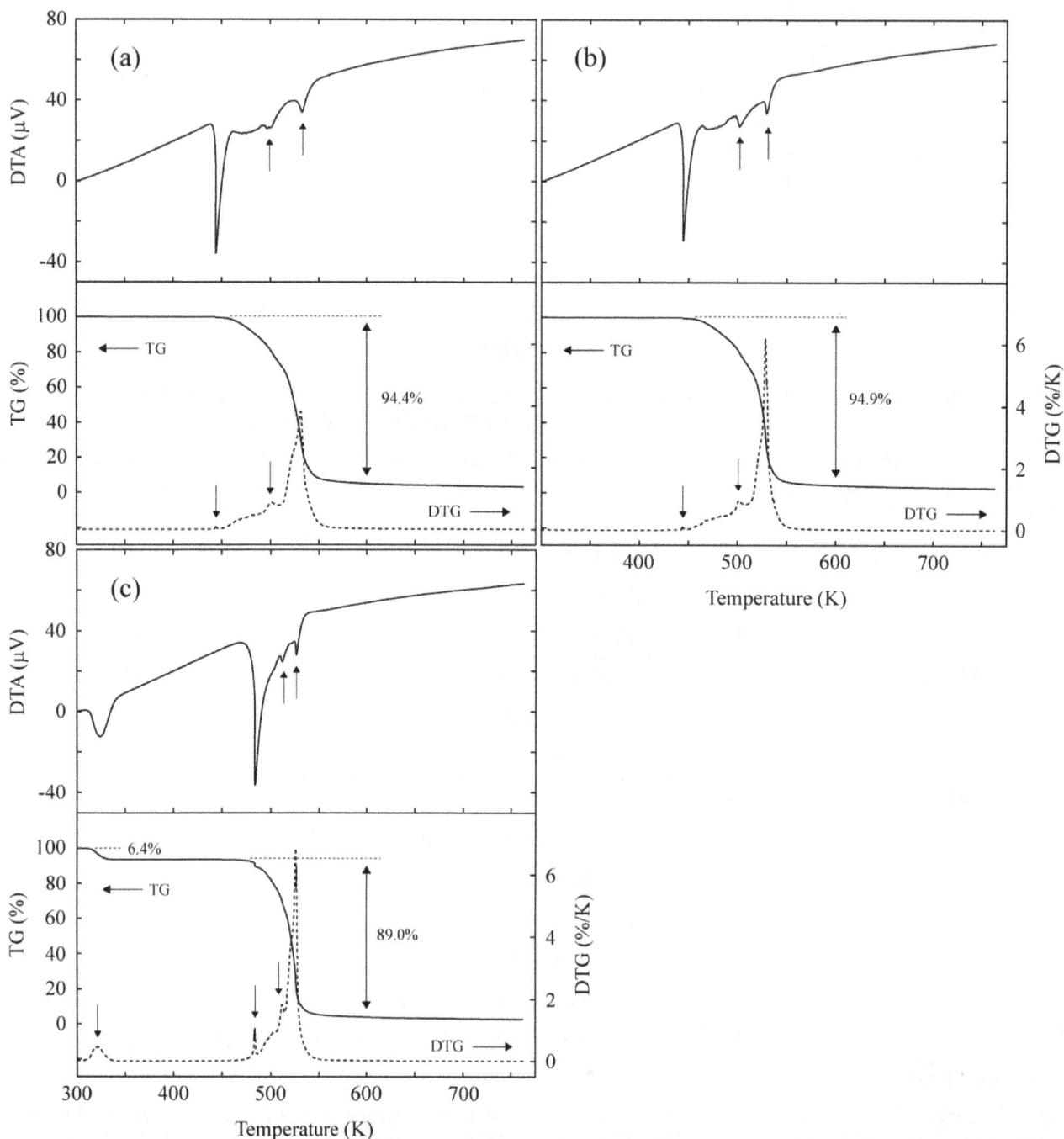

Figure 4. TG, DTG, and DTA thermograms for (a) L-tartaric acid, (b) D-tartaric acid, and (c) MDL-tartaric acid crystals on heating. The sample weights (powder) for (a), (b), and (c) curves were 6.60, 5.82, and 6.51 mg, respectively. The heating rates were 10 K min^{-1} under a dry N_2 gas flow of 300 ml min^{-1}.

The weight losses of L- and D-tartaric acid around 440 K were determined to be 94.4% and 94.9%, respectively, from the TG curves in the temperature range of 425–575 K, as shown in Fig. 4. We assume that the weight loss is caused by the evolution of $3H_2O$ and $3CO$ gases due to thermal decomposition of the sample (which is the chemical formula of $C_4H_6O_6$). The theoretical weight loss is calculated to be 92.0% [=(3×18.02 + 3×28.01) /150.09]. This value is very close to the experimental weight losses of 94.4% and 94.9%. One of four carbon atoms in the $C_4H_6O_6$ molecule remains in the sample pan by the evolution of $3H_2O$ and $3CO$ gases from the sample. After measurements, the inside of the sample pans for L- and D-tartaric acid was confirmed to change from silver to black in color.

The weight loss around 320 K of MDL-tartaric acid was determined to be 6.4% from the TG curve, and it was also determined to be 89.0% around 480 K in the temperature range of 475–575 K. As mentioned above, it is expected that the endothermic peak at 324.2 K is caused by the evaporation of bond water molecules from the sample (which is the chemical formula of $C_4H_6O_6 \cdot H_2O$). The theoretical weight loss caused by the elimination of one H_2O molecule is calculated to be 10.7% [=18.02/168.10]. We assume that the weight loss around 480 K for MDL-tartaric acid is the same as those of L- and D-tartaric acid, as mentioned above. Thus, the theoretical weight loss in the temperature range of 475–575 K is calculated to be 82.1% [=(3×18.02 + 3×28.01)/168.10]. The theoretical weight loss value of 10.7% is much larger than the experimental weight loss of 6.4%, and conversely, one of 82.1% is slightly smaller than the experimental weight loss of 89.0%. Then, we assume that one-half of bound water molecules, which is one of two crystallographically equivalent water molecules in the unit cell, evaporates around 320 K, and the remaining water contributes to the value of the weight loss by the evolution of $3H_2O$ and $3CO$ gases around 480 K. The theoretical weight loss around 320 K is calculated to be 5.4% [=0.5×18.02/168.10], and that in the temperature range of 475–575 K is 87.5% [=(0.5×18.02 + 3×18.02 + 3×28.01)/168.10]. These values calculated are very close to the experimental weight losses of 6.4% and 89.0%, respectively. One of four carbon atoms in MDL-tartaric acid at high temperature remains in the sample pan, as similar to those in L- and D-tartaric acid by decomposition. After the measurement for MDL-tartaric acid, the inside of the sample pan was confirmed to change from silver to black in color too.

3.3 Relation between Water Evaporation and Hydrogen-bonding Properties

As mentioned above, at room temperature, the O(7) atom in the bound H_2O molecule of MDL-tartaric acid forms two hydrogen bonds of O(7)–H(7)···O(6) and O(7)–H(8)···O(3) with the $C_4H_6O_6$ molecule, as shown in Table 4 and Fig. 2. The lengths of these hydrogen bonds are 2.823(1) and 2.899(1) Å, respectively, and the H atom involved in the bonds is attached to the O(7) atom. Moreover, the O(5) atom in the $C_4H_6O_6$ molecule forms the O(5)–H(4)···O(7) hydrogen bond with the O(7) atom in the H_2O molecule. The hydrogen bond length is 2.524(1) Å, and is significantly shorter than those of O(7)–H(7)···O(6) and O(7)–H(8)···O(3). It is known that the O–H–O hydrogen bond has a double-minimum potential well along the coordinate of proton motion when the O···O distance in a hydrogen bond is in the range of 2.43–2.65 Å (Fukami et al., 2010, 2014; Ichikawa, 2000). In fact, the distance of the O(5)···O(7) bond is within the range of 2.43–2.65 Å. Moreover, a difference electron density peak of 0.12 $e\text{Å}^{-3}$ at the final stage of refinement was found close to the O(7) atom on the hydrogen bond, and the peak position was observed at the distance of 0.79 Å from the O(7) atom. The H(4) atom involved in the O(5)–H(4)···O(7) hydrogen bond is attached to the O(5) atom. These findings indicate that the O(5)–H(4)···O(7) hydrogen bond has an asymmetric double-minimum potential well along the O···O bond.

Here, it is expected that the rate of proton transfer between two possible sites on the O(5)–H(4)–O(7) hydrogen bond increases with increasing temperature. When the H(4) atom in the hydrogen bond is located at the O(7) site, the H atom is attached to the O(7) atom. Then, one of hydrogen atoms (H(7) or H(8)) in the H_2O molecule is removed from the O(7) atom due to the so-called "ice rules". The H_2O molecule forms the shorter O(7)–H(4)···O(5) hydrogen bond between the H_2O and $C_4H_6O_6$ molecules by the exchange of hydrogen partners. Since the hydrogen-bonding strength is mainly influenced by O···O distance of the bond, the hydrogen bond strength between the H_2O and $C_4H_6O_6$ molecules is increased by the exchange of the hydrogen bonds. As the result, the evaporation of H_2O molecules will certainly be reduced by increasing the bonding strength. The evolution of the weight loss corresponding to 50% bound water molecules due to the evaporation around 320 K mentioned above, indicates that the exchange of the hydrogen bonds in one-half of water molecules takes place at the temperature, and that is, the relative occupancies of the two sites for the H(4) atom are estimated to be ~0.5. The occupancy of ~0.5 at each site means that the double-minimum potential well of the O(5)–H(4)–O(7) hydrogen bond is changed from asymmetric to symmetric with increasing temperature. It is concluded that the difference between the evaporation temperatures of 306.1 and 480.6 K for bound water molecules is caused by the physical properties of the O(5)–H(4)···O(7) hydrogen bond.

4. Conclusion

Single crystals of L(+)-tartaric, D(–)-tartaric, and racemic tartaric acid were grown at room temperature by slow evaporation from aqueous solutions. The exact room-temperature crystal structures of L-tartaric, D-tartaric and racemic tartaric acid, including the positions of all hydrogen atoms, respectively, were determined to be monoclinic with space group $P2_1$, monoclinic with space group $P2_1$, and triclinic with space group $P\bar{1}$, by means of single-crystal X-ray diffraction. The observed structures of L- and D-tartaric acid were almost exactly the same as each other, and very similar to those reported in the previous papers (Okaya et al., 1966; Stern & Beevers, 1950). The grown single crystals of racemic tartaric acid in this study were monohydrate racemic (MDL-) tartaric acid, $C_4H_6O_6 \cdot H_2O$, and very similar to that reported in the previous paper (Nie et al, 2001). Moreover, one of the O–H–O hydrogen bonds in the acid was 2.524(1) Å in length, and was suggested to have the asymmetric double-minimum potential well along the coordinate of proton motion at room temperature.

The weight losses due to thermal decomposition of L- and D-tartaric acid occurred at 443.0 and 443.2 K, respectively, and those in MDL-tartaric acid occurred at 306.1 and 480.6 K. The weight losses for L- and D-tartaric acid were caused by the evolution of $3H_2O$ and $3CO$ gases due to decomposition, and the losses at 306.1 and 480.6 K for MDL-tartaric acid were caused by the evaporation of one-half bound water molecules and by the remaining water molecules and the evolution of $3H_2O$ and $3CO$ gases, respectively. The weight losses of MDL-tartaric acid are related to the physical properties of the O–H–O hydrogen bond having the proton transfer between two possible sites in the double-minimum potential well along the O$\cdots$O bond, and one of four carbon atoms in the $C_4H_6O_6$ molecule of these acids remains in the sample pans upon heating to 575 K for the decomposition reactions.

References

Abdel-Kader, M. M., El-Kabbany, F., Taha, S., Abosehly, A. M., Tahoon, K. K., & El-Sharkawy, A. A. (1991). Thermal and electrical properties of ammonium tartrate. *J. Phys. Chem. Solids, 52*(5), 655-658. http://www.sciencedirect.com/science/journal/00223697/52/5

Bijvoet, J. M., Peerdeman, A. F., & Bommel, A. J. (1951). Determination of the absolute configuration of optically active compounds by means of X-rays. *Nature, 168,* 271-272. http://dx.doi.org/10.1038/168271a0

Boese, R., & Heinemann, O. (1993). Crystal structure of calcium tartrate tetrahydrate, $C_4H_4O_6Ca(H_2O)_4$. *Z. Kristallogr., 205*(1-2), 348-349. http://dx.doi.org/10.1524/zkri.1993.205.12.348

Bootsma, G. A., & Schoone, J. C. (1967). Crystal structures of mesotartaric acid. *Acta Crystallogr., 22*(4), 522-532. http://dx.doi.org/10.1107/S0365110X67001070

Burla, M. C., Caliandro, R., Camalli, M., Carrozzini, B., Cascarano, G. L., Giacovazzo, C., Mallamo, M., Mazzone, A., Polidori, G., & Spagna, R. (2012). SIR2011: a new package for crystal structure determination and refinement. *J. Appl. Crystallogr., 45*(2), 357-361. http://dx.doi.org/10.1107/S0021889812001124

Buschmann, J., & Luger, P. (1985). Structure of potassium hydrogen (+)-tartrate at 100 K, $K^+.C_4H_5O_6^-$. *Acta Crystallogr., C41*(2), 206-208. http://dx.doi.org/10.1107/S0108270185003341

Derewenda, Z. S. (2008). On wine, chirality and crystallography. *Acta Crystallogr., A64*(1), 246-258. http://dx.doi.org/10.1107/S0108767307054293

Desai, C. C., & Patel, A. H. (1988). Crystal data for ferroelectric $RbHC_4H_4O_6$ and $NH_4HC_4H_4O_6$ crystals. *J. Mater. Sci. Lett., 7*(4), 371-373. http://link.springer.com/article/10.1007/BF01730747

Farrugia, L. J. (2012). WinGX and ORTEP for Windows: an update. *J. Appl. Crystallogr., 45*(4), 849-854. http://dx.doi.org/10.1107/S0021889812029111

Firdous, A., Quasim, I., Ahmad, M. M., & Kotru, P. N. (2010). Dielectric and thermal studies on gel grown strontium tartrate pentahydrate crystals. *Mull. Mater. Sci., 33*(4), 377-382. http://www.ias.ac.in/matersci/bmsaug2010/377.pdf

Fukami, T., Miyazaki, J., Tomimura, T., & Chen, R. H. (2010). Crystal structures and isotope effect on $Na_5H_3(SeO_4)_4 \cdot 2H_2O$ and $Na_5D_3(SeO_4)_4 \cdot 2D_2O$ crystals. *Cryst. Res. Technol., 45*(8), 856-862. http://dx.doi.org/10.1002/crat.201000116

Fukami, T., Tahara, S., & Nakasone, K. (2014). Thermal properties and structures of $CsHSO_4$ and $CsDSO_4$ crystals. *Int. Res. J. Pure Appl. Chem., 4*(6), 621-637. http://dx.doi.org/10.9734/IRJPAC/2014/11331

Gal, J. (2008). The discovery of biological enantioselectivity: Louis Pasteur and the fermentation of tartaric acid, 1857 - a review and analysis 150 yr later. *Chirality., 20*(1), 5-19. http://dx.doi.org/10.1002/chir.20494

Gal, J. (2013). Citation for chemical breakthrough awards: Choosing pasteur's award-winning publication. *Bull. Hist. Chem., 38*(1), 7-12. http://www.scs.illinois.edu/~mainzv/HIST/awards/Citations/v38-1%20p7-12.pdf

Hawthorne, F. C., Borys, I., & Ferguson, R. B. (1982). Structure of calcium tartrate tetrahydrate. *Acta Crystallogr., B38*(9), 2461-2463. http://dx.doi.org/10.1107/S0567740882009042

Ichikawa, M. (2000). Hydrogen-bond geometry and its isotope effect in crystals with OHO bonds - revisited. *J. Mol. Struct., 552*(1-3), 63-70. http://dx.doi.org/doi:10.1016/S0022-2860(00)00465-8

Lowry, T. M. (1923). Pasteur as chemist. *Proc. Roy. Soc. Med., 16,* 16-20.

http://www.ncbi.nlm.nih.gov/pmc/articles/PMC2104002/

Nie, J. J., Xu, D. J., Wu, J. Y., & Chiang, M. Y. (2001). Redetermination of racemic tartaric acid monohydrate. *Acta Crystallogr.*, *E57*(5), o428-o429. http://dx.doi.org/10.1107/S1600536801006158

Okaya, Y., Stemple, N. R., & Kay, M. I. (1966). Refinement of the structure of D-tartaric acid by X-ray and neutron diffraction. *Acta Crystallogr.*, *21*(2), 237-243. http://dx.doi.org/10.1107/S0365110X66002664

Parry, G. S. (1951). The crystal structure of hydrate racemic acid. *Acta Crystallogr.*, *4*(2), 131-138. http://dx.doi.org/10.1107/S0365110X51000416

Pasteur, M. L. (1848). Sur les relations qui peuvent existre la forme cristalline, la composition chimique et Le sens de la polarisation rotatoire. *Ann. Chim. Phys.*, 24, 442-463. http://gallica.bnf.fr/ark:/12148/bpt6k65691733/f446.image.r=Annales%20de%20chimie%20et%20de%20physique.langEN

Sheldrick, G. M. (2015). Crystal structure refinement with SHELXL. *Acta Crystallogr.*, *C71*(1), 3-8. http://dx.doi.org/10.1107/S2053229614024218

Song, Q. B., Teng, M. Y., Dong, Y., Ma, C. A., & Sun, J. (2006). (2S,3S)-2,3-Dihydroxy-succinic acid monohydrate. *Acta Crystallogr.*, *E62*(8), o3378-o3379. http://dx.doi.org/10.1107/S1600536806021738

Stern, F., & Beevers, C. A. (1950). The crystal structure of tartaric acid. *Acta Crystallogr.*, *3*(5), 341-346. http://dx.doi.org/10.1107/S0365110X50000975

Tobe, Y. (2003). The reexamination of Pasteur's experiment in Japan. *Mendeleev Commun.*, *13*(3), 93-94. http://dx.doi.org/10.1070/MC2003v013n03ABEH001803

Torres, M. E., Peraza, J., Yanes, A. C., López, T., Stockel, J., López, D. M., Solans, X., Bocanegra, E., & Silgo, G. G. (2002). Electrical conductivity of doped and undoped calcium tartrate. *J. Phys. Chem. Solids*, *63*(4), 695-698. http://www.sciencedirect.com/science/journal/00223697/63/4

14

Unrestricted Floating Orbitals for the Investigation of Open Shell Systems

Eva Perlt[1], Christina Apostolidou[1], Melanie Eggers[1], Barbara Kirchner[1]

[1]Mulliken Center for Theoretical Chemistry, Rheinische Friedrich-Wilhelms-Universität Bonn, Germany

Correspondence: Prof. Dr. Barbara Kirchner, Mulliken Center for Theoretical Chemistry, Rheinische Friedrich-Wilhelms-Universität Bonn, Beringstraße 4, 53115 Bonn, Germany.
E-mail: kirchner@thch.uni-bonn.de

Abstract

The floating orbital molecular dynamics approach treats the basis functions' centers in ab initio molecular dynamics simulations variationally optimized in space rather than keeping them strictly fixed on nuclear positions. An implementation of the restricted theory for closed shell systems is already available (Perlt et al., Phys. Chem. Chem. Phys., 2014, 16, 6997–7005). In this article, the extension of the methodology to the unrestricted theory in order to cover open shell systems is introduced. The methyl radical serves as a test system to prove the correctness of the implementation and to demonstrate the scope of this method. The available spin density plots and vibrational spectra are compared to those obtained from atom-centered bases. Finally, more complex systems as well as further properties to be studied in future investigations by floating orbitals are suggested.

Keywords: floating basis functions, methyl radical, Hartree–Fock, molecular dynamics

1. Introduction

In recent years, the investigation of more and more complex systems by multi-scalar methods became feasible (Masson, Laino, Röthlisberger, & Hutter, 2009) (Ihrig, Schiffmann, & Sebastiani, 2011) (Kurzbach, Sharma, Sebastiani, Klinkhammer, & Hinderberger, 2011) (Golze, Iannuzzi, Nguyen, Passerone, & Hutter, 2013). This progress was only possible by introducing alternatives to standard methods. An example for such an alternative is the usage of floating orbitals which means that the centers of the basis functions for the construction of the wave function are not necessarily located on the nuclear positions but are optimized in space in order to minimize the total energy. A closed shell implementation of the floating orbital molecular dynamics (FOMD) approach has been presented in detail elsewhere (Perlt, Brüssel, & Kirchner, 2014). The general idea is to distinguish between centers of basis functions ρ and nuclear coordinates A when formulating the molecular integrals, which arise when applying the Hamiltonian to basis functions for the construction of the self-consistent field (SCF) equations. This is in contrast to conventional methods, where basis functions are centered on their respective nuclei. Having introduced another set of position variables, another gradient vector containing the derivative of the total energy with respect to the centers of the basis functions — the electronic gradient — is available. This one is now used to minimize the total energy with respect to the basis functions' centers which is termed floating. Afterwards the molecular gradient — the derivative of the total energy with respect to nuclear coordinates — is evaluated as usual in order to determine forces which can be applied in the frame of an integration algorithm to perform molecular dynamics simulations. FOMD is a promising tool for several reasons. At first floating functions are capable of replacing additional local functions like bond functions (Neisius & Verhaegen, Bond functions for ab initio calculations on polyatomic molecules. Hydrocarbons, 1979), (Neisius & Verhaegen, Bond functions for ab initio calculations on polyatomic molecules. Molecules containing C, N, O and H., 1981) or augmented basis sets in the case of chemically complex systems (Pitarch-Ruiz, Evangelisti, & Maynau, 2005). Furthermore, floating orbitals are potentially helpful to describe non-nuclear attractors. These are maxima in electron density which cannot be assigned to a nucleus and have been observed for solvated electron clusters from theory (Timerghazin & Peslherbe, 2007) as well as experimentally by x-ray spectroscopy (Platts, Overgaard, Jones, Iversen, & Stasch, 2011). Finally, in molecular dynamics (MD) simulations the unwanted Pulay forces arise due to incomplete local basis sets and can be eliminated by the usage of floating orbitals (Marx & Hutter, 2009).

In this contribution, the extension of the floating orbital molecular dynamics approach to unrestricted Hartree–Fock theory is presented. The article is structured as follows. In section 2, the method of the unrestricted floating orbital (UFO) molecular dynamics approach will be briefly introduced. In section 3, the case study on the methyl radical is described together with the presentation and discussion of results. The paper closes with a summary and a brief outlook on future studies.

2. Method

At this point, a short note on restricted floating orbitals is given. As already stated, the centers of the basis functions are no longer fixed on nuclear positions but optimized in space. In ab initio MD simulations, the electronic structure is determined at each MD step, where for FOMD simulations the optimization of the basis functions' centers is additionally carried out. In order to optimize the basis positions, a gradient is calculated as the first derivative of the total energy with respect to the basis functions' centers. After successful optimization, the force on the nuclei is obtained from the first derivative of the total energy with respect to their respective position coordinates. The latter one is then applied to displace the atoms according to the integration algorithm. A simplified scheme of this approach is given in Fig. 1. A detailed derivation and further references can be found in (Perlt, Brüssel, & Kirchner, 2014).

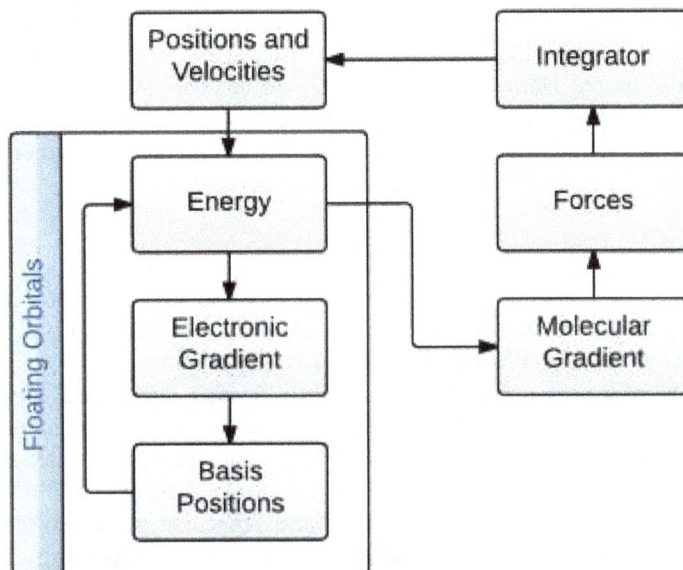

Figure 1. Simplified flow scheme of the FOMD loop.

The extension of floating orbital molecular dynamics (FOMD) to open shell systems is now straightforward. Instead of considering one spatial orbital to comprise two electrons with α and β spin, respectively, spin orbitals are defined in the unrestricted Hartree–Fock (UHF) theory:

$$\chi_i(\boldsymbol{x}) = \begin{cases} \psi_i^\alpha(\boldsymbol{r})\alpha(\omega) \\ \psi_i^\beta(\boldsymbol{r})\beta(\omega) \end{cases} \tag{1}$$

where the spin variable ω has been introduced and separated from the spatial variable $\mathbf{r}$. The variable $\mathbf{x}$ contains both spatial and spin coordinates, similarly χ_i denotes a spin orbital which is factorized to a spatial orbital ψ_i and a spin value α or β. For either value of the spin variable, the one-electron eigenvalue equation for the one-electron Fock operator $\hat{f}$ can be formulated:

$$\hat{f}(\boldsymbol{x})\psi_i^\alpha(\boldsymbol{r})\alpha(\omega) = \varepsilon_i^\alpha\psi_i^\alpha(\boldsymbol{r})\alpha(\omega) \tag{2}$$

with ε_i^α denoting the energy eigenvalues. An analogous formulation for β spin is conceived, easily. Due to normalization of spin orbitals, the spin contribution in Eq. 2 can be eliminated by multiplication with the complex conjugate and integration. Thus for both spin coordinates, a set of eigenvalue equations determining the spatial orbitals are obtained, of which one is exemplarily given for an electron with index 1 in the orbital with index j possessing α spin:

$$\hat{f}^\alpha(\boldsymbol{r}_1)\psi_j^\alpha(\boldsymbol{r}_1) = \varepsilon_j^\alpha\psi_j^\alpha(\boldsymbol{r}_1) \ . \tag{3}$$

The one-electron Fock operator $\hat{f}^\alpha$ is composed of one-electron contributions $\hat{h}$, the exchange interaction with electrons of the same spin K and two-electron Coulomb repulsive interactions J with all remaining electrons:

$$\hat{f}^\alpha(\boldsymbol{r_1}) = \hat{h}(\boldsymbol{r}_1) + \sum_{a=1}^{N^\alpha}\left[J_a^\alpha(\boldsymbol{r}_1) - K_a^\alpha(\boldsymbol{r}_1)\right] + \sum_{a=1}^{N^\beta}J_a^\beta(\boldsymbol{r}_1) \tag{4}$$

where N^α and N^β denote the number of α and β electrons, respectively. The formulation of the analogous operator for β spin is again straightforward. The spatial orbitals may now be expanded in a set of spin independent basis functions ϕ_μ, whereas the expansion coefficients $C_{\mu\nu}^\alpha$ and $C_{\mu\nu}^\beta$ are spin dependent:

$$\psi_j^\alpha = \sum_{\mu=1}^K C_{\mu j}^\alpha \phi_\mu \;,$$
(5)

$$\psi_j^\beta = \sum_{\mu=1}^K C_{\mu j}^\beta \phi_\mu \;.$$
(6)

Finally, the well-known matrix SCF equations in their unrestricted formulation read

$$F^\alpha C^\alpha = S C^\alpha \varepsilon^\alpha$$
(7)

$$F^\beta C^\beta = S C^\beta \varepsilon^\beta$$
(8)

where the matrix of overlap integrals S is spin independent. The Fock matrices F^α and F^β contain the energy integrals as obtained by the Fock operator as usual for electrons with alpha and beta spin, respectively. The vectors ε^α and ε^β contain the orbital energy eigenvalues. Eqs. 7 and 8 are mutually dependent on each other because of the dependence of both Fock operators on both spin variables. Consequently, they are solved in an iterative manner in accordance with the usual unrestricted Hartree–Fock approach. In analogy to the coefficient matrices, their product matrices — the density matrices P^α and P^β are spin dependent, as well. Their sum denotes the total density matrix P^{tot}, their difference, however, the spin density matrix P^S. Finally, the total electronic energy E_0 is obtained from

$$E_0 = \frac{1}{2}\sum_\mu \sum_\nu \left(P_{\nu\mu}^{\text{tot}} H_{\mu\nu}^{\text{core}} + P_{\nu\mu}^\alpha F_{\mu\nu}^\alpha + P_{\nu\mu}^\beta F_{\mu\nu}^\beta \right) \;.$$
(9)

To perform FOMD simulations, two distinct gradients are required: the molecular gradient for the force evaluation which is required for the integration algorithm in order to move the atoms in the molecular dynamics framework, and the electronic gradient. The latter one contains the first derivatives of the energy with respect to each basis function's center coordinates and is essential for the optimization within the floating orbital procedure. The components of both gradients can be formulated and rearranged to be composed of derivatives of molecular integrals. Those integrals comprise the components of the energy operator as well as the Gaussian type basis functions as defined by the basis set. Consequently, their derivatives are evaluated analogously to the closed shell case and afterwards combined with the spin dependent coefficient matrix elements. The formulation of the molecular gradient is textbook knowledge (Szabo & Ostlund, 1996) and the derivation of the electronic gradient is readily done according to the closed shell case. The optimization scheme used to minimize the energy with respect to orbital positions is adopted from the restricted case. That means, that as usual, the same basis set, i.e. identical functions at the same positions, is used for both α and β spin orbitals. Nevertheless, there is the possibility to separate the spin centers by occupying orbitals in different locations. This effect is obviously more pronounced for floated orbitals than for atom centered ones. In the following, the correctness of the implementation is demonstrated and certain properties are evaluated for the methyl radical.

3. Case study: Methyl Radical

3.1 Single Point Calculations

In order to demonstrate the correctness of the implementation, different calculations have been performed on the methyl radical which possesses nine electrons and therefore is necessarily an open shell system. Despite its small size, a number of theoretical studies investigating the structure and properties (Pacansky, Restricted and unrestricted Hartree-Fock calculations of the methyl radical, 1982), (Chipman, 1983) as well as spectra (Pacansky, Koch, & Miller, Analysis of the structures, infrared spectra, and Raman spectra for methyl, ethyl, isopropyl and tert-butyl radicals, 1991) of the methyl radical are available using Hartree–Fock methods. Another investigation considers the methyl radical to assess the accuracy of the spin-projected unrestricted Hartree–Fock method by investigating electron and spin densities (Glaser & Choy, 1993). Furthermore, reactions with other small molecules like carbon monoxide (Das & Lee, 2014), the recombination reaction with OH radicals (Oliveira & Bauerfeldt, 2012), and even the role in methanol synthesis (Zakharov, Ijagbuji, Tselishtev, Loriya, & Fedotov, 2015) have been studied recently. Due to its high reactivity, the methyl radical is involved in reactions which comprise the abstraction of a hydrogen atom so that many recent studies deal with the kinetic properties of these reactions (Mendes, Zhou, & Curran, 2014), (Saheb, 2015), (Tan, Yang, Krauter, Ju, & Carter, 2015). Firstly, our own FORTRAN 2003 code FORPLEASURE (Brüssel, Perlt, & Kirchner, 2012-2016) has been used to perform single point calculations with atom-centered basis functions, denoted with FP^{HF}, as well as floating orbital single point energy calculations, denoted with FP^{FO}. Both values are compared to the reference value as obtained with the program ORCA V. 3.0 (Neese, 2012) denoted as Orca^{HF}. The molecular geometry has been the same for all calculations with all $C-H$ bond lengths amounting to 108.9pm and all $\sphericalangle(HCH)$ angles being 120°. The energies as obtained by UHF calculations are given in Table 1, where also the applied basis sets are given. The electron density plot has been generated with vmd (Humphrey, Dalke, & Schulten, 1996) and the graphs with xmgrace.

Table 1. SCF energies of the methyl radical as obtained by the different methods. All values are given in atomic units of energy E_h.

Basis set	3-21G[1]	6-311G[2]
Orca^{HF}	-39.34198318	-39.55411986
FP^{HF}	-39.34198318	-39.55411986
FP^{FO}	-39.35171651	-39.56376512

It is apparent from Table 1 that the present implementation in FORPLEASURE is capable of reproducing the results of the established and widely used program ORCA. Within the chosen SCF convergence criterion of $10^{-7}E_h$, identical values for the total SCF energy are obtained. The application of a floating orbital optimization to the setup is able to reduce the SCF energy by $0.01E_h$ which may be considered as a significant stabilization. As the spin density is expected to change upon orbital displacement, spin density plots for the methyl radical as obtained with both basis sets, each with both atom-centered and floated orbitals, are provided in Figure 2.

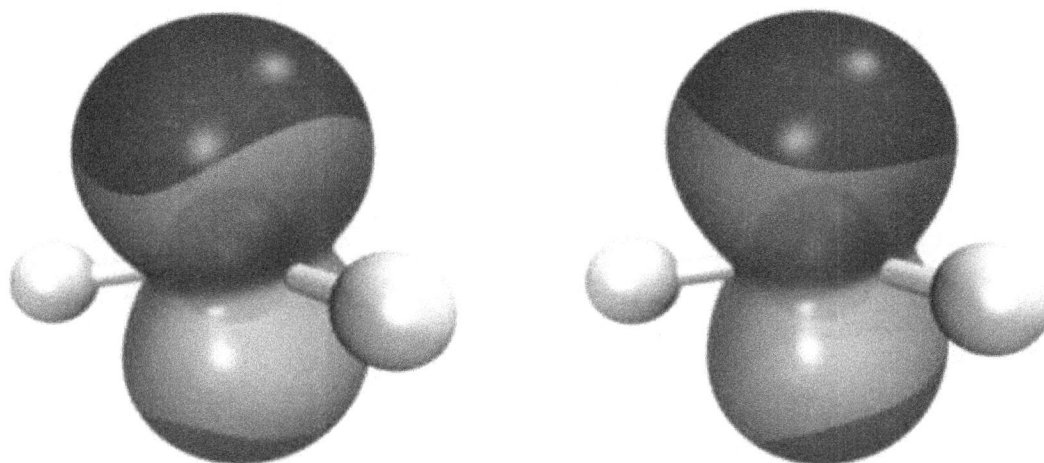

Figure 2. Spin density as obtained with (black surface) and without (light gray surface) floating orbitals for the 3-21G (Binkley, Pople, & Hehre, 1980) basis (left figure) and for the 6-311G (Krishnan, Binkley, Seeger, & Pople, 1980) basis set (right figure). Isovalue $= 0.01ea_0^{-3}$.

Figure 2 shows that the spin density as obtained by floating orbitals is shifted to larger distances from the carbon nucleus for both basis sets. Especially for systems possessing lower symmetry and more polar bonds, the differences observed for the spin density are expected to be even larger. Finally, molecular dynamics simulations have been carried out to study the impact of the displaced basis functions on the molecular gradient and the resulting trajectory. The computational details of the simulations are given in Table 2. All simulations are carried out using the program FORPLEASURE. For the HFMD simulations, the 6-311G** basis set has been augmented by polarization functions so that the effect of floating orbitals can be compared to the effect of those additional functions.

3.2 UFO Molecular Dynamics Simulations

Simulations are carried out in the microcanonical ensemble at constant particle number N, volume V and total energy E. Velocities are initialized according to a Boltzmann distribution at the temperature given in Table 2.

Table 2. Computational details of HFMD and FOMD simulations of the methyl radical.

	HFMD	FOMD
Timestep [fs]	0.3	0.3
Temperature [K]	400	400
Simulation time [ps]	30	3.4
Basis sets	3-21G	3-21G
	6-311G	
	6-311G**	

In order to study the influence of floating orbitals on dynamic properties, power and infrared (IR) spectra are studied in the following. Power spectra are obtained by TRAVIS (Brehm & Kirchner, 2011) as the Fourier transform of the velocity autocorrelation function. Therefore, the obtained data reveal the indirect effect of the floating orbitals on the trajectory of the system. Fig. 3 shows the signals at lower and higher wave numbers as well as a group of small signals

---

[1] (Binkley, Pople, & Hehre, 1980)
[2] (Krishnan, Binkley, Seeger, & Pople, 1980)

around 400cm^{-1} in the inset. The positions of the peak maxima are given in Table 3, where reference values as obtained by static HF calculations with a 6-311G** basis set (Pacansky, Koch, & Miller, Analysis of the structures, infrared spectra, and Raman spectra for methyl, ethyl, isopropyl and tert-butyl radicals, 1991) as well as experimental data (Snelson, 1970) are given. The observed peak intensities in the power and IR spectra depend on various parameters such as temperature, simulation time and the amplitude of the oscillation, i.e. the changes in polarizability and dipole moment, respectively. Since the simulation times are not identical for all simulations, the resulting peak intensities will not be discussed in the following. The wavenumbers, at which the vibrational signals occur, denote the frequency of the oscillation. Considering the classical analogs, this is dependent on the reduced mass and the force constant of the oscillation. As the masses are not affected by the basis set, the observed shifts are due to different forces. Those arise from different basis sets – by using larger basis sets, including polarization functions or applying floating basis functions. Consequently, different forces result in different trajectories showing different vibrational frequencies, which will be discussed in the following.

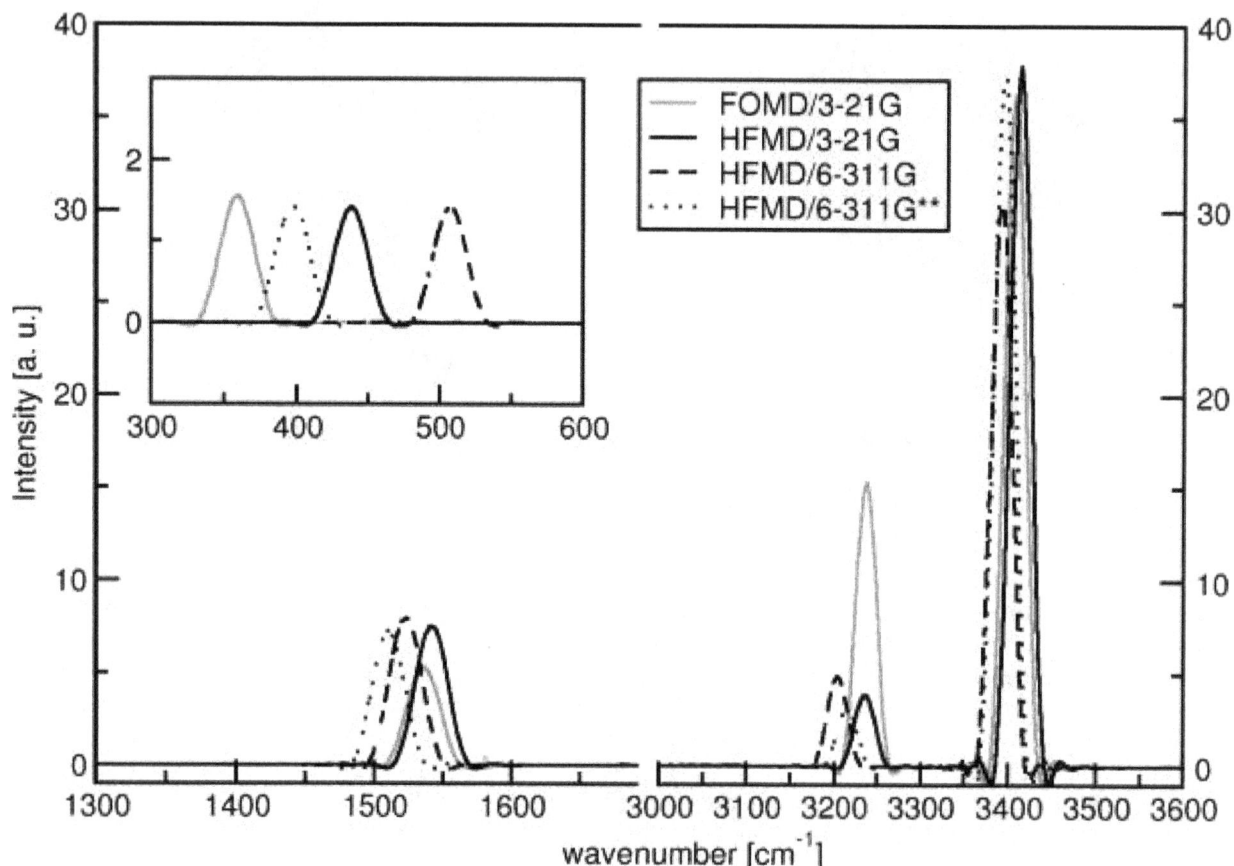

Figure 3. Power spectra as obtained by FOMD and HFMD simulations with different basis sets. The inset shows the wavenumber range from 300cm^{-1} to 600cm^{-1} enlarged.

Table 3. Positions of the peak maxima in the power spectrum of the methyl radical as observed by the different MD methods applied in this study as well as SCF (Pacansky, Koch, & Miller, Analysis of the structures, infrared spectra, and Raman spectra for methyl, ethyl, isopropyl and tert-butyl radicals, 1991) and experimental (Snelson, 1970) reference values. All values given in cm^{-1}.

	HFMD		FOMD	Calc. HF	Exp.[3]
3-21G	6-311G	6-311G**	3-21G	6-311G**[4]	
437.7	505.6	397.0	359.7	375.4	617
1540.5	1523.5	1510.0	1537.1	1512.4	1396
3237.1	3203.2	3216.7	3237.1	3274.4	
3413.6	3393.2	3396.6	3406.8	3407.4	3162

Considering the two signals above 3000cm^{-1}, a slight shift towards smaller wavenumbers is observed by increasing

[3] (Snelson, 1970)
[4] (Pacansky, Koch, & Miller, Analysis of the structures, infrared spectra, and Raman spectra for methyl, ethyl, isopropyl and tert-butyl radicals, 1991)

the basis set for HFMD simulations from 3-21G to 6-311G. Both the inclusion of polarization functions in the 6-311G** basis set as compared to the 6-311G basis and floating of the orbitals do not change the positions of the signals significantly. The observed differences are not significant within the spectral resolution which is given by TRAVIS. However, this is not astonishing since the two signals can be assigned to C-H stretching vibrations. For the description of this nonpolar bond neither polarization functions nor floating orbitals are essential so that no significant effect is observed here. This is also the case for the in-plane bending vibration observed at $\approx 1500 \text{cm}^{-1}$. Although it is observed that improving the basis set by more basis functions, polarization, or floating, yields a shift to smaller wave numbers, the differences are hardly significant. Considering the signals below 1000cm^{-1} however, large differences are obtained for the peak positions. The underlying molecular vibration is the out-of-plane bending vibration and due to the presence of electron density above and below the molecular plane, the different basis sets yield different results. In that case, the effect observed by including polarization functions in the 6-311G basis set is similar to the effect obtained by floating the orbitals of the smaller basis: The peaks are shifted to smaller wavenumbers ($\Delta v \approx 80 - 100 \text{cm}^{-1}$). This is in agreement with the observations from previous studies, where floating orbitals have also been found to cover the effect of polarization functions (Perlt, Brüssel, & Kirchner, 2014).

Furthermore, IR spectra are obtained from the autocorrelation function of the molecular dipole moment (Thomas, Brehm, Fligg, Vöhringer, & Kirchner, 2013). The dipole moment as a property directly related to the separation of charges has been found to be improved by the application of floating orbitals (Huber, 1981). Accordingly, the IR spectra are affected not only by the dynamics and their behavior with floating orbitals, but also by the dipole moment and its dependence on the basis functions' centers. The resulting IR spectra are given in Fig. 4, where again different intensities result from different simulation times and therefore will not be discussed.

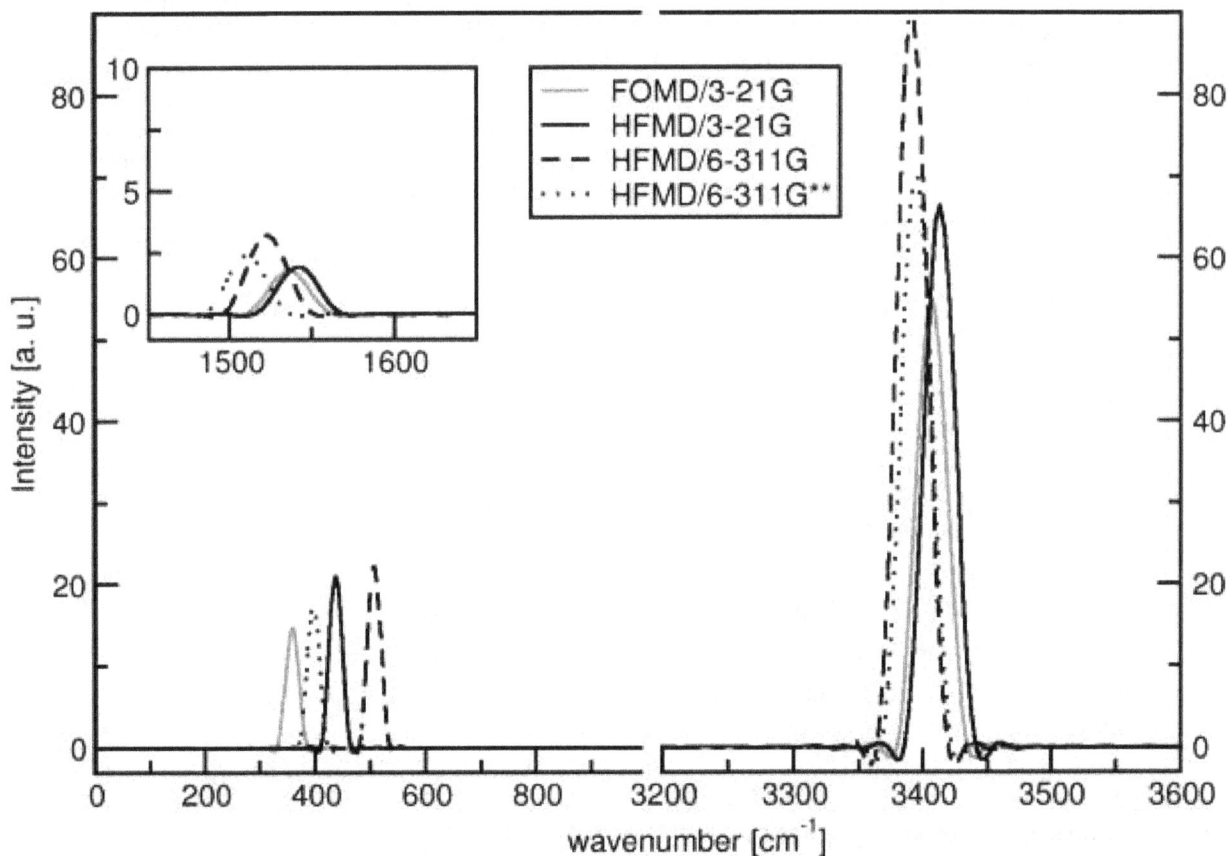

Figure 4. Infrared spectra as obtained by FOMD and HFMD simulations with different basis sets.

Table 4. Positions of the peak maxima in the IR spectrum of the methyl radical as observed by the different MD methods applied in this study as well as SCF (Pacansky, Koch, & Miller, Analysis of the structures, infrared spectra, and Raman spectra for methyl, ethyl, isopropyl and tert-butyl radicals, 1991) and experimental (Snelson, 1970) reference values. All values given in cm^{-1}.

| | HFMD | | FOMD | Calc. HF | Exp.[5] |
3-21G	6-311G	6-311G**	3-21G	6-311G**[6]	
437.7	505.6	397.0	359.7	375.4	617
1540.5	1523.5	1510.0	1537.1	1512.4	1396
3413.6	3393.2	3396.6	3406.8	3407.4	3162

For the IR spectra, results are similar to those obtained by the power spectra, see Table 4. All IR active modes observed in Refs. (Pacansky, Koch, & Miller, Analysis of the structures, infrared spectra, and Raman spectra for methyl, ethyl, isopropyl and tert-butyl radicals, 1991) and (Snelson, 1970) can be found by simulations and the corresponding peak positions are equal to those observed in the power spectra in Table 3. If systems with a permanent dipole moment are investigated, as planned in future studies, the dipole moment and consequently the IR spectra are expected to be improved by using floating orbitals. It is summarized that in agreement with the observations for the closed shell FOMD formulation (Perlt, Brüssel, & Kirchner, 2014), for all observed vibrational frequencies, the shifts as obtained by floating orbitals are similar to those obtained by additional polarization functions.

4. Summary and Outlook

The extension of floating orbitals to the unrestricted theory has been presented and the application to the methyl radical has been demonstrated. This rather small system has been chosen to rapidly evaluate the correctness of the implementation. However, in future studies, more complex systems will be examined. The great advantage of unrestricted floating orbital molecular dynamics is the possibility to investigate centers of charge, which do not necessarily need to be attributed to a nuclear center. Well-known problems of this type are solvated electrons. The floating orbital approach allows to assign a separate basis function to a solvated electron without placing it on a nucleus and optimize it. Thereby, solvated electrons may be treated by molecular dynamics and results can be compared to other theoretical studies (Uhlig & Jungwirth, Embedded Cluster Models for Reactivity of the Hydrated Electron, 2013), (Uhlig, Herbert, Coons, & Jungwirth, 2014) and experiments (Siefermann & Abel, 2011), (Abel, Buck, Sobolewski, & Domcke, 2012) which will be subject to a future study.

Acknowledgments

Financial support of the CRC 813 from the German Research Foundation is gratefully acknowledged.

References

Abel, B., Buck, U., Sobolewski, A. L., & Domcke, W. (2012). On the nature and signatures of the solvated electron in water. *Phys. Chem. Chem. Phys., 14*, 22-34.

Binkley, J. S., Pople, J. A., & Hehre, W. J. (1980). Self-consistent molecular orbital methods. 21. Small split-valence basis sets for first-row elements. *J. Am. Chem. Soc., 102*, 939-947. http://pubs.acs.org/doi/abs/10.1021/ja00523a008

Brehm, M., & Kirchner, B. (2011). TRAVIS - A Free Analyzer for Monte Carlo and Molecular Dynamics Trajectories. *J. Chem. Inf. Model., 51*, 2007-2013. http://pubs.acs.org/doi/abs/10.1021/ci200217w

Brüssel, M., Perlt, E., & Kirchner, B. (2012-2016). ForPleasure. *Version 1.1*. http://www.thch.uni-bonn.de/tc/fomd

Chipman, D. M. (1983). Theoretical study of the properties of methyl radical. *J. Chem. Phys., 78*, 3112-3132. http://scitation.aip.org/content/aip/journal/jcp/78/6/10.1063/1.445226

Das, J., & Lee, Y. P. (2014). Bimolecular reaction of CH3 + CO in solid p-H2: Infrared absorption of acetyl radical (CH3CO) and CH3-CO complex. *J. Chem. Phys., 140*, 244303. http://scitation.aip.org/content/aip/journal/jcp/140/24/10.1063/1.4883519

Glaser, R., & Choy, G. S. (1993). Electron and spin density analysis of spin-projected unrestricted Hartree--Fock density matrixes of radicals. *J. Phys. Chem., 97*, 3188-3198. http://pubs.acs.org/doi/abs/10.1021/j100115a022

Golze, D., Iannuzzi, M., Nguyen, M. T., Passerone, D., & Hutter, J. (2013). Simulation of Adsorption Processes at Metallic Interfaces: An Image Charge Augmented QM/MM Approach. *J. Chem. Theory Comput., 9*, 5086-5097. http://pubs.acs.org/doi/abs/10.1021/ct400698y

---

[5] (Snelson, 1970)
[6] (Pacansky, Koch, & Miller, Analysis of the structures, infrared spectra, and Raman spectra for methyl, ethyl, isopropyl and tert-butyl radicals, 1991)

Huber, H. (1981). Geometry optimization in ab initio SCF calculations: Part IV. Energy barriers and dipole moments obtained with floating orbital geometry optimization (FOGO). *J. Mol. Struct. THEOCHEM, 76,* 277-284. http://www.sciencedirect.com/science/article/pii/016612808185004X

Humphrey, W., Dalke, A., & Schulten, K. (1996). VMD - Visual Molecular Dynamics. *J. Mol. Graph., 14,* 33-38. http://www.sciencedirect.com/science/article/pii/0263785596000185

Ihrig, A. C., Schiffmann, C., & Sebastiani, D. (2011). Specific quantum mechanical/molecular mechanical capping-potentials for biomolecular functional groups. *J. Chem Phys., 135,* 214107. http://scitation.aip.org/content/aip/journal/jcp/135/21/10.1063/1.3664300

Krishnan, R., Binkley, J. S., Seeger, R., & Pople, J. A. (1980). Self-consistent molecular orbital methods. XX. A basis set for correlated wave functions. *J. Chem. Phys., 72,* 650-654. http://scitation.aip.org/content/aip/journal/jcp/72/1/10.1063/1.438955

Kurzbach, D., Sharma, A., Sebastiani, D., Klinkhammer, K. W., & Hinderberger, D. (2011). Dinitrogen complexation with main group radicals. *Chem. Sci., 2,* 473-479. http://pubs.rsc.org/en/content/articlelanding/2011/sc/c0sc00345j#!divAbstract

Marx, D., & Hutter, J. (2009). *Ab Initio Molecular Dynamics.* Cambridge: Cambridge University Press.

Masson, F., Laino, T., Röthlisberger, U., & Hutter, J. (2009). A QM/MM Investigation of Thymine Dimer Radical Anion Splitting Catalyzed by DNA Photolyase. *ChemPhysChem, 10,* 400-410. http://onlinelibrary.wiley.com/doi/10.1002/cphc.200800624/full

Mendes, J., Zhou, C. W., & Curran, H. J. (2014). Theoretical Chemical Kinetic Study of the H-Atom Abstraction Reactions from Aldehydes and Acids by H Atoms and OH, HO2 and CH3 Radicals. *J. Phys. Chem. A, 118,* 12089-12104. http://pubs.acs.org/doi/abs/10.1021/jp5072814

Neese, F. (2012). The ORCA Program system. *WIREs Comput. Mol. Sci., 2,* 73-78. http://wires.wiley.com/WileyCDA/WiresArticle/wisId-WCMS81.html

Neisius, D., & Verhaegen, G. (1979). Bond functions for ab initio calculations on polyatomic molecules. Hydrocarbons. *Chem. Phys. Lett., 66,* 358-362. http://www.sciencedirect.com/science/article/pii/0009261479850344

Neisius, D., & Verhaegen, G. (1981). Bond functions for ab initio calculations on polyatomic molecules. Molecules containing C, N, O and H. *Chem. Phys. Lett., 78,* 147-152. http://www.sciencedirect.com/science/article/pii/000926148185573X

Oliveira, R. C., & Bauerfeldt, G. F. (2012). Implementation of a variational code for the calculation of rate constants and application to barrierless dissociation and radical recombination reactions: CH3OH = CH3 + OH. *Int. J. Quant. Chem., 112,* 3132-3140. http://onlinelibrary.wiley.com/doi/10.1002/qua.24250/full

Pacansky, J. (1982). Restricted and unrestricted Hartree--Fock calculations of the methyl radical. *J. Phys. Chem., 86,* 485-488. http://pubs.acs.org/doi/abs/10.1021/j100393a014

Pacansky, J., Koch, W., & Miller, M. D. (1991). Analysis of the structures, infrared spectra, and Raman spectra for methyl, ethyl, isopropyl and tert-butyl radicals. *J. Am. Chem. Soc., 113,* 317-328. http://pubs.acs.org/doi/abs/10.1021/ja00001a046

Perlt, E., Brüssel, M., & Kirchner, B. (2014). Floating orbital molecular dynamics simulations. *Phys. Chem. Chem. Phys., 16,* 6997-7005. http://pubs.rsc.org/en/Content/ArticleLanding/2014/CP/c3cp54797c#!divAbstract

Pitarch-Ruiz, J., Evangelisti, S., & Maynau, D. (2005). Does a Sodium Atom bind to C60? *J. Chem. Theory Comput., 1,* 1079-1082. http://pubs.acs.org/doi/abs/10.1021/ct050035v

Platts, J. A., Overgaard, J., Jones, C., Iversen, B. B., & Stasch, A. (2011). First Experimental Characterization of a Non-nuclear Attractor in a Dimeric Magnesium(I) Compound. *J. Phys. Chem. A, 115,* 194-200. http://pubs.acs.org/doi/abs/10.1021/jp109547w

Saheb, V. (2015). Theoretical Studies on the Kinetics of Hydrogen Abstraction Reactions of H and CH3 Radicals from CH3OCH3 and Some of Their H/D Isotopologues. *J. Phys. Chem. A, 119,* 4711-4717. http://pubs.acs.org/doi/abs/10.1021/acs.jpca.5b00911

Siefermann, K., & Abel, B. (2011). The Hydrated Electron: A Seemingly Familiar Chemical and Biological Transient. *Angew. Chem. Int. Ed., 50,* 5264-5272. http://onlinelibrary.wiley.com/doi/10.1002/anie.201006521/full

Snelson, A. (1970). Infrared matrix isolation spectrum of the methyl radical produced by pyrolysis of methyl iodide and dimethyl mercury. *J. Phys. Chem., 74*, 537-544.
http://pubs.acs.org/doi/abs/10.1021/j100698a011

Szabo, A., & Ostlund, N. (1996). *Modern Quantum Chemistry: Introduction to Advanced Electronic Structure Theory.* Mineola, New York: Dover Publications.

Tan, T., Yang, X., Krauter, C. M., Ju, Y., & Carter, E. A. (2015). Ab initio Kinetics of Hydrogen Abstraction from Methyl Acetate by Hydrogen, Oxygen, Hydroxyl, and Hydroperoxy Radicals. *J. Phys. Chem. A, 119*, 6377-6390.
http://pubs.acs.org/doi/abs/10.1021/acs.jpca.5b03506

Thomas, M., Brehm, M., Fligg, R., Vöhringer, P., & Kirchner, B. (2013). Computing vibrational spectra from ab initio molecular dynamics. *Phys. Chem. Chem. Phys., 15*, 6608-6622.
http://pubs.rsc.org/en/Content/ArticleLanding/2013/CP/c3cp44302g#!divAbstract

Timerghazin, Q. K., & Peslherbe, G. H. (2007). Non-nuclear attractor of electron density as a manifestation of the solvated electron. *J. Chem. Phys., 127*, 064108.
http://scitation.aip.org/content/aip/journal/jcp/127/6/10.1063/1.2747250

Uhlig, F., & Jungwirth, P. (2013). Embedded Cluster Models for Reactivity of the Hydrated Electron. *Z. Phys. Chem., 227*, 1583-1593.
http://www.degruyter.com/view/j/zpch.2013.227.issue-9-11/zpch-2013-0402/zpch-2013-0402.xml

Uhlig, F., Herbert, J. M., Coons, M. P., & Jungwirth, P. (2014). Optical Spectroscopy of the Bulk and Interfacial Hydrated Electron from Ab Initio Calculations. *J. Phys. Chem. A, 118*, 7507-7515.
http://pubs.acs.org/doi/abs/10.1021/jp5004243

Zakharov, I. I., Ijagbuji, A. A., Tselishtev, A. B., Loriya, M. G., & Fedotov, R. N. (2015). The new pathway for methanol synthesis: Generation of methyl radical from alkanes. *J. Environ. Chem. Eng., 3*, 405-412.
http://www.sciencedirect.com/science/article/pii/S2213343714001742

15

Graft Copolymerization of Methlymethacrylate on Lignin Produced from Agricultural Wastes

Mopelola Abeke Omotoso[1], Samuel Oluwatosin Ajagun[1]

[1]Department of Chemistry, University of Ibadan, Ibadan, Nigeria

Correspondence: Mopelola Abeke Omotoso, Department of Chemistry, University of Ibadan, Ibadan, Nigeria.

E-mail: beckyomot@yahoo.com

Abstract

Lignin was isolated from banana pseudo stem, palm frond and coconut husk. The moisture content of the banana pseudo stem and palm frond was determined gravimetrically. The lignin was isolated by alkaline hydrolysis with NaOH followed by precipitation with 72% H_2SO_4. The isolated lignin was then copolymerized with methyl methacrylate via free radical mechanism. The lignin and the corresponding copolymers were characterised by FTIR analysis.

The lignin content in banana pseudo stem, palm frond and coconut husk were respectively 16%, 14% and 13% while the grafting percentage of the corresponding lignin with methlymethacrylate (MMA) are 24%, 21% and 19%. These implied that production of lignin from agricultural wastes could reduce the latter solid wastes.

The FTIR analysis of the different lignin showed important peaks such as O-H band between 3433 cm^{-1} and 3450 cm^{-1}; C=C band around 1515 cm^{-1} was absent in palm frond lignin; and C=O band was only observed in coconut husk lignin at 1696 cm^{-1}. There was a significant difference in the absorption band of the C-O (ether) of the lignin between 1107 cm^{-1} and 1198 cm^{-1}. The shift in the absorption peaks in the spectra of the isolated lignin for the different sources of the lignin and the grafting percentage corroborate the fact that the properties and composition of lignin depend on the source of the lignin. The FTIR analysis of the copolymers also showed two prominent peaks for C=O band between 1680 cm^{-1} and 1708 cm^{-1}; and C-O (ester) between 1220 cm^{-1} and 1320 cm^{-1} that were absent in the lignin. These confirm the presence of the acrylate functional groups.

Keywords: banana pseudo stem, palm frond, coconut husk, lignin, agricultural wastes, methlymethacrylate

1. Introduction

In the past century, various synthetic polymer materials have been developed in different forms, such as plastics, thermoplastics and synthetic rubbers that are used widely in a variety of fields such as packaging, construction materials, agriculture, and medical devices. Undoubtedly, those synthetic polymer materials perform very important roles in our daily lives. After rapid development for several decades, a Gordian knot is becoming increasingly serious, the continual environmental pollution is caused by non-biodegradable synthetic polymer wastes (Stéphanie and Luc 2014).

Most of the commercial products in the society are fully made of synthetic polymers. However, despite their vast usefulness, their recalcitrant nature eclipses their importance because it is at variance with the concept of green chemistry. Natural polymers which are biodegradable cannot be a direct substitute for synthetic polymers because of some of their properties. In view of this, copolymerization of the side chains of synthetic polymers onto naturally occurring backbones are being considered as a way of producing compounds that are more easily degraded in the environment. Lignin, because of its good physicochemical properties and its enormous abundance offers a promising potential for exploitation.

Lignin, as a class of biodegradable, sustainable, nontoxic and inexpensive natural amorphous hydrophobic branched heteropolymer second only to cellulose in natural abundance, has been widely used for various applications (Chung et al., 2013; Sivasankarapillai and McDonald, 2014). At present, most of lignin is discarded as waste from the pulp and paper making industry (Long et al., 2009).

Lignin is insoluble in water and stable in nature and acts as the glue that connects cellulose and hemi-cellulose. Lignin is a three-dimensional, highly cross-linked macromolecule composed of three types of substituted phenols which are monolignols (coniferyl, sinapyl, and p-coumaryl alcohols) incorporated by enzymatic polymerization yielding a vast number of functional groups and linkages (Boerjan et al., 2003; Carmen et al., 2004; Chakar and Ragauskas, 2004).

The relative amounts of the monolignol units differ considerably between plants. In softwood lignin, the network is formed primarily by coniferyl moieties (95%), the rest consisting of p-coumaryl alcohol-type units and only trace amounts of sinapyl alcohol moieties, while in hardwood and dicotyl crops like hemp and flax, various ratios of coniferyl/sinapyl have been reported (Dence and Lin, 1992).

Lignin cannot be simply removed from growing plants without causing deleterious developmental effects (Bonawitz et al., 2013). Genetic manipulation trials using natural mutants or silencing strategies have failed because they drastically reduced lignin content in a non-selective way. Nevertheless, there are cases in which mild genetic manipulations have been used to moderately reduce lignin content or modify its composition in biomass, modestly improving saccharification efficiency, forage digestibility, and pulping yield (Li et al., 2008).

Lignin from trees, plants and agricultural crops with different chemical composition and properties can be obtained by the use of several extraction methods. Over the last decade, Kraft lignin has been considered for different applications such as replacement of phenol in phenol formaldehyde resins (Mondragon, 2007 and Tejado et al., 2008), vanillin production (Arau´jo et al., 2010 and Rodrigues Pinto et al., 2011), agent in production of synthetic tannin (Soprano et al., 2005) and utilization as bio sorbents of Heavy metals (Harmita et al., 2009 and Wu et al., 2008) and dyes (Suteu et al., 2009 and Saad et al., 2012).

Graft polymerization is one of the key techniques to covalently modify the surface of lignin by introduction of various functional polymers onto the side chains of lignin, resulting in greatly enhanced mechanical, thermal resistance and UV absorption properties of the host polymers. (Xiaohuan et al., 2014).

This study aimed at isolating lignin from banana pseudo stem, palm frond, and coconut husk all of which are agricultural wastes and copolymerizing the isolated lignin with methlymethacrylate to obtain biodegradable products. This production of lignin from these wastes would also conserve its source from wood used for pulp and paper and consequently help in afforestation.

2. Materials and Method

2.1 Materials

The banana pseudo stem and palm fronds were collected from the farm behind the Chemistry Department, University of Ibadan, Ibadan. The coconut husks were purchased from Bodija market, Ibadan, Nigeria. The samples were carefully cut and washed with distilled water.

2.2 Determination of Moisture Content

The moisture content of the banana pseudo stem the palm frond and coconut husk was determined gravimetrically by loss in weight after heating to a constant weight in the oven at 105°C. This was carried out by drying crucibles to a constant weight in an oven at 105°C. Known weights of the samples were weighed into the dried crucibles and then dried to constant weight at this same temperature in the oven. The moisture content is the difference between the weight of the dried crucible with the sample and the weight of the dried crucible. This was carried out for the banana pseudostem, palm frond and the coconut husk.

2.3 Isolation of Lignin

50 g of the dried sample was boiled in 2.5 M sodium hydroxide solution on a hot plate under high pressure for two hours. It was then allowed to cool after which it was filtered with a cloth filter. The pH of the filtrate was adjusted to 3.0 with 72% sulphuric acid. The resulting precipitate was filtered with a filter paper and dried.

The precipitate of the lignin was purified by washing it with ethanol, followed by hot water after which it was dried in the oven and cooled in the desiccator.

The lignin content was determined as follows:

$$lignin\ content = \frac{weight\ of\ lignin}{weight\ of\ dried\ sample\ (50g)} \times 100\%$$

2.4 Graft Copolymerization of the Lignin and Methyl Methacrylate

The copolymerization reaction was performed as suggested by Xiaohuan Liu (2014). 1.0 g calcium chloride was added to 20 ml dimethyl sulfoxide solvent in a conical flask and then stirred until dissolved. This was followed by the addition of 1.0 g lignin and the solution was stirred for 20 min. 8 ml methyl methacrylate (MMA) was added followed by 1 ml Hydrogen peroxide. After 5 min of stirring, the conical flask was heated on a water bath set at 40°C for 24 hours. The reaction was terminated by the addition of 4.0 mg quinol (4-hydroxyphenol).

After finishing the grafting reactions, the resulting solution was added drop wise to a 20-fold excess of diluted hydrochloric acid solution (pH = 3.0), and the precipitated copolymer was filtered, and repeatedly washed with water.

The final product was dried in an oven at 80°C for 48 h.

The grafting percentage was calculated as suggested by Teena *et al.,* (2010) as follows:

$$Grafting\ percentage = \frac{W_2 - W_1}{W_1} \times 100\%$$

Where W_2 = total weight of solid mass recovered from the grafting reaction

        W_1= weight of lignin used.

2.5 Characterization of the Lignin and Its Copolymer Using FTIR

The various lignin and their corresponding copolymers were qualitatively characterized by the FTIR technique. The FTIR spectra of the lignin and copolymers were obtained on a Perkin Elmer Fourier transform infrared spectrometer. The spectra were recorded between 4000-400 cm^{-1}, using KBr as reference.

3. Result and Discussion

The moisture content of the banana pseudo stem and the palm frond are high (Table 1). This is also shown in Figure 1. The coconut husk has negligible moisture content. There was no difference in the weight of the coconut husk after drying in the oven at 105°C. The lignin content is highest in banana pseudo stem as presented in Table 1 and plotted in Figure 2.

Table 1. Physicochemical Properties of the Samples Studied

Sample	% Moisture Content	% Lignin Content on dry basis	Grafting %
Banana Pseudo stem	89	16	24
Coconut Husk	-	14	21
Palm Frond	48	13	19

Values are mean ±SD for n=3.

The grafting percentage of the various lignin is shown in Table 1. The lignin isolated from banana pseudo stem showed better grafting compared to those obtained from palm frond and coconut husk. This implies that the grafting of lignin is affected by the source since the composition also varies with the source.

The FTIR spectrum of the lignin from banana pseudo stem Figure 4 and tabulated in Table 2 showed important bands such as O-H stretching vibration at 3449 cm^{-1}, C-H stretching vibration (aromatic) at 2925 cm^{-1}, C-H stretching vibration (alkane) at 2851 cm^{-1}, C=C stretching vibration (aromatic) at 1638 cm^{-1} and 1515 cm^{-1} and C-O stretching vibration at 1107 cm^{-1}. This is similar to those observed by Carmen et al., 2004. The stretching frequencies changed in the copolymer.

The changes in the stretching frequencies of O-H, C=C and C-O and appearance of C-O (ester) as observed in the copolymer, CBPSL (Figure 5) suggests the modification of the lignin. The decrease in the stretching frequencies of O-H and C-H (deformation) is due to conjugation and hydrogen bonding in the copolymer. The reduction in the stretching frequency of C=C is due to conjugation. The emergence of the two peaks shown at 1686 cm^{-1} (C=O) and 1229 cm^{-1} (C-O) confirms the presence of the acrylate functional groups.

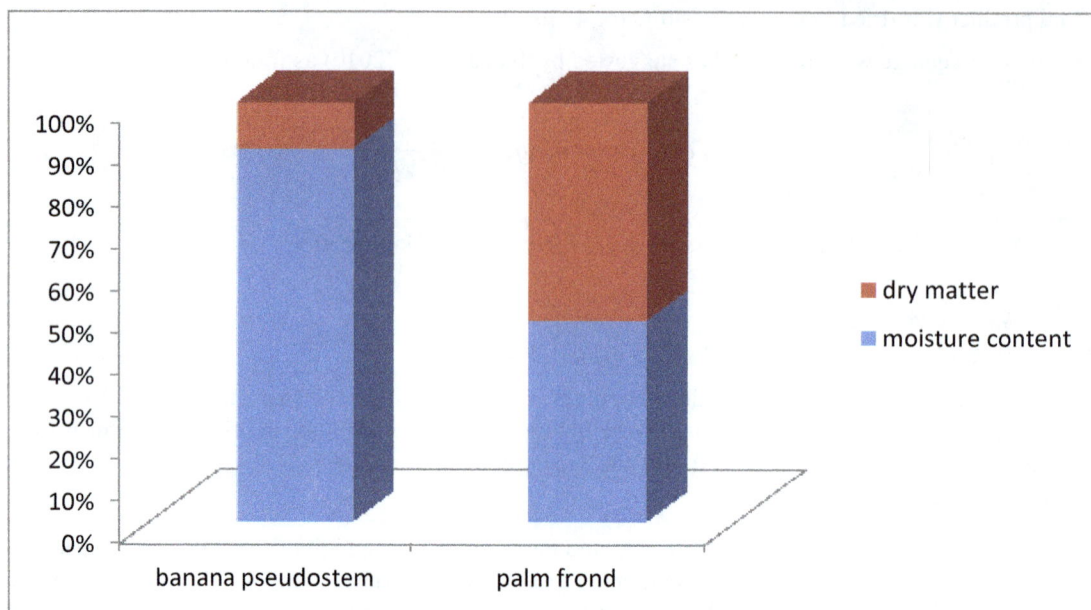

Figure 1. Moisture content of banana pseudo stem and palm frond

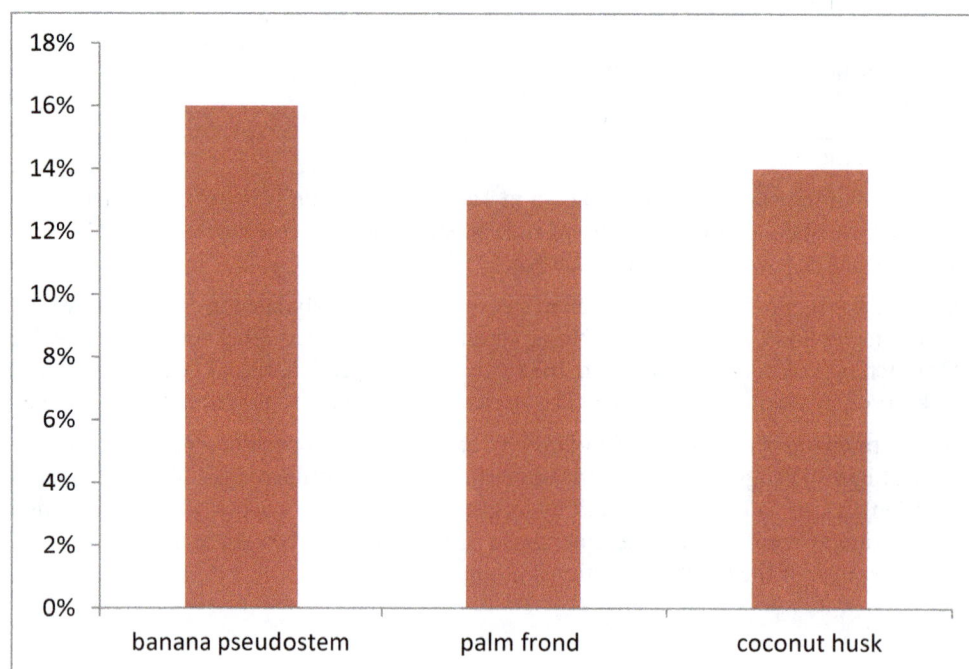

Figure 2. The lignin content of dried banana pseudo stem, coconut husk and palm frond

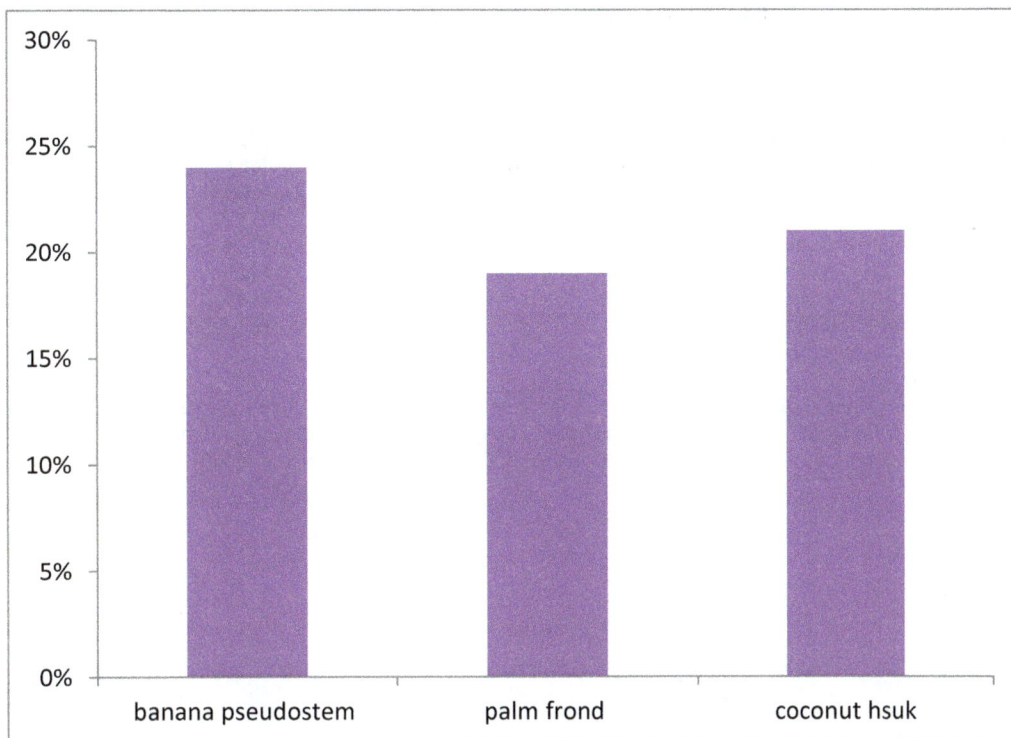

Figure 3. The grafting percentage of the various lignin

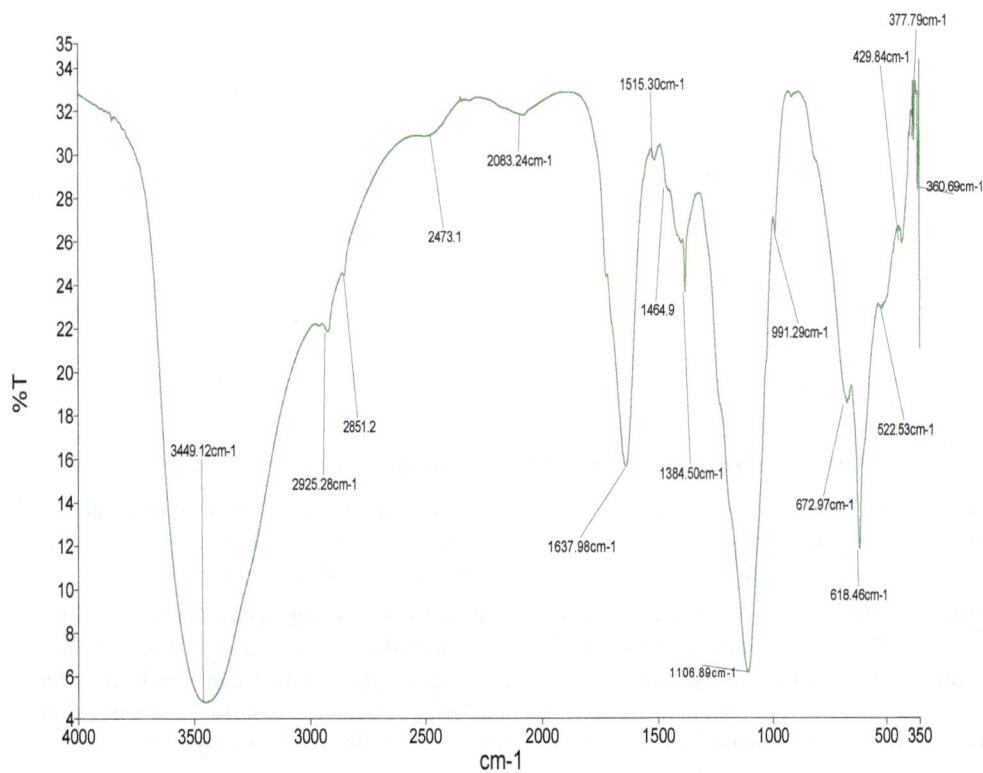

Figure 4. The FTIR spectrum of lignin isolated from banana pseudo stem

Table 2. FTIR of the Lignin from various Sources and their Corresponding Copolymer

Bond Type	Sample Absorption Band (cm-1)					
	BPSL	PFL	CHL	CBPSL	CPFL	CCHL
O-H stretching vibration	3449	3450	3433	3407	3429	3421
C-H stretching vibration (aromatic)	2925	2922	2924	2921	2919	2921
C-H stretching vibration (alkane)	2851	2851	2852	2851	2850	2852
C=O stretching vibration	--	--	1696	1686	1680	1708
C=C stretching vibration (aromatic)	1638 & 1515	1637	1626 & 1514	1621 & 1508	1621	1615 & 1514
C-H deformation	1465	--	--	1421	1463	1465
O-H deformation	1385	1384	1389	1384	1384	1384
C-O stretching vibration (ester)	--	--	--	1229	1315	1318
C-O stretching vibration (ether)	1107	1198	1195	1116	--	1123
C-O deformation	--	--	--	1027	--	--

Key: BPSL = banana pseudo stem lignin; CBPSL = copolymer of banana pseudo stem lignin and methlymethacrylate
PFL   = palm frond lignin; CPFL = copolymer of palm fond lignin and methlymethacrylate
CHL = coconut husk lignin; CCHL = copolymer of coconut husk lignin and methlymethacrylate

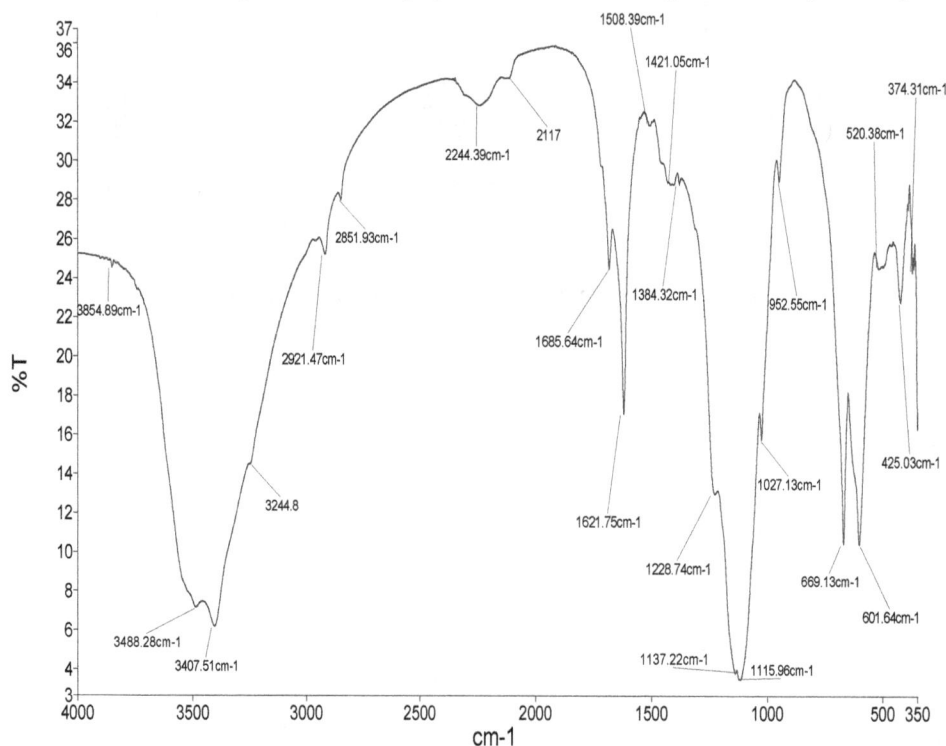

Figure 5. The FTIR spectrum of the copolymer from the lignin isolated from banana pseudo stem and MMA

The FTIR spectrum of the lignin isolated from palm frond (Figure 6) showed important bands such as O-H stretching vibration at 3450 cm^{-1}, C-H stretching vibration (aromatic) at 2922 cm^{-1}, C-H stretching vibration (alkane) at 2851 cm^{-1}, C=C stretching vibration (aromatic) at 1637 cm^{-1} and C-O stretching vibration (ether) at 1198 cm^{-1}.

The copolymerization of this lignin with methlymethacrylate that produced copolymer CPFL is evident in the changes observed in the spectra (Figure 7) notably in O-H, C-H, C=C, and C-O. The decrease in the stretching frequencies of O-H and C-H is due to conjugation and hydrogen bonding in the copolymer. The reduction in the stretching frequency of C=C is also due to conjugation. The appearance of two important peaks of C=O at 1680 cm^{-1} and C-O of ester at 1315 cm^{-1} further confirms the presence of the acrylate functional groups in CPFL copolymer.

The FTIR spectrum of the lignin obtained from coconut husk (Figure 8) showed important bands such as O-H stretching vibration at 3433 cm^{-1}, C-H stretching vibration (aromatic) at 2924 cm^{-1}, C-H stretching vibration (alkane) at 2852 cm^{-1}, C=O stretching vibration at 1696 cm^{-1}, C=C stretching vibration (aromatic) at 1626 cm^{-1} and 1514 cm^{-1} and C-O stretching vibration at 1195 cm^{-1}. The stretching frequencies changed in the copolymer.

There were changes in the stretching frequencies of O-H, C-H, C=O, C=C, and C-O in the copolymer prepared from coconut husk lignin and methlymethacrylate (Figure 9). The decrease in the stretching frequencies of O-H and C-H is due to conjugation and hydrogen bonding in the copolymer. The reduction in the stretching frequency of C=C is due to conjugation. The increase in the carbonyl stretching frequency from 1696 cm^{-1} to 1708 cm^{-1} and the emergence of the peak around 1318 cm^{-1} validates the presence of the acrylate functional groups.

Comparing the spectra of the lignin obtained from the banana pseudo stem, palm frond and coconut husk, it can be inferred that there are slight variations in their structures (Table 2). The slight variation confirms the fact that the properties and composition of lignin depend on the source (Carmen et al., 2004).

The broad band between 3433 cm^{-1} and 3450 cm^{-1} can be attributed to the hydroxyl groups in the phenolic and aliphatic structures. The bands around 2922 cm^{-1} and 2925 cm^{-1}, predominantly arising from the C-H stretching can be attributed to the aromatic methoxyl group and the aromatic moiety. The peaks between 2850 cm^{-1} and 2852 cm^{-1} arise from the C-H stretching in the aliphatic methylene group.

Unlike other lignin isolated, the spectrum of the lignin isolated from coconut husk showed a peak around 1696 cm^{-1} which is characteristic of the carbonyl group. This variance could be due to the source of the lignin.   Similar results were recorded by Carmen et al., (2004) and Li Jingjing, (2011).   Aromatic skeleton vibrations around 1620 cm^{-1} and 1515 cm^{-1} and the C-O (ether and alcohol) stretching frequencies between 1100 cm^{-1} and 1200 cm^{-1} were also prominent in all the spectra.

The copolymers all show distinctive peaks between 1680cm^{-1} and 1708cm^{-1} which is indicative of the carbonyl group. The C-O band between 1220cm^{-1} and 1320cm^{-1} was also observed in all the copolymers. This confirms the presence of the acrylate functional groups.

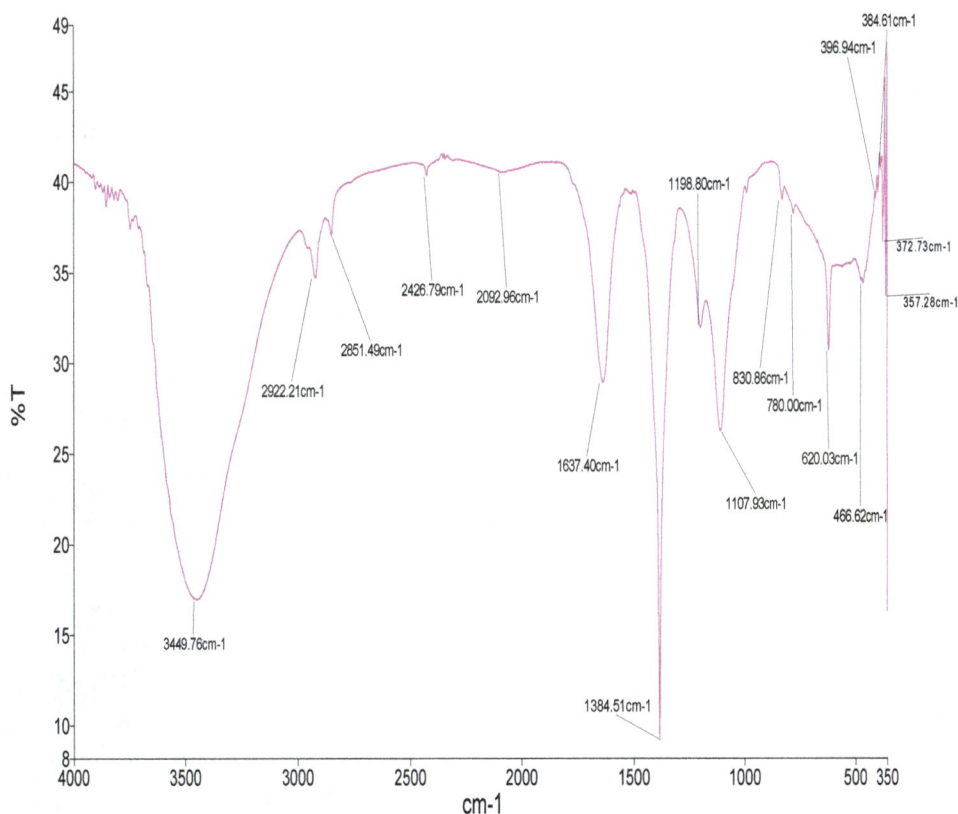

Figure 6. The FTIR spectrum of lignin isolated from palm frond

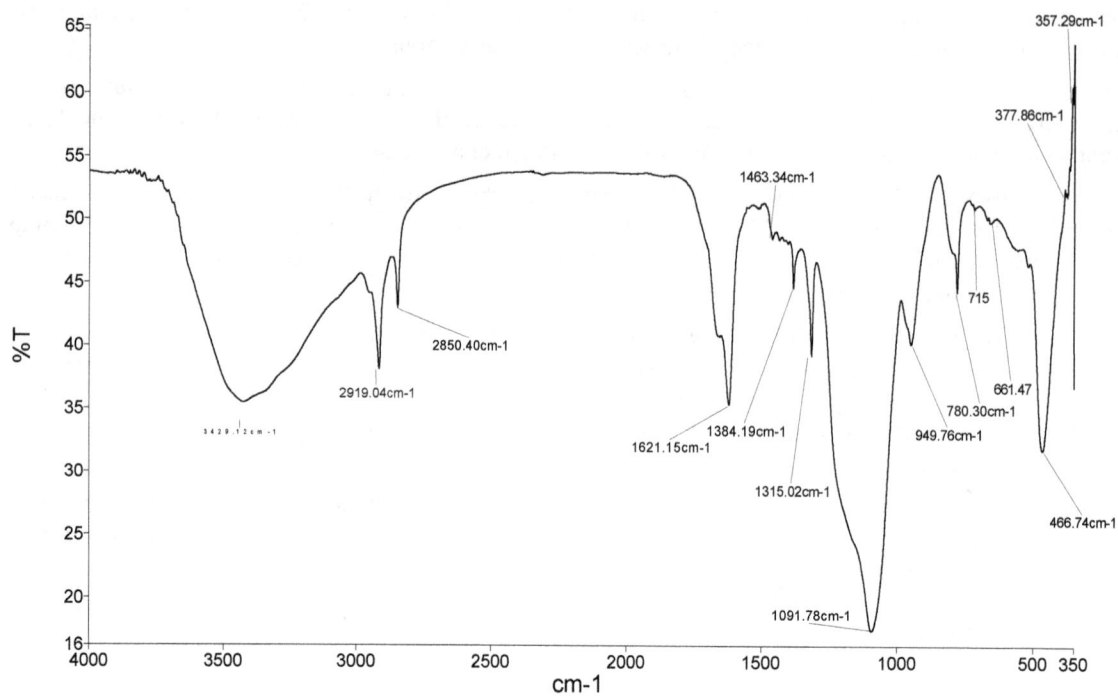

Figure 7. The FTIR spectrum of the copolymer from the lignin isolated from palm frond and MMA

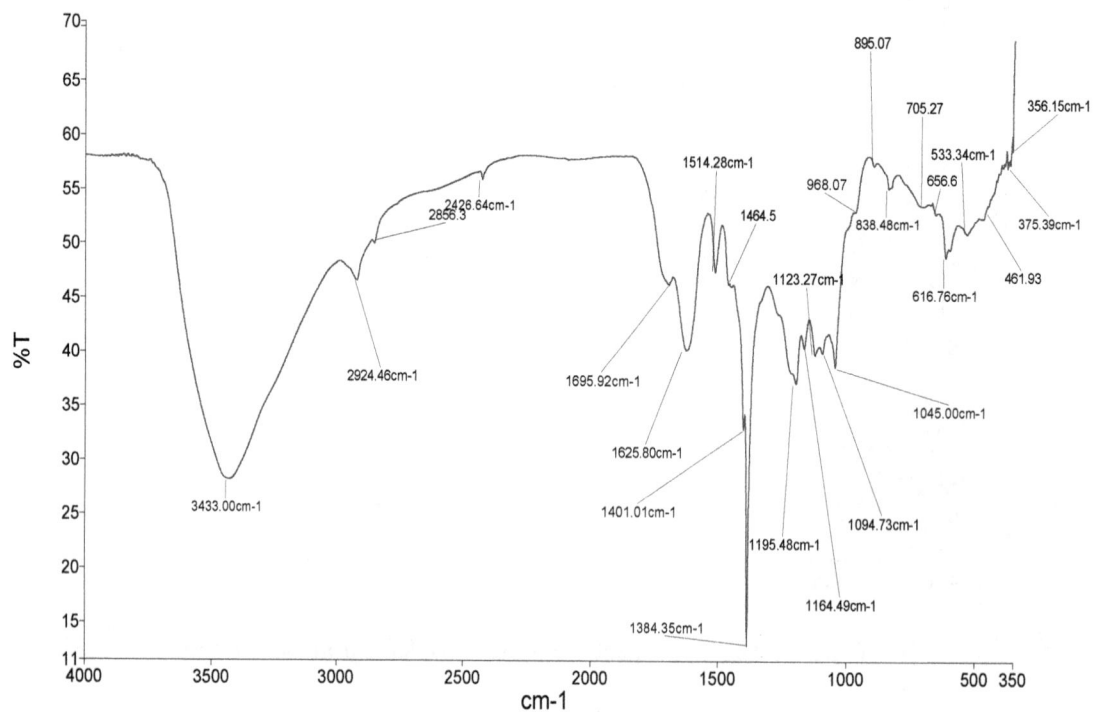

Figure 8. The FTIR spectrum of lignin isolated from coconut husk

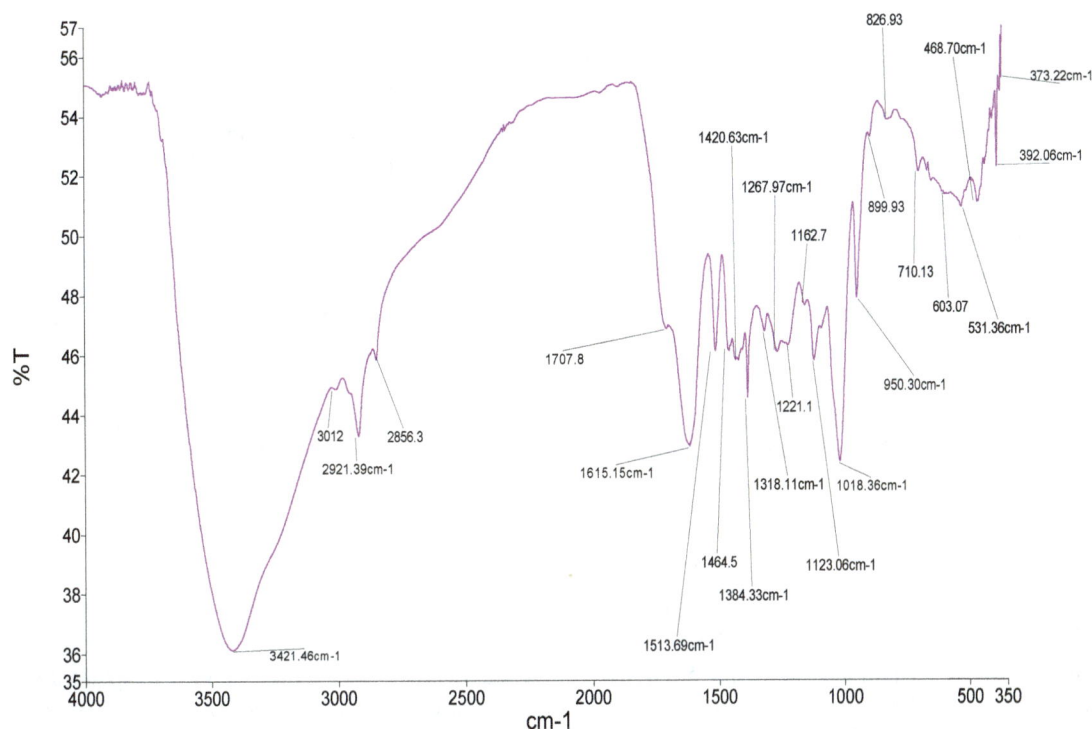

Figure 9. The FTIR spectrum of the copolymer from the lignin isolated from coconut husk

References

Arau´jo, J. D. P., Grande, C. A., & Rodrigues, A. E. (2010). Vanillin production from lignin oxidation in a batch reactor. *Chem. Eng. Res. Des.*, *88*(8), 1024-1032. http://dx.doi.org/10.1016/j.cherd.2010.01.021

Boerjan, W., Ralph, J., & Baucher, M. (2003). Lignin biosynthesis. *Annual Rev. Plant Biol.*, *54*, 519–546. http://dx.doi.org/10.1146/annurev.arplant.54.031902.134938

Bonawitz, N. D., & Chapple, C. (2013). Can genetic engineering of lignin deposition be accomplished without an unacceptable yield penalty? *Current Opinion Biotechnology*, *24*, 336-343. http://dx.doi.org/10.1016/j.copbio.2012.11.004

Carmen, B., Dominique, B., Richard, J. A. Gosselink, J. E. G., & van Dam (2004). Characterisation of structure-dependent functional properties of lignin with infrared spectroscopy. *Industrial Crops and Products, 20*, 205–218. http://dx.doi.org/10.1016/j.indcrop.2004.04.022

Chakar, F. S., & Ragauskas, A. (2004). Review of current and future softwood kraft lignin process chemistry, *Industrial Crops and Products, 20*(2), 131-141. http://dx.doi.org/10.1016/j.indcrop.2004.04.016

Chen, C. L. (1991). *Lignin, Occurrence in Woody Tissues, Isolation, Reactions, and Structure. In Wood Structure and Composition*. Lewin, M., Goldstein, I.S. New York, Marcel Dekker, Inc., 183-261.

Chung, Y. L., Olsson, J. V., Li, R. J., Frank, C. W., Waymouth, R. M., Billington, S. L., & Sattely, E. S. (2013). *ACS Sustainable Chem. Eng., 1*, 1231–1238. http://dx.doi.org/10.1021/sc4000835

Dence, C. W., & Lin, S. Y. (1992). *General structure features of lignins. Methods in Lignin Chemistry*. Springer-Verlag, Berlin-Heidelberg, 3–6.

Harmita, H., Karthikeyan, K. G., & Pan, X. (2009). Copper and cadmium sorption onto kraft and organosolv lignins. *BioResource Technol.*, *100*(24), 6183-6191. http://dx.doi.org/10.1016/j.biortech.2009.06.093

Li, X., Weng, J. K., & Chapple, C. (2008). Improvement of biomass through lignin modification. *Plant J.*, 54, 569-581. http://dx.doi.org/10.1111/j.1365-313X.2008.03457.x

Liu, X. H., Xu, Y. Z., Yu, J., Li, S. H., Wang, J. H., , Wang, C. P., & Chu, F. X. (2014). Integration of lignin and acrylic monomers towards grafted copolymers by free radical polymerization. *International Journal of Biological Macromolecules, 67*, 483–489. http://dx.doi.org/10.1016/j.ijbiomac.2014.04.005

Long, C., Chang, Y. T., Nan, Y. N., Chao, Y. W., Qiang, F., & Qin, Z. (2009). Preparation and Properties of

Chitosan/Lignin Composite Films. *Chinese Journal of Polymer Science,* *27*(5), 739–746. http://dx.doi.org/10.1142/S0256767909004448

Mousavioun, P., & Doherty, W. O. S. (2009). Chemical and thermal properties of fractionated bagasse soda lignin. *Ind. Crops Prod.,* *31*, 52-58. http://dx.doi.org/10.1016/j.indcrop.2009.09.001

Rodrigues, P. P. C., Borges, D. S. E. A., & Rodrigues, A. E. (2011). Insights into oxidative conversion of lignin to high-added value phenolic aldehydes. *Ind. Eng. Chem. Res.,* *50*(2), 741-748. http://dx.doi.org/10.1021/ie102132a

Saad, R., Radovic-Hrapovic, Z., Ahvazi, B., Thiboutot, S., Ampleman, G., & Hawari, J. (2012). Sorption of 2,4-dinitroanisole (DNAN) on lignin. *J. Environ. Sci.,* *24*(5), 808-813. http://dx.doi.org/10.1016/S1001-0742(11)60863-2

Sivasankarapillai, G., & McDonald, A. G. (2014). Lignin valorization by forming thermally stimulated shape memory copolymer elastomers – partially crystalline hyper branched polymer crosslinks, *J. of Applied Polymer Science,* *131*(22). http://dx.doi.org/10.1002/app.41103

Stéphanie, L., & Luc, A. (2014). Chemical modification of lignins, Towards biobased polymers. Université de Strasbourg. *Progress in Polymer Science,* *39*, 1266–1290. http://dx.doi.org/10.1016/j.progpolymsci.2013.11.004

Suparno, O., Covington, A. D., Phillips, P. S., & Evans, C. S. (2005). An innovative new application for waste phenolic compounds, use of Kraft lignin and naphthols in leather tanning. *Resource Conservation. Recycle.,* *45*(2), 114-127. http://dx.doi.org/10.1016/j.resconrec.2005.02.005

Suteu, D., Zaharia, C., Muresan, A., Muresan, R., & Popescu, A. (2009). Using of industrial waste materials for textile wastewater treatment. *Environ. Eng. Managers J.,* *8*(5), 1097-1102.

Tejado, A., Kortaberria, G., Pen˜a, C., Blanco, M., Labidi, J., & Echeverrı J. M. (2008). Lignins for phenol replacement in novolactype phenolic formulations. II. Flexural and compressive mechanical properties. *J. Appl. Polymer. Sci.,* *107*(1), 159-165. http://dx.doi.org/10.1002/app.27003

Wu, Y., Zhang, S., Guo, X., & Huang, H. (2008). Adsorption of chromium(III) on lignin. *BioResource. Technol.,* *99*(16), 7709-7715. http://dx.doi.org/10.1016/j.biortech.2008.01.069

16

Noble Metal Nanoparticles and Their (Bio) Conjugates. II. Preparation

Ignác Capek

Correspondence: Ignác Capek, Slovak Academy of Sciences, Polymer Institute, Institute of Measurement Science, Dúbravská cesta, Bratislava and Faculty of Industrial Technologies, TnUni, Púchov, Slovakia

Abstract

Hybrid nanoparticles of gold and silver can not only retain the beneficial features of both nanomaterials, but also possess unique advantages (synergism) over the other two types. Novel pseudospherical and anisotropic nanoparticles, bimetallic triangular nanoparticles, and core@shell nanoparticles were prepared by the different procedures for various applications and understanding both the particle evolution (nucleation) and nanoparticle anisotropy. Hybrid nanoparticles of gold and silver are considered to be low in toxicity, and exhibit facile surface functionalization chemistry. Furthermore, their absorption peaks are located in visible and near-infrared region. These nanoparticles provide significant plasmon tunability, chemical and surface modification properties, and significant advances in the growth into anisotropic nanostructures. The photoinduced synthesis can be used to prepare various (sub) nanoparticles and OD and 1D nanoparticles. Ostwald and digestive ripening provided narrower particle size distribution.

Keywords: functional, hybrid, core@shell, bimetallic and (an) isotropic noble metal nanoparticles, Oswaldt and digestive ripening process

1. Introduction

The noble metal nanoparticle colloids were reported to exhibit a wide range of various applications from different fluids to in vitro and vivo as a carrier for drug delivery into cells and tissues to optical and magnetic manipulations in biomedical systems as well (Sun et al., 2000). The unique physical and chemical properties of nanoparticles are given by their large surface area or a large surface-to-volume ratio and high reactivity. The unique optical and magnetic properties of nanoparticles are controlled by the type of atoms (metals), composition of nanoparticles, particle shape, type and concentration of functional groups and so on (Feldmann, 2003). For example, the composite nanoparticles with the iron atoms domains generate the magnetism and with the gold atoms exhibit the unique optical features. Therefore such nanoparticles exhibit unique properties which mostly strongly differ from the bulk or classical materials.

The future generation of environmental nanotechnologies based on noble metal nanoparticles is supposed to replace conventional environmental technologies by novel nano(bio)technologies (Modi, et al., 2003). Zero- and one-dimensional (0D and 1D) nanoparticles and two- and three (2D and 3D) nanoparticle assays belong to these unique nanomaterials. Especially the nanostructures decorated with different (bio) molecules are exceptional materials with nonconventional such as biomedical harvesting properties.

Polymer- or biopolymer decorated metal nanoparticles are prepared by several procedures with and without templates. Most existing procedures explore the strong affinity of alkylthiol ligands and surfactants to gold and silver (Brust et al., 1994), the use of alkyl or aryl derivatives of disulfides (Shon et al., 2001), and thioethers (Li et al., 2001) as well. Furthermore, organic compounds with polar groups (Boal et al., 2000), thioacetate groups (Brousseau et al., 1999), amino- and carboxylate groups and tetradentate thioether ligands (Maye et al., 2002) have recently been used to mediate the formation of 0D or 1D nanoparticles. In this second part of the review (Preparation II) we summarize and discuss the preparation hybrid, functional and bimetallic noble metal nanoparticles including their particle nucleation, ripening and optical properties.

2. Hybrid Nanoparticles

Hybrid nanoparticles are composed of different inorganic and organic components that can not only retain the beneficial features of both nanomaterials, but also possess unique advantages (synergism) over the other two types. For example, the ability to combine a multitude of organic and inorganic components in a modular fashion allows for systematic tuning of the properties of the resultant hybrid nanomaterial. Currently, there are four common synthetic strategies to prepare hybrid nanoparticles. The first strategy involves the preparation of each component particle separately and then both (or all) components were incorporated into micrometer size porous bead (Sathe et al., 2006), polymeric micelle (Park et al., 2008), silica core or shell matrix (Gole et al., 2008) or alternate polymeric layers (Kulakovich et al., 2002). In the second approach one of the presynthesized particles is coated with polymer of desired thickness and then attached with other presynthesized nanoparticles (Zhai et al., 2009). In the third approach the seed nanoparticle is prepared first, and then it is used to grow with other atoms (Selvan et al., 2007; Jiang et al., 2008). In the last approach each presynthesized component particle is functionalized first, and then hybrid particles are prepared by chemical/biochemical linkage (Pons et al., 2007; Fu et al., 2009). In the first two methods the large composite nanoparticles (>50 nm to micrometer size) with a resultant poor colloidal stability are formed. The third approach, on the contrarty, provides small nanoparticles (Selvan et al., 2007). The last procedure provides small nanoparticles with the low yield and often mixed with nanoparticle side products and requires some separation methods (Fu et al., 2004).

Hybrid nanoparticles of gold and silver are considered to be low in toxicity, and exhibit facile surface functionalization chemistry (Jain et al., 2008). Furthermore, their absorption peaks are located in visible and near-infrared region. Hybrid gold/silver nanoparticles provide significant plasmon tunability, chemical and surface modification properties, and significant advances in the growth into anisotropic nanostructures (Xiang et al., 2008). Novel pseudospherical and anisotropic nanoparticles, bimetallic triangular nanoparticles, and core@shell nanoparticles were prepared by the different (above mentioned) procedures for various applications and understanding both the particle evolution (nucleation) and nanoparticle anisotropy (Xue et al., 2007). The photoinduced synthesis of noble metal nanoparticles is broadly used approach to initiate photochemical reduction processes leading to oligomers, subnanoparticles and OD and 1D nanoparticles. Xue et al. have used the photochemical approach for the preparation of Au@Ag OD and 1D (nanorods, nanoprisms) nanoparticles and explained the mechanism of the photochemical growth of silver nanoprisms using spherical metal seeds (Xue et al., 2007).

Anisotropic hybrid nanoparticles can be also prepared by several approaches. One of them is a microemulsion. Herein the mixing of single microemulsions can tune the final structure and composition of final nanoparticles (Capek, 2004; Glomm, 2005; Hunter, 2004; Murphy & Jana, 2002). The second approach is based on the (co)deposition of atoms or primary metal nanoparticles onto seeds and their encapsulation within the various templates. Co-deposition of small gold nanoparticles onto nanowires or encapsulation of gold nanoparticles or their atoms within metal colloids was used to prepare desired anisotropic nanostructures (Hu et al., 2006). This approach can be used to vary the colloidal and collective properties of 1D nanomaterials, for example, the transformation of gold spherical nanoparticles onto 1D nanoparticles (Nam et al., 2006). Core@shell nanoparticles belong to the unique composite nanostructures. They can be prepared by various procedures, for example, by the reduction of two or more metal salts with different reactivity in solutions. The reduction of most reactive metal salts leads to the formation of particle core and the postponed reduction of other salt can form the atoms or primary nanoparticles that deposit onto the preformed nanoparticles. The seed nanoparticles can also be used as a solid matrix or template for the deposition of other present atoms. This procedure was used to prepare various ferromagnetic or plasmonic/magnetic nanostructures such as gold@cobalt, gold@Fe_xO_y, and so on, nanoparticles with a nondipolar gold core and a ferromagnetic cobalt (iron oxide) shell (Yu et al., 2005; Korth et al., 2008). The gold@cobalt nanoparticles were prepared by the thermolysis of $Co_2(CO)_8$ in the presence of the oleylamine-capped gold nanoparticles as seeds. The amine-termined oligopolystyrene (OPS) surfactants were used for the stabilization of these hybrid nanostructures. TEM and HRTEM confirmed the presence of OPS-capped Au@Co nanostructures by the formation of larger (d ~ 22 nm) nanoparticles compared to the AuNP seeds. These particle nanostructures self-assembled into the mesoscopic nanoparticle chains spanning hundreds of nanometers to micrometers in length. HRTEM further confirmed the formation of hybrid OPS-capped Au@CoNPs by the imaging of a discrete core@shell morphology composed of the darker high-atomic-number gold nanoparticle core (d_{core} ~ 13 nm) and a lighter shell from the metallic cobalt (thickness ~ 4-5 nm).

A heterogeneous nucleation of composite noble metal nanoparticle is based on the mixing of a precursor (salt, reductant and additives) solution with a micelle solution saturated with the seed metal nanoparticles (Lin et al., 1999a). This procedure was used to prepare sets of metal composite nanoparticles. Herein the inverse cationic micelles were used to grow the seeds as the feedstock. These hybrid nanoparticles of silver and gold were also prepared by Wilcoxon and Provencio (Wilcoxon & Provencio, 2004). The red colored initial solution with gold nanoparticles with an absorbance peak at 518 nm blue shifts as thicker shells of silver are formed around the core (gold seed) and the core@shell nanostructure appeared. The formation of silver shell is accompanied by yellowing the gold@silver nanoparticles solution with a narrow symmetrical plasmon typical for the silver nanocluster. In the reverse approach (gold deposition onto silver seed nanoparticles - silver@gold nanoparticles), the absorbance peak shifts in the opposite manner. It is noteworthy to mention that even when the composite nanoparticles have much larger fraction of atoms of silver (particle core – seed), the damping of the gold plasmon is much stronger than is observed in a pure gold nanoparticle of the same size.

Silica-based nanomaterials and nanoscale metal–organic frameworks (NMOFs) belong to other interesting hybrid nanomaterials (Taylor-Pashow et al., 2010). The former group can be divided into two nanostructures - silica-based hybrid nanomaterials and solid silica particles and mesoporous silica nanoparticles (MSNPs). Both exhibit unique properties such as high surface areas, tunable pore sizes, and large pore volumes and therefore they are used as the drug delivery nanomaterials, imaging agents or therapeutics. Required cargoes can be either directly incorporated in the silica pores or grafted to the outer surface of the solid silica particles via various functional groups. Thus, the MSNPs must be first functionalized and then coupled with imaging or therapeutic agents. Then these conjugates via loading of cargo into the pores, covalent grafting, and co-condensation of siloxy-derived groups are formed. Silica matrixes can serve as a core or shell template for more complex composite nanostructures with noble metal nanoparticles.

These silica-based nanostructures are mainly synthesized by two approaches: sol–gel and reverse microemulsion. The former approach provides monodisperse solid silica particles ranging in size from 50 nm to 2 μm (Stöber et al., 1968). This synthesis consists of two steps: the hydrolysis and condensation of a silica monomer (tetraethyl orthosilicate (TEOS)) in ethanol solution and as a catalyst is ammonia used (Scheme 5a). The size of the particles can be tuned by variations of precursor concentrations, the type of reaction solvent and temperature. For example, increased the TEOS concentration from 0.05 M to 0.67 M while keeping the other reactant concentrations and conditions constant affords silica particles from 20 to 880 nm in size (Wang et al., 2010). The high stability of silica colloid results from the electrostatic repulsion among the negatively charged silica particles. The second common method for the preparation of silica nanoparticles with narrow particle size distribution is based on the reverse microemulsion that controls simply the kinetics of precursor TEOS hydrolysis, particle nucleation and growth (Scheme 5b) (Rieter et al., 2007). For example, the size and number of final nanoparticles can be tuned by varying the water to surfactant ratio (ω).

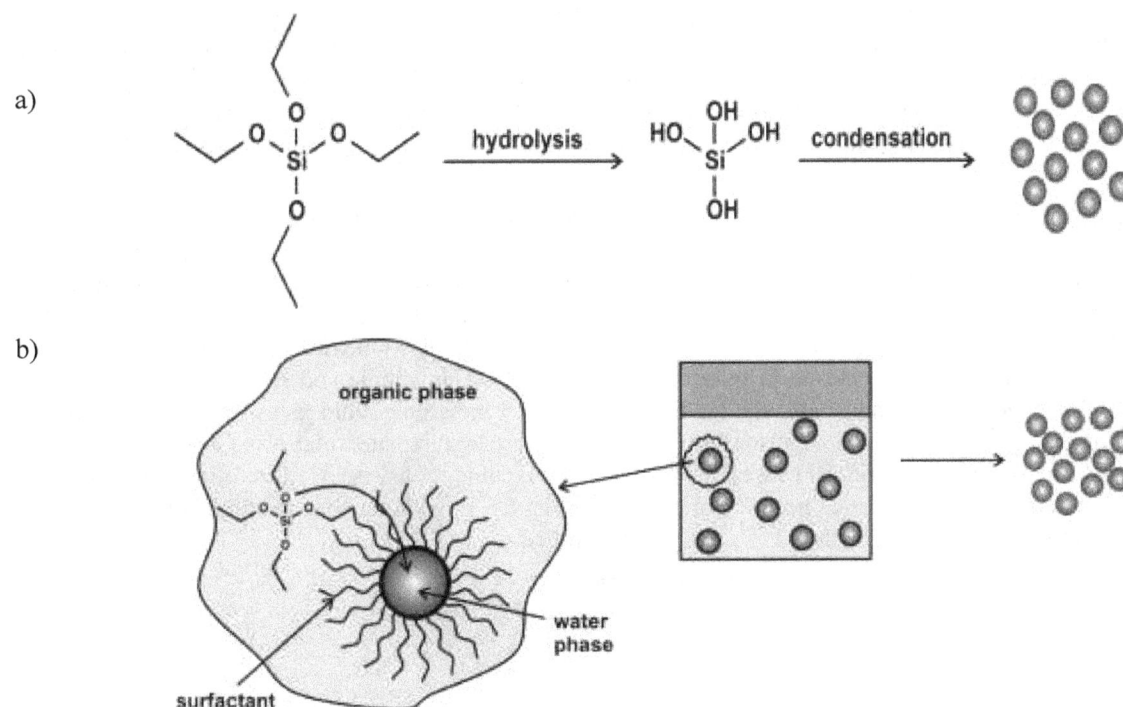

Scheme 5. Methods for synthesizing solid silica nanoparticles. (a) The Stöber method, (b) The reverse phase microemulsion (Taylor-Pashow et al., 2010).

Chak et al. have discussed (Chak et al., 2009) the synthesis of composite nanoparticles consisting of small gold nanoparticles (2 nm), dibenzo(24)crown-8 (DB24C8) and dibenzylammonium (DBA). DB24C8 and DBA fragments were connected with the gold nanoparticle via the amine group links, acted as supramolecular recognition motifs and formed form a 1:1 supramolecular complex, namely, (DBA⊂DB24C8) pseudorotaxane (Badjic et al., 2006). The conjugate of amine monofunctional gold nanoparticles (1-AuNPs, decorated with dibenzyl ammonium (DBA)-disulfide ligand) together with a crown-ether-decorated polystyrene resin (PS-DB24C8) and crownether-decorated superparamagnetic iron oxide nanoparticles (SPIO-DB24C8) formed interesting recognition nanomotifs. This research was devoted to the control of the number of active entities per nanoparticle (Grainger & Castner, 2008). Herein, the 1-AuNPs–based smart nanomotifs were formed by self-assembling of supramolecular dimers and trimers bisDB24C8 and trisDB24C8 (Scheme 6). Both the gold-based suprastructures and monofunctional 1-AuNP exhibit the same plasmonic band at λ ~ 520 nm (Mock et al., 2008). These smart nanoparticle-biomolecule suprastructures with the controlled number of bioentities per nanoparticle, would allow more accurate detection of molecules associated with particular diseases, offering drastic improvements in disease detection, therapy, and prevention (Jain et al., 2008).

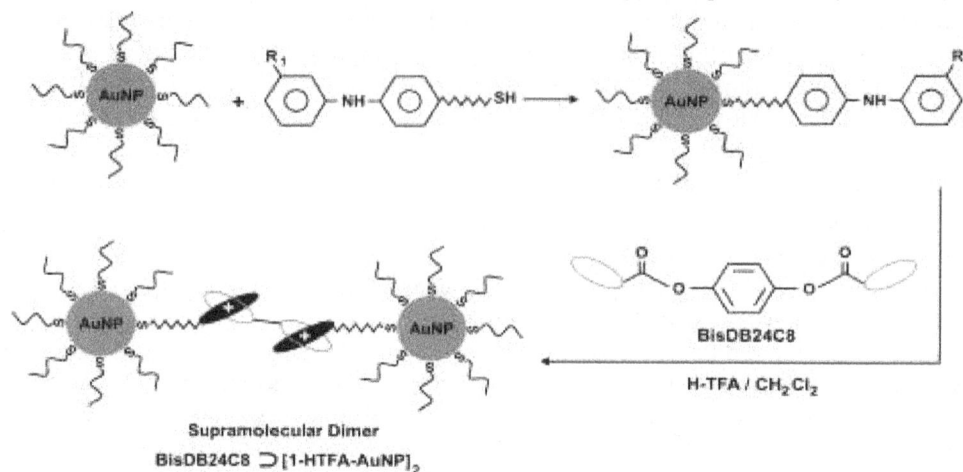

Scheme 6. Supramolecular self-assembly of the 1-AuNPs with bisDB24C8 and trisDB24C8 to afford the dimers bisDB24C8⊃(1·HTFA-AuNP)₂ and trimers trisDB24C8⊃(1·HTFA-AuNP)₃, respectively (Chak et al., 2009).

The further smart suprastructures (hydrogen-bonded supramolecular pseudorotaxane structures) were formed by the conjugation of SPIO-DB24C8 spheres with the functionalized AuNPs. The shell of the suprastructure was formed by gold nanoparticles and its nature was a function of gold particle size. Smaller AuNPs are observed to be well deposited on the periphery of the SPIO-DB24C8 microspheres. The SPIO-DB24C8 microspheres are composed of SPIO nanocrystals confined in thin silica shells, which are decorated with the crown ethers by amide linkages (Chak et al., 2009).

Yang et al. (Yang et al., 2004) have described the fabrication of advanced 1-D metallic organic nanocomposites by loading a vacant lipid nanotubes (LNTs) hollow cylinder with water-soluble gold (AuNPs) and silver nanoparticles (AgNPs) using capillary force (Scheme 7). They achieved the fabrication of gold or silver nanoparticles encapsulated in the cylindrical hollow of the glycolipid nanotube with high-axial ratio by loading the vacant LNT hollow cylinder with aqueous gold or silver nanoparticles using. Aqueous gold or silver nanoparticles (1-3 nm wide) are favorable to form the AuNPs@LNT or AgNPs@LNT nanocomposite in relatively higher yields. This simple and mild approach led to the fabrication of a 1-D metallic-organic nanocomposite loading well defined metallic colloid particles inside the organic nanotube hollow. In addition, the 1-D nanocomposite functions as a convergent template to fabricate a gold nanowire by removing the LNT shell through a firing process. Thus, the fabrication consists of the following steps: (1) Lyophilization of LNTs to empty the internal hollow volume, (2) filling LNTs with aqueous solution containing metal (Au or Ag) nanoparticles, and (3) removing the LNT shell by firing process.

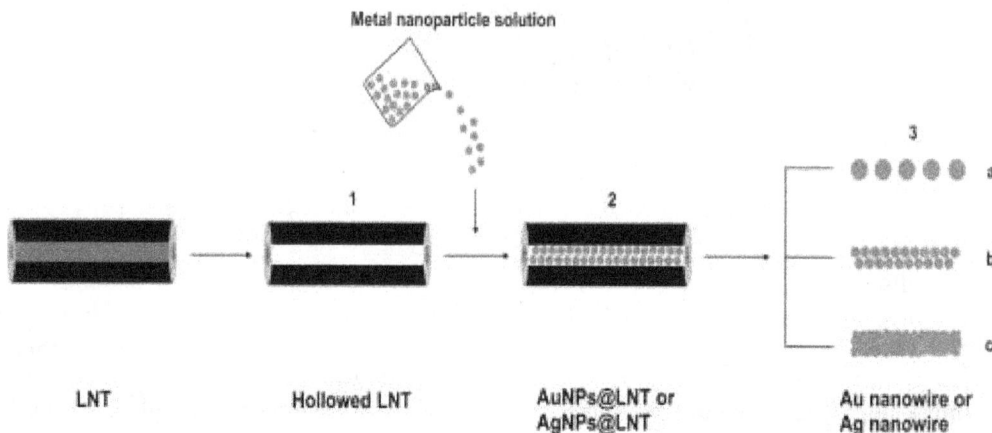

Scheme 7. Schematic diagram for the fabrication of AuNPs@LNT or AgNPs@LNT and a gold nanowire. (1) Lyophilization of LNTs to empty the internal hollow volume, (2) filling LNTs with aqueous solution containing metal (Au or Ag) nanoparticles, and (3) removing the LNT shell by firing process: (a) discrete and (b, c) continuous gold nanowires (Yang et al., 2004).

3. Other Nanoparticles

Among valuable noble metal nanoparticles one can find also platinum nanoparticles. The syntheses of platinum nanoparticles of different sizes and shapes such as tetrahedral (Ahmadi, et al., 1996), cubic (Ahmadi, et al., 1996; Fu et al., 2002), nanowire (Chen et al., 2004) and tetrapods (Herricks et al., 2004) have been reported. Platinum nanoparticles were prepared in inverse microemulsions. By mixing of two microemulsions (microdroplets saturated either by precursor such as the platinum salt ($PtCl_6^{2-}$)) or by the reducing agent) started the nucleation and platinum nanoparticles appeared (Ingelsten et al., 2001). The nonionic stabilizers (poly(ethylene glycol)monododecyl ethers ($C_{12}E_4$ (Brij 30), $C_{12}E_5$, $C_{12}E_6$)), anionic stabilizer (sodium bis(2-ethylhexyl)sulfosuccinate (AOT)), and their mixtures were used to formulate the nanoreactors (water-solubilized microdroplets). The reduction of platinum salt is a function of droplet fusion which is given by the type of the palisade stabilizer layer at the droplet surface. The hydrophilic alcohol ethoxylate favors a flux of stabilizer and other reactants back and forth between the bulk hydrocarbon domain and the droplets interface. In contrast, the hydrophobic stabilizer (AOT) favors more elastic collisions and slower rate of fusion of droplets and the formation of smaller nanoparticles (Fletcher et al., 1987). The fusion of droplets is regulated by the composition of interface layer formed by both stabilizers.

Mixtures of hydrophobic AOT and nonionic emulsifiers (based on PEG) were reported to be more efficient than their homo-mixtures in the preparation of platinum nanoparticles (Shelimov et al., 2000). The mechanism of particle formation is also based on the formation mixed micelles or microdroplets and the accumulation of salt precursor in the micelles (microdroplets) via interaction of the polar headgroups of PEG chains and the sulfonate group of AOT with the platinum salt. In the second step the precursor (hexachloroplatinic complex) is reduced by the reducing agent via the droplet scission. The chemisorption of functional surfactants on the particle surface or on the atom aggregates leads to the inhibition of both particle growth and agglomeration and the stable nanoparticles can be fabricated. Platinum nanoparticles were also synthesized by reducing H_2PtCl_6 with hydrazine in the w/o microemulsion consisting of $C_{12}E_4$(Brij 30) alone, n-heptane and water (Rivadulla et al., 1997). Absorption spectrum of the aqueous solution of H_2PtCl_6 shows two different bands (at 220 nm and 260 nm). After mixing of two different microdroplets, absorption spectrum showed only one peak centered at 236 nm which is attributed to the formation of Pt nanoparticles. The position of the observed band differs from the expected one appeared at ~215 nm (Creighton & Eadon, 1991). This difference was discussed in terms of the formation of anisotropic nanoparticles (nanorods (NRs)) and the interaction between the particles (formation of aggregates). The conformation and packing of the PEG can strongly influence the accessibility of the polymer chains for interaction (the complex formation, hybridization,...) and the particle mobility (Scheme 8). The mobility of the PEG-coated gold nanoparticles is always retarded by the increased ligand density which is directly related to the particle retention and distribution between organs.

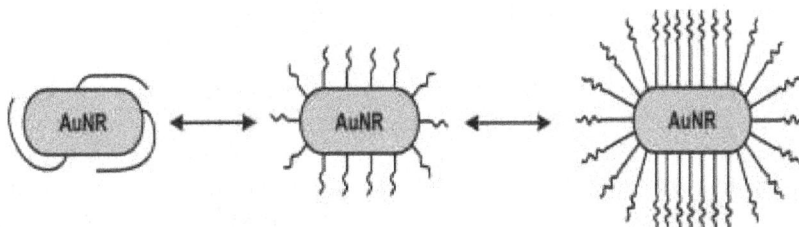

Scheme 8. Different possible configurations of PEG molecules attached to the surface of gold nanoparticles.

The photoinitiation mechanism of particle nucleation is complex because some or all components of the reaction system can absorb light. These molecules such as nonionic surfactant (its terminal hydroxyl groups), reducing agents or additives (dyes) are then transferred to the excited entities and participate in the charge transfer reactions and reduction processes. This complex approach was discussed by Rivadulla at al. (Rivadulla et al., 1997) who prepared platinum nanoparticles at different temperatures with/without exposure to light irradiation. They used AOT instead of PEG-based stabilizer because AOT has no terminal hydroxyl groups and only small changes appeared. However, at temperature 50 °C the reaction produces a brown color that is observed also by platinum particles obtained by hydrazine reduction, confirming the formation of colloidal Pt particles without addition of the reducing agent. Sulfonate groups of the excited ionic surfactant AOT are supposed to donate electrons, causing reduction of silver salt during irradiation and nucleate nanoparticles. Excited reactants might also take part in reduction mechanism via donation of electrons or radicals. Indeed, UV irradiation of surfactant micelles saturated with monomer initiated the microemulsion polymerization of vinyl monomers (Capek, 2014). Thus, the radicals formed within the microdroplets or the continuous phase at higher temperature or by irradiation might start the reduction of metal precursors. Furthermore, impurities in reactants may also donate electrons (the radicals) by excitation caused by the absorption of light or transformed to radical intermediates at higher temperatures (Sheu & Chen, 1988).

The inverse microemulsion (water/Triton X-100/propanol-2/cyclohexane) saturated with Pt and Ru salts provided the hybrid nanoparticles of Pt/Ru (Zhang & Chan, 2003). The particle sizes were varied in the range 2.5 - 4.6 nm by varying the precursor salt and hydrazine concentrations. At low Pt-Ru salt concentration, below 20 mM, the particles with the diameter around 2.5 nm appeared. When the Pt-Ru concentration increased above 20 mM (up to 60 mM), the size of Pt-Ru alloy nanoparticles increased to 4.0 – 4.5 nm. Beyond 60 mM the average nanoparticle size remained constant at 4.5 nm. Similar results were obtained for microdroplets stabilized with cationic surfactants (Chen & Wu, 2000). The common feature of the present studies is that the size of initial microdroplets is much larger than the size of metal nanoparticles. The particle size of the latter varied from 26 nm to 83 nm with increasing concentrations of Ru and Pt salts from 2 to 100 mM. This behavior was discussed in the following two terms: 1) For the {(Ru} + (Pt)} salts mixture below 20 mM the generation of one metal particle requires several Pt-Ru microdroplets. 2) Beyond 20 mM, a single precursor droplet saturated with Pt-Ru requires the nucleation to start the entrance of hydrazine droplets. The particle nucleation and growth are controlled by the collision among Pt-Ru-and hydrazine containing droplets. Furthermore the collision among the formed is very low because they are outnumbered by the hydrazine droplets by a factor of $>10^4$.

Pure palladium nanoparticles have been prepared by the reverse microemulsion consisting of {bis(N-octylethylenediamine)palladium(II) chloride ((Pd(oct-en)$_2$)Cl$_2$), water, chloroform and NaBH$_4$ as a reducing agent (Iida et al., 2002; Hamada et al., 2000). Monodisperse metal nanoparticles appeared at a low precursor (Pd(oct-en)$_2$)Cl$_2$ concentration with the (water)/((Pd(oct-en)$_2$)Cl$_2$) ratio (ω) = 7.6. At a very low amount of water $\omega < 10$, the structure of the aggregates confirms a reverse micelle type. At a somewhat higher level of water $\omega \geq 20$, the classical microdroplets appeared. At $\omega = 50$ an onset of the bicontinuous structure creates. The further variation in droplet size and shape one can obtain by using different solvents and reaction conditions. For example, the addition of methanol changed spherical to anisotropic nanoparticles. This is attributed to the variation of structures of the stabilizer/palladium complexes and microdroplets and the participation these complexes on the reaction and stabilization processes. The alkylethylenediamine/palladium complex, for example, acts as a metal precursor and a metal-particle stabilizer as well.

The spherical nanoparticles of Pt were nucleated and grew in the microdroplets stabilized by nonionic surfactant polyvinylpyrollidone (PVP) and saturated with potassium hexachloroplatinate salt in the presence of hydrogen as a reductant (Yu & Xu, 2003). The similar procedure was used to prepare cubic platinum nanoparticles capped with oxalate (Fu et al., 2002) and tetrahedral platinum nanoparticles capped with PVP (Yu & Xu, 2003; Narayanan & El-Sayed, 2005; Narayanan & El-Sayed, 2004a; Narayanan & El-Sayed, 2004b). Polyacrylate-based stabilizer was used in the modified version of the hydrogen reduction method to produce polymer capped cubic platinum nanoparticles (Ahmadi et al., 1996; Narayanan & El-Sayed, 2004a; Narayanan & El-Sayed, 2004b; Narayanan & El-Sayed, 2004c). Capping agents based on PEG and PVP were also used to prepare and stabilize the platinum nanowires and tetrapods (Iida et al., 2002).

Min et al. have used PEGylated gold nanorods (PEG-covered AuNR, AuNR@PEG) as a carrier for platinum drug complexes (Scheme 9) (Min et al., 2010). The prodrug Pt(IV) complex was conjugated to the surface of PEG-AuNRs. On entering cells, the high reactivity of Pt(IV) prodrug via entering the cells was modified (reduced by cellular reductants, e.g. glutathione) to the divalent platinum form. Simultaneously the cisplatin can be released from the carrier matrix and/or interacts with cell biopolymers (proteins, enzymes,..). PEG-based particle stabilizer and linkage are widespread in the clinical use. For example, PEG-AuNR conjugates are highly stable, relatively noncytotoxic in vivo, and the PEGylated drugs, nanoparticles and various therapeutics can be „masked" from the human's immune system (Veronese & Pasut, 2005). Thus as shown earlier, AuNR@PEG – platinum drug ligand conjugates enhance both the cell uptake of platinum drugs and the cytotoxicity of the conjugated cisplatin compared with cisplatin alone.

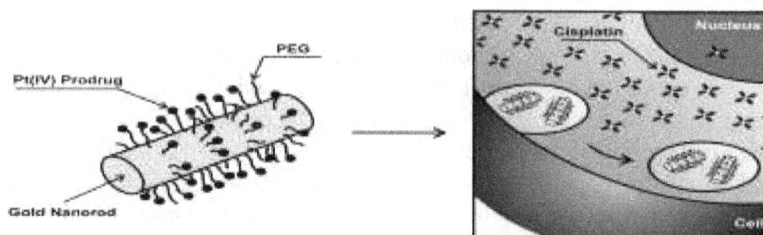

Scheme 9. Schematic illustration of AuNR@PEG-platinum ligand conjugates for drug delivery (Min et al., 2010)

4. Ripening Processes

The continuous degradation of monomer dispersion mostly increases the particle size distribution. The large droplets growth on the expense of the small ones and this process is called Ostwald ripening (De Smet et al., 1997). In this process the high surface energy of the small droplets promotes their dissolution and the dissolved material is then re-deposited on the large particles. The "Ostwald ripening" proceeds also in the colloids with metal nanoparticles or nanocrystals. This process involves the growth of large particles (crystals) on the expense of the small ones by the adsorption of atoms from the atom-saturated continuous phase. The surface metal atoms from small nanoparticles have a higher solubility than those from large nanoparticles. A group of free standing crystallites with unequal sizes in nonequilibrium form will further differentiate and redistribute themselves through the above solid–solution–solid process to achieve a more uniform size distribution (Scheme 10).

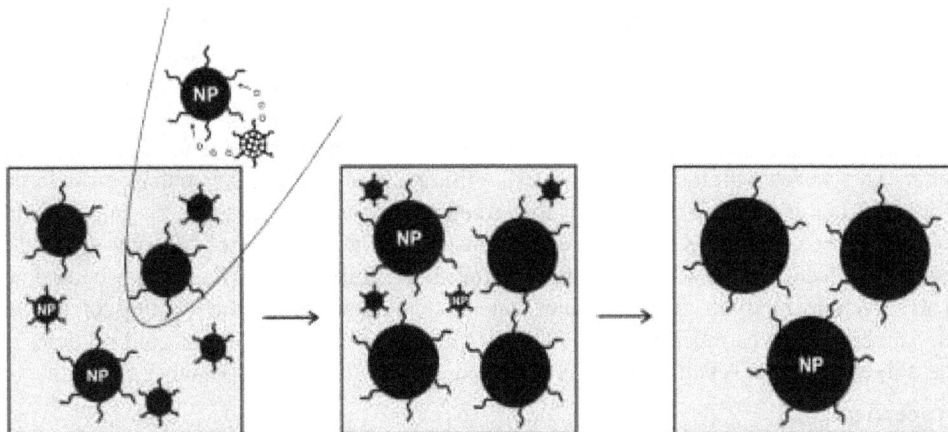

Scheme 10. Oswald ripening process.

The particle size distribution of metal nanoparticles can be narrower by the heating of nanoparticles solution under reflux known as the "digestive ripening" (Lin et al., 1999b). This simple procedure dramatically approves the particle size distribution. Herein this approach was first successfully applied for the formation of uniform gold nanoparticles (Lin et al., 2000). The alkylthiols-decoration of nanoparticles improve the particle size distribution of gold nanoparticles, that is, for example a very polydisperse colloid by ligating the nanoparticles with dodecanethiol followed by a digestive ripening process gives a monodisperse colloid. Continuous deaggregation of the surface oligomers occurs via the incorporation of additional stabilizer molecules that helps to release the surface embryos from the particle surface. Furthermore, these oligomers may also disassemble and collide with further large particles, finally forming an irreversible particle. The energy barrier for this oligomer deaggregation from large nanoparticles (spontaneous and random deaggregation occurring throughout the solution) is high and hence requires higher degree of metastability (e.g., supersaturation) compared to that for the small nanoparticles. In the case of large nanoparticles, growth occurs on surfaces of pre-existing large nanoparticles, hence, some energy is released by the partial elimination of those pre-existing interface by their interaction with primary oligomers. This energy gain decreases the overall free energy barrier and favors the growth of large nanoparticles. Therefore, the particle growth via small-large nanoparticles interaction proceeds at the lower (super) saturation level of large nanoparticles compared to that of small nanoparticles.

A digestive ripening process provided narrower particle distribution in the following study. The gold colloid with sizes ranging from 1 to 40 nm by heating tuned the particle size to the range of 4-4.5 nm (TEM) (Stoeva et al., 2002). The appearance of a narrow plasmon absorption peak (UV/vis absorption) with the maximum at 513 nm confirmed the presence of uniform gold nanoparticles. Without the digestive ripening a broad plasmon absorption band with no definite maximum appeared (Ferrell et al., 1985). A similar trend in the narrowing of the particle size distribution by heating of gold/toluene dispersion was reported in the next study (Lin et al., 1986). Herein the fraction of small spherical nanoparticles with sizes of about 4 nm strongly increased as well as their tendency to self-assemble into both 2D layers and 3D structures. Even the mechanism of this process is not entirely clear the surfactant-coated nanoparticles are the necessary starting material. The gold powder itself or its mixtures with various surfactant does not undergo to digestive ripening. The surfactant-capped nanoparticles take part in the diffusion of "free atoms" in solid-solvent-solid steps and this transform is much more active with the smaller nanoparticles. The surface atoms of small nanoparticles are speculated to be "much more free" to desorb into the solution than those of the large nanoparticles. The surface metal atoms of very small nanoparticles are "nearly free" to be desorbed into the continuous phase and adsorbed by solid phase of large nanoparticles (Klabunde & Mulukutla, 2001). The surfaces of both small and large nanoparticles (water- or oil-soluble metal nanoparticles) are passivated or wetted with the solute and surfactant molecules and the solubility of these surface particle entities is higher at small particles. This behavior is besides of above mentioned discussion also probably due to the fact that these nanoparticles were prepared by a "wet chemistry" procedure and therefore the solution is saturated with surfactant and free atoms. We speculate that the encapsulation of surfactant molecules into the particle matrix might favor the discussed dissolution events. The predominant adsorption of atoms by large nanoparticles shift the equilibrium in favor of the growth of large nanoparticles. The continuous solubilization of surface particle entities leads to the saturation of solution with surfactant and free atoms. Furthermore, the free surfactant micelles can serve as vesicles for the transfer of atoms from small nanoparticles via solution to the large particles (a la multiemulsion).

The heating of gold colloids can change not only their size but also their shape. This is reason why new formed colloid may appear in different colors. For example, the red gold colloid changed to a brown color one (Lin et al., 1986; Bain et al., 1989). The color change is also a function of solvent type (acetone, toluene or their mixtures). In great excess of acetone the gold nanoparticles are strongly solvated by acetone and the bond of thiol (RSH) molecules on the particles' surface becomes weaker. Acetone (or similar solvents), with its nonbonding electron pairs, can serve as a reasonably good ligand for gold but can only compete with RSH at high acetone concentrations. Under such a condition the particle agglomeration is more favored as well as the color change. As larger volume of acetone is removed, the thiol competes better and better. This varies with the solubility of alkylthiols in solvent and it is enhanced with the long-chained thiol because it is less soluble in acetone than in toluene. Acetone acts as a weak stabilizing agent, which is substituted by alkylthiol molecules when acetone is evaporated. This ensures good dispersity of the long-chained thiol (apolar)-ligated gold particles in the toluene medium. On the contrary, the short-chained thiol (less polar)-ligated gold particles are well dispersed in the acetone medium. These studies can be also devoted to the wetting of surfactant-decorated gold nanoparticles. Toluene achieves much better wetting of the thiol molecules on the gold particles' surface compared to the more polar solvent acetone. A similar trend is expected also for the colloids in other solvents.

5. Optical Properties

The field of noble metal nanoparticles research has received tremendous attention due to their unique collective properties. The properties of these nanoparticles arise from their size confinement effect (Kubo, 1962). A distinct feature of small gold nanoparticles is the red color of their colloids that is caused by the surface plasmon resonance (SPR) absorption (Link & El-Sayed, 1999a). For a spherical nanoparticle much smaller than the wavelength of light, an electromagnetic field at a certain frequency induces a resonant, coherent oscillation of metal free electrons across the nanoparticle. This oscillation is known as the surface plasmon resonance (SPR) (Scheme 11). For the gold and silver nanoparticles the SPR maximum lies at visible frequencies (Link & El-Sayed, 2003; Mulvaney, 1996). The surface Plasmon oscillation of the metal electrons results in a strong enhancement of absorption and scattering of electromagnetic radiation in resonance with the SPR frequency of the studied nanoparticles, giving them intensive colors.

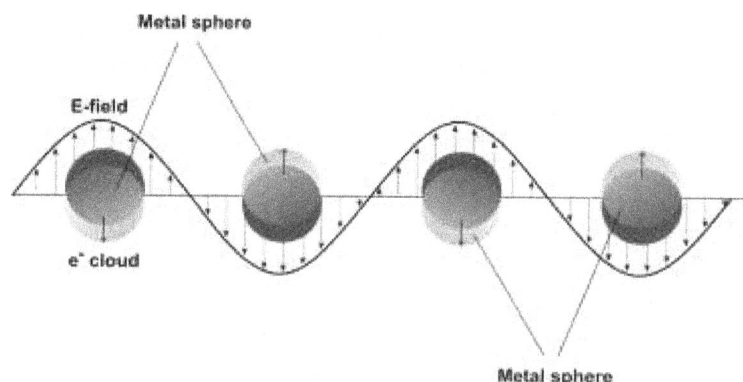

Scheme 11. Schematic view of the surface plasmon resonance absorption by noble metal nanoparticles (Jain et al., 2007).

Colloidal gold nanoparticles sample is ruby red in color. Mie explained this phenomenon theoretically by solving Maxwell's equation for the absorption and scattering of electromagnetic radiation by spherical particles (Mie, 1908). Gold nanoparticles with diameters ~ 10-20 nm exhibit a red color and absorption peak around 520 nm (Bohren & Huffman, 1983). The frequency and cross/section of SPR absorption and scattering depend on several parameters such as the size and shape of nanoparticles, atom type, surface particle composition, particle concentration, dielectric properties of the surrounding media, and inter-particle interactions (Jain et al., 2006a). The extreme sensitivity of the bandwidth, the peak height and shape, and the position of the absorption (or scattering) maximum of SPR spectra to environmental changes has initiated the development of gold nanoparticle-based sensors (Thanh & Rosenzweig, 2002; Reynolds et al., 2000). Small gold nanoparticles with a diameter ≤ 20 nm show essentially absorption (Jain et al., 2006b). The relative contribution of light scattering to the total extinction of the particle increases with increasing the particle diameter above 20 nm. This behavior of the gold nanoparticles is applied for molecular labeling. The Plasmon band shifts to longer wavelengths with increasing particle size, the change in the particle shape, the interparticle interactions and particle associations (Brause et al., 2002; Kerker, 1985; Sosa et al., 2003).

Gold nanoparticles have much higher absorption coefficients than conventional dyes. Furthermore, absorption spectra of gold nanoparticle do not undergo bleaching, that is, absorption coefficients of gold nanoparticles do not decrease with UV irradiation period. This is not the case for organic dyes that undergo bleaching and degradation. The gold nanoparticles has been already applied in colorimetric detection of analytes due to the following features: 1) the induced aggregation of gold nanoparticles attributed to the interaction with some probes changes the color of colloids and 2) , the refractive index of the AuNPs environment changes due to adsorption of biological analytes. The great enhancement of electromagnetic field at the surface of AuNPs by interaction of electrons with electromagnetic radiation offers other interesting physical properties with great potential for biodiagnostic assays (Baptista et al., 2008).

This behavior is presented in the spherical (zero-dimensional, 0D) gold nanoparticles coated with glutathione (GSH), lipoic acid (LA), DNA, bovine serum albumin (BSA) and chitosan (Figure 1). The decorated nanoparticles are monodisperse which is confirmed by a single peak in the absorbance spectra (Figure 1). The maximum of absorbance λ_{max} of gold nanoparticles in the absence of a capping agent was observed to be at around 530 nm with the mean hydrodynamic diameter in the range of 20-30 nm. In the presence of all additives the absorption peaks are shifted to higher wavelength. The strongest shift in the λ_{max} is observed for LA conjugate. Thus the λ_{max} appeared around 540-580 nm for GSH- and 560-620 nm for LA- decorated gold nanoparticles. The nanoparticles with different sizes led to the colloid with different colors, that is, the wine red color for the bare nanoparticles changed to blue for GSH- and dark blue for LA- decorated gold nanoparticles (Ahirwal & Mitra, 2009; Bakshi et al., 2010). The color of both GSH- and LA-stabilized nanoparticles did not change with time compared to the color of bare nanoparticles.

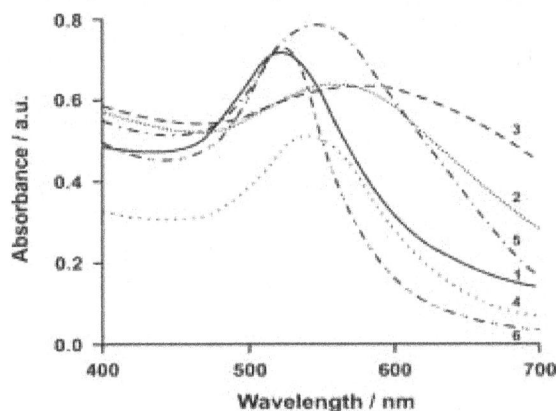

Figure 1. UV-Visible spectrum of (1) gold nanoparticles (AuNP), (2) GSH-AuNP, (3) LA-AuNP, (4) DNA-AuNP, (5) BSA-AuNP and (6) chitosan (Ahirwal & Mitra, 2009; Bakshi et al., 2010).

The further interesting feature of gold nanoparticles is their ability to efficiently quench emission from excited organic dyes located within a few nanometers of the particle surface (Lakowicz, 2001). When fluorophore probes are attached to gold surfaces, the gold surface can provide fluorescence quenching, and no organic quencher dye is required. The presence of gold or silver provides alternative nonradiative energy decay paths that can change the fluorescence quantum yield of a fluorophore. At close distances (< 5 nm) fluorescence is quenched and at longer distances (7.5–10.0 nm), it is enhanced (Lakowicz, 2001). Perez-Luna et al. (Perez-Luna et al., 2002) have used a gold surface to quench fluorescence of bound molecules and detected emission after displacement in a competitive immunoassay. Fluorescently labeled thio-oligonucleotides were attached to gold surfaces to demonstrate proof of principle for nucleic acid assays. Fluorescence from the unstructured probes can be quenched with single-stranded oligonucleotides due to their high flexibility and forming looped structures with the gold surface (Nie & Emory, 1997). Upon hybridization the double-stranded DNA is rigid and can't form looped structures to the particle surface. The major shortcoming of both solution-based and particle -based molecular beacons are the limited multiplexability (Varma-Basil et al., 2004).

Numerous assays were developed for a wide range of analytes (e.g., DNA, protein, metal ions, enzyme, and small molecular drugs) to follow photochemical and photophysical events within the studied systems. For example, Du et al. (Du et al., 2003) demonstrated quenching of hairpin DNA sequences attached to a planar gold surface, to mimic a microarray experiment, and successfully distinguished two DNA sequences, and they also started to investigate the thermodynamic and kinetic response of the sensor (Du et al., 2005). Dubretret et al. (Dubertret et al., 2001) used a hairpin loop beacon probe structure on small gold nanoparticles, while Maxwell et al. (Maxwell et al., 2002) showed that even unstructured oligonucleotide probes could be employed.

Spherical gold nanoparticles have been already discussed as the distinct colorimetric probes. The one-dimensional (nonspherical) nanoparticles have been also used to construct plasmon coupling-based colorimetric assays with novel optical characteristics (Huang et al., 2007; Link & El-Sayed, 2005). Gold nanorods possess two bands: a weaker band around 530 nm that corresponds to the transverse Plasmon oscillation and a stronger band at longer wavelengths arising from the Plasmon oscillation of electrons along the longitudinal axis of the nanorods (Link et al., 1999b). The latter band can be even shifted into the near-IR region by an increase in the nanorod aspect ratio (Huang et al., 2006). Particle association and/or agglomeration (or interparticle interaction) is accompanied with the redshift of the transverse peak and the blueshift of the longitudinal peak.

Mock and coworkers (Mock et al., 2002) studied the absorption spectra of silver nanoparticles with different sizes and shapes in the whole range of visible region. The prepared spherical and nonspherical with anisotropical silver nanoparticles with different sizes (40–120 nm) and shapes (spheres, decahedrons, triangular truncated pyramids and platelets) absorbed light in the blue, green and red part of the spectrum, respectively. The SPR peak intensity increased with size and the shift to longer wavelengths is observed at one-dimensional nanoparticles.

6. Conclusions

Novel hybrid pseudospherical and anisotropic nanoparticles, bimetallic triangular nanoparticles, and core@shell nanoparticles were prepared by the different procedures for various applications. Hybrid nanoparticles of gold and silver are considered to be low in toxicity, and exhibit facile surface functionalization chemistry. Furthermore, their absorption peaks are located in visible and near-infrared region. These nanoparticles provide significant plasmon tunability, chemical and surface modification properties, and significant advances in the growth into anisotropic nanostructures. Ostwald and digestive ripening provided narrower particle size distribution. The unique collective properties of noble metal nanoparticles have generated considerable interest for their use in a wide range of diverse applications. Their properties results from their large surface-to-volume ratio that strongly differ from the bulk materials. Silver and gold nanoparticles have much higher absorption coefficients than conventional dyes and their absorption spectra do not undergo bleaching. The gold nanoparticles has been already applied in colorimetric detection of analytes due to the following features: 1) the induced aggregation of gold nanoparticles attributed to the interaction with some probes changes the color of colloids and 2) , the refractive index of the AuNPs environment changes due to adsorption of biological analytes. The heating of silver gold colloids can change not only their size but also their shape. This is reason why new formed colloid may appear in different colors. The color change is also a function of solvent type. The passivation of metal particles with polar and apolar organic solvents might lead to the gold-solvent colloid which has particles well dispersed in polar or apolar solvents.

Acknowledgements

This research is supported by the VEGA grants 2/0040/14 and 2/0152/13 and APVV-0125-11 grant.

Nomenclature

0D	zero-dimensional
1D	one-dimensional
AFM	atomic force microscopy
AgNPs	silver nanoparticles
AOT	sodium bis(2-ethylhexyl)sulfosuccinate
AuNPs	gold nanoparticles
CA	citric acid
$C_{12}E_4$ (Brij 30), $C_{12}E_5$, $C_{12}E_6$, $C_{12}E_4$(Brij 30)	poly(ethylene glycol)monododecyl ethers
CTAB	cetyltrimethyl ammonium bromide
DB24C8	dibenzo(24)crown-8
DBA	dibenzylammonium
DDAB	didodecyldimethylammonium bromide
Dx	dextran
H_2PtCl_6	hexachloroplatinum acid
HRTEM	high resolution TEM
LNTs	lipid nanotubes

MSNPs	mesoporous silica nanoparticles
NMOFs	nanoscale metal–organic frameworks
NRs	nanorods
OPS	oligopolystyrene
OR	Oswald ripening
PEG	poly(ethylene glycol)
PMAA	poly(methacrylic acid)
$PtCl_6^{2-}$	platinum salt
PVP	polyvinylpyrollidone
SPIO	superparamagnetic iron oxide
SPIO-DB24C8	crownether-decorated superparamagnetic iron oxide nanoparticles
TEM	transmission electron microscopy
TEOS	tetraethyl orthosilicate
ω	ratio of (water)/(surfactant

References

Ahirwal, G. K., & Mitra, C. K. (2009). Direct electrochemistry of horseradish peroxidase-gold nanoparticles conjugate. *Sensors, 9*(2), 881-894. http://dx.doi.org/10.3390/s90200881

Ahmadi, T. S., Wang, Z. L., Green, T. C., Henglein, A., & El-Sayed, M. A. (1996). Shape-Controlled Synthesis of Colloidal Platinum Nanoparticles. *Science, 272*, 1924-1926. http://dx.doi.org/10.1126/science.272.5270.1924

Badjic, J. D., Ronconi, C. M., Stoddart, J. F., Balzani, V., Silvi, S., & Credi, A. (2006). Operating molecular elevators. *Journal of the American Chemical Society, 128*, 1489-1499. http://dx.doi.org/10.1021/ja0543954

Bain, C. D., Evall, J., & Whitesides, G. M. (1989). Formation of Monolayers by the Coadsorption of Thiols on Gold: Variation in the Head Group, Tail Group, and Solven. *Journal of the American Chemical Society, 111*, 7155-7164. http://dx.doi.org/10.1021/ja00200a039

Bakshi, M. S., Jaswal, V. S., Possmayer, F., & Petersen, N. O. (2010). Solution Phase Interactions Controlled Ordered Arrangement of Gold Nanoparticles in Dried State. *Journal of Nanoscience and Nanotechnology, 10*, 1747-1756. http://dx.doi.org/10.1166/jnn.2010.2050

Baptista, P., Pereira, E., Eaton, P., Doria, G., Miranda, A., Gomes, I., Quaresma, P., & Franco, R. (2008). Gold nanoparticles for the development of clinical diagnosis methods. *Analytical and Bioanalytical Chemistry, 391*, 943-950. http://dx.doi.org/10.1007/s00216-007-1768-z

Boal, A. K., Ilhan, F., DeRouchey, J. E., Thurn-Albrecht, T., Russell, T. P., & Rotello, V. M. (2000). Self-assembly of nanoparticles into structured spherical and network aggregates. *Nature, 404*, 746-748. http://dx.doi.org/10.1038/35008037

Bohren, C. F., & Huffman, D. R. (1983). Absorption and Scattering of Light by Small Particles; John Wiley & Sons: New York, 544.

Brause, R., Moeltgen, H., & Kleinermanns, K. (2002). Characterization of laser-ablated and chemically reduced silver colloids in aqueous solution by UV/VIS spectroscopy and STM/SEM microscopy. *Applied Physics B: Lasers and Optics, 75*, 711-716. http://dx.doi.org/10.1007/s00340-002-1024-3

Brousseau, L. C., Novak, J. P., Marinakos, S. M., & Feldheim, D. L. (1999). Assembly of phenylacetylene-bridged gold nanocluster dimers and trimers. *Advanced Materials, 11*, 447-449. http://dx.doi.org/10.1002/(SICI)1521-4095(199904)11:6<447::AID-ADMA447>3.0.CO;2-I

Brust, M., Walker, M., Bethell, D., Schiffrin, D. J., & Whyman, R. (1994). Synthesis of thiol-derivatised gold nanoparticles in a two-phase Liquid–Liquid system. *Journal of the Chemical Society, Chemical Communications,* 801-802. http://dx.doi.org/10.1039/c39940000801

Capek, I. (2004). Preparation of metal nanoparticles in water-in-oil (w/o) microemulsions. *Advances in Colloid and Interface Science, 110*, 49-74. http://dx.doi.org/10.1016/j.cis.2004.02.003

Capek, I. (2014). On photoinduced polymerization of acrylamide. *Designed Monomers and Polymers, 17*, 356-363.

http://dx.doi.org/10.1080/15685551.2013.840510

Chak, C. P., Xuan, S., Mendes, P. M., Yu, J. C., Cheng, C. H. K., & Leung, K. C. F. (2009). Discrete functional gold nanoparticles: hydrogen bond-assisted synthesis, magnetic purification, supramolecular dimer and trimer formation. *ACS Nano, 3*, 2129-2138. http://dx.doi.org/10.1021/nn9005895

Chen, D. H., & Wu, S. H. (2000). Synthesis of Nickel Nanoparticles in Water-in-Oil Microemulsions. *Chemistry of Materials, 12*, 1354-1360. http://dx.doi.org/10.1021/cm991167y

Chen, J., Herricks, T., Geissler, M., & Xia, Y. (2004). Single-crystal nanowires of platinum can be synthesized by controlling the reaction rate of a polyol process. *Journal of the American Chemical Society, 126*, 10854-10855. http://dx.doi.org/10.1021/ja0468224

Creighton, A., & Eadon, D. (1991). Ultraviolet–visible absorption spectra of the colloidal metallic elements. *Journal of the Chemical Society, Faraday Transactions, 87*, 3881-3891. http://dx.doi.org/10.1039/ft9918703881

De Smet, Y., Deriemaeker, L., & Finsy, R. (1997). A Simple Computer Simulation of Ostwald Ripening. *Langmuir, 13*, 6884-6888. http://dx.doi.org/10.1021/la970379b

Du, H., Disney, M. D., Miller, B. L., & Krauss, T. D. (2003). Hybridization-based unquenching of DNA hairpins on Au surfaces: prototypical "molecular beacon" biosensors. *Journal of the American Chemical Society, 125*, 4012-4013. http://dx.doi.org/10.1021/ja0290781

Du, H., Strohsahl, C. M., Camera, J., Miller, B. L., & Krauss, T. D. (2005). Sensitivity and specificity of metal surface-immobilized "molecular beacon" biosensors. *Journal of the American Chemical Society, 127*, 7932-7940. http://dx.doi.org/10.1021/ja042482a

Dubertret, B., Calame, M., & Libchaber, A. J. (2001). Single-mismatch detection using gold-quenched fluorescent oligonucleotides. *Nature Biotechnology, 19*, 365-370. http://dx.doi.org/10.1038/86762

Feldmann, C. (2003). Polyol-Mediated Synthesis of Nanoscale Functional Materials. *Advanced Functional Materials, 13*, 101-107. http://dx.doi.org/10.1002/adfm.200390014

Ferrell, T. L., Callcott, T. A., & Warmark, R. (1985). Plasmons and surfaces. *American Scientist, 73*, 344-353.

Fletcher, P. D. I., Howe, A. M., & Robinson, B. H. (1987). The kinetics of solubilisate exchange between water droplets of a water-in-oil microemulsion. *Journal of the Chemical Society, Faraday Transactions, 1*(83), 985-1006. http://dx.doi.org/10.1039/f19878300985

Fu, A. H., Micheel, C. M., Cha, J., Chang, H., Yang, H., & Alivisatos, A. P. (2004). Discrete nanostructures of quantum dots/Au with DNA. *Journal of the American Chemical Society, 126*, 10832-10833. http://dx.doi.org/10.1021/ja046747x

Fu, X., Wang, Y., Wu, N., Gui, L., & Yang, Y. (2002). Shape-Selective Preparation and Properties of Oxalate-Stabilized Pt Colloid. *Langmuir, 18*, 4619-4624. http://dx.doi.org/10.1021/la020087x

Fu, Y., Zhang, J., & Lakowicz, J. R. (2009). Silver-enhanced fluorescence emission of single quantum dot nanocomposites. *Chemical Communications*, 313-315. http://dx.doi.org/10.1039/B816736B

Glomm, W. R. J. (2005). Functionalized Gold Nanoparticles for Applications in Bionanotechnology. *Journal of Dispersion Science and Technology, 26*, 389-414. http://dx.doi.org/10.1081/DIS-200052457

Gole, A., Agarwal, N., Nagaria, P., Wyatt, M. D., & Murphy, C. J. (2008). One-pot synthesis of silica-coated magnetic plasmonic tracer nanoparticles. *Chemical Communications*, 6140-6142. http://dx.doi.org/10.1039/b814915a

Grainger, D. W., & Castner, D. G. (2008). Nanobiomaterials and Nanoanalysis: Opportunities for Improving the Science to Benefit Biomedical Technologies. *Advanced Materials, 20*, 867-877. http://dx.doi.org/10.1002/adma.200701760

Hamada, K., Hatanaka, K., Kawai, T., & Konno, K. (2000). Synthesis of Nanosize Pt Particles by Direct Addition of Hydrazine in Ionic Reversed Micelle Systems. *Shikizai Kyokaishi, 73*, 385-390. http://dx.doi.org/10.4011/shikizai1937.73.385

Herricks, T., Chen, J., & Xia, Y. (2004). Polyol synthesis of platinum nanoparticles: control of morphology with sodium nitrate. *Nano Letters, 4*(12), 2367-2371. http://dx.doi.org/10.1021/nl048570a

Hu, J., Wen, Z., Wang, Q., Yao, Q., Zhang, Q., Zhou, J., & Li, J. (2006). Controllable Synthesis and Enhanced Electrochemical Properties of Multifunctional $Au_{core}Co_3O_{4shell}$ Nanocubes. *Journal of Physical Chemistry B, 110*, 24305-24310. http://dx.doi.org/10.1021/jp063216h

Huang, X., El-Sayed, I. H., Qian, W., & El-Sayed, M. A. (2006). Cancer Cell Imaging and Photothermal Therapy in the Near-Infrared Region by Using Gold Nanorods. *Journal of the American Chemical Society, 128*, 2115-2120. http://dx.doi.org/10.1021/ja057254a

Huang, Y. F., Lin, Y. W., & Chang, H. T. (2007). Control of the surface charges of Au-Ag nanorods: selective detection of iron in the presence of poly(sodium 4-stryenesulfonate). *Langmuir, 23*, 12777-12781. http://dx.doi.org/10.1021/la701668e

Hunter, R. J. (2004). Foundations of Colloid Science. Oxford University Press: New York.

Iida, M., Ohkawa, S., Er, H., Asaoka, N., & Yoshikawa, H. (2002). Formation of Palladium(0) Nanoparticles from a Microemulsion System Composed of Bis (N-octylethylenediamine)palladium(II) Chloride Complex. *Chemistry Letters, 10*, 1050-1051. http://dx.doi.org/10.1246/cl.2002.1050

Ingelsten, H. H., Bagwe, R., Palmqvist, A., Skoglundh, M., Svanberg, C., Holmberg, K., & Shah, D. O. (2001). Kinetics of the formation of nano-sized platinum particles in water-in-oil microemulsions. *Journal of Colloid and Interface Science, 241*, 104-111. http://dx.doi.org/10.1006/jcis.2001.7747

Jain, P. K., El-Sayed, I. H., & El-Sayed, M. A. (2007). Au nanoparticles target cancer. *Nano Today, 2*, 18-29. http://dx.doi.org/10.1016/S1748-0132(07)70016-6

Jain, P. K., Eustis, S., & El-Sayed, M. A. (2006a). Plasmon Coupling in Nanorod Assemblies: Optical Absorption, Discrete Dipole Approximation Simulation, and Exciton-Coupling Model. *Journal of Physical Chemistry B, 110*, 18243-18253. http://dx.doi.org/10.1021/jp063879z

Jain, P. K., Huang, X., El-Sayed, I. H., & El-Sayed, M. A. (2008). Noble metals on the nanoscale: optical and photothermal properties and some applications in imaging, sensing, biology, and medicine. *Accounts of Chemical Research*, 41, 1578-1586. http://dx.doi.org/10.1021/ar7002804

Jain, P. K., Lee, K. S., El-Sayed, I. H., & El-Sayed, M. A. (2006b). Calculated absorption and scattering properties of gold nanoparticles of different size, shape, and composition: applications in biological imaging and biomedicine. *Journal of Physical Chemistry B, 110*, 7238-7248. http://dx.doi.org/10.1021/jp0571170o

Jiang, J., Gu, H., Shao, H., Devlin, E., Papaefthymiou, G. C., & Ying, J. Y. (2008). Bifunctional Fe3O4-Ag heterodimer nanoparticles for two-photon fluorescence imaging and magnetic manipulation. *Advanced Materials, 20*, 4403-4407. http://dx.doi.org/10.1002/adma.200800498

Kerker, M. (1985). The optics of colloidal silver: something old and something new. *Journal of Colloid and Interface Science*, 105, 297-314. http://dx.doi.org/10.1016/0021-9797(85)90304-2

Klabunde, K. J., & Mulukutla, R. S., (2001). Nanoscale Materials in Chemistry; K. J. Klabunde, ed. Wiley Interscience: New York Chapt, 7, 223. http://dx.doi.org/10.1002/0471220620

Korth, B. D., Keng, P., Shim. I. B., Tang, C., Kowalewski, T., & Pyun, J. (2008). Synthesis, Assembly and Functionalization of Polymer Coated Ferromagnetic Nanoparticles. *ACS Nano, 996*, 272-285. http://dx.doi.org/10.1021/bk-2008-0996.ch020

Kubo, R. (1962). Electronic properties of metallic fine particles I. *Journal of the Physical Society of Japan, 17*, 975-986. http://dx.doi.org/10.1143/JPSJ.17.975

Kulakovich, O., Strekal, N., Yaroshevich, A., Maskevich, S., Gaponenko, S., Nabiev, I., Woggon, U., & Artemyev, M. (2002). Enhanced Luminescence Of Cdse Quantum Dots On Gold Colloids. *Nano Letters, 2*, 1449-1452. http://dx.doi.org/10.1021/nl025819k

Lakowicz, J. R. (2001). Radiative decay engineering: biophysical and biomedical applications. *Analytical Biochemistry, 298*, 1-24. http://dx.doi.org/10.1006/abio.2001.5377

Li, X. M., de Jong, M. R., Inoue, K., Shinkai, S., Huskens, J., & Reinhoudt, D. N. (2001). Formation of gold colloids using thioether derivatives as stabilizing ligands. *Journal of Materials Chemistry, 11*, 1919-1923. http://dx.doi.org/10.1039/b101686p

Lin, S., Franklin, M. T., & Klabunde, K. J. (1986). Nonaqueous colloidal gold. Clustering of metal atoms in organic media. *Langmuir, 2*, 259-260. http://dx.doi.org/10.1021/la00068a027

Lin, X. M., Sorensen, C. M., & Klabunde, K. J. (2000). Digestive ripening, nanophase segregation and superlattice formation in gold nanocrystal colloids. *Journal of Nanoparticle Research, 2*, 157-164. http://dx.doi.org/10.1023/A:1010078521951

Lin, X. M., Sorensen, C. M., Klabunde, K. J., & Hajipanayis, G. C. (1999a). Control of Cobalt Nanoparticle Size by the

Germ Growth Method in Inverse Micelle System: Size Dependent Magnetic Properties. *Journal of Materials Research, 14*, 1542-1547. http://dx.doi.org/10.1557/JMR.1999.0207

Lin, X. M., Wang, G. M., Sorensen, C. M., & Klabunde, K. J. (1999b). Formation and dissolution of gold nanocrystal superlattices in a colloidal solution. *Journal of Physical Chemistry B, 103*(26), 5488-5492. http://dx.doi.org/10.1021/jp990729y

Link, S., & El-Sayed, M. A. (1999a). Size and temperature dependence of the plasmon absorption of colloidal gold nanoparticles. *Journal of Physical Chemistry B, 103*, 4212-4217. http://dx.doi.org/10.1021/jp984796o

Link, S., & El-Sayed, M. A. (2003). Optical properties and ultrafast dynamics of metallic nanocrystals. *Annual Review of Physical Chemistry, 54*, 331-366. http://dx.doi.org/10.1146/annurev.physchem.54.011002.103759

Link, S., & El-Sayed, M. A. (2005). Simulation of the optical absorption spectra of gold nanorods as a function of their aspect ratio and the effect of the medium dielectric constant. *Journal of Physical Chemistry B, 109*, 10531-10532. http://dx.doi.org/10.1021/jp058091f

Link, S., Mohamed, M. B., & El-Sayed, M. A. (1999b). Simulation of the optical absorption spectra of gold nanorods as a function of their aspect ratio and the effect of the medium dielectric constant. *Journal of Physical Chemistry B, 103*, 3073-3077. http://dx.doi.org/10.1021/jp990183f

Maxwell, D. J., Taylor, J. R., & Nie, S. (2002). Self-assembled nanoparticle probes for recognition and detection of biomolecules. *Journal of the American Chemical Society, 124*, 9606-9612. http://dx.doi.org/10.1021/ja025814p

Maye, M. M., Chun, S. C., Han, L., Rabinovich, D., & Zhong, C. J. (2002). Novel spherical assembly of gold nanoparticles mediated by a tetradentate thioether. *Journal of the American Chemical Society, 124*, 4958-4959. http://dx.doi.org/10.1021/ja025724k

Mie, G. (1908). Contributions to the optics of turbid media, particularly of colloidal metal solutions. *Annals of Physics, 25*, 377-445. http://dx.doi.org/10.1002/andp.19083300302

Min, Y., Mao, C., Xu, D., Wang, J., & Liu, Y. (2010). Gold nanorods for platinum based prodrug delivery. *Chemical Communications, 46*, 8424-8426. http://dx.doi.org/10.1039/c0cc03108a

Mock, J. J., Barbic, M., Smith, D. R., Schultz, D. A., & Schultz, S. (2002). Shape effects in plasmon resonance of individual colloidal silver nanoparticles. *Journal of Chemical Physics, 116*, 6755-6759. http://dx.doi.org/10.1063/1.1462610

Mock, J. J., Hill, R. T., Degiron, A., Zauscher, S., Chilkoti, A., & Smith, D. R. (2008). Distance-dependent plasmon resonant coupling between a gold nanoparticle and gold film. *Nano Letters, 8*, 2245-2252. http://dx.doi.org/10.1021/nl080872f

Modi, A., Koratkar, N., Lass, E., Wei, B. Q., & Ajayan, P. M. (2003). Miniaturized gas ionization sensors using carbon nanotubes. *Nature, 424*, 171-174. http://dx.doi.org/10.1038/nature01777

Mulvaney, P. (1996). Surface Plasmon Spectroscopy of Nanosized Metal Particles. *Langmuir, 12*, 788-800. http://dx.doi.org/10.1021/la9502711

Murphy, C. J., & Jana, N. R. (2002). Controlling the Aspect Ratio of Inorganic Nanorods and Nanowires. *Advanced Materials, 14*, 80-82. http://dx.doi.org/10.1002/1521-4095(20020104)14:1<80::AID-ADMA80>3.0.CO;2-#

Nam, K. T., Kim, D. W., Yoo, P. J., Chiang, C. Y., Meethong, N., Hammond, P. T., Chiang, Y. M., & Belcher, A. M. (2006). Virus-enabled synthesis and assembly of nanowires for lithium ion battery electrodes. *Science, 312*, 885-888. http://dx.doi.org/10.1126/science.1122716

Narayanan, R., & El-Sayed, M. A. (2004a). Shape-Dependent Catalytic Activity of Platinum Nanoparticles in Colloidal Solution. *Nano Letters, 4*(7), 1343-1348. http://dx.doi.org/10.1021/nl0495256

Narayanan, R., & El-Sayed, M. A. (2004b). Effect of Nanocatalysis in Colloidal Solution on the Tetrahedral and Cubic Nanoparticle SHAPE: Electron-Transfer Reaction Catalyzed by Platinum Nanoparticles. *Journal of Physical Chemistry B, 108*(18), 5726-5733. http://dx.doi.org/10.1021/jp0493780

Narayanan, R., & El-Sayed, M. A. (2004c). Changing Catalytic Activity during Colloidal Platinum Nanocatalysis Due to Shape Changes: Electron-Transfer Reaction. *Journal of the American Chemical Society, 126*(23), 7194-7195. http://dx.doi.org/10.1021/ja0486061

Narayanan, R., & El-Sayed, M. A. (2005). Effect of colloidal nanocatalysis on the metallic nanoparticle shape: the Suzuki reaction. *Langmuir, 21*(5), 2027-2033. http://dx.doi.org/10.1021/la047600m

Nie, S., & Emory, S. R. (1997). Probing Single Molecules and Single Nanoparticles by Surface-Enhanced Raman

Scattering. *Science, 275*, 1102-1106. http://dx.doi.org/10.1126/science.275.5303.1102

Park, J. H., von Maltzahn, G., Ruoslahti, E., Bhatia, S. N., & Sailor, M. (2008). Micellar hybrid nanoparticles for simultaneous magnetofluorescent imaging and drug delivery. *Angewandte Chemie International Edition, 47*, 7284-7288. http://dx.doi.org/10.1002/anie.200801810

Perez-Luna, V. H., Yang, S., Rabinovich, E. M., Buranda, T., Sklar, L. A., Hampton, P. D., & López, G. P. (2002). Fluorescence biosensing strategy based on energy transfer between fluorescently labeled receptors and a metallic surface. *Biosensors and Bioelectronics, 17*, 71-78. http://dx.doi.org/10.1016/S0956-5663(01)00260-3

Pons, T., Medintz, I. L., Sapsford, K. E., Higashiya, S., Grimes, A. F., English, D. S., & Mattoussi, H. (2007). On the quenching of semiconductor quantum dot photoluminescence by proximal gold nanoparticles. *Nano Letters, 10*, 3157-3164. http://dx.doi.org/10.1021/nl071729+

Reynolds, R. A., Mirkin, C. A., & Letsinger, R. L. (2000). A gold nanoparticle/latex microsphere-based colorimetric oligonucleotide detection method. *Pure and Applied Chemistry, 72*, 229-235. http://dx.doi.org/10.1351/pac200072010229

Rieter, W. J., Kim, J. S., Taylor, K. M. L., An, H., Tarrant, T., & Lin, W. (2007). Hybrid silica nanoparticles for multimodal imaging. *Angewandte Chemie International Edition, 46*, 3680-3682. http://dx.doi.org/10.1002/anie.200604738

Rivadulla, J. F., Vergara, M. C., Blanco, M. C., Lopez-Quintela, M. A., & Rivas, J. (1997). Optical Properties of Platinum Particles Synthesized in Microemulsions. *Journal of Physical Chemistry B, 101*, 8997-9004. http://dx.doi.org/10.1021/jp970528z

Sathe, T. R., Agrawal, A., & Nie, S. (2006). Mesoporous silica beads embedded with semiconductor quantum dots and iron oxide nanocrystals: dual-function microcarriers for optical encoding and magnetic separation. *Analytical Chemistry, 78*, 5627-5632. http://dx.doi.org/10.1021/ac0610309

Selvan, S. T., Patra, P. K., Ang, C. Y., & Ying, J. Y. (2007). Synthesis of silica-coated semiconductor and magnetic quantum dots and their use in the imaging of live cells. *Angewandte Chemie International Edition, 46*, 2448-2452. http://dx.doi.org/10.1002/anie.200604245

Shelimov, B. N., Lambert, J. F., Che, M., & Didillon, B. (2000). Molecular-level studies of transition metal-support interactions during the first steps of catalysts preparation: platinum speciation in the hexachloroplatinate/alumina system. *Journal of Molecular Catalysis A: Chemical, 158*, 91-99. http://dx.doi.org/10.1016/S1381-1169(00)00047-9

Sheu, E. Y., & Chen, S. H. (1988). Thermodynamic analysis of polydispersity in ionic micellar systems and its effect on small-angle neutron scattering data treatment. *Journal of Physical Chemistry, 92*, 4466-4474. http://dx.doi.org/10.1021/j100326a044

Shon, Y. S., Mazzitelli, C., & Murray, R. W. (2001). Unsymmetrical Disulfides and Thiol Mixtures Produce Different Mixed Monolayer-Protected Gold Clusters. *Langmuir, 17*, 7735-7741. http://dx.doi.org/10.1021/la015546t

Sosa, I. O., Noguez, C., & Barrera, R. G. (2003). Optical Properties of Metal Nanoparticles with Arbitrary Shapes. *Journal of Physical Chemistry B, 107*, 6269-6275. http://dx.doi.org/10.1021/jp0274076

Stöber, W., Fink, A., & Bohn, E. (1968). Controlled growth of monodisperse silica spheres in the micron size range. *Journal of Colloid and Interface Science, 26*, 62-69. http://dx.doi.org/10.1016/0021-9797(68)90272-5

Stoeva, S., Klabunde, K. J., Sorensen, C. M., & Dragieva, I. (2002). Gram-Scale Synthesis of Monodisperse Gold Colloids by the Solvated Metal Atom Dispersion Method and Digestive Ripening and Their Organization into Two- and Three-Dimensional Structures. *Journal of the American Chemical Society, 124*, 2305-2311. http://dx.doi.org/10.1021/ja012076g

Sun, S., Murray, C. B., Weller, D., Folks, L., & Moser, A. (2000). Monodisperse FePt nanoparticles and ferromagnetic FePt nanocrystal superlattices. *Science, 287*, 1989-1992. http://dx.doi.org/10.1126/science.287.5460.1989

Taylor-Pashow, K. M. L., Della Rocca, J., Huxforda, R. C., & Lin, W. (2010). Hybrid nanomaterials for biomedical applications. *Chemical Communications, 46*, 5832-5849. http://dx.doi.org/10.1039/c002073g

Thanh, N. T. K., & Rosenzweig, Z. (2002). Development of an Aggregation-Based Immunoassay for Anti-Protein A Using GoldNanoparticles. *Analytical Chemistry, 74*, 1624-1628. http://dx.doi.org/10.1021/ac011127p

Varma-Basil, M., El-Hajj, H., Marras, S. A., Hazbon, M. H., Mann, J. M., Connell, N. D., Kramer, F. R., & Alland, D. (2004) Molecular beacons for multiplex detection of four bacterial bioterrorism agents. *Clinical Chemistry, 50*,

1060-1062. http://dx.doi.org/10.1373/clinchem.2003.030767

Veronese, F. M., & Pasut, G. (2005). PEGylation, successful approach to drug delivery. *Drug Discovery Today, 10*, 1451-1458. http://dx.doi.org/10.1016/S1359-6446(05)03575-0

Wang, X. D., Shen, Z. X., Sang, T., Cheng, X. B., Li, M. F., Chen, L. Y., & Wang, Z. S. (2010). Preparation of spherical silica particles by Stober process with high concentration of tetra-ethyl-orthosilicate. *Journal of Colloid and Interface Science, 341*, 23-29. http://dx.doi.org/10.1016/j.jcis.2009.09.018

Wilcoxon, J. P., & Provencio, P. P. (2004). Heterogeneous Growth of Metal Clusters from Solutions of Seed Nanoparticles. *Journal of the American Chemical Society, 126*, 6402-6408. http://dx.doi.org/10.1021/ja031622y

Xiang, Y., Wu, X., Liu, D., Li, Z., Chu, W., Feng, L., Zhang, K., Zhou, W., & Xie, S. (2008). Gold nanorod-seeded growth of silver nanostructures: from homogeneous coating to anisotropic coating. *Langmuir, 24*, 3465-3470. http://dx.doi.org/10.1021/la702999c

Xue, C., Millstone, J. E., Li, S., & Mirkin, C. A. (2007). Plasmon-Driven Synthesis of Triangular Core-Shell Nanoprisms from Gold Seeds. *Angewandte Chemie International Edition, 46*, 8436-8439. http://dx.doi.org/10.1002/anie.200703185

Yang, B., Kamiya, S., Shimizu, Y., Koshizaki, N., & Shimizu, T. (2004). Glycolipid nanotube hollow cylinders as substrates: Fabrication of one-dimensional metallic-organic nanocomposites and metal nanowires. *Chemistry of Materials, 16*, 2826-2831. http://dx.doi.org/10.1021/cm049695j

Yu, H., Chen, M., Rice, P. M., Wang, S. X., White, R. L., & Sun, S. (2005). Dumbbell-like bifunctional Au-Fe3O4 nanoparticles. *Nano Letters, 5*, 379-382. http://dx.doi.org/10.1021/nl047955q

Yu, Y., & Xu, B. (2003). Selective formation of tetrahedral Pt nanocrystals from K_2PtCl_6/PVP. *Chinese Science Bulletin, 48*(23), 2589-2593. http://dx.doi.org/10.1360/03wb0098

Zhai, Y., Zhai, J., Wang, Y., Guo, S., Ren, W., & Dong, S. (2009). Fabrication of iron oxide core/gold shell submicrometer spheres with nanoscale surface roughness for efficient surface-enhanced Raman scattering. *Journal of Physical Chemistry C, 113*, 7009-7014. http://dx.doi.org/10.1021/jp810561q

Zhang, X., & Chan, K. Y. (2003). Water-in-Oil Microemulsion Synthesis of Platinum−Ruthenium Nanoparticles, Their Characterization and Electrocatalytic Properties. *Chemistry of Materials, 15*, 451-459. http://dx.doi.org/10.1021/cm0203868

17

Experimental Investigation of the Hyperfine Structure of Neutral Praseodymium Spectral Lines and Discovery of New Energy Levels

Shamim Khan[1,2], Imran Siddiqui[1,3], Syed Tanweer Iqbal[1,4], Zaheer Uddin[1,3], G. H. Guthöhrlein[5], L. Windholz[1]

[1]Institute for Experimental Physics, Graz University of Technology, Petersgasse 16, A-8010 Graz, Austria

[2]Department of Physics, Islamia College Peshawar, Peshawar, Pakistan

[3]Department of Physics, University of Karachi, 75270 Karachi, Pakistan

[4]Department of Physics, The NED University of Engineering and Technology, University Rd, 75270 Karachi, Pakistan

[5]Laboratorium für Experimentalphysik, Helmut-Schmidt-Universität, Universität der Bundeswehr Hamburg, Holstenhofweg 85, 22043 Hamburg, Germany

Correspondence: L. Windholz, Institute for Experimental Physics, Graz University of Technology, Petersgasse 16, A-8010 Graz, Austria

Abstract

Experimental investigations of Pr I spectral lines were performed by means of laser induced fluorescence spectroscopy, using a hollow cathode discharge lamp as source of free atoms. The wavelengths for the laser excitation were found by the help of a highly resolved Fourier transform spectrum. Altogether we excited 236 unclassified lines and analysed their hyperfine structure, which led, together with the measured wavelengths of the observed fluorescence lines, to the discovery of 32 new even parity and 38 odd parity fine structure energy levels. These levels allow to classify more than 670 spectral lines of Pr I. The wave number calibrated Fourier transform spectrum allowed us to determine the energies of most of these newly discovered levels with an uncertainty of 0.015 cm^{-1}. Angular momenta, parity, and magnetic and electric hyperfine interaction constants (A and B) of the new levels were also determined.

Keywords: praseodymium, new energy levels, laser spectroscopy

1. Introduction

Praseodymium is a rare earth element belonging to the lanthanides group. Its atomic number is 59, and it has naturally occurring only one stable isotope ^{141}Pr with nuclear spin quantum number I = 5/2. The ground state $^4I^{\circ}_{9/2}$ belongs to the electronic configuration [Xe] $4f^3$ $6s^2$, having five valence electrons outside a Xenon-like electron core. Its nuclear magnetic dipole moment amounts to $\mu_I = 4.2754(5)$ μ_N (Macfarlane, Burum & Shelby 1982) and its nuclear electric quadrupole moment is Q = -0.0024b (Böklen, Bossert & Foerster 1975). Due to the five valence electrons, praseodymium has a very large number of fine structure (fs) levels and consequently an extremely line rich spectrum.

The first investigation of hyperfine (hf) structure in the spectrum of praseodymium can be traced back as early as 1929, when H.E. White (White 1929) investigated the wave number separation of hf structure components of 173 spectral lines. He was the first in determining the nuclear spin quantum number of ^{141}Pr. Since 1929 and over the span of five decades, several authors investigated the spectrum of praseodymium and discovered several electronic levels which were then collected and published in 1978 (Martin, Zalubas & Hagan 1978). A major contribution to the understanding and development of the spectrum of praseodymium (atom and singly charged ion) came from A. Ginibre (Ginibre 1981, Ginibre-Amery 1988, Ginibre 1989a, Ginibre 1989b). Her published (and also not yet published) work is based on the study of the fine and hyperfine structures in the configurations $4f^25d6s^2$ and $4f^25d^26s$ of neutral praseodymium. In this work, she calculated the wave numbers for a large number of new fs levels and also determined their hyperfine interaction constants. In 1981, Childs and Goodmann, using an atomic-beam - laser-rf double-resonance technique, determined the hyperfine (hf) constants of several low-lying metastable electronic levels with high accuracy (Childs & Goodman 1981). Kuwamoto et al. (Kuwamoto et al. 1996) investigated the hf structure of Pr I lines by using an atomic beam of neutral praseodymium, crossed perpendicular with an exciting laser beam, and determined A and B constants for 57 levels. The energies of 11 fs levels were determined for the first time. Krzykowski et al. (Krzykowski, Furmann, Stefanska, Jarosz & Kajoch 1997) accurately determined the values of the hf structure constants of some lower levels belonging to the configuration $4f^35d6s$ and for the upper levels of the investigated Pr-I lines, excited by laser light. The group of G. H.

Guthöhrlein (Guthöhrlein 2005) investigated a big number of Pr I lines using laser-induced fluorescence spectroscopy. Unfortunately, most of the results are still not published. Based on the available experimental results, in 2003 J. Ruczkowski et al. applied a semi-empirical method to describe configuration and designation of the known Pr I levels (Ruczkowski, Stachowska, Elantkowska, Guthöhrlein & Dembczyński 2003). In 2006 Furmann et al. (Furmann, Krzykowski, Stefańska & Dembczyński 2006) reported hf structure investigations of Pr I lines and the discovery of 57 new electronic levels with odd parity.

Since 2004, the group in Graz is concerned with experimental investigations of the hf structure of praseodymium lines using laser induced fluorescence (LIF) spectroscopy (Uddin 2006). In parallel to the laser spectroscopic work highly resolved Fourier-transform spectra were also recorded. These spectra led to the discovery of more than 9000 previously unknown Pr I and Pr II spectral lines, from which more than 1100 could been classified (Gamper et al. 2011). Later on, we focused on searching for new fs levels of praseodymium and were able to discover a significant number of levels, some of which have already been published, see refs. (Uddin et al. 2012a, Shamim, Siddiqui & Windholz 2011, Syed et al. 2011, Uddin et al. 2012b, Siddiqui, Shamim & Windholz 2014, Uddin et al. 2012c, Uddin 2014, Gamper, Khan, Siddiqui & Windholz 2013, Siddiqui, Shamim & Windholz 2016, Uddin, Siddiqui, Tanweer, Jilani & Windholz 2015).

The present investigation of the spectrum of praseodymium not only broadens the understanding of the Pr I level scheme but is also of interest to astrophysicists (Ryabchikova, Savanov, Malanushenko & Kudryavtsev 2001, Dolk et al. 2002) in analysing the Pr abundance in various stars.

All wavelengths given in this paper are in Å in standard air. Conversion from wavelength to wave number was performed using the formula given by Peck and Reeder (Peck & Reeder 1972) for the refractive index of air.

2. Experimental Setup

A laser-induced fluorescence (LIF) technique in combination with Fourier transform (FT) spectroscopy is used for the investigation of the hf structure of lines in the spectrum of praseodymium. The experimental setup is shown in Fig. 1, and a detailed description of each component used is given in ref. (Shamim, Siddiqui & Windholz 2011). Excitations of Pr lines were performed by means of a tunable ring dye laser, pumped by a single frequency semiconductor diode pumped Nd-Vanadate (Verdi V-18) laser. The investigations were carried out in the spectral regions of the dyes Rhodamine 6G (5700-6100 Å) , Sulforhodamine B (6100-6300 Å), DCM (6300-6800 Å) and Stilbene 3 (4200-4700 Å). The source of free praseodymium atoms is a dc hollow cathode discharge lamp (HCL), filled with Ar as buffer gas, in which free praseodymium atoms were produced by cathode sputtering. One of the advantages of using a HCL is, that, as a result of collisional processes within the Ar-Pr plasma, free Pr atoms are not only produced in its ground state but also in high lying excited states up to about 25000 cm^{-1}.

Of course the HCL emits all spectral lines of Pr and Ar (and their first ions). In order to distinguish the light emitted from the HCL from LIF intensity, the exciting laser beam is modulated with the help of a mechanical chopper 1 (see Fig. 1). The light emitted by the discharge is dispersed by a monochromator whose output is detected by a photomultiplier. Only lines influenced by the laser light show the same modulation and are detected with the help of Lock-In amplifier 1. The LIF signal together with the transmission signal of a temperature-stabilized marker etalon are digitally stored in a computer for further analysis.

The investigation of a line begins by tuning the laser light wavelength into resonance with the wavelength of a hf component of the investigated line. This wavelength is extracted from the FT spectrum. As a result of the excitation one expects laser-induced fluorescence lines, which are then detected. If the upper level of the excited transition is known, the monochromator transmission wavelength can be set to an expected decay line of this level and the hf structure of the line can be recorded by scanning the laser light frequency over the transition. If the upper level is previously unknown, we first have to set the laser frequency to a strong hf component of the line (wavelength known from the FT spectrum) and to search for LIF lines by tuning the transmission wavelength of the monochromator. The function of chopper 2 and Lock-In amplifier 2 will be explained when discussing the discovery of the level at 19111 cm^{-1}.

3. Data evaluation and Analysis

The analysis of the recorded hf pattern can lead to the knowledge of parameters of the combining fs levels such as total angular momentum J and magnetic dipole and electric quadrupole hf constants A and B, respectively. In case of praseodymium, which has a very small nuclear quadrupole moment, the value of B is very small for most of the levels and is often neglected. Exceptions are high precision Doppler free laser spectroscopic methods like experiments on an atomic beams, where the hf constants can usually be determined with high precision, and so also small B factors are determined reliably (Krzykowski, Furmann, Stefanska, Jarosz & Kajoch 1997, Childs & Goodman 1981). The evaluation of the recorded hf patterns also gives an estimated center-of-gravity (cg) wave number of the excited line. Together with the values of J and A (and sometimes B) and the wavelengths of the experimentally observed fluorescence lines, the identification of the involved fs levels is possible.

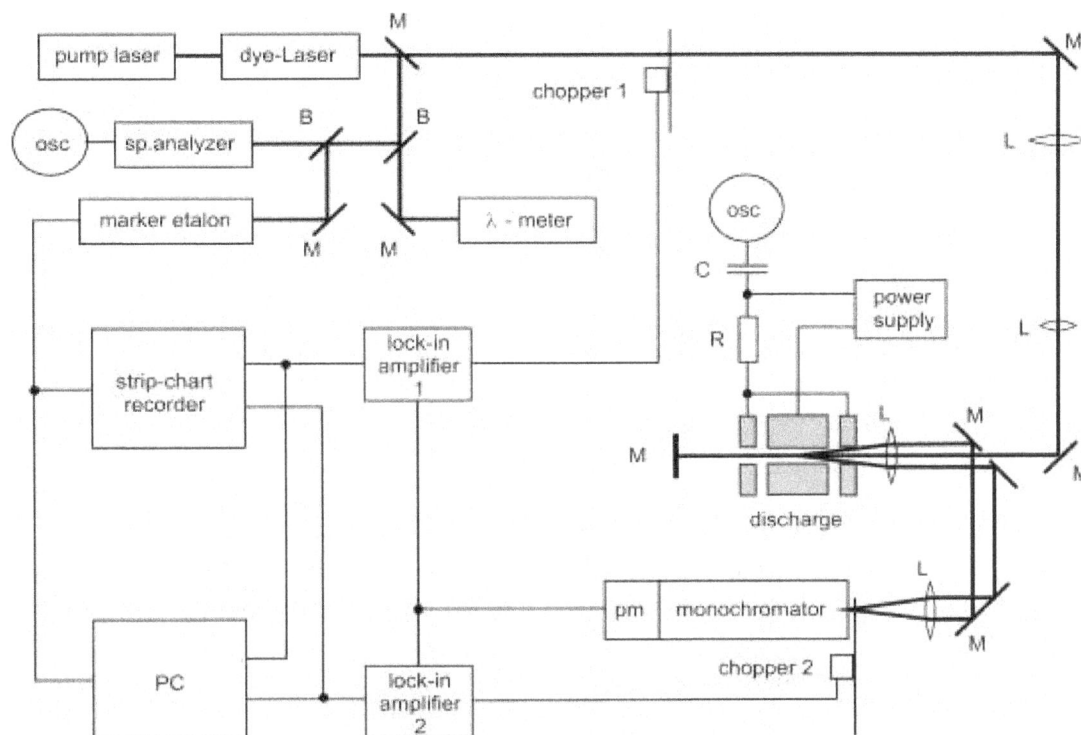

Figure 1. Experimental setup. M mirror, L lens, B beam splitter, R resistor, C capacitor, pm photomultiplier, osc oscilloscope, PC computer

Of special interest is the excitation of lines which cannot be classified as transitions between already known energy levels. In such cases the wavelengths of the fluorescence lines are previously unknown and have to be found by tuning the monochromator which disperses the light emitted by the discharge.

If now one of the combining levels is already known, let us say, the lower level, the energy of the upper combining unknown level is determined by adding the cg wave number of the line to the energy of the lower level. On the other hand, if an upper level is already known, the energy of lower combining unknown level is determined by subtraction of the cg wave number from the energy of the upper level. In the worst case, where both the combining levels are not known, only the observed fluorescence wavelengths can be used for the determination of the energies of the combining levels. For a detailed description of the method see ref. (Uddin et al. 2012a).

In an otherwise slightly simpler situation, where one (or several) of the observed fluorescence lines are in the wavelength range of our laser, laser excitation of a former fluorescence line can be performed and the hf pattern can be recorded on the previously excited line. Since the probability that a previously unknown energy level decays to an already known level is high, the energy of the new upper level can be determined. In each case, at least one of the levels involved in the excited transition must be identified.

4. Discovery of the New Level at 19111.80 cm^{-1}, Even Parity, J = 13/2, A = 762 MHz

A quite prominent line at 5636.940 Å, appearing in the FT spectrum with a high signal-to noise-ratio (SNR) of 140, was experimentally investigated using laser spectroscopy. The line exhibits a well-defined hf structure with clearly visible hf pattern, but with unresolved hf components, see Fig. 1. The line could not be classified using already known level energies, thus we had to assume that a previously unknown energy level is involved in the transition.

Laser excitation was performed by tuning the laser wavelength to the highest peak of the pattern at 5636.92 Å. A single fluorescence line at 5228 Å was observed when tuning the transmission wavelength of the monochromator from 3000 Å to 7000 Å. Observation of only one LIF line can happen when the upper level has only a few number of decay channels or has a very low transition probability for the other lines.

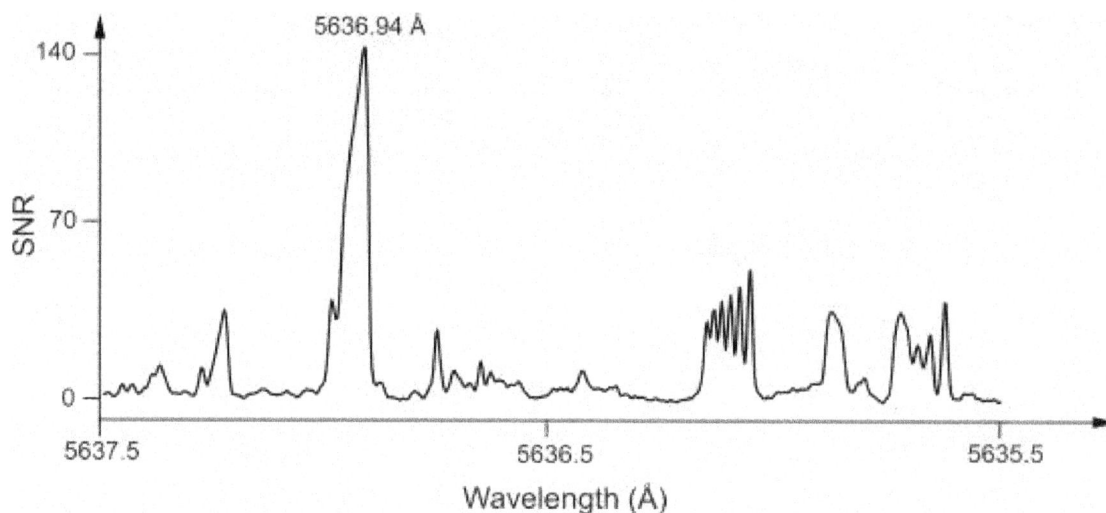

Figure 2. Line 5636.940 Å in the FT spectrum of Praseodymium

The hf structure of the line was recorded by scanning the laser frequency across the line. The LIF spectrum had a good signal-to-noise ratio and showed the hf structure pattern as appearing in FT spectrum, but with higher resolution (Figure 2).

The recorded hf structure was then fitted assuming different pairs of J-values for upper and lower level. For this, the program "Fitter" was used (Guthöhrlein 1998) which minimizes the sum of the squares of the differences between measured and calculated intensity values (ESS). The best fit situation with the lowest ESS value was obtained for $J_{up} = 13/2$, $J_{lo} = 11/2$, $A_{up} = 762$ MHz and $A_{lo} = 733$ MHz, see Figure 3. The cg wavelength obtained from the fitting procedure was 5636.940 Å.

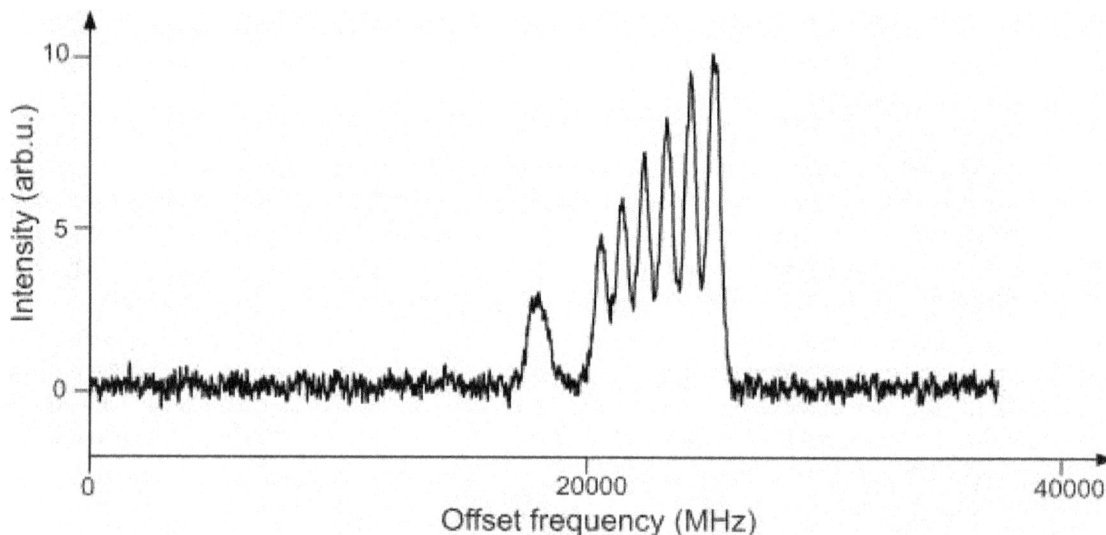

Figure 3. Recorded hf pattern of the line 5636.940 Å (LIF at line 5228 Å)

First we assumed that an unknown upper level is involved in the excitation of the investigated line. Therefore we searched in our database of known energy levels for a level having J = 11/2 and approximately the A-value from the fit. Several levels, having both even or odd parity, fulfilled these requirements (see Table 1). Hypothetical new levels were calculated by adding the cg wave number of the excited line at 5636.940 Å (17735.198 cm^{-1}) to the energies of these levels. Then we calculated possible decay lines of the hypothetical levels, but none of them could explain the observed LIF line at 5228 Å. Thus we repeated the procedure assuming that a new lower level is involved, but still without any success.

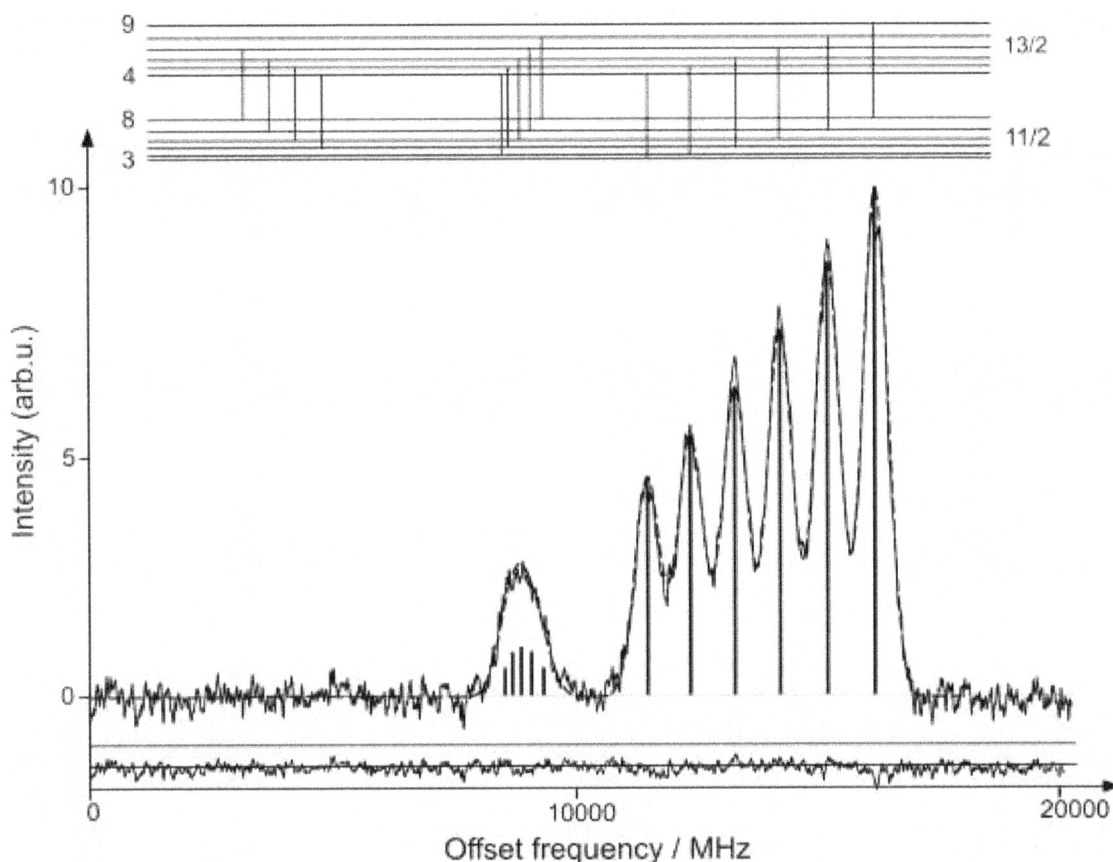

Figure 4. Best fit situation of the recorded hyperfine structure of line 5636.940 Å. The lower trace shows the residual between experimental and fitted hf pattern (x 0.5)

Next we thought about a possible scenario in which both the lower and upper of the combining levels are not known and were yet to be discovered. But to solve such a situation the wavelengths of at least two LIF lines have to be determined with high precision. The method is described e.g. in ref. (Windholz 2016). But here we obtained only one LIF line, so this scenario cannot be resolved.

Nevertheless, we tried to find a more accurate value of the fluorescence wavelength. The light emitted by the HCL was modulated using a second mechanical chopper, working with much higher frequency than the first chopper, modulating the intensity of the exciting laser light. The output of the photomultiplier was given now to two Lock-In amplifiers, one set to the chopping frequency of the laser light and the second one to that of the light emitted by the HCL. The laser frequency was fixed to the strongest hyperfine component of the line 5636.940 Å and the monochromator transmission wavelength was tuned around the line 5228 Å (from 5210 to 5240 Å). Both output signals of the LI-amplifiers were recorded simultaneously (see Figure 4). One trace now shows the LIF signal, obtained when the transmission wavelength of the monochromator agrees with the LIF line. The second trace shows the emission spectrum of the HCL. This spectrum now can be wavelength calibrated with help of the FT spectrum. In this way the high wavelength precision of the FT spectrum can be used to determine the LIF wavelength with an uncertainty of lower than 0.1 Å, despite of the fact that the monochromator used has a resolution of only 1 Å. The determined value of the fluorescence wavelength was 5227.97(10) Å.

In the FT spectrum there is a very strong line (SNR = 5300) at 5227.968 Å, see figure 5. This line is already classified as a transition from the ground level to a known low lying upper level at 19122.567 cm^{-1}, even parity, J = 11/2. Thus we had to believe that the observed LIF line is this strong line.

Thus again we came back to searching for a new lower energy level. But a fit of the recorded structure (Figure 2) is not possible with low ESS taking the hf constants of the level 19122.567 cm^{-1} as fixed values in the fitting procedure. This suggested that the upper level involved in the excitation of line at 5636.940 Å is not 19122.567 cm^{-1}.

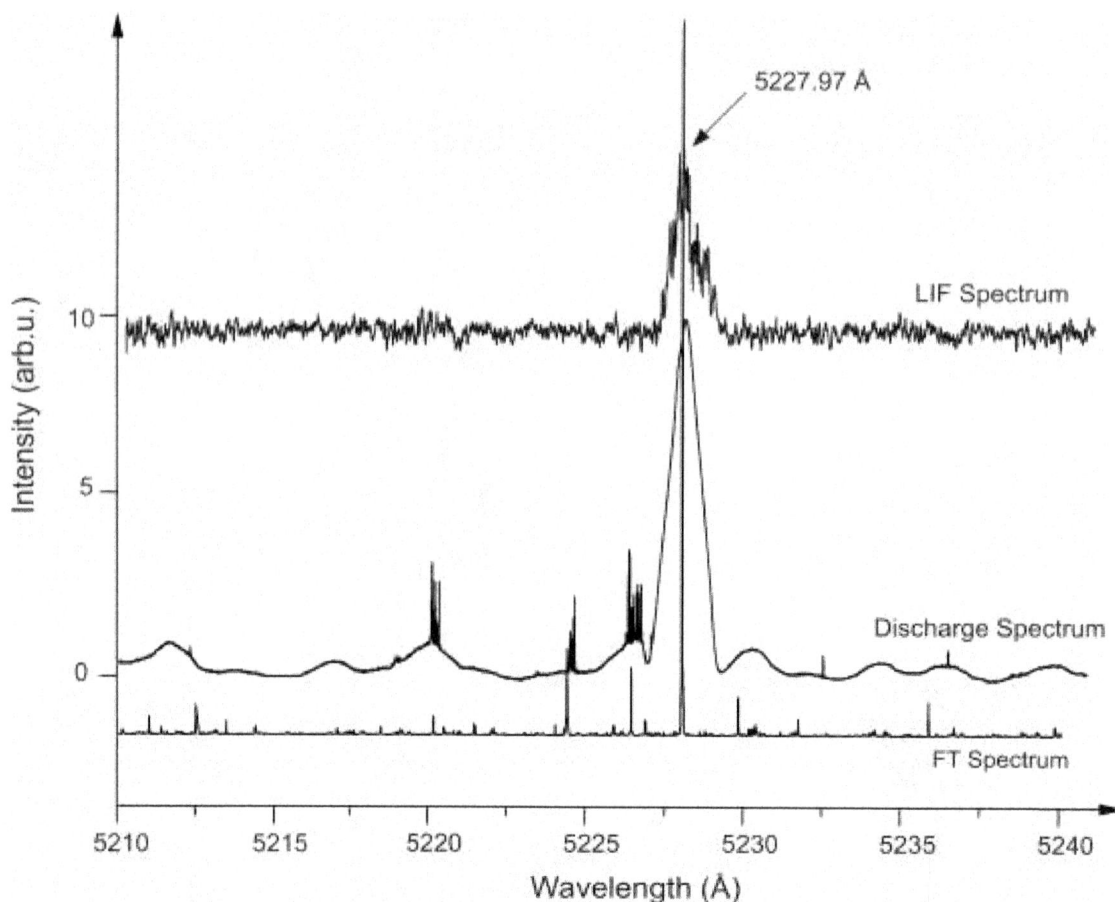

Figure 5. Determination of the fluorescence wavelength 5227.968 Å

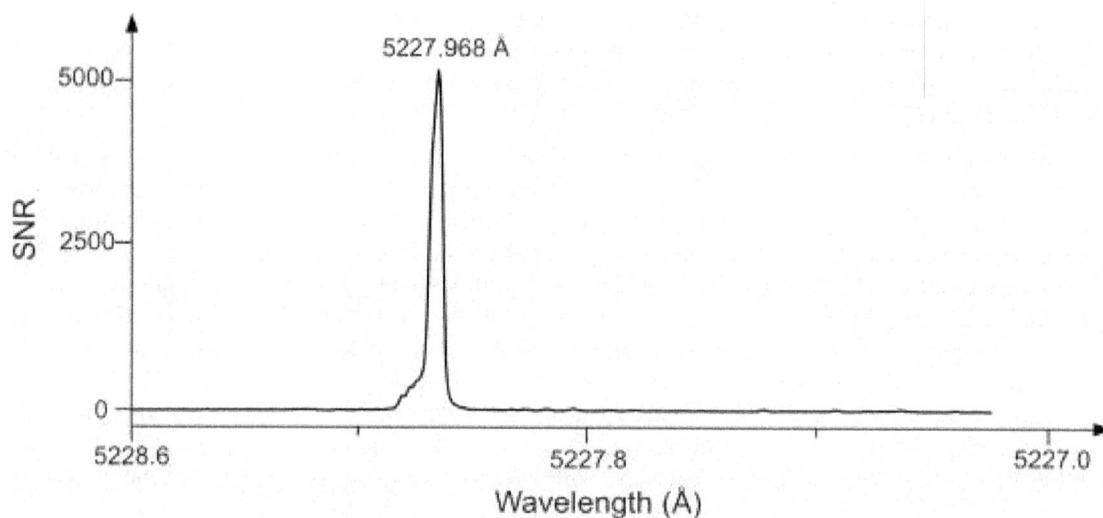

Figure 6. Spectral line 5227.968 Å as appearing in the FT spectrum

Thus we had to conclude that the observed LIF line at 5636.940 Å is not a direct decay line of the new energy level. In this case we have to assume that the new level has an energy close to $19122.567 \text{ cm}^{-1}$ and that the laser-induced enhancement of the population of the new level is (at least partly) transferred to $19122.567 \text{ cm}^{-1}$ by collisions.

We now treat again the list of lower levels having $J = 11/2$ and A close to 733 MHz as determined by the fit procedure. For the energy of the level we expect $19122 \pm 1000 \text{ cm}^{-1}$ due to the assumed collisional coupling. The wave number of the excited line is 17735 cm^{-1}. This gives $19122 \pm 1000 - 17735 = 1387 \pm 1000 \text{ cm}^{-1}$ for the lower level of the excitation. Indeed, as can be seen from Table 1, only one level, 1376.602 cm^{-1}, even parity, is located in the estimated energy region.

Moreover, when fitting the pattern shown in Fig. 2 with A- and B-values for the lower level, the best fit (lowest ESS) is obtained for 1376.602 cm^{-1} as lower level.

The odd parity level at 1376.602 cm^{-1}, J_{lo} = 11/2 and A_{lo} = 730.393 MHz, B_{lo} = -11.877 MHz is the lowest metastable level of Pr I, the nearest neighbour of the ground level. Taking now this level as the lower one of the laser excitation, we obtain a new even parity level at 19111.80 cm^{-1}, having J = 13/2, A = 762(2) MHz. B is assumed to be zero.

The newly calculated hypothetical level was then confirmed by a second laser excitation. First we calculated possible decay lines of the newly introduced level to lower levels. From this list we tried to excite lines in the wavelength range of our lasers. We were successful with the line 6146.45 Å, which appears in the FT spectrum with SNR = 24. Again we observed a strong LIF signal at the collisionally coupled line 5228 Å. The correct cg wavelengths of both excited lines were determined from the FT spectrum, i.e. 5636.940 Å and 6146.447 Å. The energy of the lower levels were already corrected earlier, therefore the energy of the newly discovered level could be determined with less uncertainty to be 19111.800(10) cm^{-1}. The magnetic and electric hf constants for the lower combining levels 1376.605 cm^{-1} and 2846.741 cm^{-1} were determined with high precision using laser-atomic beam spectroscopy [31, 9], therefore from a repeated recording of both laser-excited lines, the electric quadrupole hf constant for the newly discovered upper level could also been determined. Therefore the hf constants of the new upper level are finally A = 760(1) MHz, B = -20(10) MHz.

Table 1 shows the energies of all the possible combinations of known lower levels and calculated upper levels for the line 5636.940 Å. It also displays the ESS value of the best fit situation for each combination of J- and A-values of the combining levels (A- and B-values of the lower levels fixed during the fitting procedure).

Table 1. All the possible combinations of known lower levels and calculated upper levels for the line 5636.940 Å

Assumed lower level, J = 11/2			Calculated upper level, J = 13/2		
Energy (cm^{-1})	A (MHz)	B (MHz)	Energy (cm^{-1})	A (MHz)	ESS (arb.u.)
1	2	3	4	5	6
odd			even		
1376.602	730.393	-11.877	19111.8	761.83	4.98
10431.716	701.7(5)	-10(7)	28166.91	735.84	6.27
22577.235	685.4(5)		40312.43	721.91	7.77
even			odd		
6313.24	756.3		24048.42	782.67	6.16
8829.063	769.4	-31	26564.26	793.72	7.63
9483.518	731.3(2)	-15.4(80)	27218.72	761.15	5.00
9675.029	683.2(5)		27410.23	720.02	8.01
14505.065	777.2		32240.26	800.65	8.72
14981.5	687(1)		32716.7	723.29	7.60

With the help of a calculated list of decay lines of higher lying odd levels to the new level we could classify two further lines in the FT spectrum (7586 and 9632 Å, see Table 2). One additional excitation was also successful (at 7086.57 Å, transition to the upper level 32289.455 cm^{-1}, odd parity). The complete scheme of all transitions in which the new level is involved is shown in Figure 7.

5. Results and Discussion

In similar way as decribed for the level 19111 cm^{-1} in chapter 4, or using methods described in ref. (Windholz 2016), we found altogether 70 previously unknown energy levels of Pr I.

The data of all newly discovered levels are given in Table 2. The levels are listed separately for even and odd parity and are sorted by their value of J and their energy. In col. 1 the value of J is given, in col. 2 the level energy together with its estimated uncertainty (one standard deviation) is given. In col. 3 the magnetic dipole hf constant A is given. With the exception of the level at 19111 cm^{-1}, the electric quadrupole hf constant B could not be determined reliably and therefore was assumed to be zero. In the next column the cg wavelengths at which laser excitation was performed, are given. Values having three figures after the decimal point are determined with help of the FT spectrum, the other values with help of the lambdameter measuring the laser wavelength (uncertainty ± 0.01 Å). Wavelengths of observed fluorescence lines are given in Col. 5. Lines in the FT spectrum classified due to their cg wavelength and hf pattern are given in col.6. Some levels were discovered as lower levels of the excited transitions, and their existence was detected by decay lines of the combining upper levels. In such cases one finds in col. 5 the entry "see Table 3".

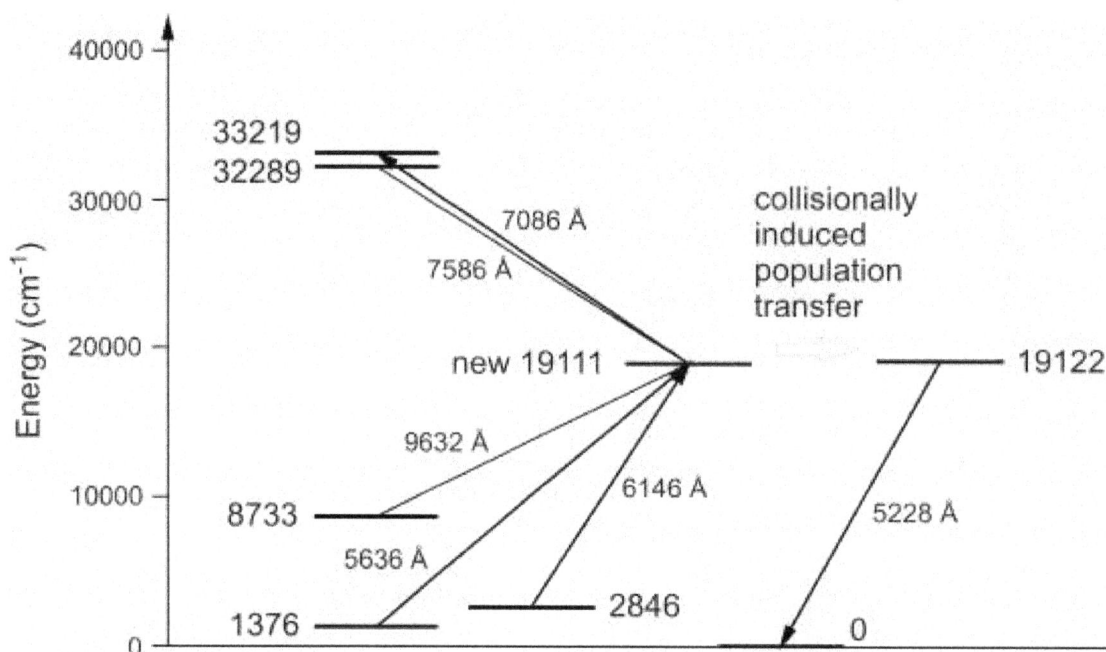

Figure 7. Level scheme of all observed lines classified by the new level at 19111 cm^{-1}. Laser excitations are shown as arrows; lines classified in the FT spectrum as thin full lines. The observed LIF at 5228 Å is an indirect line caused by collisionally induced population transfer. Wave numbers in cm^{-1}

In Table 3 all lines classified by new levels, which were discovered as lower levels of the excited transitions, are listed. Once such (hypothetical) level was introduced, its existence was checked by trying excitation from this level to other known upper levels. In such case we expect laser-induced decay lines from the known upper level. Thus in Table 3 also the observed fluorescence channels are mentioned. In cols. 1-3 the data from Table 2 are repeated (J, energy and A). In cols. 4-6 the excited line is given. If the line appears in the FT spectrum, the cg wavelength is given with three figures after the decimal point, and in col. 6 the SNR is given. If the cg wavelength is given with only two figures after decimal point, the cg wavelength is determined with the lambdameter measuring the laser wavelength, and in col. 6 SNR is set to 1. In col. 5 (C.) a comment is given: nl means a previously unknown spectral line not contained in commonly used spectral tables (e.g. Harrison 1969), cl means that the line was already known (in most cases from the work of A.Ginibre (Ginibre-Amery 1988)), but not classified. e means the line was excited by laser light. The data of the combining upper level are given in cols 7-9. Some of these are previously unknown levels but discovered within this work. If a decay of the upper level was observed, wavelength and SNR in the FT spectrum are given. nl means again a previously unknown line. If in col. 12 SNR = 1 is given, the LIF on this line was observed, but the line does not show up in the FT spectrum. In cols. 13 and 14 J and energy of the lower levels of the LIF lines are given. Finally in col. 15 a reference to the source of hf constant A (given in col. 9), used when determining the A-value of the new level (given in col. 3), is given.

Table 4 contains the lines classified by the newly found even levels, which were discovered as upper levels of the investigated lines. The structure of the table is similar to Table 3. A comment (C.) f means the line was observed as LIF line. The lines classified by the discovered upper odd levels are given in Table 5.

Table 2. Previously unknown energy levels of Pr I discovered within this work

J	Energy (cm⁻¹)	A (MHz)	Excitation wavelength (Å)	Fluorescence wavelength (Å)	Lines in FT spectrum. Wavelength (Å)
1	2	3	4	5	6
Even Parity					
7/2	14429.047(15)	911(5)	5704.123, 5820.611	see Table 3	9762.631
7/2	22486.816(20)	948.34	7022.178	4445.802	8448.802
7/2	26121.377(15)	805(5)	5594.030,6792.357	5594.030, 6792.357	
7/2	27123.623(15)	684(3)	6070.12, 6255.292	3685.774, 5296.965, 6494.68	7209.096, 7273.985
7/2	27231.854(15)	773(3)	6135.075, 7153.27, 7217.149	5266.763, 6213.212, 6449.335	
7/2	27381.792(15)	662(3)	6256.568, 6387.551, 6545.725	5668.50, 5976.44	7139.865, 7968.417
7/2	27679.200(20)	654(5)	5802.49, 6045.14	5145.50, 5574.49, 6991.366	
7/2	27998.766(20)	520(4)	6145.299, 6586.785, 6629.709	6145.299, 6586.785	
7/2	30654.508(20)	599(3)	5778.53, 5869.29, 6551.97	4781.251, 5173.255, 5373.53, 5605.879, 6006.045	
7/2	30920.290(20)	420(5)	6439.79	4883.762, 5103.07	6101.124
7/2	31531.300(15)	619(4)	6195.93	3170.53, 5691.024, 5881.793, 5987.943	
7/2	31593.97(3)	641(4)	6050.73, 6171.958	4575.66, 5670.80, 6050.73	
7/2	31664.74(3)	732(4)	6145.11, 5940.46	4916.246, 5940.46	
7/2	31937.43(3)	641(3)	4851.192, 5927.51, 6043.80	4652.58, 5256.68, 5351.44, 5562.418, 6043.80	
9/2	22825.611(20)	570(5)	7145.88	4379.813, 4660.916	
9/2	22996.562(20)	779(5)	7059.610	4624.060	4347.255
9/2	24606.934(10)	777(2)	6111.977, 7052.618	4303.507	
9/2	25744.156(10)	823(3)	5659.740, 5912.451	3883.276	6751.467
9/2	26231.182(10)	645(3)	6327.578, 6705.624	6327.578, 6894.405	6536.480, 8762.130
9/2	27109.750(20)	783(3)	6075.244, 6181.40, 6500.543, 7271.13	3687.661, 5253.579, 5300.862	
9/2	27282.002(15)	750(3)	5647.454, 6630.474	5419.539, 6266.953, 6428.537	8023.198
9/2	27913.895(15)	496(3)	6025.531, 5888.603	3767.21, 5240.04, 6363.744	
9/2	30938.737(15)	634(3)	5981.884, 6094.26	4879.36, 5685.186, 5772.959, 5981.884	7138.169, 7449.325
9/2	31362.709(15)	622(3)	6043.863	4589.47, 4988.56, 5162.365, 5553.338	6010.168, 6261.353, 6497.932
9/2	31721.60(5)	574(3)	6123.70	4900.80, 5180.83, 5442.87, 5523.27, 5644.21	
9/2	32556.286(15)	679(5)	4862.656, 5825.84	3206.29, 4709.75, 5547.27, 5665.50, 6030.118, 6079.293	
9/2	33877.134(15)	609(4)	5727.48, 6111.25	4843.484, 5056.89	
9/2	35607.370(15)	665(5)	4830.293	3973.85, 4744.09, 4946.36, 5244.98, 5526.70	
11/2	23654.406(10)	613(2)	6419.18, 6489.916, 6700.129	6489.916	7156.112, 7860.865, 8615.006
11/2	23865.975(15)	752(2)	6997.854, 7049.349	4445.297	
11/2	25357.817(10)	669(3)	6050.701, 6336.191, 6697.824	3942.441, 4168.755	
11/2	25855.320(10)	683(2)	5838.861, 5873.834	3866.580	6481.778, 6879.052, 7077.868, 7085.725, 7241.556
11/2	26222.282(15)	590(3)	5749.862, 6007.065, 6044.970	4023.706, 4276.771, 6540.285	7131.724, 7140.794
11/2	26467.250(15)	857(2)	5669.98, 7326.027	3984.422, 5956.74	
11/2	26937.739(15)	720(5)	5794.30, 7359.74, 7363.24	3711.208, 5491.679	
11/2	27888.407(20)	570(4)	5726.871, 6193.275	3770.83, 6374.11	7800.123
11/2	27985.186(15)	618(2)	5688.50, 5695.298	3572.299, 5462.66, 6156.364	6844.894
11/2	29524.677(15)	493(3)	6175.40, 6192.193, 7075.92	3551.63, 5012.43, 6534.800, 6891.518	3386.025, 5230.327
11/2	31332.402(15)	628(3)	5844.222, 6273.258, 6761.49	3337.29, 4595.869, 4997.932, 5104.245, 5555.083	
11/2	32672.078(15)	529(3)	5770.52, 5988.29, 6071.057, 6352.019	4684.206, 4835.418, 5675.71, 5786.790, 5874.56	
11/2	32891.473(15)	399(4)	5593.36, 5648.07, 5714.22,	4634.98, 4784.648	

			5799.785, 7202.48		
11/2	32989.188(20)	478(5)	5682.49, 5956.362	4615.624, 4762.376	
11/2	33096.940(20)	660(5)	5647.893, 5995.465	3304.812, 4738.054, 5385.69, 5546.210	
11/2	33254.220(15)	583(5)	5598.150	5340.444, 5498.235, 5582.924	5939.440
11/2	33781.556(15)	517(3)	5687.833, 5759.01, 5933.723	5437.584, 5759.010	3231.67
13/2	19111.800(10)	760(2)	5636.940, 6146.447, 7086.57	5228[a]	9632.785, 7586.509
13/2	22592.830(15)	500(5)	6888.745	5062.883	5489.434, 8384.228
13/2	24722.464(15)	703(2)	4914.712, 6252.560	4282.21, 4570.00	7113.695, 7460.619
13/2	27020.835(15)	632(3)	4415.767, 6026.379, 6368.318, 6747.350, 7305.278	3898.408, 4135.493, 5754.13, 6755.464	7318.457
13/2	27150.550(15)	536(3)	5711.488, 6065.693, 6688.793	3878.79, 4113.421, 5428.225, 6696.766	7236.686, 7249.619, 7603.473
13/2	27340.868(15)	557(2)	5662.01, 5996.449	5650.053, 6438.939	6612.468
13/2	27519.089(15)	487(3)	5933.025, 6172.411, 7520.760	3824.103, 4051.974, 5593.71	6365.865, 7048.643
13/2	27626.232(15)	443(3)	6489.97, 6836.087, 7007.878	6131.847, 6295.526,	
13/2	27821.680(25)	644(5)	6745.932	5828.36, 6218.985, 6408.65	
13/2	33840.270(20)	540(3)	5660.918, 5705.48, 5739.60, 5780.94, 5913.12	3225.548	6230.529

Odd Parity

7/2	15396.135(15)	719(2)	5682.49, 5714.22, 5825.84, 5878.40, 5995.548, 6043.80, 6123.70, 6145.287, 6171.956, 6195.93, 6273.258, 6439.79, 6551.97, 6616.346, 7075.92	see Table 3	4946.362, 5598.150, 5647.893, 5786.790, 6261.353, 6372.878, 8732.880, 9118.367
7/2	20012.862(15)	1133(5)	7045.286	4976[b]	8793.405, 9903.501
7/2	21975.458(10)	344(3)	5698.62, 6440.004	6189.254	7498.891, 7551.988, 8604.434, 9588.908, 9678.695
7/2	22451.973(15)	1125(4)	6248.204, 7240.105, 7490.283	4743[c]	5547.926, 7289.586, 7976.244, 8156.221, 8265.447, 8943.890, 9169.796, 9637.438
7/2	23293.356(15)	759(3)	5722.36, 5936.053, 5965.723	5300.434	8256.077, 8457.236, 8751.047, 9834.368
7/2	24443.790(10)	1046(3)	7016.088, 7496.383	5582.478	6785.520, 7591.178, 7950.416, 8357.456, 8477.849, 9091.907
7/2	25350.082(15)	1074(3)	5984.123, 6478.297, 7526.755	4779.268, 5290.019, 5313.568	7057.366, 7307.241, 7872.786
7/2	25788.548(15)	138(6)	5624.172, 5831.080	5007.20, 5192.56, 5501.713, 5863.14, 6215.714	7590.434
7/2	26915.490(15)	1122(3)	5613.118, 5808.71, 6547.334, 7055.51	4446.504, 4739.674, 4905.42, 5288.863	7853.830
7/2	27852.358(15)	868(4)	5661.72, 7469.521	4268.629	4671.467, 5043.218
7/2	27890.013(15)	877(5)	5649.665, 6504.25	4261.78, 4530.358, 4663.26, 4681.56, 4931.40, 5033.66, 5219.81	
9/2	17518.399(20)	874(4)	6111.25, 7566.54	see Table 3	7449.325
9/2	21063.222(15)	1093(2)	6967.185, 7055.071, 7435.23	6011.225	8171.588
9/2	22395.393(15)	724(4)	7269.896, 7369.16	5703.289, 6216.347, 6303.498	8194.017, 8304.292, 8917.363, 8996.381, 9839.067
9/2	22524.055(10)	1129(2)	6252.773, 6706.579	6252.773, 6706.579	7542.062, 7666.289, 8603.474, 8816.190, 8893.417, 9045.942, 9109.571, 9247.454, 9449.344
9/2	23008.838(10)	1227(2)	6666.715, 6673.919	5381.616, 5510.442	7275.959, 8525.734, 8850.571, 9035.326, 9278.867, 9531.065
9/2	23934.491(10)	785(3)	5673.386, 5745.890 6126.861	5242.935, 6618.31	7362.963, 7672.215, 8337.793, 9504.223, 9846.492
9/2	24196.152(10)	714(5)	5058.313, 5660.757 6177.590, 7521.18	5171.962, 5441.178, 6297.108, 6428.128	6795.001, 8561.896, 9273.537
9/2	24485.172(10)	1105(3)	6069.198, 6570.021	4985.407, 5095.769, 5569.608	8453.073, 9176.123, 9339.919, 9511.740, 9699.660
9/2	24830.361(10)	940(4)	5807.98, 5944.622	4901.040, 5259.627, 5398.903, 6055.21	6176.266, 8476.085
9/2	26035.395(10)	877(2)	5962.563, 6121.76	4627.654, 4722.595, 4946.052, 5069.024,	7819.677

9/2	26748.513(15)	912(5)	5865.622, 6423.26, 7406.55, 7507.67	5174.218, 5427.976 4568.687, 4990.040, 5034.96, 5340.614, 5424.938	4945.946
9/2	27292.757(20)	768(2)	5613.504, 5684.116	4656.388, 4858.07, 5360.746	
9/2	27659.987(15)	927(2)	4303.982, 5966.38, 5972.20, 6411.697	4385.986, 4683.240, 4987.993, 5257.221, 5567.856	
9/2	27980.556(15)	768(3)	5987.19, 6787.042, 6871.86, 7418.838, 7564.350	4245.390, 4511.846, 4661.788, 4740.793, 4909.47	5461.314, 7690.756
11/2	16399.404(20)	541(2)	5913.11, 5933.723, 6352.019	see Table 3	
11/2	21926.369(10)	882(3)	5714.613, 6992.917	5714.613, 6992.917	7898.193, 8022.236, 8574.069, 8640.943, 8691.213, 9634.268, 9646.510
11/2	23042.780(10)	891(4)	7033.518, 7362.643	5371.801, 5500.152, 5805.63	6122.530, 7172.779, 7258.024, 7824.854, 8430.534, 9500.323, 9620.895, 9709.173, 9990.187
11/2	24816.383(10)	744(2)	5927.775, 5955.302, 6060.343, 7387.03	4904.40, 5402.980, 5522.655	6181.599, 6363.059, 6602.604, 7185.876, 8643.863, 8660.299
11/2	25009.019(10)	983(3)	5860.83, 5990.391, 6439.24	5347.309, 5464.505	7709.545, 8997.499
11/2	25746.884(10)	784(2)	5617.818, 5845.29, 6544.171	4690.294, 4787.848, 5017.675, 5144.278, 5222.310, 5252.653, 5909.289	8081.917
11/2	26848.512(15)	590(5)	5634.305, 5686.772	4459.79, 4547.902, 4965.254 5290.319, 5395.656	5312.230, 6272.458
11/2	27944.838(15)	654(4)	5964.78, 6537.026	4519.130, 4921.167, 5352.94, 5415.225	7262.018
13/2	25443.665(15)	491(3)	4858.40, 5715.20, 5994.217	5225.816, 5306.358, 5389.124, 5612.150, 6655.946, 6675.071	6263.873, 7459.513, 7577.607, 7909.719
13/2	25765.460(15)	582(3)	5902.808, 6140.074, 7325.375, 7358.89	4783.593, 5217.247, 5297.234, 5512.568, 5611.961	6132.888, 6213.154, 6536.224, 7284.603, 7627.381
13/2	26810.061(15)	860(3)	5699.239, 5763.547, 5769.892, 7442.26	4877.439, 4947.529, 5301.107	5212.333, 5419.676, 5540.310, 6100.929
13/2	27139.468(15)	776(3)	5662.24, 5747.48, 7264.132, 7413.65 7535.314	5124.323, 5459.858, 5594.186	5440.982
13/2	27500.390(20)	883(2)	5608.43, 5800.911, 7349.556	4416.914, 5548.813	5630.651

[*] For the level at 19111 cm^{-1} the hf constant B could be determined to be B = -20(10) MHz

[a] Indirect fluorescence due to collisional transfer of population from the upper excited level to the even level at 19122.567 cm^{-1} which decay via fluorescence channel 5228 Å to the odd ground level

[b] Indirect fluorescence due to collisional transfer of population from the upper excited level to the even level at 20089.269 cm^{-1} which decay via fluorescence channel 4976 Å to the odd ground level

[c] Indirect fluorescence due to collisional transfer of population from the upper excited level to the even level at 22451.755 cm^{-1} which decay via fluorescence channel 4743 Å to the odd level at 1376.602 cm^{-1}

Abbreviations in col. 16: tw ... this work, G96 (Guthöhrlein 1996-2005), U12 (Uddin 2012b), SSW16 (Siddiqui, Shamim & Windholz 2016)

Table 3. New energy levels discovered as lower levels of the excited transitions

	New lower level			Line			Combining upper level			Observed fluorescence line from upper level			Lower level of LIF line		Ref. to col. 9
J	Energy (cm^{-1})	A (MHz)	Lambda (Å)	C.	SNR	J	Energy (cm^{-1})	A (MHz)	Lambda (Å)	C.	SNR	J	Energy (cm^{-1})	
1	2	3	4	5	6	7	8	9	10	11	12	13	14	15
	Even						**Odd**							
7/2	14429.047(15)	911(5)	5704.123	nl,e	12	9/2	31955.363	760(2)	3932.829	nl	5	7/2	6535.572	tw
			5820.611	nl,e	20	9/2	31604.60	742(2)	3738.921	nl	6	11/2	4866.515	tw
			9762.631	cl	5	7/2	24669.405	745(1)						G96
	Odd						**Even**							
9/2	15396.135(15)	719(2)	4946.362	nl	4	9/2	35607.370	665(5)						tw
			5598.150	cl	26	11/2	33254.220	583(5)						tw
			5647.893	nl	12	11/2	33096.929	660(5)						tw
			5682.49	nl,e	1	11/2	32989.188	478(5)	4762.376	nl	8	11/2	11997.137	tw
			5714.22	nl,e	1	11/2	32891.473	399(4)	4784.648		6	11/2	11997.137	tw
			5786.793	nl	4	11/2	32672.078	529(3)						tw
			5825.84	nl,e	1	9/2	32556.286	679(5)	4862.656	nl	12	11/2	11997.137	tw
			5878.396	nl,e	12	9/2	32402.860	544(2)	4744.046	nl	25	9/2	11329.696	U12
			5995.548	nl,e	5	7/2	32070.559	668(2)	4820.053		20	9/2	11329.696	U12
			6043.80	nl,e	1	7/2	31937.43	641(3)	4851.192	nl	4	9/2	11329.696	tw
			6123.70	nl,e	1	9/2	31721.60	574(3)	4900.80	nl	1	9/2	11322.443	tw
			6145.11	nl	1	7/2	31664.74	732(4)	4916.246	nl	3	9/2	11329.696	tw
			6145.287	nl,e	3	11/2	31664.262	601(2)	4916.362		12	9/2	11329.696	U12
			6171.956	nl,e	5	7/2	31593.97	641(4)	6050.73	nl	1	9/2	15071.618	tw
			6195.93	nl,e	1	7/2	31531.300	619(4)	5691.024	nl	1	5/2	13964.645	tw
			6261.353	nl	6	9/2	31362.709	622(3)						tw
			6273.258	nl,e	9	11/2	31332.402	628(3)	5555.083		25	13/2	13335.868	tw
			6372.878	nl	7	11/2	31083.290	620(2)						SSW16
			6439.79	nl,e	1	7/2	30920.300	420(5)	4883.761		17	7/2	10449.997	tw
			6551.97	nl,e	1	7/2	30654.508	599(3)	4781.251	nl	5	5/2	9745.334	tw
			6616.346	nl,e	5	9/2	30506.050	675(2)	5213.30	nl	1	9/2	11329.696	G96
			7075.92	nl,e	1	11/2	29524.677	493(3)	3386.024	nl	20	9/2	0.000	tw
			8732.880	cl	6	11/2	26843.963	624(2)						tw
			9118.367	nl	7	7/2	26360.014	877(5)						G96
9/2	17518.399(15)	847(4)	6111.25	nl,e	1	9/2	33877.134	609(4)	4843.484		8	7/2	13236.606	tw
			7449.325	nl	5	9/2	30938.737	634(3)						
			7566.54	nl,e	1	9/2	30730.841	668(3)	4929.385		15	7/2	10449.997	G96
11/2	16933.404(15)	541(2)	5913.12	nl,e	1	6.5	33840.270	540(3)	3225.548	nl	5	13/2	2846.741	tw
			5933.723	cl,e	20	11/2	33781.556	517(3)	5759.010	nl	5	11/2	16422.282	tw
			6352.019	nl,e	5	11/2	32672.078	529(3)	6071.057		10	13/2	16205.041	tw

Table 4. Lines classified by the newly discovered even levels

Abbreviations in col. 11: CG81 (Childs & Goodman 1981), G96 (Guthöhrlein 1996-2005)

	New upper even level			Line			Combining known lower odd levels			Ref. to cols. 9,10
J	Energy (cm^{-1})	A (MHz)	Lambda (Å)	C.	SNR	J	Energy (cm^{-1})	A (MHz)	B (MHz)	
1	2	3	4	5	6	7	8	9	10	11
7/2	22486.816	948(5)	4445.802	nl,f	7	4.5	0.000	926.209	-11.878	CG81
			7022.178	cl,e	53	4.5	8250.141	213.531	-4.136	CG81
			8448.802	nl	3	3.5	10654.060	169(2)		G96
7/2	26121.377	805(5)	5594.030	nl,e,f	22	4.5	8250.141	213.531	-4.136	CG81
			6792.357	nl,e,f	18	3.5	11403.011	1142(3)		G96
7/2	27123.623	684(3)	3685.77	nl,f	3	4.5	0.000	926.209	-11.878	CG81
			5296.965	nl,f	20	4.5	8250.141	213.531	-4.136	CG81
			6070.127	nl,e	8	3.5	10654.053	169(2)		G96
			6255.292	nl,e	21	2.5	11141.576	169(2)		G96
			6494.682	nl,f	10	4.5	11730.668	1365(5)		G96
			7209.096	cl	8	2.5	13256.082	1074(3)		G96
			7273.985	cl	8	4.5	13379.788	932(1)	10(20)	G96
	and so on									

Full version of Table 4: see Appendix

Table 5. Lines classified by the newly discovered odd levels

Abbreviations in col. 11: K96 (Kuwamoto et al. 1996), G96 (Guthöhrlein 1996-2005)

	New upper odd level			Line			Combining known lower even levels			Ref. to cols. 9,10
J	Energy (cm^{-1})	A (MHz)	Lambda (Å)	C.	SNR	J	Energy (cm^{-1})	A (MHz)	B (MHz)	
1	2	3	4	5	6	7	8	9	10	11
7/2	20012.862	1133(5)	7045.286	nl,e	10	4.5	5822.890	855.8(4)	-17(7)	K96
			8793.405	nl	12	4.5	8643.824	797(2)		G96
			9903.501	cl	55	3.5	9918.190	1057.4(5)	22(6)	K96

7/2	21975.458	344(3)	5698.622	nl,e	10	4.5	4432.225	923.2(4)	-22(7)	K96
			6189.254	cl,f	10	4.5	5822.890	855.8(4)	-17(7)	K96
			6440.004	nl,e	28	2.5	6451.808	1189.6(6)	-5(5)	K96
			7498.891	nl	11	4.5	8643.824	797(2)		G96
			7551.988	nl	12	2.5	8737.556	1149(5)		G96
			8604.434	nl	7	4.5	10356.737	1406(1)		G96
			9588.908	cl	12	4.5	11549.602	1064(2)		G96
			9678.690	cl	10	2.5	11646.312	1317(10)		G96
and so on										

Full version of Table 5: see Appendix

6. Conclusion

Experimental investigation of the hyperfine structure of Pr I lines is reported in this paper. The investigation resulted in the discovery of 70 Pr I fs levels having even and odd parity. The magnetic dipole interaction constants A of all new levels could be determined, and for one level also the very small electric quadrupole interaction constant B. The investigation is carried out in the spectral regions of Rhodamine 6G, Sulforhodamine B, DCM and Stilbene 3. Furthermore the discovery of these levels led to the classification of ca. 670 praseodymium spectral lines.

7. Acknowledgements

The authors are thankful to the staff of the Institute of Experimental Physics, TU Graz for providing technical assistance at various occasions. Shamim Khan and Imran Siddiqui are also thankful to the Higher Education Commission Pakistan and University of Karachi, Pakistan for providing financial support during extended stays in Graz.

References

Böklen, K. D., Bossert, T., Foerster, W., Fuchs, H. H., & Nachtsheim, G. (1975). Hyperfine structure measurements in the $^4I_{9/2}$ ground state of ^{141}Pr. *Z. Physik. A, 274*, 195-201. https://doi.org/10.1007/BF01437730

Childs, W. J., & Goodman, L. S. (1981). Double resonance, fluorescence spectroscopy, and hyperfine structure in Pr I. *Phys. Rev. A, 24*, 1342. https://doi.org/10.1103/PhysRevA.24.1342

Dolk, L., Wahlgren, G. M., Lundberg, H., Li, Z. S., Litzen, U., Ivarsson, S., ... & Hubrig, S. (2002). The presence of Nd and Pr in HgMn stars. *Astron. Astrophys, 385*, 111-130. https://doi.org/10.1051/0004-6361:20020118

Furmann, B., Krzykowski, A., Stefańska, D., & Dembczyński, J. (2006). New levels and hyperfine structure evaluation in neutral praseodymium. *Physica Scripta, 74*, 658. https://doi.org/10.1088/0031-8949/74/6/010

Gamper, B., Khan, S., Siddiqui, I., & Windholz, L. (2013). Modelling of emission spectra of Pr I by summarizing hyperfine patterns of overlapping spectral lines. *Eur. Phys. J. Special Topics, 222*, 2171-2178. https://doi.org/10.1140/epjst/e2013-01993-9

Gamper, B., Uddin, Z., Jahangir, M., Allard, O., Knöckel, H., Tiemann, E., & Windholz, L. (2011). Investigation of the hyperfine structure of Pr I and Pr II lines based on highly resolved Fourier transform spectra. *J. Phys. B, 44*, 045003. https://doi.org/10.1088/0953-4075/44/4/045003

Ginibre-Emery, A. (1988). *Classification et étude paramétrique des spectres complexes á l`aide de l`interprétation des structures hyperfines: spectres I et II du praséodyme* (Unpublished doctoral dissertation). Université de Paris-Sud, Centre d'Orsay

Ginibre, A. (1981). Fine and hyperfine structures in the configurations $4f^2\,5d\,6s^2$ and $4f^2\,5d^2\,6s$ of neutral praseodymium. *Physica Scripta, 23*, 260-267. https://doi.org/10.1088/0031-8949/23/3/008

Ginibre, A. (1989). Fine and hyperfine structures of singly ionized praseodymium: I. energy levels, hyperfine structures and Zeeman effect, classified lines. *Physica Scripta, 39*, 694-709. https://doi.org/10.1088/0031-8949/39/6/005

Ginibre, A. (1989). Fine and hyperfine structures of singly ionized praseodymium: II. parametric interpretation of fine and hyperfine structures for the even levels of singly ionised praseodymium. *Physica Scripta 39*, 710-721. https://doi.org/10.1088/0031-8949/39/6/006

Guthöhrlein, G. H. (1996-2005). Helmut Schmidt-Universität, Universität der Bundeswehr Hamburg, Laboratorium für Experimental physik; unpublished material taken from several diploma theses (supervisor G.H. Guthöhrlein)

Guthöhrlein, G. H. (1998). Program package "Fitter", unpublished, Helmut-Schmidt-Universität, Universität der Bundeswehr Hamburg, Holstenhofweg 85, D-22043 Hamburg, Germany. The program code can be obtained from G.H. Guthöhrlein (g.h.g@hsu-hh.de) or L.W. (windholz@tugraz.at)

Guthöhrlein, G. H. (2005). Private communication, Helmut-Schmidt-Universität, Universität der Bundeswehr Hamburg, Holstenhofweg 85, D-22043 Hamburg, Germany

Krzykowski, A., Furmann, B., Stefanska, D., Jarosz, A., & Kajoch, A. (1997). Hyperfine structures in the configuration $4f^3$ 5d 6s of the praseodymium atom. *Optics Communications 140*, 216. https://doi.org/10.1016/S0030-4018(97)00165-X

Lew, H. (1953). The ground state hyperfine structure and nuclear magnetic moment of praseodymium. *Phys. Rev. 89*, 530. https://doi.org/10.1103/PhysRev.89.530.2

Macfarlane, R. M., Burum. D. P., & Shelby, R. M. (1982). New Determination of the Nuclear Magnetic Moment of [141]Pr. *Phys. Rev. Lett., 49*, 636. https://doi.org/10.1103/PhysRevLett.49.636

Martin, W. C., Zalubas, R., & Hagan, L. (1978). Atomic Energy Levels – The Rare Earth Elements. *National Bureau of Standards NSRDS-NBS 60* (Washington, DC: US GPO)

Peck, E. R., & Reeder, K. (1972). Dispersion of air. *J. Opt. Soc. Am., 62*, 958-962. https://doi.org/10.1364/JOSA62.000958

Ruczkowski, J., Stachowska, E., Elantkowska, M., Guthöhrlein, G. H., & Dembczyński, J. (2003). Interpretation of the hyperfine structure of the even configuration system of Pr I. *Physica Scripta, 68*, 133. https://doi.org/10.1238/Physica.Regular.068a00133

Ryabchikova, T. A., Savanov, I. S., Malanushenko, V. P., Kudryavtsev, D. O. (2001). A study of rare earth elements in the atmospheres of chemically peculiar stars. Pr III and Nd III lines. *Astronomy Reports, 45*, 382-388. https://doi.org/10.1051/0004-6361:20020118

Shamim, K., Siddiqui, I., & Windholz, L. (2011). Experimental investigation of the hyperfine spectra of Pr I – lines: discovery of new fine structure levels with low angular momentum. *Eur. J. Phys. D, 64*, 209. https://doi.org/10.1140/epjd/e2011-20250-9

Siddiqui, I., Shamim, K., & Laurentius, W. (2016). Experimental investigation of the hyperfine spectra of Pr I – lines: discovery of new fine structure energy levels with medium angular momentum quantum number between 7/2 and 13/2. *Eur. Phys. J. D, 70*, 44. https://doi.org/10.1140/epjd/e2016-60485-2

Siddiqui, I., Shamim, K., & Windholz, L. (2014). Experimental investigation of the hyperfine spectra of Pr I – lines: discovery of new fine structure levels with high angular momentum. *Eur. Phys. J. D, 68*, 122. https://doi.org/1140/epjd/e2014-50025-7

Syed, T. I., Siddiqui, I., Shamim, K., Uddin, Z., Guthöhrlein, G. H., & Windholz, L. (2011). New even and odd parity levels of neutral praseodymium. *Physica Scripta 84*, 065303. https://doi.org/10.1088/0031-8949/84/06/065303

Takeshi, K., Ichita, E., Atsushi, F., Takashi, H., Takayoshi, H., Yoshihisa, I., … & Tohru, T. (1996). Systematic study of fine and hyperfine structures in Pr I by Doppler-free atomic-beam laser spectroscopy. *J. Phys. Soc. Japan, 65*, 3180. https://doi.org/10.1143/JPSJ.65.3180

Uddin, Z. (2006). *Hyperfine structure studies of tantalum and praseodymium* (Unpublished doctoral dissertation). Graz University of Technology, Graz, Austria.

Uddin, Z. (2014). New odd levels of Pr I with low angular momentum. *Chin. J. Phys., 52*, 770. https://doi.org/10.6122/CJP.52.770

Uddin, Z., El Bakkali, D., Gamper, B., Shamim, K., Imran, S., Guthöhrlein, G. H., & Windholz, L. (2012). Laser spectroscopic investigations of praseodymium I transitions: New energy levels. *Advances in Optical Technologies*, 639126. https://doi.org/10.1155/2012/639126

Uddin, Z., Siddiqui, I., Shamim, K., Gamper, B., Abdul-Hafidh, E. H., & Windholz, L. (2012). New levels of the Pr atom with almost simiar energies. *Journal of Physical Science and Application, 2*, 88-94.

Uddin, Z., Siddiqui, I., Tanweer, J., Jilani, S. U., & Windholz, L. (2015). Classification of some blended spectral lines of praseodymium. *J. Phys. B, 48*, 135001. https://doi.org/10.1088/0953-4075/48/13/135001

Uddin, Z., Zafar, R., Sikander, R., Siddiqui, I., Shamim, K., & Windholz, L. (2012). Investigation of Pr I lines by a simulation of their hyperfine patterns: discovery of new levels. *J. Phys. B, 45*, 205001. https://doi.org/10.1088/0953-4075/45/20/205001

Wavelength Tables, edited by Harrison G.R. (Massachusetts Institute of Technology, The M.I.T. Press, (1969).

White, H. E. (1929). Hyperfine structure in singly ionized praseodymium. *Phys. Rev., 34*, 1397. https://doi.org/10.1103/PhysRev.34.1397

Windholz, L. (2016). Finding of previously unknown energy levels using Fourier-transform and laser spectroscopy. *Physica Scripta, 91*, 114003 http://dx.doi.org/10.1088/0031-8949/91/11/114003

Appendix: Full versions of Tables 4 and 5

Table 4. Lines classified by the newly discovered even levels.

Abbreviations in col. 11: tw ... this work, CG81 (Childs & Goodman 1981), G96 (Guthöhrlein 1996-2005), K96 (Kuwamoto et al. 1996), KF97 (Krzykowski, Furmann, Stefanska, Jarosz & Kajoch, 1997), SSW16 (Siddiqui, Shamim & Windholz 2016), U12a (Uddin et al. 2012a), SSW14 (Siddiqui, Shamim & Windholz 2014)

New upper even level			Line				Combining known lower odd levels			Ref. to cols. 9,10
J	Energy (cm⁻¹)	A (MHz)	Lambda (Å)	C.	SNR	J	Energy (cm⁻¹)	A (MHz)	B (MHz)	
1	2	3	4	5	6	7	8	9	10	11
7/2	22486.816(20)	948(5)	4445.802	nl,f	7	9/2	0.000	926.209	-11.878	CG81
			7022.178	cl,e	53	9/2	8250.141	213.531	-4.136	CG81
			8448.802	nl	3	7/2	10654.060	169(2)		G96
7/2	26121.377(15)	805(5)	5594.030	nl,e,f	22	9/2	8250.141	213.531	-4.136	CG81
			6792.357	nl,e,f	17	7/2	11403.011	1142(3)		G96
7/2	27123.623(15)	684(3)	3685.77	nl,f	3	9/2	0.000	926.209	-11.878	CG81
			5296.965	nl,f	20	9/2	8250.141	213.531	-4.136	CG81
			6070.127	nl,e	8	7/2	10654.053	169(2)		G96
			6255.292	nl,e	21	2.5	11141.576	169(2)		G96
			6494.682	nl,f	10	9/2	11730.668	1365(5)		G96
			7209.096	cl	8	2.5	13256.082	1074(3)		G96
			7273.985	cl	8	9/2	13379.788	932(1)	10(20)	G96
7/2	27231.854(15)	773(3)	5266.763	nl,f	14	9/2	8250.141	213.531	-4.136	CG81
			6135.075	nl,e	6	9/2	10936.652	930(1)	0(10)	G96
			6213.214	nl,f	10	2.5	11141.576	169(2)		G96
			6449.336	nl,f	12	9/2	11730.668	1365(5)		G96
			7153.27	nl,e	1	2.5	13256.082	1074(3)		G96
			7217.149	nl,e	9	9/2	13379.788	932(1)	10(20)	G96
7/2	27381.792(15)	662(3)	5668.50	nl,f	20	2.5	9745.334	626.4(3)	5(3)	K96
			5976.44	cl.f	6	7/2	10654.053	169(2)		G96
			6256.568	cl,e	30	7/2	11403.011	1142(3)		G96
			6387.551	cl,e	6	9/2	11730.668	1365(5)		G96
			6545.725	nl,e	8	2.5	12108.867	1275(2)		G96
			7139.865	nl	7	9/2	13379.788	932(1)	10(20)	G96
			7968.417	nl	4	2.5	14835.699	1414(5)		G96
7/2	27679.200(20)	654(5)	3611.792	nl	3	9/2	0	926.209	-11.878	CG81
			5145.496	nl,f	8	9/2	8250.141	213.531	-4.136	CG81
			5574.496	nl,f	22	2.5	9745.334	626.4(3)	5(3)	K96
			5802.49	nl,e	1	7/2	10449.997	541(2)		G96
			6045.14	nl,e	1	2.5	11141.576	169(2)		G96
			6991.366	cl,f	18	9/2	13379.788	932(1)	10(20)	G96
7/2	27998.766(20)	520(4)	6145.299	nl,e,f	9	9/2	11730.668	1365(5)		G96
			6586.785	cl,e,f	8	9/2	12821.044	1127(3)		G96
			6629.709	nl,e	6	7/2	12919.316	180(2)		G96
7/2	30654.508(20)	599(3)	4781.251	nl,f	5	5/2	9745.334	626.4(3)	5(3)	K96
			5173.255	nl,f	12	9/2	11329.696	530(3)		G96
			5373.53	nl,f	1	9/2	12049.942	275(2)		G96
			5605.879	nl,f	10	9/2	12821.044	1127(3)		G96
			5778.53	nl,e	1	9/2	13354.043	1273(3)		G96
			5869.29	nl,e	24	7/2	13621.400	879(1)		G96
			6006.045	nl,f	4	7/2	14009.225	1100(1)		G96
			6551.97	nl,e	1	9/2	15396.135	719(2)		tw
7/2	30920.290(20)	420(5)	4684.624	nl	2	9/2	9579.820	789(1)		G96
			4883.762	cl,f	17	7/2	10449.997	541(2)		G96
			5103.07	nl,f	1	9/2	11329.696	530(3)		G96
			6101.124	nl	6	9/2	14534.393	100(2)		G96
			6439.79	nl,e	1	9/2	15396.135	719(2)		tw
7/2	31531.300(15)	619(4)	3170.53	nl,f	1	9/2	0.000	926.209	-11.878	CG81
			5691.024	nl,f	20	5/2	13964.645	185(2)		G96
			5881.793	nl,f	3	9/2	14534.393	100(2)		G96
			5987.943	nl,f	15	5/2	14835.699	1414(5)		G96
			6195.93	nl,e	1	9/2	15396.135	719(2)		tw
7/2	31593.97(3)	642(4)	4575.66	nl,f	1	5/2	9745.334	626.4(3)		K96
			5670.796	nl,f	10	5/2	13964.645	185(2)		G96
			6050.73	nl,e,f	1	9/2	15071.618	635(3)		G96
			6171.958	nl,e	5	9/2	15396.135	719(2)		tw
7/2	31664.74(3)	732(4)	4916.246	nl,f	3	9/2	11329.696	530(3)		G96
			5940.46	nl,e	1	5/2	14835.699	1414(5)		G96
			6145.11	nl,e	1	9/2	15396.135	719(2)		tw
7/2	31937.43(3)	641(3)	4652.58	nl,f	1	7/2	10449.997	541(2)		G96
			4851.192	nl,e	3	9/2	11329.696	530(3)		G96
			5256.68	nl,f	1	7/2	12919.316	180(2)		G96

			5351.44	nl,f	1	5/2	13256.082	1074(3)		G96
			5562.418	nl,f	20	5/2	13964.645	185(2)		G96
			5927.51	nl,e	1	9/2	15071.618	635(3)		G96
			6043.80	nl,f,e	1	9/2	15396.135	719(2)		tw
9/2	22825.611(20)	570(5)	4379.813	nl,f	5	9/2	0.000	926.209	-11.878	CG81
			4660.916	cl,f	270	11/2	1376.602	730.393	-11.877	CG81
			7145.88	nl,e	1	11/2	8835.389	949.091	-13.721	CG81
9/2	22996.562(20)	779(5)	4347.255	nl	5	9/2	0.000	926.209	-11.878	CG81
			4624.060	nl,f	95	11/2	1376.602	730.393	-11.877	CG81
			7059.610	cl,e	76	11/2	8835.389	949.091	-13.721	CG81
9/2	24606.934(10)	777(2)	4303.507	nl,f	41	11/2	1376.602	730.393	-11.877	CG81
			6111.977	cl,e	62	9/2	8250.141	213.531	-4.136	CG81
			7052.620	nl,e	7	11/2	10431.716	701.7(5)	-10(7)	KF97
9/2	25744.156(10)	823(3)	3883.276	nl,f	10	9/2	0.000	926.209	-11.878	CG81
			5659.740	nl,e	15	11/2	8080.402	238.352	-22.961	CG81
			5912.451	e	45	11/2	8835.389	949.091	-13.721	CG81
			6751.469	cl	20	9/2	10936.652	930(1)	0(10)	G96
9/2	26231.182(10)	645(3)	6327.578	e,f	30	11/2	10431.716	701.7(5)	-10(7)	KF97
			6536.480	nl	8	9/2	10936.652	930(1)	0(10)	G96
			6705.623	nl,e	10	11/2	11322.443	1272(1)	75(50)	G96
			6894.405	cl	20	9/2	11730.668	1365(5)		G96
			8762.130	cl	22	11/2	14821.565	544/2)		SSW16
9/2	27109.750(20)	783(3)	3687.661	nl,f	14	9/2	0.000	926.209	-11.878	CG81
			5253.579	nl,f	14	11/2	8080.402	238.352	-22.961	CG81
			5300.862	cl,f	42	9/2	8250.141	213.531	-4.136	CG81
			6075.244	cl,e	132	7/2	10654.053	169(2)		G96
			6181.40	nl,e	35	9/2	10936.652	930(1)	0(10)	G96
			6500.543	cl,e	7	9/2	11730.668	1365(5)		G96
			7271.13	nl,e	1	11/2	13360.511	151(3)		G96
9/2	27282.002(15)	750(3)	5419.539	nl,f	10	11/2	8835.389	949.091	-13.721	CG81
			5647.454	nl,e	16	9/2	9579.820	789(1)		G96
			6266.953	cl,f	4	9/2	11329.696	530(3)		G96
			6428.537	cl,f	7	9/2	11730.668	1365(5)		G96
			6630.473	cl,e	11	11/2	12204.286	1010(2)		G96
			8023.198	nl	5	11/2	14821.565	544(2)		G96
9/2	27913.895(15)	496(3)	3767.21	cl,f	1	11/2	1376.602	730.393	-11.877	CG81
			5240.04	nl,f	1	11/2	8835.389	949.091	-13.721	CG81
			5888.603	cl,e	18	9/2	10936.652	930(1)	0(10)	G96
			6025.531	cl,e	22	11/2	11322.443	1272(1)	75(50)	G96
			6363.774	cl,f	22	11/2	12204.286	1010(2)		G96
9/2	30938.737(15)	634(3)	4879.36	nl,f	1	7/2	10449.997	541(2)		G96
			5685.186	nl,f	5	9/2	13354.043	1272(3)		G96
			5772.959	nl,f	16	7/2	13621.400	879(1)		G96
			5981.884	nl,f,e	6	11/2	14226.220	869(3)		G96
			6094.26	nl,e	1	9/2	14534.393	100(2)		G96
			7138.169	nl	4	11/2	16933.404	541(2)		tw
			7449.325	nl	5	9/2	17518.399	847(4)		tw
9/2	31362.709(15)	622(3)	4589.47	nl,f	1	9/2	9579.820	789(1)		G96
			4988.56	nl,f	1	11/2	11322.443	1272(1)	75(50)	G96
			5162.365	nl,f	12	11/2	11997.137	585(3)		G96
			5553.338	nl,f	10	11/2	13360.511	151(3)		G96
			6010.168	cl	10	11/2	14728.843	811(2)		G96
			6043.863	nl,e	5	11/2	14821.570	544(2)		SSW16
			6261.353	nl	6	9/2	15396.135	719(2)		tw
			6497.932	cl	5	11/2	15977.45	66(3)		G96
9/2	31721.60(5)	574(3)	4900.80	nl,f	1	11/2	11322.443	1272(1)	75(50)	G96
			5442.87	nl,f	1	9/2	13354.043	1273(3)		G96
			5523.27	nl,f	1	7/2	13621.400	879(1)		G96
			5644.21	nl	1	7/2	14009.225	1100(1)		G96
			6123.70	nl,e	1	9/2	15396.135	719(2)		tw
9/2	32556.286(15)	679(5)	3206.29	nl,f	1	11/2	1376.602	730.393	-11.877	CG81
			4709.75	nl,f	3	9/2	11329.696	530(3)		G96
			4862.656	cl,a	11	11/2	11997.137	585(3)		G96
			5547.27	nl,f	1	9/2	14534.393	100(2)		G96
			5665.50	nl,f	1	9/2	14910.476	611(2)		G96
			5825.84	nl,e	1	9/2	15396.135	719(2)		tw
			6030.118	nl,f	10	11/2	15977.45	66(3)		G96
			6079.29	nl,f	1	7/2	16111.538	1128(3)		G96
9/2	33877.134(15)	609(4)	4843.484	cl,f	8	7/2	13236.606	726(1)		G96
			5056.89	cl,f	1	7/2	14107.700	860(3)		G96
			5727.48	nl,e	2	11/2	16422.282	605(3)		G96
			6111.25	nl,e	1	9/2	17518.394	847(4)		tw
9/2	35607.370(15)	665(5)	3973.85	nl,f	1	7/2	10449.997	541(2)		G96
			4744.09	nl,f	1	9/2	14534.393	100(2)		G96
			4830.293	cl,e	11	9/2	14910.476	611(2)		G96

			4946.36	nl,f	2	9/2	15396.135	719(2)		tw
			5244.98	nl,f	1	11/2	16546.842	429(2)		SSW16
			5526.70	nl,f	1	9/2	17518.399	847(4)		tw
11/2	23654.406(10)	613(2)	6419.185	cl,e	20	11/2	8080.402	238.352	-22.961	CG81
			6489.916	cl,e,f	78	9/2	8250.141	213.531	-4.136	CG81
			6700.129	nl,e	5	13/2	8733.440	854.297	-31.807	CG81
			7156.112	cl	20	13/2	9684.184	991.907	-7.246	CG81
			7860.865	cl	12	9/2	10936.652	930(1)	0(10)	G96
			8615.006	nl	6	9/2	12049.942	275(2)		G96
11/2	23865.975(15)	752(2)	4445.297	nl,f	45	11/2	1376.602	730.393	-11.877	CG81
			6997.854	cl,e	22	9/2	9579.820	789(1)		G96
			7049.349	nl,e	5	13/2	9684.184	991.907	-7.246	CG81
11/2	25357.817(10)	669(3)	3942.441	nl,f	8	9/2	0.000	926.209	-11.878	CG81
			4168.755	nl,f	11	11/2	1376.602	730.393	-11.877	CG81
			6050.701	cl,e	24	11/2	8835.389	949.091	-13.721	CG81
			6336.191	cl,e	20	9/2	9579.820	789(1)		G96
			6697.824	cl,e	40	11/2	10431.716	701.7(5)	-10(7)	KF97
11/2	25855.320(10)	683(2)	3866.580	nl,f	8	9/2	0.000	926.209	-11.878	CG81
			5838.861	nl,e	7	13/2	8733.440	854.297	-31.807	CG81
			5873.834	nl,e	31	11/2	8835.389	949.091	-13.721	CG81
			6481.778	cl	48	11/2	10431.716	701.7(5)	-10(7)	KF97
			6879.052	cl	22	11/2	11322.443	1272(1)	75(50)	G96
			7077.868	nl	8	9/2	11730.668	1365(5)		G96
			7085.725	cl	10	13/2	11746.328	401(1)	50(20)	G96
			7241.556	nl	6	9/2	12049.942	275(2)		G96
11/2	26222.282(15)	590(3)	4023.707	nl,f	13	11/2	1376.602	730.393	-11.877	CG81
			4276.771	nl,f	8	13/2	2846.741	613.240	-12.850	CG81
			5749.862	nl,e	12	11/2	8835.389	949.091	-13.721	CG81
			6007.065	cl,e	9	9/2	9579.820	789(1)		G96
			6044.970	cl,e	62	13/2	9684.184	991.907	-7.246	CG81
			6540.285	nl,f	16	9/2	10936.652	930(1)	0(10)	G96
			7131.724	nl	5	11/2	12204.286	1010(2)		G96
			7140.794	nl	5	13/2	12222.091	1133(4)		G96
11/2	26467.250(15)	857(2)	3984.422	nl,f	7	11/2	1376.602	730.393	-11.877	CG81
			5669.98	cl,e	12	11/2	8835.389	949.091	-13.721	CG81
			5956.74	nl,f	1	13/2	9684.184	991.907	-7.246	CG81
			7326.027	nl,e	8	9/2	12821.044	1127(3)		G96
11/2	26937.739(15)	720(5)	3711.208	nl,f	30	9/2	0.000	926.209	-11.878	CG81
			5491.679	cl,f	95	13/2	8733.440	854.297	-31.807	CG81
			5794.30	nl,e	1	13/2	9684.184	991.907	-7.246	CG81
			7359.74	nl,e	1	9/2	13354.043	1272(3)		G96
			7363.24	nl,e	1	11/2	13360.511	151(3)		G96
11/2	27888.407(20)	570(4)	3770.83	nl,f	1	11/2	1376.602	730.393	-11.877	CG81
			5726.871	nl,e	8	11/2	10431.716	701.7(5)	-10(7)	KF97
			6193.275	cl,e	23	13/2	11746.328	401(1)	50(20)	G96
			6374.113	nl,f	12	11/2	12204.286	1010(2)		G96
			7800.120	nl	5	9/2	15071.618	635(3)		G96
11/2	27985.186(15)	618(2)	3572.299	nl,f	20	9/2	0.000	926.209	-11.878	CG81
			5462.663	nl,f	6	13/2	9684.184	991.907	-7.246	CG81
			5688.503	nl,e	55	13/2	10410.745	655.9(3)	-29(7)	K96
			5695.298	cl,e	51	11/2	10431.716	701.7(5)	-10(7)	KF97
			6156.364	nl,f	8	13/2	11746.328	401(1)	50(20)	G96
			6844.894	cl	12	9/2	13379.788	932(1))	10(20)	G96
11/2	29524.677(15)	493(3)	3386.025	nl	20	9/2	0.000	926.209	-11.878	CG81
			3551.63	nl,f	1	11/2	1376.602	730.393	-11.877	CG81
			5012.43	nl,f	1	9/2	9579.820	789(1)		G96
			5230.327	cl	40	13/2	10410.736	655.9(3)	-29(7)	K96
			6175.40	nl,e	1	13/2	13335.868	895(1)	100(50)	G96
			6192.193	nl,e	17	9/2	13379.788	932(1)	10(20)	G96
			6534.800	nl,f	3	11/2	14226.220	869(3)		G96
			6891.518	nl,f	3	13/2	15018.088	108(3)		G96
			7075.92	nl,e	1	9/2	15396.135	719(2)		tw
11/2	31332.402(15)	628(3)	3337.290	nl	2	11/2	1376.602	730.393	-11.877	CG81
			4595.869	cl,f	9	9/2	9579.820	789(1)		G96
			4997.932	nl	4	9/2	11329.696	530(3)		G96
			5104.245	nl	7	13/2	11746.328	401(1)	50(20))	G96
			5555.083	cl,f	25	13/2	13335.868	895(1)	100(50)	G96
			5844.222	nl	6	11/2	14226.220	869(3)		G96
			6273.258	nl	9	9/2	15396.135	719(2)		tw
			6376.54	nl	1	13/2	15654.235	577(1)		SSW16
			6761.49	nl	1	11/2	16546.842	429(2)		SSW16
11/2	32672.078(15)	529(3)	4684.206	nl,f	4	9/2	11329.696	530(3)		G96
			4835.418	cl,f	17	11/2	11997.137	585(3)		G96
			5675.71	nl,f	1	13/2	15058.01	693(3)		tw
			5770.52	nl,e	1	13/2	15347.431	674(2)		U12a

			5786.790	nl,f	5	9/2	15396.135	719(2)		tw
			5874.56	nl,f	5	13/2	15654.235	577(1)		SSW16
			5988.29	nl,e	1	11/2	15977.45	66(3)		G96
			6071.057	cl,e	10	13/2	16205.041	733(2)		U12a
			6352.019	nl,e	5	11/2	16933.404	541(2)		tw
11/2	32891.473(15)	399(4)	4634.98	nl,f	1	11/2	11322.443	1272(1)	75(50)	G96
			4784.648	cl,f	6	11/2	11997.137	585(3)		G96
			5593.36	nl,e	1	13/2	15018.088	108(3)		G96
			5648.07	nl,e	1	13/2	15191.218	666(5)		U12a
			5714.22	nl,e	1	9/2	15396.135	719(2)		tw
			5799.785	nl,e	7	13/2	15654.235	577(1)		SSW16
			7202.48	nl,e	1	11/2	19011.200	844(5)		tw
11/2	32989.188(20)	478(5)	4615.624	nl,f	4	9/2	11329.696	530(3)		G96
			4762.376	nl,f	8	11/2	11997.137	585(3)		G96
			5682.49	nl,e	1	9/2	15396.135	719(2)		tw
			5956.362	cl,e	5	13/2	16205.041	733(2)		U12a
11/2	33096.940(20)	660(5)	3304.812	nl,f	3	13/2	2846.741	613.240	-12.850	CG81
			4738.054	nl,f	3	11/2	11997.137	585(3)		G96
			5385.69	nl,f	1	9/2	14534.393	100(2)		G96
			5546.210	nl,f	15	9/2	15071.618	635(3)		G96
			5647.893	nl,e	12	9/2	15396.135	719(2)		tw
			5995.465	nl,e	6	11/2	16422.282	605(3)		G96
11/2	33254.220(15)	583(5)	5340.444	nl,f	5	9/2	14534.393	100(2)		G96
			5498.235	nl,f	3	9/2	15071.618	635(3)		G96
			5582.924	nl,f	3	13/2	15347.431	674(2)		U12a
			5598.150	cl,e	26	9/2	15396.135	719(2)		tw
			5939.440	cl	9	11/2	16422.282	605(3)		G96
11/2	33781.556(15)	517(3)	3231.67	nl,f	1	13/2	2846.741	613.240	-12.850	CG81
			5437.584	nl,f	9	9/2	15396.135	719(2)		tw
			5687.833	nl,e	4	13/2	16205.041	733(2)		U12a
			5759.01	nl,e	5	11/2	16422.282	605(3)		G96
			5933.723	cl,e	20	11/2	16933.404	541(2)		tw
13/2	19111.800(10)	760(2)	5636.940	cl,e	140	11/2	1376.602	730.393	-11.877	CG81
			6146.447	nl,e	24	13/2	2846.741	613.240	-12.850	CG81
			7086.57	nl,e	1	13/2	33219.119	671(1)		SSW16
			7586.510	cl	10	11/2	32289.455	665(1)		U12a
			9632.785	cl	4	13/2	8733.440	854.297	-31.807	CG81
13/2	22592.830(15)	500(5)	5062.883	cl,f	205	13/2	·2846.741	613.240	-12.850	CG81
			5489.434	nl	18	15/2	4381.072	541.575	-14.558	CG81
			6888.746	cl,e	22	11/2	8080.402	238.352	-22.961	CG81
			8384.228	cl	9	15/2	10668.950	951.310	-2.670	CG81
13/2	24722.464(15)	703(2)	4282.21	nl,f	1	11/2	1376.602	730.393	-11.877	CG81
			4570.00	nl,f	1	13/2	2846.741	613.240	-12.850	CG81
			4914.712	cl,e	160	15/2	4381.072	541.575	-14.558	CG81
			6252.560	cl,e	9	13/2	8733.440	854.297	-31.807	CG81
			7113.695	cl	9	15/2	10668.950	951.310	-2.670	CG81
13/2	27020.835(15)	632(3)	3898.408	nl,f	28	11/2	1376.602	730.393	-11.877	CG81
			4135.493	nl,f	5	13/2	2846.741	613.240	-12.850	CG81
			4415.767	nl,e	4	15/2	4381.072	541.575	-14.558	CG81
			5754.13	nl,f	1	15/2	9646.830	907.515	-23.132	CG81
			6026.379	cl,e	70	11/2	10431.716	701.7(5)	-10(7)	KF97
			6368.318	nl,f	50	11/2	11322.443	1272(1)	75(50)	G96
			6747.350	nl,f	36	11/2	12204.286	1010(2)		G96
			6755.465	cl	20	13/2	12222.091	1133(4)		G96
			7305.278	nl,f	5	13/2	13335.868	895(1)	100(50)	G96
			7318.457	nl	11	11/2	13360.511	151(3)		G96
13/2	27150.550(15)	536(3)	3878.79	nl,f	2	11/2	1376.602	730.393	-11.877	CG81
			4113.421	nl,f	8	13/2	2846.741	613.240	-12.850	CG81
			5428.225	nl,f	5	13/2	8733.440	854.297	-31.807	CG81
			5711.488	nl,e	12	15/2	9646.830	907.515	-23.132	CG81
			6065.693	cl,e	32	15/2	10668.950	951.310	-2.670	CG81
			6688.793	cl,e	14	11/2	12204.286	1010(2)		G96
			6696.766	cl,f	7	13/2	12222.091	1133(4)		G96
			7236.686	cl	11	13/2	13335.868	895(1)	100(50)	G96
			7249.619	cl	12	11/2	13360.511	151(3)		G96
			7603.473	cl	30	11/2	14002.294	566(2)		G96
13/2	27340.868(15)	557(2)	5650.054	nl,f	15	15/2	9646.830	907.515	-23.132	CG81
			5662.01	nl,e	1	13/2	9684.184	991.907	-7.246	CG81
			5996.449	cl,e	28	15/2	10668.950	951.310	-2.670	CG81
			6438.939	cl,f	12	15/2	11814.647	355(2)	0(10)	G96
			6612.468	cl	12	13/2	12222.091	1133(4)		G96
13/2	27519.089(15)	487(3)	3824.103	nl,f	5	11/2	1376.602	730.393	-11.877	CG81
			4051.974	nl,f	4	13/2	2846.741	613.240	-12.850	CG81
			5593.710	nl,f	10	15/2	9646.830	907.515	-23.132	CG81
			5933.025	cl,e	48	15/2	10668.950	951.310	-2.670	CG81

			6172.410	cl,e	47	11/2	11322.443	1272(1)	75(50)	G96
			6365.865	cl	12	15/2	11814.647	355(2)	0(10)	G96
			7048.643	cl	13	13/2	13335.868	895(1)	100(50)	G96
			7520.76	nl,e	4	11/2	14226.220	869(3)		G96
13/2	27626.232(15)	443(3)	6131.847	cl,f	22	11/2	11322.443	1272(1)	75(50)	G96
			6295.526	cl,f	24	13/2	11746.328	401(1)	50(20)	G96
			6489.97	nl,e	5	13/2	12222.091	1133(4)		G96
			6836.090	nl,e	17	15/2	13002.023	317(2)	30(50)	G96
			7007.878	cl,e	6	11/2	13360.511	151(3)		G96
13/2	27821.680(25)	644(5)	5828.360	nl,f	8	15/2	10668.950	951.310	-2.670	CG81
			6218.985	nl,f	6	13/2	11746.328	401(1)	50(20)	G96
			6408.654	cl,f	22	13/2	12222.091	1133(4)		G96
			6745.932	cl,e	15	15/2	13002.023	317(2)	30(50)	G96
13/2	33840.270(20)	540(3)	3225.548	nl,f	5	13/2	2846.741	613.240	-12.850	CG81
			5660.918	nl,e	5	15/2	16180.200	883(2)		SSW14
			5705.48	nl,e	1	13/2	16318.118	281(1)		G96
			5739.60	nl,e	1	11/2	16422.282	605(3)		G96
			5780.94	nl,e	1	11/2	16546.842	429(2)		SSW16
			5913.12	nl,e	1	11/2	16933.404	541(2)		tw
			6230.529	nl	3	15/2	17794.708	221(2)		SSW14

Table 5. Lines classified by the newly discovered odd levels

Abbreviations in col. 11: tw ... this work, CG81 (Childs & Goodman 1981), G96 (Guthöhrlein 1996-2005), K96 (Kuwamoto et al. 1996), U12a (Uddin et al. 2012a)

New upper even level			Line				Combining known lower odd levels			Ref. to cols. 9,10
J	Energy (cm^{-1})	A (MHz)	Lambda (Å)	C.	SNR	J	Energy (cm^{-1})	A (MHz)	B (MHz)	
1	2	3	4	5	6	7	8	9	10	11
7/2	20012.862(15)	1133(5)	7045.286	nl,e	10	9/2	5822.890	855.8(4)	-17(7)	K96
			8793.405	nl	12	9/2	8643.824	797(2)		G96
			9903.501	cl	55	7/2	9918.190	1057.4(5)	22(6)	K96
7/2	21975.458(10)	344(3)	5698.622	nl,e	20	9/2	4432.225	923.2(4)	-22(7)	K96
			6189.254	cl,f	10	9/2	5822.890	855.8(4)	-17(7)	K96
			6440.004	nl,e	28	5/2	6451.808	1189.6(6)	-5(5)	K96
			7498.891	nl	11	9/2	8643.824	797(2)		G96
			7551.988	nl	12	5/2	8737.556	1149(5)		G96
			8604.434	nl	7	9/2	10356.737	1406(1)		G96
			9588.908	cl	12	9/2	11549.602	1064(2)		G96
			9678.690	cl	10	5/2	11646.312	1317(10)		G96
7/2	22451.973(15)	1125(4)	5547.926	nl	34	9/2	4432.225	923.2(4)	-22(7)	K96
			6248.203	nl,e	70	5/2	6451.808	1189.6(6)	-5(5)	K96
			7240.105	cl,e	45	9/2	8643.824	797(2)		G96
			7289.586	nl	5	5/2	8737.556	1149(5)		G96
			7490.283	cl,e	6	9/2	9105.021	689.7(3)	-3(5)	K96
			7976.244	nl	8	7/2	9918.190	1057.4(5)	22(6)	K96
			8156.221	nl	8	7/2	10194.768	855(1)		G96
			8265.447	cl	12	9/2	10356.737	1406(1)		G96
			8943.890	cl	34	7/2	11274.229	1286(1)	-10(20)	G96
			9169.796	cl	19	9/2	11549.602	1064(2)		G96
			9637.438	cl	24	5/2	12078.621	1566(1)		G96
7/2	23293.356(15)	759(3)	5300.434	cl	60	9/2	4432.225	923.2(4)	-22(7)	K96
			5722.360	nl,e	19	9/2	5822.890	855.8(4)	-17(7)	K96
			5936.053	nl,e	55	5/2	6451.808	1189.6(6)	-5(5)	K96
			5965.723	cl,e	12	7/2	6535.572	979(1)	25(30)	G96
			8256.077	cl	12	9/2	11184.396	692(1)	15(30)	G96
			8457.236	cl	8	7/2	11472.410	273(3)		G96
			8751.048	nl	3	7/2	11869.290	210(1)		G96
			9834.371	nl	5	5/2	13127.722	156(1)	0(10)	G96
7/2	24443.790(10)	1046(3)	5582.478	cl,f	90	7/2	6535.572	979(1)	25(30)	G96
			6785.521	nl	7	5/2	9710.600	164(2)		G96
			7016.088	cl,e	8	7/2	10194.768	855(1))		G96
			7496.383	nl,e	6	5/2	11107.696	658(2)		G96
			7591.178	nl	7	7/2	11274.229	1286(1)	-10(20)	G96
			7950.415	nl	3	7/2	11869.290	210(1)		G96
			8357.456	cl	6	5/2	12481.714	937(1)	20(20)	G96
			8477.849	nl	4	9/2	12651.586	723(3)		G96
			9091.907	cl	9	5/2	13448.016	825(1)	25(30)	G96
7/2	25350.082(15)	1074(3)	4779.268	cl,f	45	9/2	4432.225	929(1)	-22(7)	K96
			5290.019	cl,f	35	5/2	6451.808	1189.6(6)	-5(5)	K96
			5313.568	nl,f	25	7/2	6535.572	979(1)	25(30)	G96
			5984.123	cl,e	38	9/2	8643.824	797(2)		G96

			6478.293	cl,e	14	7/2	9918.190	1057.4(5)	22(6)	G96
			7057.366	nl	6	9/2	11184.396	692(1)	15(30)	G96
			7307.241	nl	3	7/2	11668.794	805(2)		G96
			7526.755	nl,e	3	9/2	12067.802	873(6)		G96
			7872.786	nl	12	9/2	12651.586	723(3)		G96
7/2	25788.548(15)	138(6)	5007.20	nl,f	1	9/2	5822.890	855.8(4)	-17(7)	K96
			5192.556	nl,f	7	7/2	6535.572	979(1)	25(30)	G96
			5501.713	nl,f	40	7/2	7617.440	866.9(5)	-4(5)	G96
			5624.172	cl,e	70	7/2	8013.089	168(1)		G96
			5723.065	nl	10	9/2	8320.240	255(2)		G96
			5831.080	nl,e	11	9/2	8643.824	797(2)		G96
			5863.136	nl,f	11	5/2	8737.556	1149(5)		G96
			6215.714	nl,f	15	7/2	9704.744	779(1)	-50(30)	G96
			7590.434	nl	5	7/2	12617.700	883(2)		G96
7/2	26915.490(15)	1122(3)	4446.504	nl,f	12	9/2	4432.225	923.2(4)	-22(7)	K96
			4739.673	nl,f	4	9/2	5822.890	855.8(4)	-17(7)	K96
			4905.42	cl,f	4	7/2	6535.572	979(1)	25(30)	G96
			5288.863	cl,f	40	7/2	8013.089	168(1)		U12a
			5613.118	nl,e	7	9/2	9105.021	689.7(3)	-3(5)	K96
			5808.71	nl,e	1	7/2	9704.744	779(1)	-50(30)	G96
			6547.334	nl,e	6	5/2	11646.312	1317(10)		G96
			7055.51	nl,e	1	9/2	12746.067	982(1)	10(10)	G96
			7853.830	cl	7	9/2	14186.352	910(1)	-60(30)	G96
7/2	27852.358(15)	868(4)	4268.629	nl,f	12	9/2	4432.225	923.2(4)	-22(7)	K96
			4671.467	cl	8	5/2	6451.808	1189.6(6)	-5(5)	K96
			5043.218	cl	38	9/2	8029.275	797(2)		G96
			5661.72	nl,e	1	7/2	10194.768	855(1)		G96
			7469.521	nl,e	3	9/2	14468.303	762(2)	20(20)	G96
7/2	27890.013(15)	877(5)	4261.78	nl,f	1	9/2	4432.225	923.2(4)	-22(7)	K96
			4530.358	nl,f	8	9/2	5822.890	855.8(4)	-17(7)	K96
			4663.26	nl,f	1	5/2	6451.808	1189.6(6)	-5(5)	K96
			4681.56	nl,f	1	7/2	6535.572	979(1)	25(30)	G96
			4931.396	cl.f	4	7/2	7617.440	866.9(5)	-4(5)	K96
			5033.66	nl,f	1	9/2	8029.275	797(2)		G96
			5219.809	nl,f	7	5/2	8737.556	1149(5)		G96
			5649.665	nl,e	21	7/2	10194.768	855(1)		G96
			6504.25	nl,e	1	9/2	12519.705	693(3)		G96
9/2	21063.222(15)	1093(2)	6011.225	cl,f	30	9/2	4432.225	923.2(4)	-22(7)	K96
			6172.386	nl	11	11/2	4866.515	867.997	-50.319	CG81
			6777.789	cl	5	11/2	6313.224	756(1)		G96
			6967.185	nl,e	7	11/2	6714.184	474.692	-29.633	CG81
			7055.071	cl,e	15	11/2	6892.934	551.934	-24.736	CG81
			7435.228	nl,e	5	7/2	7617.440	866.9(5)	-4(5)	K96
			8171.588	nl	8	11/2	8829.063	769(1)	-30(20)	G96
9/2	22395.393(15)	724(4)	5703.289	nl,f	7	11/2	4866.515	867.997	-50.319	CG81
			6216.347	nl,f	10	11/2	6313.224	756(1)		G96
			6303.498	nl,f	21	7/2	6535.572	979(1)	25(30)	G96
			6448.806	cl	42	11/2	6892.934	551.934	-24.736	CG81
			7269.895	cl,e	44	9/2	8643.824	797(2)		G96
			7369.161	nl,e	6	11/2	8829.063	769(1)	-30(20)	G96
			8194.047	cl	10	7/2	10194.768	855(1)		G96
			8304.292	nl	8	9/2	10356.737	1406(1)		G96
			8917.363	nl	7	9/2	11184.396	692(1)	15(30)	G96
			8996.381	cl	18	11/2	11282.865	1049(2)		G96
			9839.067	nl	3	11/2	12234.616	1164(2)		G96
9/2	22524.055(10)	1129(2)	6252.774	cl,e,f	75	7/2	6535.572	979(1)	25(30)	G96
			6706.579	cl,e,f	20	7/2	7617.440	866.9(5)	-4(5)	K96
			7542.062	cl	15	11/2	9268.726	977(1)	-24(20)	G96
			7666.289	nl	3	11/2	9483.518	731(1)	-15(10)	G96
			8603.474	cl	12	11/2	10904.034	301(1)	-20(10)	G96
			8816.190	nl	8	9/2	11184.396	692(1)	15(30)	G96
			8893.417	cl	14	11/2	11282.865	1049(2)		G96
			9045.942	nl	3	7/2	11472.410	273(3)		G96
			9109.571	cl	20	9/2	11549.602	1064(2)		G96
			9247.454	cl	16	9/2	11713.236	818(2)		G96
			9449.344	cl	26	11/2	11944.207	1003(1)		G96
9/2	23008.838(10)	1227(2)	5381.616	cl,f	105	9/2	4432.225	929(1)	-22(7)	K96
			5510.442	nl,f	21	11/2	4866.515	867.997	-50.319	CG81
			6666.715	cl,e	22	7/2	8013.089	168(1)		U12a
			6673.919	cl,e	41	9/2	8029.275	797(2)		G96
			7275.959	nl	7	11/2	9268.726	977(1)	-24(20)	G96
			8525.734	cl	55	11/2	11282.865	1049(2)		G96
			8850.571	cl	16	9/2	11713.236	818(2)		G96
			9035.326	cl	50	11/2	11944.207	1003(1)		G96
			9278.867	cl	32	11/2	12234.616	1169(2)		G96

			9531.065	cl	8	9/2	12519.705	693(3)		G96
9/2	23934.491(10)	785(3)	5242.935	nl,f	14	11/2	4866.515	867.997	-50.319	CG81
			5519.790	cl	20	9/2	5822.890	855.8(4)	-17(7)	K96
			5673.386	cl,e	116	11/2	6313.224	756(1)		G96
			5745.890	cl,e	50	7/2	6535.572	979(1)	25(30)	G96
			6126.861	cl,e	27	7/2	7617.440	866.9(5)	-4(5)	K96
			6618.310	cl,f	7	11/2	8829.063	769(1)	-30(20)	G96
			6816.719	cl	6	11/2	9268.726	977(1)	-24(20)	G96
			7362.963	nl	5	9/2	10356.737	1406(1)		G96
			7672.215	nl	3	11/2	10904.034	301(1)	-20(10)	G96
			8337.793	cl	7	11/2	11944.207	1003(1)		G96
			9504.223	cl	9	7/2	13415.739	797(2)		G96
			9846.493	cl	7	7/2	13781.374	807(1)		G96
9/2	24196.152(10)	714(4)	5058.313	cl,e	105	9/2	4432.225	923.2(4)	-22(7)	K96
			5171.962	nl,f	18	11/2	4866.515	867.997	-50.319	CG81
			5441.178	cl,f	28	9/2	5822.890	855.8(4)	-17(7)	K96
			5660.757	nl,e	28	7/2	6535.572	979(1)	25(30)	G96
			6177.590	cl,e	128	7/2	8013.089	168(1)		U12a
			6297.108	cl,f	22	9/2	8320.240	255(2)		G96
			6428.128	cl,f	21	9/2	8643.824	797(2)		G96
			6795.001	nl	3	11/2	9483.518	731(1)	-15(10)	G96
			6884.620	nl	12	11/2	9675.029	683(1)		G96
			7521.185	nl,e	7	11/2	10904.034	301(1)	-20(10)	G96
			8561.896	cl	11	9/2	12519.705	693(3)		G96
			9273.537	cl	13	7/2	13415.739	797(2)		G96
9/2	24485.172(10)	1110(2)	4985.407	nl,f	10	9/2	4432.225	923.2(4)	-22(7)	K96
			5095.769	cl,f	60	11/2	4866.515	867.997	-50.319	CG81
			5569.608	nl,f	33	7/2	6535.572	979(1)	25(30)	G96
			6069.198	cl,e	19	7/2	8013.089	168(1)		G96
			6570.021	nl,e	6	11/2	9268.726	977(1)	-24(25)	G96
			8453.073	nl	6	11/2	12658.401	995(5)		U12a
			9176.123	cl	27	7/2	13590.311	1151(2)		G96
			9339.919	cl	5	7/2	13781.374	807(1)		G96
			9511.740	cl	15	9/2	13974.732	854(1)	-15(15)	G96
			9699.660	nl	6	11/2	14178.365	986(1)	-20(10)	G96
9/2	24830.361(10)	940(4)	4901.040	cl,f	38	9/2	4432.225	923.2(4))	-22(7)	K96
			5259.627	nl,f	62	9/2	5822.890	855.8(4))	-17(7)	K96
			5398.903	cl,f	90	11/2	6313.224	756(1)		G96
			5807.980	nl,e	10	7/2	7617.440	866.9(5)	-4(5)	K96
			5944.622	cl,e	130	7/2	8013.089	168(1)		G96
			6055.214	nl,f	50	9/2	8320.240	255(2)		G96
			6176.266	nl	8	9/2	8643.824	797(2)		G96
			8476.085	cl	5	11/2	13035.697	796(3)		G96
9/2	26035.395(10)	877(2)	4627.654	nl,f	19	9/2	4432.225	923.2(4))	-22(7)	K96
			4722.595	cl,f	13	11/2	4866.515	867.997	-50.319	CG81
			4946.052	cl,f	22	9/2	5822.890	855.8(4)	-17(7)	K96
			5069.024	cl,f	60	11/2	6313.224	756(1)		G96
			5174.218	nl,f	11	11/2	6714.184	474.692	-29.633	CG81
			5427.976	cl,f	27	7/2	7617.440	866.9(5)	-4(5)	K96
			5962.563	cl,e	14	11/2	9268.726	977(1)	-24(20)	G96
			6121.762	cl,e	8	7/2	9704.744	779(1)	-50(30)	G96
			7819.677	cl	5	11/2	13250.662	424(2)		G96
9/2	26748.513(15)	912(5)	4568.687	nl,f	8	11/2	4866.515	867.997	-50.319	CG81
			4945.946	nl	3	7/2	6535.572	979(1)	25(30)	G96
			4990.040	cl,f	14	11/2	6714.184	474.692	-29.633	CG81
			5034.96	nl,f	1	11/2	6892.934	551.934	-24.736	CG81
			5340.614	nl,f	28	9/2	8029.275	797(2)	-	G96
			5424.938	nl,f	7	9/2	8320.240	255(2)		G96
			5865.622	nl,e	8	7/2	9704.744	779(1	-50(30)	G96
			6423.26	nl,e	1	9/2	11184.396	692(1	15(30)	G96
			7406.55	nl,e	1	11/2	13250.662	424(2)		G96
			7507.67	nl,e	1	9/2	13432.468	559(2)		G96
9/2	27292.757(20)	768(2)	4656.39	nl,f	3	9/2	5822.890	855.8(4)	-17(7)	G96
			4858.07	nl,f	1	11/2	6714.184	474.692	-29.633	CG81
			5360.746	cl,f	70	9/2	8643.824	797(2)		G96
			5613.504	nl,e	22	11/2	9483.518	731(1)	-15(10)	G96
			5684.116	cl,e	41	7/2	9704.744	779(1)	-50(30)	G96
9/2	27659.987(15)	927(2)	4303.982	nl,e	11	9/2	4432.225	923.2(4)	-22(7)	K96
			4385.986	nl,f	10	11/2	4866.515	867.997	-50.319	CG81
			4683.240	cl,f	22	11/2	6313.224	756(1)		G96
			4987.993	nl,f	7	7/2	7617.440	866.9(5)	-4(5)	K96
			5257.221	cl,f	60	9/2	8643.824	797(2)		G96
			5567.856	nl,f	5	7/2	9704.744	779(1)	-50(30)	G96
			5966.38	n,e	1	11/2	10904.034	301(1)	-20(10)	G96
			5972.195	nl,e	10	9/2	10920.365	632(2)		G96

			6411.696	nl,e	10	9/2	12067.802	873(6)		G96
9/2	27980.556(15)	768(2)	4245.390	nl,f	9	9/2	4432.225	923.2(4)	-22(7)	K96
			4511.846	nl,f	3	9/2	5822.890	855.8(4)	-17(7)	K96
			4661.788	nl,f	3	7/2	6535.572	979(1)	25(30)	G96
			4740.793	cl,f	4	11/2	6892.934	551.934	-24.736	CG81
			4909.47	nl,f	1	7/2	7617.440	866.9(5)	-4(5)	K96
			5170.068	nl	8	9/2	8643.824	797(2)		G96
			5461.314	cl	13	11/2	9675.029	683(1)		G96
			5987.193	nl,e	8	11/2	11282.865	1049(2)		G96
			6787.044	nl,e	4	11/2	13250.662	424(2)		G96
			6871.862	nl,e	10	9/2	13432.468	559(2)		G96
			7398.655	nl	10	9/2	14468.303	762(2)	20(20)	G96
			7418.838	nl,e	6	11/2	14505.065	777(3)		G96
			7564.350	nl,e	6	9/2	14764.288	867(2)		G96
			7690.756	cl	14	11/2	14981.5	687(1)		G96
			7767.927	cl	19	7/2	15110.642	947(2)		G96
11/2	21926.369(10)	882(3)	5714.613	cl,e,f	170	9/2	4432.225	923.2(4)	-22(7)	K96
			6992.917	cl,e,f	28	13/2	7630.132	776.286	-43.592	CG81
			7898.193	cl	13	11/2	9268.726	977(1)	-24(20)	G96
			8022.236	cl	7	13/2	9464.440	1056(1)	-15(10)	G96
			8640.941	nl	4	9/2	10356.737	1406(1)		G96
			8691.213	cl	14	13/2	10423.654	869(1)	25(30)	G96
			9634.268	nl	17	9/2	11549.602	1064(2)		G96
			9646.510	cl	9	13/2	11562.762	819(2)		G96
11/2	23042.780(10)	891(4)	5371.801	cl,f	90	9/2	4432.225	923.2(4)	-22(7)	K96
			5500.152	cl,f	170	11/2	4866.515	867.997	-50.319	CG81
			5805.628	nl,f	30	9/2	5822.890	855.8(4)	-17(7)	K96
			6122.530	nl	9	11/2	6714.184	474.692	-29.633	CG81
			7033.518	cl,e	20	11/2	8829.063	769(1)	-30(20)	G96
			7172.779	cl	11	9/2	9105.021	689.7(3)	-3(5)	K96
			7258.024	cl	9	11/2	9268.726	977(1)	-24(20)	G96
			7362.643	cl,e	22	13/2	9464.440	1056(1)	-15(10)	G96
			7824.854	cl	6	13/2	10266.501	972(2)		G96
			7951.713	cl	7	13/2	10470.329	628(3)		G96
			8430.534	nl	5	9/2	11184.396	692(1)	15(30)	G96
			9500.323	cl	7	9/2	12519.705	693(3)		G96
			9620.895	cl	6	9/2	12651.586	723(3)		G96
			9709.173	cl	14	9/2	12746.067	982(1)	10(10)	G96
			9990.187	cl	10	11/2	13035.697	796(3)		G96
11/2	24816.383(10)	744(2)	4904.40	nl,f	1	9/2	4432.225	923.2(4)	-22(7)	K96
			5402.980	cl,f	155	11/2	6313.224	756(19		G96
			5522.655	cl,f	175	11/2	6714.184	474.692	-29.633	CG81
			5927.775	cl,e	16	13/2	7951.323	644(1)	-30(20)	G96
			5955.302	cl,e	120	9/2	8029.275	797(2)		G96
			6060.343	cl,e	45	9/2	8320.240	255(2)		G96
			6181.599	cl	12	9/2	8643.824	797(2)		G96
			6363.059	cl	6	9/2	9105.021	689.7(3)	-3(5)	K96
			6602.604	cl	10	11/2	9675.029	683(1)		G96
			7185.875	nl	5	11/2	10904.034	301(1)	-20(10)	G96
			7387.032	nl,e	5	11/2	11282.865	1049(2)		G96
			8643.863	nl	2	11/2	13250.662	424(2)		G96
			8660.299	nl	7	9/2	13272.613	900(2)		G96
11/2	25009.019(10)	983(3)	5347.309	nl,f	28	11/2	6313.224	756(1)		G96
			5464.505	cl,f	32	11/2	6714.184	474.692	-29.633	CG81
			5860.83	nl,e	1	13/2	7951.323	644(1)	-30(20)	G96
			5990.391	cl,e	10	9/2	8320.240	255(2)		G96
			6439.235	cl,e	6	11/2	9483.518	731(1)	-15(10)	G96
			7434.966	nl	10	13/2	11562.762	819(2)		G96
			7709.545	nl	6	13/2	12041.655	1049(2)		G96
			8997.499	nl	7	13/2	13897.874	900(2)		G96
11/2	25746.884(10)	784(2)	4690.294	cl,f	60	9/2	4432.225	923.2(4)	-22(7)	K96
			4787.848	cl,f	8	11/2	4866.515	867.997	-50.319	CG81
			5017.675	cl,f	65	9/2	5822.890	855.8(4)	-17(7)	K96
			5144.278	cl,f	75	11/2	6313.224	756(1)		G96
			5222.311	nl,f	16	13/2	6603.591	755.456	-48.633	CG81
			5252.653	nl,f	8	11/2	6714.184	474.692	-29.633	CG81
			5617.818	cl,e	80	13/2	7951.323	644(1)	-30(20)	G96
			5845.29	nl,e	20	9/2	8643.824	797(2)		G96
			5909.289	cl,f	23	11/2	8829.063	769(1)	-30(20)	G96
			6544.171	nl,e	7	13/2	10470.329	628(3)		G96
			8081.917	nl	5	11/2	13376.992	868(2)		G96
11/2	26848.512(15)	590(5)	4459.79	nl,f	1	9/2	4432.225	923.2(4)	-22(7)	K96
			4547.902	nl,f	2	11/2	4866.515	867.997	-50.319	CG81
			4965.254	cl,f	6	11/2	6714.184	474.692	-29.633	CG81
			5290.322	nl,f	8	13/2	7951.323	644(1)	-30(20)	G96

J	Level	A	λ	Cl.	I	J'	Upper	A	B	Ref
			5312.230	nl	18	9/2	8029.275	797(2)		G96
			5395.656	nl,f	24	9/2	8320.240	255(2)		G96
			5634.305	cl,e	40	9/2	9105.021	689.7(3)	-3(5)	K96
			5686.772	nl,e	14	11/2	9268.726	977(1)	-24(20)	G96
			6276.458	nl	5	9/2	10920.365	632(2)		G96
11/2	27944.838(15)	654(4)	4519.130	nl,f	18	9/2	5822.890	855.8(4)	-17(7)	K96
			4921.167	cl,f	5	13/2	7630.132	776.286	-43.592	CG81
			5352.944	nl,f	20	11/2	9268.726	977(1)	-24(20)	G96
			5415.225	nl,f	20	11/2	9483.518	731(1)	-15(10)	G96
			5964.777	cl,e	8	9/2	11184.396	692(1)	15(30)	G96
			6537.026	nl,e	4	9/2	12651.586	723(3)		G96
			7262.018	nl	4	11/2	14178.365	986(1)	-20(10)	G96
13/2	25443.665(15)	491(3)	4858.40	nl,e	1	11/2	4866.515	867.997	-50.319	CG81
			5225.816	cl,f	75	11/2	6313.224	756(1)		G96
			5306.358	cl,f	100	13/2	6603.591	755.456	-48.633	CG81
			5389.124	nl,f	100	11/2	6892.934	551.934	-24.736	CG81
			5612.151	cl,f	40	13/2	7630.132	776.286	-43.592	CG81
			5715.20	nl,e	5	13/2	7951.323	644(1)	-30(20)	G96
			5994.217	cl	30	15/2	8765.542	763.557	-45.805	CG81
			6263.873	cl	8	11/2	9483.518	731(1)	-15(10)	G96
			6655.946	cl,f	15	13/2	10423.654	869(1	25(30)	G96
			6675.071	cl,f	9	15/2	10466.689	1042(2)	-20(30)	G96
			7459.513	nl	4	13/2	12041.655	1049(2)		G96
			7577.607	cl	4	15/2	12250.519	608(1)		G96
			7909.719	cl	10	15/2	12804.468	732(1)		G96
13/2	25765.460(15)	582(3)	4783.593	cl,f	16	11/2	4866.515	867.997	-50.319	CG81
			5217.247	nl,f	14	13/2	6603.591	755.456	-48.633	CG81
			5297.234	cl,f	12	11/2	6892.934	551.934	-24.736	CG81
			5512.568	cl,f	22	13/2	7630.132	776.286	-43.592	CG81
			5611.961	cl,f	145	13/2	7951.323	644(1)	-30(20)	G96
			5902.808	cl,e	27	11/2	8829.063	769(1)	-30(20)	G96
			6132.888	cl	11	13/2	9464.440	1056(1)	-15(10)	G96
			6140.074	cl,e	100	11/2	9483.518	731(1)	-15(10)	G96
			6213.154	nl	28	11/2	9675.029	683(1)		G96
			6536.224	nl	8	13/2	10470.329	628(3)		G96
			7284.604	nl	8	13/2	12041.655	1049(2)		G96
			7325.375	nl,e	15	13/2	12118.039	554(1)	-45(30)	G96
			7358.854	nl,e	8	11/2	12180.207	679(3)		G96
			7627.381	nl	4	11/2	12658.401	995(5)		G96
13/2	26810.061(15)	860(3)	4877.439	cl,f	7	11/2	6313.224	756(1)		G96
			4947.529	nl,f	35	13/2	6603.591	755.456	-48.633	CG81
			5212.333	nl	16	13/2	7630.132	776.286	-43.592	CG81
			5301.107	cl,f	30	13/2	7951.323	644(1)	-30(20)	G96
			5419.676	cl	35	15/2	8363.901	763.306	-48.253	CG81
			5540.310	nl	11	15/2	8765.542	763.557	-45.805	CG81
			5699.239	nl,e	17	11/2	9268.726	977(1)	-24(20)	G96
			5763.547	nl,e	6	13/2	9464.440	1056(1)	-15(10)	G96
			5769.892	nl,e	7	11/2	9483.518	731(1)	-15(10)	G96
			6100.929	cl	12	13/2	10423.654	869(1)	25(30)	G96
			7442.26	nl,e	4	11/2	13376.992	868(2)		G96
13/2	27139.468(15)	776(3)	5124.323	cl,f	110	13/2	7630.132	776.286	-43.592	CG81
			5440.982	cl	40	15/2	8765.542	763.557	-45.805	CG81
			5459.859	cl,f	90	11/2	8829.063	769(1)	-30(20)	G96
			5594.186	cl,f	52	11/2	9268.726	977(1)	-24(20)	G96
			5662.24	nl,e	1	11/2	9483.518	731(1)	-15(10)	G96
			5747.48	nl,e	1	15/2	9745.376	540(2)		G96
			7264.132	nl,e	5	11/2	13376.992	868(2)		G96
			7413.65	nl,e	1	13/2	13654.555	501(1)	20(20)	G96
			7535.314	nl,e	4	11/2	13872.266	872(3)		G96
13/2	27500.390(20)	883(2)	4416.914	nl,f	3	11/2	4866.515	867.997	-50.319	CG81
			5548.813	cl,f	16	11/2	9483.518	731(1)	-15(10)	G96
			5608.428	nl,e	15	11/2	9675.029	683(1)		G96
			5630.651	nl	11	15/2	9745.376	540(2)		G96
			5800.912	cl,e	18	13/2	10266.501	972(2)		G96
			7349.556	nl,e	6	13/2	13897.874	900(3)		G96

18

Synthesis and Characterization of New 8-trifluloromethyl Quinazolin-2,4-(*3H*)-Dione Nucleosides

Laila M. Break

Correspondence: Laila M. Break, Taif University, Al-Haweiah, P.O.Box 888 Zip Code 21974. Taif, Saudi Arabia. E-mail: lailabreek@gmail.com

Abstract

Synthesis of 8-trifluloromethyl quinazolin-2,4-(*1H,3H*)-dione **2**. which have been ribosylated by coupling with 1-*O*-acetyl-2,3,5-tri-*O*-benzoyl-β-D-ribofuranose **4** by using the silylation method, afforded mixture β-and α-anomeric of the benzoylated nucleoside derivatives **5** and **6**, respectively. Debenzoylation of each of **5** and **6** by sodium metal in dry methanol to afford the corresponding free nucleosides **7** and **8** respectively. The structures of the newly synthesis compounds have been confirmed on the basis of elemental analyses, IR, ^{1}HNMR, ^{13}CNMR and Mass spectral data.

Keywords:1-*O*-Acetyl-2,3,5-trihydroxy-β-D-ribofuranose, Nucleosides, Quinazolin-2,4-(*1H,3H*)-dione, Trifluoromethyl

1. Introduction

Quinazolinone is a heterocyclic compound that occupies a distinct and place in the field of medicinal chemistry. Many of them were showed antimicrobial, anti-inflammatory, anticonvulsant, analgesic and anticancer agents (Safinaz E.Abbas et al, 2013; A. Kumar et al, 2011; K.M. Amin et al, 2010; M.M. Aly, 2010; A. Kumar, 2003; S.T. Al-Rashood, 2006 and N. Mulakayala, 2012).

Quinazoline nucleosides were first synthesized by Stout and Robins in 1968 as pyrimidine nucleoside analogs (Stout M. G and Robins R. K., 1968) and consequent synthetic studies were contributed by Dunkel and Pfleiderer in the 1990s. (Dunkel M and Pfleiderer W, 1991, 1992 and1993). More recently, several quinazoline-2,4-dione nucleosides have been incorporated into oligonucleotides as pyrimidine nucleoside substitutes to study the binding affinity and base pairing selectivity(Michel J et al, 1997; Diwan A. R., 1969; T. C. Chien, 2005; F. E. M. El-Baih, 2004; Tun-Cheng CHIEN, 2004).

Many familiar drugs and pharmacological studies contain trifluoromethyl groups. Quinazoline-2,4-diones bearing a trifluoromethyl group derivatives were an inhibitor of human immunodeficiency virus-1 reverse transcriptase, antagonists at ionotropic glutamate receptors (Hao Chen et al, 2003; Vittoria Colotta, 2012) and anticancer compound trifluoromethyl-substituted pyrazole N-nucleoside(Ayman M. Saleh et al, 2016).

In this review, quinazolin-2,4-(*3H*)-dione nucleosides containing trifluoromethyl group were designed as part of our continuing interest in the synthesis of new nucleosides as expected their biological activity.

2. Material and Methods

Melting points were measured on Gallenkamp melting point apparatus (UK) and are uncorrected. The purity of the compounds was checked by thin layer chromatography (TLC). Thin layer chromatography (TLC) was performed on silica gel sheets F1550 LS 254 of Schleicher & Schull and column chromatography on Merck silica gel 60 (particle size 0.063–0.20. Elemental analyses were obtained on an Elementary Vario EL 1150C analyzer. IR spectra were recorded on KBr discs on Fourier Transform infrared and Pie Unicom SP 300 Infrared Spectrophotometers at Taif University. ^{1}H NMR and ^{13}C NMR spectra were obtained on a Varian (850 MHz) EM 390 USA instrument at King Abdel-Aziz University by using TMS as the internal reference. Mass spectra were recorded on a JEOL-JMS-AX500 at King Abdel-Aziz University, Saudi Arabia.

3. Experimental

8-trifluoromethyl quinazolin-2,4-(*1H,3H*)-dione 2

2-Amino-3-trifluloromethyl benzoic acid **1** (Aldrich; 0.019mol, 4g) was added drops of acetic acid in (100 ml) water, the solution of KNCO (0.049mol, 4g) was dropped to the mixture was stirred in an ice bath at 1h. The reaction mixture was added sodium hydroxide (10 g) and overnight at room temperature and then filtered. The precipitate was neutralized with dilute sulfuric acid (1:1) and washed (3×20 ml) water. The compound purified by column chromatography on silica gel with (Ethylacetate: Acetone 9:1) to afford white crystals.

Yield (63.23%), w. 6.1g, m.p. <300°C white ; v (cm^{-1}) (KBr) 3400, 3075,1740, 1640; ^{1}HNMR (850MHz); (DMSO-D$_6$): 11.62 (s, 1H) NH-1, 11.20 (s,1H) NH-3, 8.22 (s, 1H) H$_5$, 7.79 (d, 1H, J = 7.5 Hz) H$_7$, 7.18 (d, 1H, J = 7.5 Hz) H$_6$. ^{13}CNMR (850MHz): δ 167.27, 154.77, 133.08, 130.82,128.39, 125.71, 124.15, 123.09, 109.80; MS m/z: 230 (M$^+$, 45%). Anal. Calcd. for C$_9$H$_5$F$_3$N$_2$O$_2$; M.wt: 230.14; C,46.97; H,2.19; F,24.77; N, 12.17 (%); Found: C, 46.26; H, 2.89; F,24.01; N,11.05 (%).

Synthesis of protection nucleoside of 8-trifluloromethyl quinazolin-2,4-(*1H,3H*)-dione 2

General Procedure

Silyation of 8-trifluloromethyl quinazolin-2,4-(*1H,3H*)-dione **2** (0.021 mol) with hexamethyldisilazane (HMDS) (20 ml) was refluxed for 3days with a catalytic a few crystals of ammonium sulfate under exclusion of moisture. Excess of HMDS was removed in vacuo by co-evaporation with dry dichloroethane gave the silyated derivative **3**, using the Vorbruggen's silylation method (Vorbruggen et al, 1981). The residue was dissolved in 20 ml of dry 1,2-dichloroethane and then 1-*O*-acetyl-2,3,5-tri-*O*-benzoyl-β-D-ribofuranose **4** (10.8 g, 0.021 mol) was added. The mixture was added dropwise onto a mixture (4.5ml) of (10 ml trimethylsilyl trifluoromethane sulfonate (TMSOTf) in dry 1,2-dichloroethane (50 ml)). The mixture was stirred at room temperature for 24 h, and then washed with a saturated solution of aqueous sodium bicarbonate (3 × 50 ml), washed with water (3 × 50 ml), and dried over anhydrous sodium sulfate. the organic phase was extracted by CH$_2$Cl$_2$, dried over MgSO$_4$ and evaporated. The solvent was removed under vacuum gave an anomeric mixture of β and α-1-(2,3,5-tri-*O*-benzoyl-D-ribofuranosyl)-8-trifluloromethylquinazolin-2,4-(*3H*)-dione. The protected nucleoside was separated by column chromatography on silica gel with dichloromethane: acetone (9:1) as eluent to afford a white crystal pure β-anomeric **5** and α-anomeric **6** respectively, in good yields.

β-1-(2,3,5-Tri-*O*-benzoyl-D-ribofuranosyl)-8-trifluloromethyl quinazolin-2,4-(*3H*)-dione 5

Yield (52.12%), w. 2.7 g, m.p. 119°C white; IR v (cm^{-1}) (KBr) 3042, 1725, 1680; ^{1}HNMR (850MHz); (CDCl$_3$): δ 9.47(s, 1H) H$_{3\,Amide}$, 8.26 (d, 1H, J = 7.5 Hz) H$_5$, 8.03 (d, 1H, J = 7.2 Hz) H$_7$, 7.92 (dd, 1H, J = 15.7 Hz) H$_6$, 7.51-7.25 (m, 15H) H$_{(Ar-H)}$, 6.48 (d, 1H, J = 7.5 Hz) H$_{1'}$, 6.16 (dd, 1H, J = 8.4 Hz) H$_{2'}$, 6.09 (t, 1H, J = 13. 4 Hz) H$_{3'}$, 4.82-4.77 (dd, 1H, J = 4.6 Hz) H$_{5'}$ 4.69-4.56 (m, 1H) H$_{4'}$. ^{13}CNMR (850MHz) (CDCl$_3$): δ 166.30, 165.46, 164.87,165.53$_{C=O's\ groups}$, 148.17C$_4$, 146.78 C$_2$, 133.61-128.33 $_{Ar-carbons}$, 119.11 CF$_3$, 88.02 C$_{1'}$, 79.47 C$_{2'}$,73.93 C$_{3'}$, 70.98 C$_{4'}$, 63.82 C$_{5'\ sugar\ carbons}$. Anal. Calcd. for C$_{35}$H$_{25}$F$_3$N$_2$O$_9$; M.wt: 674.58; C,62.32; H,3.74; F,8.45; N, 4.15; (%); Found: C, 62.26; H, 3.89; F,8.21; N,3.95 (%).

α-1-(2,3,5-Tri-*O*-benzoyl-D-ribofuranosyl)-8-trifluloromethyl quinazolin-2,4-(*3H*)-dione 6

Yield (47.87%), w. 2.48 g, m.p. 105-107°C white color; IR v (cm^{-1}) (KBr) 3020, 1725, 1685; ^{1}HNMR (850MHz) (CDCl$_3$): δ 9.16 (s, 1H) H$_{Amide}$, 8.06 (d, 1H, J = 7.4 Hz) H$_5$, 7.98 (d, 1H, J = 7.2 Hz) H$_7$, 7.89 (d, 1H, J = 7.4 Hz) H$_6$, 7.55-7.25 (m, 15H) H$_{(Ar-H)}$, 6.73 (d, 1H, J = 2.2 Hz) H$_{1'}$, 6.14 (dd, 1H, J = 8.8 Hz)H$_{2'}$, 6.03-5.89 (m, 1H) H$_{3'}$, 4.81-4.57 (m, 1H) H$_{4'}$ 3.73 (m, 1H) H$_{5'}$. ^{13}CNMR (850MHz): δ 166.46,165.82, 165.40 $_{C=O's\ groups}$, 147.95C$_4$, 147.26 C$_2$, 13.73-128.40 $_{Aromatic\ carbons}$, 118.01 CF$_3$, 87.46 C$_{1'}$, 79.37 C$_{2'}$,73.90 C$_{3'}$, 71.69 C$_{4'}$, 63.83 C$_{5'}$. (CH$_3$Cl : CH$_3$COOCH$_2$CH$_3$) (9:1). Anal. Calcd. for C$_{35}$H$_{25}$F$_3$N$_2$O$_9$; M.wt: 674.58; C,62.32; H,3.74; F,8.45; N, 4.15; (%); Found: C, 62.15; H, 3.45; F,8.91; N,4.09 (%).

Deprotection of **5** and **6**. Synthesis of free nucleosides **7** and **8** respectively

General Procedure

The pure anomer of each β **5** and α **6** (0.001 mol for each), dry absolute methanol (20 ml) and sodium metal (0.055 g, 0.001mol) was stirred at room temperature for 48h. The solvent was evaporated under vacuum to give a colorless solid, which was dissolved in hot water and neutralized with few drops acetic acid. Purification of each compound by TLC chromatographic on silica gel with chloroform: ethyl acetate (9: 1) to afford colorless and white crystals of the following Zemplen et al.'s method (Zemplen et al, 1939) to afford the free nucleosides **7** and **8**, respectively.

β-1-(2,3,5-Trihydroxy-D-ribofuranosyl)-8-trifluloromethyl quinazolin-2,4-(*3H*)-dione 7

Yield (81.67%), w. 0.307g. m.p. 185°C white color; IR v (cm^{-1}) (KBr) 3450, 3032, 1715, 1685; ^{1}HNMR (600MHz)(DMSO-D$_6$): δ 11.59 (s, 1H) H$_{1\,Amide}$, 8.06 (d, 1H, J = 5.5 Hz) H$_5$, 7.80-7.79 (d, 1H, J = 8.7 Hz) H$_7$, 7.77-7.76 (d, 1H, J = 5.2 Hz) H$_6$, 6.17 (d, 1H, J = 7.5 Hz) H$_{1'}$, 5.27 (s, 1H) H$_{2'}$, 5.07 (m, 1H) H$_{3'}$, 4.45 (t, 1H) H$_{5'}$ 4.13 (s, 1H) H$_{4'}$, 3.80-3.76 (m, 1H) H$_{2'OH}$, 3.66-3.61 (m, 1H) H$_{3'OH}$, 3.58-3.41 (m, 1H) H$_{3'OH}$. ^{13}C NMR: 155.07 C$_4$, 153.63 C$_2$, 139.6, 137.59, 133.73, 129.92 , 128.86, 120.00 , 118.96 CF$_3$, 89.89 C$_{1'}$, 85.31 C$_{2'}$, 69.41 C$_{3'}$, 69.29 C$_{4'}$, 61.63 C$_{5'}$.

$(CH_3COOCH_2CH_3$: Acetone) (9:1); MS m/z: 362 (M^+, 21%). Anal. Calcd. for $C_{14}H_{13}F_3N_2O_6$; M.wt: 362.26; C,46.42; H,3.62; F,15.73; N, 7.73; (%); Found: C, 46.12; H, 3.75; F,15.91; N,7.24 (%).

α-1-(2,3,5-Trihydroxy-D-ribofuranosyl)-8-trifluloromethyl quinazolin-2,4-(*3H*)-dione 8

Yield (94.73%), w. 0.41 g, m.p. 220°C white color; IR v (cm^{-1}) (KBr) 3450, 1714, 1690; ^{1}HNMR (600MHz)(DMSO-D_6): δ 11.85 (s, 1H) H_{1Amide}, 8.06 (d,1H, J = 1.46 Hz)H_5, 7.80 (t, 1H, J = 11.37 Hz) H_7, 7.77 (d, 1H, J = 1.83 Hz) H_8, 6.53 (d, 1H, J = 4.5 Hz) $H_{1'}$, 5.25 (d, 1H, J = 5.87 Hz) $H_{2'}$, 5.05 (t, 1H, J = 21.65 Hz) $H_{3'}$, 4.46 (q, 1H) $H_{5'}$, 4.13 (q, 1H) $H_{4'}$, 3.8 (m, 2H) $H_{4'a}$, 3.60 (s, 1H) OH_2 3.58 (s, 1H) OH_3 3.57 (s, 1H) OH_5. ^{13}C NMR: 161.07 C_4, 150.63 C_2, 139.67, 137.59, 133.73, 129.93, 125.34, 120.00, 118.82 CF_3, 89.91 $C_{1'}$, 85.32$C_{2'}$, 69.42 $C_{3'}$, 69.29 $C_{4'}$, 61.63 $C_{5'}$. ($CH_3COOCH_2CH_3$: Acetone) (9:1); MS m/z: 362 (M^+, 36%). Anal. Calcd. for $C_{14}H_{13}F_3N_2O_6$; M.wt: 362.26; C,46.42; H,3.62; F,15.73; N, 7.73; (%); Found: C, 46.12; H, 3.75; F,15.57; N,7.13 (%).

4. Results and Discussion

The structures of the products **2-8** were established and confirmed on the bases of their elemental analyses and spectral data (IR, ^{1}H and ^{13}C NMR) (see the Experimental section)(Scheme 1). Thus, their ^{1}H NMR spectra of compound **2** showed doublet signals at assigned to the aromatic protons of H-5 H-6 and H-7 and two a singlet signal of amide NH-3 and NH-1.

^{1}H NMR spectra of **5** and **6** showed in each case a doublet signals at δ 6.48 (d, 1H, J = 7.5 Hz) $H_{1'}$ for compound **5** and at δ 6.73 (d, 1H, J = 2.2 Hz) $H_{1'}$ for compound **6** assigned to the anomeric proton of the ribose moiety with spin–spin coupling constant ($J_{1',2'}$) equal to 7.5 Hz, which confirms the β-anomeric configuration. While confirms the -anomeric configuration showed spin–spin coupling constant ($J_{1',2'}$) equal to 2.2 Hz, which confirms the -anomeric configuration for compound **8** (Break et al, 2014; Break et al, 2013; Break & Mosselhi, 2012; Mosselhi & Break, 2011and Break et al, 2010 and Abdullah Hijazi, 1988). The ^{1}H NMR spectra of nucleosides free showed a doublet signals at δ 6.17for compound **7** spin– spin coupling constant ($J_{1',2'}$) equal to 7.5 Hz -anomeric configuration and at δ 6.53 for compound **8** assigned to spin–spin coupling constant ($J_{1',2'}$) equal to 4.5 Hz, which confirms the -anomeric configuration. The ^{1}H NMR of compounds **5** and **6** showed the expected base moiety protons in addition to the sugar moiety protons (see the Experimental section).

The ^{13}C NMR of nucleoside products revealed the signals are due to the three benzoyl carbonyl groups at and for compound **5**, and 166.46, 165.82 and 165.40 for compound **6**, while showed the two signals of amide carbons at 148.17, 146.78 for compound **5**, and at 147.95C_4, 147.26 C_2 for compound **6**, The twenty one signals at 133.61-128.33 and at 133.73-128.40 Aromatic carbons for compound **5** and **6** respectively.

The five signals were assigned to C-1', C-2', C-3' , C-4', and C-5' of the sugar moiety, at δ 88.02 $C_{1'}$, 79.47 $C_{2'}$,73.93 $C_{3'}$, 70.98 $C_{4'}$ and 63.82 $C_{5'}$ for compound **5**, at δ 87.46 $C_{1'}$, 79.37 $C_{2'}$,73.90 $C_{3'}$, 71.69 $C_{4'}$, 63.83 $C_{5'}$ for compound **6**, at δ 89.89 $C_{1'}$, 85.31 $C_{2'}$, 69.41 $C_{3'}$, 69.29 $C_{4'}$, 61.63 $C_{5'}$ for compound **7** and at δ 89.91 $C_{1'}$, 85.32$C_{2'}$, 69.42 $C_{3'}$, 69.29 $C_{4'}$, 61.63 $C_{5'}$ for compound **8**.

The ^{13}C NMR of CF_3 group showed at δ 119.11, 118.01, 118.96 and 118.82 of compounds (**5, 6, 7** and **8**) respectively Break and Break. The IR spectrum of compounds **5** and **6** showed the stretching vibration frequencies of the carbonyl C=O groups at 1725 cm^{-1}. IR spectra of compounds **7** and **8** showed absorptions around 3450 cm^{-1} for (OH) and 1715 cm^{-1} for (C=O).

5. Conclusion

Quinazolinone nucleosides are scientific importance in many biologically active compounds. So synthesis and characterization of 8-trifluloromethyl quinazolin-2,4-(*1H,3H*)-dione **2**. Ribosylation of compound **3** with 1-*O*-acetyl-2,3,5-tri-*O*-benzoyl-β-D-ribofuranose **4** afforded mixture β-and α-anomeric of the benzoylated nucleoside derivatives **5** and **6**, respectively. Debenzoylation of the latter affording the corresponding new free N-nucleosides **7** and **8**, respectively. Nucleosides obtained have been identified by their spectral analysis.

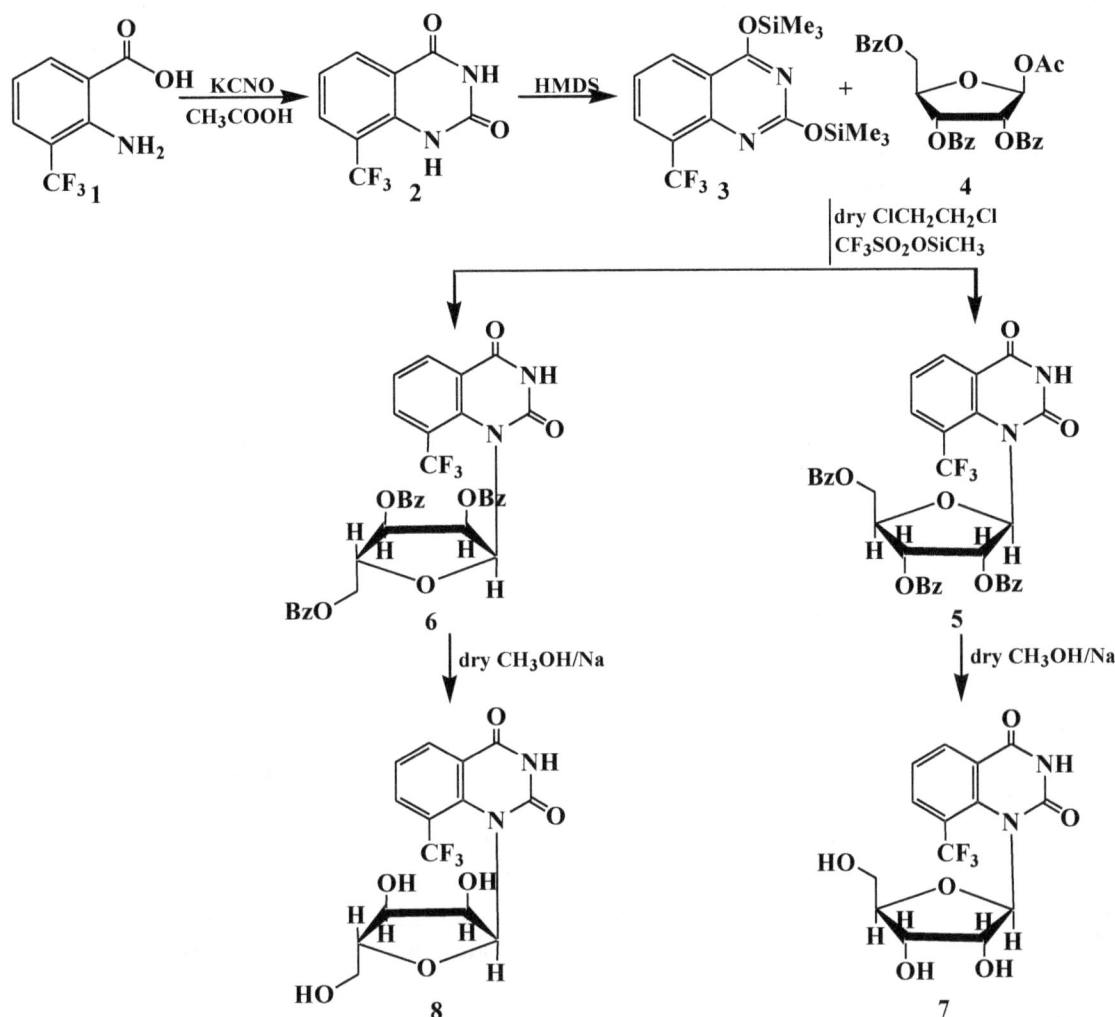

Schem (1). 8-Trifuloro quinazolin-2,4-(3H)-dione Nucleosides

References

AbdullahM H. (1988). Nucleosides, IV Synthesis and Properties of 2-Metylthio-Naphthimidazole-Ribonucleoside. *Nucleosides & nucleotides, 7*(4), 537-547. https://doi.org/10.1080/07328318808075395

Al-Rashood, S. T., Aboldahab, I. A., Nagi, M. N., Abouzeid, L. A., Abdel-Aziz, A. A. M., Abdel-hamide, ... & El-Subbagh, H. I. (2006). Synthesis, dihydrofolate reductase inhibition, antitumor testing, and molecular modeling study of some new 4(3H)-quinazolinone analogs., *Bioorg. Med. Chem., 14*, 8608–8621. https://doi.org/10.1016/j.bmc.2006.08.030

Aly, M. M., Mohamed, Y. A., El-Bayouki, K. A. M., Basyouni, W. M., & Abbas, S. Y. (2010). Synthesis of some new 4(3H)-quinazolinone-2-carboxaldehyde thiosemicarbazones and their metal complexes and a study on their anticonvulsant, analgesic, cytotoxic and antimicrobial activities – Part-1., *Eur. J. Med. Chem., 45*, 3365–3373. https://doi.org/10.1016/j.ejmech.2010.04.020

Amin, K.M., Kamel, M.M., Anwar, M. M., Khedr, M., & Syam, Y. M. (2010). Synthesis, biological evaluation and molecular docking of novel series of spiro [(2H,3H) quinazoline-2,10-cyclohexan]-4(1H)-one derivatives as anti-inflammatory and analgesic agents., *Eur. J. Med. Chem., 45*, 2117–2131. https://doi.org/10.1016/j.ejmech.2009.12.078

Ayman, M. S., Taha, M. O., Aziz, M. A., Mahmoud A. Al-Qudah, AbuTayehx, R. F. & Syed A. Rizvi. (2016). Novel anticancer compound [trifluoromethyl-substituted pyrazole N-nucleoside] inhibits FLT3 activity to induce differentiation in acute myeloid leukemia cells, *Cancer Letters., 375*(2), 199–208. https://doi.org/10.1016/j.canlet.2016.02.028

Break, L. M., & Mosselhi, A. N. M. (2012). Synthesis, structure and Antimicrobial activity of new 3- and 2- arylmethyl and arylacyl-3H [1,2,4] triazino [3,2-b]-quinazoline-2,6 (1H) diones as expect as DNA fluorphores. *Research Journal of Chemical Science, 2*(5), 23-28.

Break, L. M., Shmiss, N. A. M. M., & Mosselhi, M. A. N. (2010). Synthesis of some news-nucleoside derivatives of 2-thioxo and (2,4-Dithioxo)-5,6,7,8-Tetrahydrobenzo-Thieno[2,3-d]Pyrimi-din-4-(3H) Ones. Phosphorus, *Sulfur and Silicon and the Related Elements, 185*(8), 1615–1622. https://doi.org/10.1080/10426500903147159

Chen, H., Chen, W.Q., Gan, L. S., & Mutlib, A. E. (2003). Metabolism of (S)-5,6-DifluoroI-4-Cyclopropylethynyl-4-Trifluorormethyl-3,4-dihydro-2(1H)-Quinazolinone, A non-Nucleoside Reverse Transcriptase Inhibitor, In Human Liver Microsomes. Metabolic Activation and Enzyme Kinetics. *Drug Metabolism and Disposition, 31*(1), 122–132. https://doi.org/10.1124/dmd.31.1.122

Chien, T. C., Chen, C. S., & Chern, J. W. (2005). Nucleosides XIII. Facile synthesis of 4-amino-1-(2-deoxy-β-D-ribofuranosyl)quinazolin-2-one as a 2-deoxycytidine analog for oligonucleotide synthesis. *Journal of the Chinese Chemical Society, 52*(6), 1237–1244. https://doi.org/10.1002/jccs.200500178

Chien, T. C., Chen, C. S., Yu, F. H., & Chern, J. W.(2004). Nucleosides XI.1) Synthesis and Antiviral Evaluation of 5-Alkylthio-5deoxy Quinazolinone Nucleoside Derivatives as S-Adenosyl-Lhomocysteine Analog, *Chem. Pharm. Bull., 52*(12) 1422-1426. https://doi.org/10.1248/cpb.52.1422

Diwan, A. R., Robins, R. K., & Prusoff, W. H. (1969). Antiviral activity of certain substituted purine and pyrimidine nucleosides. *Experientia, 25,* 98-100. PMID: 4304043. https://doi.org/10.1007/BF01903922

Dunkel, M., & Pfleiderer, W. (1991). *Nucleosides Nucleotides, 10,* 799—817. https://doi.org/10.1080/07328319108046663

Dunkel, M., & Pfleiderer, W. (1992). *Nucleosides Nucleotides, 11,* 787—819. https://doi.org/10.1080/07328319208021742

Dunkel, M., & Pfleiderer, W. (1993). *Nucleosides Nucleotides, 12,* 125—374. https://doi.org/10.1080/07328319308021199

Dunkel, M., & Pfleiderer, W. (1993). Nucleosides. LII. Synthesis and properties of quinazoline-3'-azidonucleosides. *Nucleosides and Nucleotides, 12*(2), 125–137. https://doi.org/10.1080/07328319308021199

El-Baih, F. E. M., Bakari, S. B. A., & Hijazi, A. A. (2004). Synthesis and spectroscopic properties of quinazolinedione derivatives. *Journal of King Abdelaziz University, 16,* 41–53. https://doi.org/10.4197/Sci.16-1.5

Kumar, A., Sharma, P., Kumari, P., & Kalal, B. L.(2011). Exploration of antimicrobial and antioxidant potential of newly synthesized 2,3-disubstituted quinazoline-4(3H)-ones., *Bioorg. Med. Chem. Lett., 21.* 4353–4357 https://doi.org/10.1016/j.bmcl.2011.05.031

Kumar, A., Sharma, S., Archana, Bajaj, K., Sharma, S., Panwar, H., ... Srivastava, V. K. (2003). Some new 2,3,6-trisubstituted quinazolinones as potent anti-inflammatory, analgesic and COX-II inhibitors., Bioorg Med Chem, 11, 5293–5299. https://doi.org/10.1016/S0968-0896(03)00501-7

Laila, M. B. (2015). Synthesis of the Novel 3-Benzotriazole-5-yl difluoromethyl-5-trifluoromethyl benzotriazole Nucleosides. *International Journal of Chemistry, 7*(2), 99. https://doi.org/10.5539/ijc.v7n2p99

Laila, M. B. (2016). Synthesis of Some of Fluorinated Benzimidazole Nucleosides. *International Journal of Chemistry. 8*(1), 188. https://doi.org/10.5539/ijc.v8n1p188

Laila, M. B., Mohamed, M., & Abdel-Hafez, S. H. (2014). Synthesis of New Organoselenium Compounds Containing Nucleosides as Antioxidant. Orient. *J. Chem., 30*(4), 1639-1645. https://doi.org/10.13005/ojc/300423

Laila, M. B., Mosselhi, A. M., & Elshafai, N. M. (2013). Nucleosides 8 [18]: Ribosylation of Fused Quinazolines—Synthesis of New [1,2,4]Triazolo[5,1-b]- and [1,2,4]Triazino[3,2-b]quinazoline Nucleosides of Fluorescence Interest. *Journal of Chemistry.* https://doi.org/10.1155/2013/612756

Michel, J., Gueguen, G., Vercauteren, J., & Moreau, S. (1997). *Tetrahedron, 53,* 8457-8478. https://doi.org/10.1016/S0040-4020(97)00564-4

Mosselhi, A., Mosselhi, N., & Laila, M. B. (2011). Nucleosides 79: Synthesis, structure, and biological, activity of new 6-arylidenamino-2-thio- and 2-benzylthiopyrimidine N-nucleosides. *Nucleosides, Nucleotides and Nucleic Acids, 30,* 681–695. https://doi.org/10.1080/15257770.2011.597628

Mulakayala, N., Kandagatla, B., Ismail, R., Rapolu,K., Rao, P., Mulakayala, C., Kumar, S., Iqbal, J., & Oruganti, S. (2012). InCl$_3$-catalyscd synthesis of 2-aryl quinazolin-4(3H)-ones and 5-aryl pyrazolo[4,3-d]pyrimidin-7(6H)-ones and their evaluation as potential anticancer agents., *Bioorg. Med. Chem. Lett., 22,* 5063–5066.

https://doi.org/10.1016/j.bmcl.2012.06.003

Safinaz E. Abbas, Flora, F. B., Hanan, H. G., & Eman, R. M. (2013). Synthesis and antitumor activity of certain 2,3,6-trisubstituted quinazolin-4(3H)-one derivatives., Bulletin of Faculty of Pharmacy, Cairo University. Volume 51, Issue 2, 273–282. https://doi.org/10.1016/j.bfopcu.2013.08.003

Stout, M. G., & Robins, R. K. (1968). The synthesis of some quinazoline nucleosides., *J. Org. Chem., 33*, 1219—1225. https://doi.org/10.1021/jo01267a061

Vittoria, C., Ombretta, L., Daniela, C., Flavia, V., Lucia, S., Chiara, C., ... Stefano, M.(2012). 3-Hydroxy-1*H*-quinazoline-2,4-dione derivatives as new antagonists at ionotropic glutamate receptors: Molecular modeling and pharmacological studies. *European Journal of Medicinal Chemistry 54*, 470–482. https://doi.org/10.1016/j.ejmech.2012.05.036

Vorbruggen, H, Krolikiewicz, K., & Bennua, B. (1981). *Chem. Ber., 114*, 1234. https://doi.org/10.1002/cber.19811140404

Zemplen, G., Gerecs, A., & Hadacsy, I. (1939). Ber. Dtsch. *Chem. Ges., 69*, 1827. https://doi.org/10.1002/cber.19360690807

Permissions

All chapters in this book were first published in IJC, by Canadian Center of Science and Education; hereby published with permission under the Creative Commons Attribution License or equivalent. Every chapter published in this book has been scrutinized by our experts. Their significance has been extensively debated. The topics covered herein carry significant findings which will fuel the growth of the discipline. They may even be implemented as practical applications or may be referred to as a beginning point for another development.

The contributors of this book come from diverse backgrounds, making this book a truly international effort. This book will bring forth new frontiers with its revolutionizing research information and detailed analysis of the nascent developments around the world.

We would like to thank all the contributing authors for lending their expertise to make the book truly unique. They have played a crucial role in the development of this book. Without their invaluable contributions this book wouldn't have been possible. They have made vital efforts to compile up to date information on the varied aspects of this subject to make this book a valuable addition to the collection of many professionals and students.

This book was conceptualized with the vision of imparting up-to-date information and advanced data in this field. To ensure the same, a matchless editorial board was set up. Every individual on the board went through rigorous rounds of assessment to prove their worth. After which they invested a large part of their time researching and compiling the most relevant data for our readers.

The editorial board has been involved in producing this book since its inception. They have spent rigorous hours researching and exploring the diverse topics which have resulted in the successful publishing of this book. They have passed on their knowledge of decades through this book. To expedite this challenging task, the publisher supported the team at every step. A small team of assistant editors was also appointed to further simplify the editing procedure and attain best results for the readers.

Apart from the editorial board, the designing team has also invested a significant amount of their time in understanding the subject and creating the most relevant covers. They scrutinized every image to scout for the most suitable representation of the subject and create an appropriate cover for the book.

The publishing team has been an ardent support to the editorial, designing and production team. Their endless efforts to recruit the best for this project, has resulted in the accomplishment of this book. They are a veteran in the field of academics and their pool of knowledge is as vast as their experience in printing. Their expertise and guidance has proved useful at every step. Their uncompromising quality standards have made this book an exceptional effort. Their encouragement from time to time has been an inspiration for everyone.

The publisher and the editorial board hope that this book will prove to be a valuable piece of knowledge for researchers, students, practitioners and scholars across the globe.

List of Contributors

Athina Mardhatillah
Faculty of Pharmacy, Universitas Padjadjaran, Jl. Raya Bandung-Sumedang Km.21 Jatinangor, West Java 45363, Indonesia
Faculty of Pharmacy, University of Jenderal Achmad Yani, Jl. Terusan Jenderal Sudirman Cimahi, West Java, Indonesia

Mutakin Mutakin and Jutti Levita
Faculty of Pharmacy, Universitas Padjadjaran, Jl. Raya Bandung-Sumedang Km.21 Jatinangor, West Java 45363, Indonesia

Enos Masheija Rwantale Kiremire
Department of Chemistry and Biochemistry, University of Namibia, Private Bag 13301, Windhoek, Namibia

Bin Yao
1Materials Engineering and Science Program, South Dakota School of Mines and Technology, Rapid City, SD, 57701

Praveen Kolla
Chemistry and Applied Biological Sciences Department, South Dakota School of Mines and Technology, Rapid City, SD, 57701

Ranjit Koodali and Chia-Ming Wu
Chemistry Department, University of South Dakota, Vermillion, SD, 57065

Alevtina Smirnova
Materials Engineering and Science Program, South Dakota School of Mines and Technology, Rapid City, SD, 57701
Chemistry and Applied Biological Sciences Department, South Dakota School of Mines and Technology, Rapid City, SD, 57701

Jonathan H. Loftus
Applied Biochemistry Group, School of Biotechnology, Dublin City University, Dublin 9, Ireland
Immunomodulation Group, School of Biotechnology, Dublin City University, Dublin 9, Ireland

Gregor S. Kijanka
Biomedical Diagnostics Institute, Dublin City University, Dublin 9, Ireland

Richard O'Kennedy
Applied Biochemistry Group, School of Biotechnology, Dublin City University, Dublin 9, Ireland
Biomedical Diagnostics Institute, Dublin City University, Dublin 9, Ireland

Christine E. Loscher
Immunomodulation Group, School of Biotechnology, Dublin City University, Dublin 9, Ireland

Chetansing K. Rajput
Correspondence: Dr. Chetansing K. Rajput, M.B.B.S., T.N.M.C., Mumbai University, ACST, 306, Vikrikar Bhavan, Nashik-422010, India

Kasem K. Kasem, Christopher Santuzzi, Nick Daanen and Kortany Baker
School of Sciences, Indiana University Kokomo, Kokomo, IN, 46904, USA

Jun-Zai Yu and Xue-Hai Ju
Key Laboratory of Soft Chemistry and Functional Materials of MOE, School of Chemical Engineering, Nanjing University of Science and Technology, Nanjing 210094, P. R. China

Feng-Qi Zhao and Si-Yu Xu
Laboratory of Science and Technology on Combustion and Explosion, Xi'an Modern Chemistry Research Institute, Xi'an 710065, P. R. China

Lu-Jing Sun
School of Chemistry and Materials Science, Nanjing Normal University, Nanjing 210094, P. R. China

Naofumi Naga, Yukie Uchiyama and Yuri Takahashi
1Department of Applied Chemistry, Materials Science Course, College of Engineering, Shibaura Institute of Technology, 3-7-5 Toyosu, Koto-ku, Tokyo 135-8548, Japan

Hidemitsu Furukawa
Department of Mechanical Systems Engineering, Graduate School of Science and Engineering, Yamagata University, 4-3-16 Jonan, Yonezawa, Yamagata992-8510, Japan

Valter A. Nascimento, Petr Melnikov, André V. D. Lanoa, Anderson F. Silva and Lourdes Z. Z. Consolo
School of Medicine of the Federal University of Mato Grosso do Sul/UFMS, Brazil

Jiao Yang, Zhenzhen Liu, Huan Ge and Sufang Sun
College of Chemistry and Environmental Science, Hebei University, Baoding, China

Takanori Fukami, Shuta Tahara, Chitoshi Yasuda and Keiko Nakasone
Department of Physics and Earth Sciences, Faculty of Science, University of the Ryukyus, Japan

Eva Perlt, Christina Apostolidou, Melanie Eggers and Barbara Kirchner
Mulliken Center for Theoretical Chemistry, Rheinische Friedrich-Wilhelms-Universität Bonn, Germany

Mopelola Abeke Omotoso and Samuel Oluwatosin Ajagun
Department of Chemistry, University of Ibadan, Ibadan, Nigeria

Ignác Capek
Slovak Academy of Sciences, Polymer Institute, Institute of Measurement Science,
Dúbravská cesta, Bratislava and Faculty of Industrial Technologies, TnUni, Púchov, Slovakia

Shamim Khan
Institute for Experimental Physics, Graz University of Technology, Petersgasse 16, A-8010 Graz, Austria
Department of Physics, Islamia College Peshawar, Peshawar, Pakistan

Imran Siddiqui and Zaheer Uddin
Institute for Experimental Physics, Graz University of Technology, Petersgasse 16, A-8010 Graz, Austria
Department of Physics, University of Karachi, 75270 Karachi, Pakistan

Syed Tanweer Iqbal
Institute for Experimental Physics, Graz University of Technology, Petersgasse 16, A-8010 Graz, Austria
Department of Physics, The NED University of Engineering and Technology, University Rd, 75270 Karachi, Pakistan

G. H. Guthöhrlein
Laboratorium für Experimentalphysik, Helmut-Schmidt-Universität, Universität der Bundeswehr Hamburg, Holstenhofweg 85, 22043 Hamburg, Germany

L. Windholz
Institute for Experimental Physics, Graz University of Technology, Petersgasse 16, A-8010 Graz, Austria

Laila M. Break
Taif University, Al-Haweiah, P.O.Box 888 Zip Code 21974. Taif, Saudi Arabia

Index

www.ingramcontent.com/pod-product-compliance
Lightning Source LLC
Chambersburg PA
CBHW080530200326
41458CB00012B/4388